# Student Guide with
## Solutions and Sample Tests

# Student Guide
## with Solutions and Sample Tests

TO ACCOMPANY

## Intermediate Algebra
FOURTH EDITION

**M. A. Munem and W. Tschirhart**

**Macomb Community College**

WORTH PUBLISHERS, INC.

**STUDENT GUIDE
WITH SOLUTIONS AND SAMPLE TESTS**

to accompany

**INTERMEDIATE ALGEBRA**
Fourth Edition

Copyright © 1988 by Worth Publishers, Inc.

All rights reserved. No part of this publication may be reproduced, stored in a retrieval system, or transmitted in any form or by any means, electronic, mechanical, photocopying, recording, or otherwise, without the prior written permission of the publisher.

Printed in the United States of America
ISBN: 0-87901-379-6
2 3 4 5 – 92 91 90 89

**WORTH PUBLISHERS, INC.**
33 Irving Place
New York, New York 10003

# Preface

This *Student Guide with Solutions and Sample Tests* is designed to help you master the material presented in *Intermediate Algebra* by Munem and Tschirhart. There are many ways to make good use of the information provided here.

To ensure that you can identify the key concepts covered in each chapter of the textbook, review the **list of objectives** in this Guide—both before and after reading the chapter.

As you work through each section of the textbook, you will find it very helpful to complete the **semiprogrammed problems** in the corresponding section of the Guide. Difficult problems are broken down into a sequence of steps that will help you to follow the logic of the solution and to understand the concepts and techniques fully. Be sure to cover the answers in the left column with a strip of paper—you are only shortchanging yourself if you peek! The value of this exercise is that it alerts you to any weakness in your problem-solving ability *before* it is revealed in a test. If you do have trouble with a problem in the Guide, review that topic in the textbook, carefully following the examples given.

When you have worked through a section of the textbook and completed the semiprogrammed problems in the Guide, you are ready to turn back to the textbook and tackle the section problem set. First solve two or three problems at the beginning of each group—they are usually the easiest. Compare what you have done with the **solutions** provided here to **every other odd-numbered problem** in the textbook. Have you left out any steps? Does your review alert you to anything you have not understood? You might also want to check your answers against those provided in the textbook appendix—to all the odd-numbered problems.

Once you've followed these steps for every assigned section in a chapter, you should be ready for any test. To be sure you are, take **Chapter Test A** in the Guide, writing out every step of the solution to each problem. (Wrong answers are more often due to carelessness than to lack of understanding.) Then compare your solutions with those given, making sure you understand any discrepancies. If you are likely to be given a multiple-choice test, take **Chapter Test B** and check your answers against those provided.

It's easy to get rusty with newly learned mathematical skills. The **Cumulative Review Problem Sets** are designed to prevent that from happening. In each chapter of the Guide, you will find a cumulative review problem set (with answers) covering all the chapters in the textbook up to that point. If you take the time as you complete each chapter of the textbook to solve these 20 or so problems, you will be very glad you did: this review will serve to refresh and reinforce all the abilities you have gained from the course.

To review, the resources available to you in this Guide are:

1. Chapter Objectives
2. Semiprogrammed Problems—about 15 for each section of each chapter.
3. Solutions to Selected Odd Problems—step-by-step solutions to every other odd-numbered problem in the textbook.

4. A Cumulative Review Problem Set—about 20 problems reviewing the important concepts and skills covered in the textbook through the current chapter, with answers provided.
5. Chapter Test A—about 20 problems, with step-by-step solutions to each.
6. Chapter Test B—about 20 multiple-choice problems, with answers provided.

Good luck!

Warren, Michigan  
July 1988

M. A. Munem  
W. Tschirhart

# Contents

| Chapter 1 | Numbers and Their Properties | 1 |
|---|---|---|
| 1.1 | Symbols and Notation of Algebra | 1 |
| 1.2 | Sets of Numbers | 3 |
| 1.3 | Properties of Real Numbers | 5 |
| 1.4 | Operations with Signed Numbers | 6 |
| 1.5 | Operations on Rational Numbers | 7 |
| 1.6 | Calculators and Approximations | 10 |
| 1.7 | The Language of Algebra | 11 |
| | Cumulative Review Problem Set | 13 |
| | Chapter Tests | 14 |

| Chapter 2 | The Algebra of Polynomials | 18 |
|---|---|---|
| 2.1 | Polynomials | 18 |
| 2.2 | Addition and Subtraction of Polynomials | 19 |
| 2.3 | Properties of Positive Integral Exponents | 21 |
| 2.4 | Multiplication of Polynomials | 23 |
| 2.5 | Factoring Polynomials | 25 |
| 2.6 | Factoring by Recognizing Special Products | 26 |
| 2.7 | Factoring Trinomials of the Form $ax^2 + bx + c$ | 28 |
| 2.8 | Division of Polynomials | 29 |
| | Cumulative Review Problem Set | 32 |
| | Chapter Tests | 33 |

| Chapter 3 | The Algebra of Fractions | 37 |
|---|---|---|
| 3.1 | Rational Expressions | 37 |
| 3.2 | Multiplication and Division of Fractions | 39 |
| 3.3 | Addition and Subtraction of Fractions | 42 |
| 3.4 | Complex Fractions | 46 |
| | Cumulative Review Problem Set | 49 |
| | Chapter Tests | 50 |

| Chapter 4 | Linear Equations and Inequalities | 55 |
|---|---|---|
| 4.1 | Equations | 55 |
| 4.2 | Equations Involving Fractions | 58 |

| | | | |
|---|---|---|---|
| | 4.3 | Literal Equations and Formulas | 62 |
| | 4.4 | Translating Verbal Expressions into Algebraic Expressions | 65 |
| | 4.5 | Applications of Linear Equations—Word Problems | 67 |
| | 4.6 | Inequalities | 73 |
| | 4.7 | Linear Inequalities | 75 |
| | 4.8 | Equations Involving Absolute Values | 78 |
| | 4.9 | Inequalities Involving Absolute Values | 80 |
| | | Cumulative Review Problem Set | 83 |
| | | Chapter Tests | 84 |
| **Chapter 5** | **Exponents, Radicals, and Complex Numbers** | | **88** |
| | 5.1 | Zero and Negative Integer Exponents | 88 |
| | 5.2 | Roots and Rational Exponents | 90 |
| | 5.3 | Radicals | 93 |
| | 5.4 | Addition and Subtraction of Radical Expressions | 94 |
| | 5.5 | Multiplication and Division of Radical Expressions | 96 |
| | 5.6 | Complex Numbers | 98 |
| | | Cumulative Review Problem Set | 100 |
| | | Chapter Tests | 102 |
| **Chapter 6** | **Nonlinear Equations and Inequalities** | | **107** |
| | 6.1 | Solving Quadratic Equations by Factoring | 107 |
| | 6.2 | Solving Quadratic Equations by Roots Extraction and Completing the Square | 109 |
| | 6.3 | Using the Quadratic Formula to Solve Quadratic Equations | 114 |
| | 6.4 | Applications of Quadratic Equations | 118 |
| | 6.5 | Equations in Quadratic Form | 122 |
| | 6.6 | Equations Involving Radicals | 125 |
| | 6.7 | Equations Involving Rational Exponents | 129 |
| | 6.8 | Nonlinear Inequalities | 133 |
| | | Cumulative Review Problem Set | 140 |
| | | Chapter Tests | 141 |
| **Chapter 7** | **Graphing Linear Equations and Inequalities in Two Variables** | | **147** |
| | 7.1 | The Cartesian Coordinate System | 147 |
| | 7.2 | Distance Between Two Points | 151 |
| | 7.3 | Graphs of Linear Equations | 154 |
| | 7.4 | The Slope of a Line | 160 |
| | 7.5 | Equations of Lines | 164 |
| | 7.6 | Graphs of Linear Inequalities | 169 |

| | | Cumulative Review Problem Set | 173 |
| --- | --- | --- | --- |
| | | Chapter Tests | 175 |
| **Chapter 8** | **Systems of Linear Equations and Inequalities** | | **180** |
| | 8.1 | Systems of Linear Equations in Two Variables | 180 |
| | 8.2 | Systems of Linear Equations in Three Variables | 186 |
| | 8.3 | Determinants | 189 |
| | 8.4 | Cramer's Rule | 191 |
| | 8.5 | Applications Involving Linear Systems | 195 |
| | 8.6 | Systems of Linear Inequalities | 198 |
| | | Cumulative Review Problem Set | 201 |
| | | Chapter Tests | 203 |
| **Chapter 9** | **Logarithms** | | **210** |
| | 9.1 | Exponential Equations and Logarithms | 210 |
| | 9.2 | Basic Properties of Logarithms | 212 |
| | 9.3 | Common Logarithms | 215 |
| | 9.4 | Using a Calculator to Evaluate Logarithmic and Exponential Expressions | 219 |
| | 9.5 | Applications | 221 |
| | | Cumulative Review Problem Set | 224 |
| | | Chapter Tests | 226 |
| **Chapter 10** | **Functions and Related Curves** | | **230** |
| | 10.1 | Functions | 230 |
| | 10.2 | Variation | 234 |
| | 10.3 | Linear and Quadratic Functions | 236 |
| | 10.4 | Exponential and Logarithmic Functions | 244 |
| | 10.5 | Graphs of Special Curves—Conic Sections | 249 |
| | 10.6 | Systems Containing Quadratic Equations | 256 |
| | | Cumulative Review Problem Set | 263 |
| | | Chapter Tests | 265 |
| **Chapter 11** | **Topics in Algebra** | | **272** |
| | 11.1 | Sequences | 272 |
| | 11.2 | Series | 275 |
| | 11.3 | The Binomial Theorem | 278 |
| | | Cumulative Review Problem Set | 282 |
| | | Chapter Tests | 284 |

| | | | |
|---|---|---|---|
| **Chapter 12** | **Geometry** | | **289** |
| | 12.1 | Basic Elements of Geometry | 289 |
| | 12.2 | Polygons | 290 |
| | 12.3 | Areas of Plane Figures | 292 |
| | 12.4 | Perimeters of Plane Figures | 294 |
| | 12.5 | Volumes and Surface Areas | 296 |
| | | Cumulative Review Problem Set | 300 |
| | | Chapter Tests | 302 |
| **Appendices** | **Appendix A** | **Tables** | **307** |
| | | Table I   Common Logarithms | 307 |
| | | Table II   Powers and Roots | 309 |
| | **Appendix B** | **Formulas from Geometry** | **310** |

# 1 Numbers and Their Properties

In this chapter we deal with numbers and their properties. We also include material on the use of calculators, approximations, and the language of algebra. After completing the appropriate sections in this chapter, the student should be able to:

1. Use the symbols and notations of algebra.
2. Locate sets of numbers on a number line.
3. Justify the basic operations of real numbers.
4. Perform operations with signed numbers.
5. Perform operations on rational numbers.
6. Round off numbers.
7. Use exponential notation and formulas.

## 1.1 Symbols and Notation of Algebra

### SEMIPROGRAMMED PROBLEMS

In problems 1–4, translate each statement into symbols.

1. The sum of $x$ and 8 is 24. The word "sum" refers to the result of the operation of _____. Thus we have:
   _____ = 24.

addition

$x + 8$

2. The difference of 11 and $y$ is 13. The word "difference" refers to the result of the operation of _____. Thus we have:
   _____ = 13.

subtraction

$11 - y$

3. The product of 5 and $z$ equals the sum of $z$ and 40. The word "product" refers to the result of the operation of _____.
   Thus we have:
   ____ = ____.

multiplication

$5z, z + 40$

4. The quotient of $t$ and 3 equals the difference of $t$ and 1. The word "quotient" refers to the result of the operation of _____.
   Thus we have:
   ____ = ____.

division

$\dfrac{t}{3}, t - 1$

## CHAPTER 1 NUMBERS AND THEIR PROPERTIES

In problems 5 and 6, write an equivalent word statement for each given statement represented by the symbols.

**5** $x + 5 \leq 14$

sum

$x + 5$ means the _____ of $x$ and 5.

less than or equal to

$\leq$ means _____.

Thus we have:

sum, less than or equal to

The _____ of $x$ and 5 is _____ 14.

**6** $4y - 3 \neq 7$

$4y$ means the _____ of 4 and $y$.

product

$4y - 3$ means the _____ of $4y$ and 3.

difference

$\neq$ means _____.

is not equal to

Thus we have:

difference, $4y$ and 3, not equal to

The _____ of _____ is _____ 7.

In problems 7 and 8, identify the base and exponent, then write each expression in expanded form and find its value.

**7** $3^6$

3, 6

The base is _____. The exponent is _____.

$3 \cdot 3 \cdot 3 \cdot 3 \cdot 3 \cdot 3$, 729

The expanded form is _____, so that $3^6 =$ _____.

**8** $8^3$

8, 3

The base is _____. The exponent is _____.

$8 \cdot 8 \cdot 8$, 512

The expanded form is _____, so that $8^3 =$ _____.

In problems 9–12, find the value of each expression by using the rule for the order of operations.

**9** $3 \cdot 5 + 7$

15, 22

$3 \cdot 5 + 7 =$ _____ $+ 7 =$ _____

**10** $2[4 + (7 - 2)] - 11$

5

$2[4 + (7 - 2)] - 11 = 2[4 +$ _____$] - 11$

9

$= 2[$_____$] - 11$

18, 7

$=$ _____ $- 11 =$ _____

**11** $3^2[5^2 - (3^2 + 2^2)]$

9, 4

$3^2[5^2 - (3^2 + 2^2)] = 9[25 - ($_____ $+$ _____$)]$

13

$= 9[25 -$ _____$]$

12, 108

$= 9[$_____$] =$ _____

**12** $\dfrac{30 + 24 \div 2^2}{2^3 + 8 \div 2}$

We evaluate the numerator and denominator separately. Thus we have:

4

$30 + 24 \div 2^2 = 30 + 24 \div$ _____

6, 36

$= 30 +$ _____ $=$ _____

8

$2^3 + 8 \div 2 =$ _____ $+ 8 \div 2$

8, 4, 12

$=$ _____ $+$ _____ $=$ _____

Therefore,

$\frac{36}{12}$, 3

$\dfrac{30 + 24 \div 2^2}{2^3 + 8 \div 2} =$ _____ $=$ _____

## SOLUTIONS TO SELECTED ODD PROBLEMS   Section 1.1, text pages 6–8

**1**    $y + 3 = 15$      **5**    $x - 2 > 5$      **9**    $y \div 3 = y - 5$      **13**    $3(8y) \leq 7(y + 8)$

**17**    The product of 3 and $x$ is less than or equal to 15.

**21**    Two less than the product of 10 and $x$ does not equal 15.

**25**    Three more than the quotient of $u$ and 4 equals 7.

**29**    Base = 5; exponent = 2; $5^2 = 5 \cdot 5 = 25$

**33**    Base = 2; exponent = 7; $2^7 = 2 \cdot 2 \cdot 2 \cdot 2 \cdot 2 \cdot 2 \cdot 2 = 128$

**37**    Base = 11; exponent = 2; $11^2 = 11 \cdot 11 = 121$

**41**    $5 \cdot 7 - 4$      **45**    $4 + 4 \cdot 4 - 4 \div 4$      **49**    $6 + 2 - 7 + 9 \cdot 2 \div 3$
     $= 35 - 4$            $= 4 + 16 - 1$           $= 6 + 2 - 7 + 18 \div 3$
     $= 31$                $= 19$                 $= 6 + 2 - 7 + 6$
                                                             $= 7$

**53**    $8 - (12 - 2) \div (4 + 1)$    **57**    $5 \cdot 6 + 11$      **61**    $(9 + 6)^2 \div 5$      **65**    $3^2 + 4^2$
     $= 8 - 10 \div 5$             $= 30 + 11$           $= (15)^2 \div 5$          $= 9 + 16$
     $= 8 - 2$                   $= 41$                  $= 225 \div 5$            $= 25$
     $= 6$                                              $= 45$

**69**    $2^2 + 3^2 + 4^2$      **73**    $9 \cdot 2^2 + 10 \div 5 + 4^3$      **77**    $15^2 - [11 \cdot 5 + 4(2 \cdot 4 - 3)]$
     $= 4 + 9 + 16$           $= 9 \cdot 4 + 2 + 64$           $= 225 - [55 + 4(8 - 3)]$
     $= 29$                     $= 36 + 2 + 64$             $= 225 - [55 + 4(5)]$
                           $= 102$                    $= 225 - [55 + 20]$
                                                           $= 225 - 75$
                                                            $= 150$

**81**    $\dfrac{24 + 6^2 \div 3}{7 + 6 \cdot 3}$      **85**    $\left[\dfrac{(5 + 3)6}{2} - 9\right] \div 3$      $\left[\dfrac{(7 + 3)6}{2} - 9\right] \div 3$

     $= \dfrac{24 + 36 \div 3}{7 + 18}$           $= \left[\dfrac{(8)(6)}{2} - 9\right] \div 3$        $= \left[\dfrac{(10)(6)}{2} - 9\right] \div 3$

     $= \dfrac{24 + 12}{25}$                $= \left[\dfrac{48}{2} - 9\right] \div 3$            $= \left[\dfrac{60}{2} - 9\right] \div 3$

     $= \dfrac{36}{25}$                     $= [24 - 9] \div 3$            $= [30 - 9] \div 3$
                             $= 15 \div 3$                    $= 21 \div 3$
                             $= 5$                           $= 7$

## 1.2   Sets of Numbers

### SEMIPROGRAMMED PROBLEMS

In problem 1, use set notation to describe the given set of numbers.

**1**   Use set notation to describe the set $A$ of counting numbers greater than 3 but less than 8.

braces

Using _____ to enclose the elements of the set, we have:

$\{4, 5, 6, 7\}$

$A = $ _____.

# CHAPTER 1 NUMBERS AND THEIR PROPERTIES

In problems 2 and 3, express the points on the number lines as sets.

$\{-3, -1, 2, 3\}$

2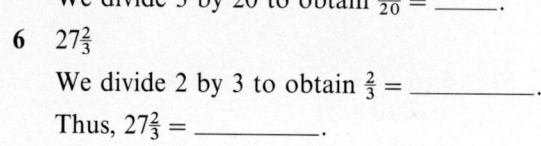

$\{-\frac{7}{2}, -2, -\frac{1}{2}, \frac{5}{2}\}$

3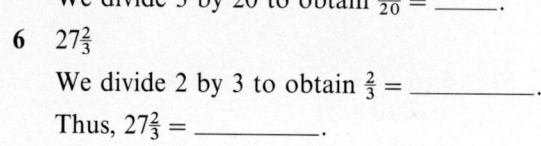

4 The subsets of $\{2, 4, 6\}$ are $\emptyset$, $\{2\}$, $\{4\}$, $\{6\}$, $\{2, 4\}$,

$\{2, 6\}, \{4, 6\}, \{2, 4, 6\}$ _____, _____, and _____.

In problems 5 and 6, express each rational number as a decimal.

5 $\frac{3}{20}$

0.15

We divide 3 by 20 to obtain $\frac{3}{20} = $ _____.

6 $27\frac{2}{3}$

$0.\overline{6}$

We divide 2 by 3 to obtain $\frac{2}{3} = $ _____.

$27.\overline{6}$

Thus, $27\frac{2}{3} = $ _____.

In problems 7 and 8, express each decimal as a quotient of integers.

7 0.4

$0.4 = \dfrac{4}{\rule{1cm}{0.4pt}} = \rule{1cm}{0.4pt}$

$10, \frac{2}{5}$

8 $-0.0036$

$-0.0036 = -\dfrac{36}{\rule{1cm}{0.4pt}} = \rule{1cm}{0.4pt}$

$10{,}000, -\frac{9}{2{,}500}$

In problems 9 and 10, identify each number as being rational or irrational.

9 $\sqrt{2}$

square

$\sqrt{2}$ is the square root of a positive integer that is not a perfect _____.

irrational

Thus, $\sqrt{2}$ is _____.

10 $\frac{11}{7}$

integers

$\frac{11}{7}$ is the quotient of two _____.

rational

Thus, $\frac{11}{7}$ is _____.

## SOLUTIONS TO SELECTED ODD PROBLEMS Section 1.2, text pages 13-14

1 $5 \in A$; $5 \notin B$; $10 \in A$; $4 \notin B$; $10 \in B$; $12 \notin A$; $8 \in B$; $9 \in A$

5 $C = \{x \mid x \text{ is a counting number and } 4 < x < 17\}$. The set is finite.

9 $\emptyset, \{5\}, \{6\}, \{7\}, \{8\}, \{5,6\}, \{5,7\}, \{5,8\}, \{6,7\}, \{6,8\}, \{7,8\}, \{5,6,7\}, \{5,6,8\}, \{5,7,8\}, \{6,7,8\}, \{5,6,7,8\}$

13 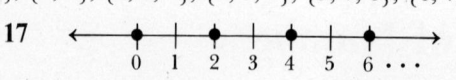    17 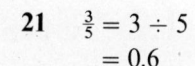    21 $\frac{3}{5} = 3 \div 5 = 0.6$

25 $\frac{4}{5} = 4 \div 5 = 0.8$    29 $-\frac{7}{3} = -7 \div 3 = -2.\overline{3}$    33 $2.64 = 2\frac{64}{100} = 2\frac{16}{25} = \frac{66}{25}$    37 $0.0527 = \frac{527}{10{,}000}$

41 rational    45 rational    49 rational    53 rational

## 1.3 Properties of Real Numbers

### SEMIPROGRAMMED PROBLEMS

In problems 1 and 2, complete the statements by applying the commutative properties.

6 + 7

8 · 5

**1** 7 + 6 = _____

**2** 5 · 8 = _____

In problems 3 and 4, complete the statements by applying the associative properties.

2

6

**3** 5 + (2 + 7) = (5 + _____) + 7

**4** _____ · (4 · 3) = (6 · 4) · 3

In problems 5 and 6, complete the statements by applying the distributive properties.

5 · 7

8 · 2

**5** 5 · (4 + 7) = 5 · 4 + _____

**6** (3 + 8) · 2 = 3 · 2 + _____

In problems 7–10, state the property that justifies each equality.

additive identity

multiplicative identity

additive inverse

multiplicative inverse

**7** $\sqrt{5} + 0 = \sqrt{5}$ _____

**8** $\sqrt{7} \cdot 1 = \sqrt{7}$ _____

**9** $\sqrt{11} + (-\sqrt{11}) = 0$ _____

**10** $9 \cdot \frac{1}{9} = 1$ _____

In problems 11 and 12, fill in the blanks to make the statement an illustration of the given property.

y

2 − y

**11** Transitive Property:

If $7 = x$ and $x = y$, then $7 = $ _____.

**12** Substitution Property:

If $4x - 3y = 14$ and $x = 2 - y$, then $4($_____$) - 3y = 14$.

### SOLUTIONS TO SELECTED ODD PROBLEMS   Section 1.3, text pages 19–20

**1**  15 + 17 = 32     17 + 15 = 32

**5**  0.6 + 0.5 = 1.1     0.5 + 0.6 = 1.1

**9**  2 + (7 + 6)     (2 + 7) + 6
    = 2 + 13         = 9 + 6
    = 15             = 15

**13**  4.1 + (6.8 + 3.3)     (4.1 + 6.8) + 3.3
    = 4.1 + 10.1         = 10.9 + 3.3
    = 14.2               = 14.2

**17**  3 · (5 + 8)     3 · 5 + 3 · 8
    = 3(13)         = 15 + 24
    = 39            = 39

**21**  2.1(1.1 + 3.4)     2.1 · 1.1 + 2.1 · 3.4
    = 2.1(4.5)          = 2.31 + 7.14
    = 9.45              = 9.45

**25**  Associative property for multiplication

**29**  Multiplicative inverse

**33**  Zero-factor property (ii)

**37**  Zero-factor property (i)

**41**  Identity property for addition

**43**  (a)  Identity property for multiplication and the negative property;
    (b)  distributive property;

**6** CHAPTER 1 NUMBERS AND THEIR PROPERTIES

    **(c)** additive inverse;
    **(d)** zero-factor property (i)

**47** $3t + 2t = 5t$

## 1.4 Operations with Signed Numbers

### SEMIPROGRAMMED PROBLEMS

In problems 1–3, find the sums.

|  |  |
|---|---|
|  | **1** $(-13) + (-5)$ |
| $-18$ | $(-13) + (-5) = -(13 + 5) = $ _____ |
|  | **2** $5 + (-21)$ |
| $-16$ | $5 + (-21) = -(21 - 5) = $ _____ |
|  | **3** $(-9) + 17$ |
| $8$ | $(-9) + 17 = +(17 - 9) = $ _____ |

In problems 4–6, find the differences.

|  |  |
|---|---|
|  | **4** $5 - (-4)$ |
| $4, 9$ | $5 - (-4) = 5 + (\_\_\_) = \_\_\_$ |
|  | **5** $(-3) - (-11)$ |
| $11, 8$ | $(-3) - (-11) = (-3) + (\_\_\_) = \_\_\_$ |
|  | **6** $(-9) - 15$ |
| $-15, -24$ | $(-9) - 15 = (-9) + (\_\_\_) = \_\_\_$ |

In problems 7–10, find the products.

|  |  |
|---|---|
|  | **7** $7 \cdot (-6)$ |
| $6, -42$ | $7 \cdot (-6) = -(7 \cdot \_\_\_) = \_\_\_$ |
|  | **8** $(-5) \cdot (-4)$ |
| $20$ | $(-5) \cdot (-4) = 5 \cdot 4 = \_\_\_$ |
|  | **9** $(-1) \cdot 2 \cdot (-3) \cdot (-1)$ |
| $-, -6$ | $(-1) \cdot 2 \cdot (-3) \cdot (-1) = \_\_\_ (1 \cdot 2 \cdot 3 \cdot 1) = \_\_\_$ |
|  | **10** $(-2) \cdot (-3) \cdot (-1) \cdot (-4)$ |
| $+, 24$ | $(-2) \cdot (-3) \cdot (-1) \cdot (-4) = \_\_\_ (2 \cdot 3 \cdot 1 \cdot 4) = \_\_\_$ |

In problems 11 and 12, find the quotients.

|  |  |
|---|---|
|  | **11** $15 \div (-3)$ |
| $-5$ | $15 \div (-3) = -(15 \div 3) = \_\_\_$ |
|  | **12** $(-63) \div (-7)$ |
| $9$ | $(-63) \div (-7) = 63 \div 7 = \_\_\_$ |

In problems 13 and 14, evaluate each expression.

|  |  |
|---|---|
|  | **13** $(-5)[(-3) + 7]$ |
| $(-5)(7)$ | $(-5)[(-3) + 7] = (-5)(-3) + \_\_\_$ |
| $-35, -20$ | $\qquad\qquad\quad = 15 + (\_\_\_) = \_\_\_$ |

$-2$
$-8$
$19$

14  $3^2 \cdot 6 \div 2 + (6-8)^3$
    $3^2 \cdot 6 \div 2 + (6-8)^3 = 9 \cdot 6 \div 2 + (\underline{\phantom{xx}})^3$
    $= 54 \div 2 + (\underline{\phantom{xx}})$
    $= \underline{\phantom{xx}}$

## SOLUTIONS TO SELECTED ODD PROBLEMS   Section 1.4, text pages 29–31

1  $|3| = 3$

5  $-|16| = -(16) = -16$

9  $-\left|-\frac{5}{7}\right| = -\left(\frac{5}{7}\right) = -\frac{5}{7}$

13  $(-6) + (-8) = -(6+8) = -14$

17  $(-8) + 8 = 0$

21  $(-21) + 9 = -(21-9) = -12$

25  $(15.2) + (-13.7) = 15.2 - 13.7 = 1.5$

29  $3 + (-5) + (-7)$
    $= 3 + (-12)$
    $= -(12 - 3)$
    $= -9$

33  $6 + (-9) + (-1)$
    $= 6 + (-10)$
    $= -(10 - 6)$
    $= -4$

37  $(-25) - (-45)$
    $= -25 + 45$
    $= 45 - 25$
    $= 20$

41  $(-8) - 8$
    $= (-8) + (-8)$
    $= -(8 + 8)$
    $= -16$

45  $0.052 - (-0.007)$
    $= 0.052 + 0.007$
    $= 0.059$

49  $(-111) - (-17)$
    $= -111 + 17$
    $= -94$

53  $3 \cdot (-5)$
    $= -(3 \cdot 5)$
    $= -15$

57  $(-9) \cdot 0.3$
    $= -[(9) \cdot (0.3)]$
    $= -2.7$

61  $(-3) \cdot (-2) \cdot (-4)$
    $= -(3 \cdot 2 \cdot 4)$
    $= -24$

65  $(-1) \cdot (-2) \cdot (-3) \cdot (-5)$
    $= 1 \cdot 2 \cdot 3 \cdot 5$
    $= 30$

69  $(-10) \div 5$
    $= -(10 \div 5)$
    $= -2$

73  $(-57) \div (-19)$
    $= 57 \div 19$
    $= 3$

77  $(-75) \div (-5)$
    $= 75 \div 5$
    $= 15$

81  $(-350) \div (-70)$
    $= 350 \div 70$
    $= 5$

85  $(-2)[9 + (-11)]$
    $= (-2)(-2)$
    $= 4$

89  $(-5)[(-15) \div 3]$
    $= (-5)(-5)$
    $= 25$

93  $(-36) \div [(-8) \div (-4)]$
    $= (-36) \div 2$
    $= -18$

97  $17 - 5 \cdot 2 \div 2 - 5 + 4$
    $= 17 - 10 \div 2 - 5 + 4$
    $= 17 - 5 - 5 + 4$
    $= 11$

101  $5(-2)^3 \cdot (-3)^2 \div (-8)$
     $= 5(-8)(9) \div (-8)$
     $= -360 \div (-8)$
     $= 45$

105  $(-5)^2 + 4 \cdot [(-5) + 3] \div (-2)$
     $= 25 + 4[-2] \div (-2)$
     $= 25 + (-8) \div (-2)$
     $= 25 + 4$
     $= 29$

109  $5 + 20 \div (-2)^2 - 3^4$
     $= 5 + 20 \div 4 - 81$
     $= 5 + 5 - 81$
     $= -71$

113  $|(-8) - 3|^2 \neq |(-8)|^2 - |3|^2$
     $|(-8) - 3|^2 = |-11|^2 = (11)^2 = 121$

117  Let $-4$ and $-5$ represent the losses. Then $(-4) + 7 + (-5) = -2$ represents a loss of 2 yards in relation to the original line of scrimmage.

## 1.5  Operations on Rational Numbers

### SEMIPROGRAMMED PROBLEMS

In problems 1 and 2, reduce each fraction to lowest terms.

$7 \cdot 7, 7$

1  $\dfrac{49}{56}$

   $\dfrac{49}{56} = \dfrac{\phantom{xxx}}{7 \cdot 8} = \dfrac{\phantom{xxx}}{8}$

# CHAPTER 1 NUMBERS AND THEIR PROPERTIES

$-3 \cdot 7, -3$

2 $\dfrac{-21}{35}$

$$\dfrac{-21}{35} = \dfrac{\rule{1cm}{0.4pt}}{5 \cdot 7} = \dfrac{\rule{1cm}{0.4pt}}{5}$$

In problems 3–7, combine the fractions and simplify.

3 $\dfrac{3}{8} + \dfrac{4}{8}$

$4, \dfrac{7}{8}$

$$\dfrac{3}{8} + \dfrac{4}{8} = \dfrac{3 + \rule{0.6cm}{0.4pt}}{8} = \rule{1cm}{0.4pt}$$

4 $\dfrac{9}{16} - \dfrac{3}{16}$

$3, \dfrac{6}{16}, \dfrac{3}{8}$

$$\dfrac{9}{16} - \dfrac{3}{16} = \dfrac{9 - \rule{0.6cm}{0.4pt}}{16} = \rule{1cm}{0.4pt} = \rule{1cm}{0.4pt}$$

5 $\dfrac{1}{2} + \dfrac{3}{5}$

$2, 11$

$$\dfrac{1}{2} + \dfrac{3}{5} = \dfrac{1(5) + 3(\rule{0.6cm}{0.4pt})}{(2)(5)} = \dfrac{\rule{0.8cm}{0.4pt}}{10}$$

6 $\dfrac{5}{8} - \dfrac{2}{5}$

$8, 9$

$$\dfrac{5}{8} - \dfrac{2}{5} = \dfrac{5(5) - 2(\rule{0.6cm}{0.4pt})}{(8)(5)} = \dfrac{\rule{0.8cm}{0.4pt}}{40}$$

7 $\dfrac{5}{8} + \dfrac{7}{12}$

$24$

The LCD of the given fractions is _____, so that

$14, 14, \dfrac{29}{24}$

$$\dfrac{5}{8} + \dfrac{7}{12} = \dfrac{15}{24} + \dfrac{\rule{0.6cm}{0.4pt}}{24} = \dfrac{15 + \rule{0.6cm}{0.4pt}}{24} = \rule{1cm}{0.4pt}$$

In problems 8–12, perform the indicated operations and simplify.

8 $\dfrac{3}{4} \cdot \dfrac{8}{9}$

$3, \dfrac{2}{3}$

$$\dfrac{3}{4} \cdot \dfrac{8}{9} = \dfrac{3 \cdot 8}{4 \cdot 9} = \dfrac{24}{36} = \dfrac{2 \cdot 12}{(\rule{0.5cm}{0.4pt}) \cdot 12} = \rule{1cm}{0.4pt}$$

9 $\dfrac{21}{44} \cdot \left(\dfrac{-8}{35}\right)$

$-168, -6, -\dfrac{6}{55}$

$$\dfrac{21}{44} \cdot \left(\dfrac{-8}{35}\right) = \dfrac{21 \cdot (-8)}{44 \cdot 35} = \dfrac{\rule{0.8cm}{0.4pt}}{1540} = \dfrac{(\rule{0.5cm}{0.4pt}) \cdot 28}{55 \cdot 28} = \rule{1cm}{0.4pt}$$

10 $\dfrac{5}{6} \div \dfrac{1}{4}$

$\dfrac{4}{1}, 20, \dfrac{10}{3}$

$$\dfrac{5}{6} \div \dfrac{1}{4} = \dfrac{5}{6} \cdot \left(\rule{0.6cm}{0.4pt}\right) = \dfrac{\rule{0.8cm}{0.4pt}}{6} = \rule{1cm}{0.4pt}$$

11 $\dfrac{4}{15} \div \left(\dfrac{-2}{3}\right)$

$12, -\dfrac{2}{5}$

$$\dfrac{4}{15} \div \left(\dfrac{-2}{3}\right) = \dfrac{4}{15} \cdot \left(\dfrac{3}{-2}\right) = \dfrac{\rule{0.8cm}{0.4pt}}{-30} = \rule{1cm}{0.4pt}$$

−10

$\frac{-10}{36}$

−180, −1

12 $\frac{2}{3} \cdot \left(\frac{-5}{12}\right) \div \frac{5}{18}$

$\frac{2}{3} \cdot \left(\frac{-5}{12}\right) \div \frac{5}{18} = \frac{\underline{\quad}}{36} \div \frac{5}{18}$

$= \frac{\underline{\quad}}{\underline{\quad}} \cdot \frac{18}{5}$

$= \frac{\underline{\quad}}{180} = \underline{\quad}$

## SOLUTIONS TO SELECTED ODD PROBLEMS  Section 1.5, text pages 38–40

**1** $\frac{15}{45} = \frac{1 \cdot \cancel{15}}{3 \cdot \cancel{15}} = \frac{1}{3}$

**5** $\frac{-65}{91} = -\frac{65}{91}$
$= -\frac{5 \cdot \cancel{13}}{7 \cdot \cancel{13}}$
$= -\frac{5}{7}$

**9** $\frac{-45}{-72} = \frac{45}{72}$
$= \frac{5 \cdot 9}{8 \cdot 9}$
$= \frac{5}{8}$

**13** $\frac{2}{5} + \frac{1}{5} = \frac{2+1}{5}$
$= \frac{3}{5}$

**17** $\frac{7}{12} + \frac{3}{12}$
$= \frac{7+3}{12}$
$= \frac{10}{12}$
$= \frac{5}{6}$

**21** $\frac{1}{2} + \frac{3}{8}$
$= \frac{1 \cdot 4}{2 \cdot 4} + \frac{3}{8}$
$= \frac{4}{8} + \frac{3}{8}$
$= \frac{7}{8}$

**25** $\frac{11}{12} + \frac{7}{16}$
$= \frac{11 \cdot 4}{12 \cdot 4} + \frac{7 \cdot 3}{16 \cdot 3}$
$= \frac{44}{48} + \frac{21}{48}$
$= \frac{65}{48}$

**29** $\frac{7}{12} + \frac{4}{15}$
$= \frac{7 \cdot 5}{12 \cdot 5} + \frac{4 \cdot 4}{15 \cdot 4}$
$= \frac{35}{60} + \frac{16}{60}$
$= \frac{51}{60} = \frac{17}{20}$

**33** $\frac{2}{3} + \frac{7}{6} - \frac{3}{4} = \frac{2 \cdot 4}{3 \cdot 4} + \frac{7 \cdot 2}{6 \cdot 2} - \frac{3 \cdot 3}{4 \cdot 3}$
$= \frac{8}{12} + \frac{14}{12} - \frac{9}{12}$
$= \frac{8 + 14 - 9}{12}$
$= \frac{13}{12}$

**37** $\frac{7}{12} - \frac{3}{2} + \frac{4}{9} = \frac{7 \cdot 3}{12 \cdot 3} - \frac{3 \cdot 18}{2 \cdot 18} + \frac{4 \cdot 4}{9 \cdot 4}$
$= \frac{21}{36} - \frac{54}{36} + \frac{16}{36}$
$= \frac{21 - 54 + 16}{36}$
$= \frac{-17}{36} = -\frac{17}{36}$

**41** $\frac{4}{5} \cdot \frac{10}{12} = \frac{40}{60} = \frac{2 \cdot \cancel{20}}{3 \cdot \cancel{20}} = \frac{2}{3}$

**45** $\left(\frac{-57}{34}\right) \cdot \left(\frac{51}{19}\right) = -\left(\frac{3 \cdot \cancel{19}}{2 \cdot \cancel{17}}\right) \cdot \left(\frac{3 \cdot \cancel{17}}{\cancel{19}}\right) = -\frac{9}{2}$

**49** $\left(\frac{-7}{5}\right) \cdot \left(\frac{15}{21}\right) = -\left(\frac{\cancel{7}}{\cancel{5}}\right)\left(\frac{\cancel{5} \cdot \cancel{3}}{\cancel{7} \cdot \cancel{3}}\right) = -1$

**53** $\left(\frac{-18}{35}\right) \cdot \left(\frac{-14}{27}\right) = \left(\frac{2 \cdot \cancel{9}}{5 \cdot \cancel{7}}\right) \cdot \left(\frac{2 \cdot \cancel{7}}{3 \cdot \cancel{9}}\right) = \frac{4}{15}$

**57** $\left(\frac{-25}{24}\right) \div \left(\frac{15}{16}\right) = -\left(\frac{5 \cdot \cancel{5}}{\cancel{8} \cdot 3}\right) \cdot \left(\frac{\cancel{8} \cdot 2}{\cancel{5} \cdot 3}\right) = -\frac{10}{9}$

**61** $\frac{20}{21} \div \left(\frac{-15}{14}\right) = \frac{\cancel{5} \cdot 4}{3 \cdot \cancel{7}} \cdot \left[-\left(\frac{2 \cdot \cancel{7}}{\cancel{5} \cdot 3}\right)\right] = -\frac{8}{9}$

**65** $\left(\frac{-18}{35}\right) \div \left(\frac{18}{42}\right) = -\left(\frac{\cancel{18}}{5 \cdot \cancel{7}}\right) \cdot \left(\frac{6 \cdot \cancel{7}}{\cancel{18}}\right) = -\frac{6}{5}$

**69** $\left(\frac{-39}{51}\right) \div \left(\frac{26}{34}\right) = -\left(\frac{\cancel{3} \cdot \cancel{13}}{\cancel{3} \cdot \cancel{17}}\right) \cdot \left(\frac{\cancel{2} \cdot \cancel{17}}{\cancel{2} \cdot \cancel{13}}\right) = -1$

**73** $\left(\frac{9}{16}\right) \cdot \left(\frac{-2}{3}\right) \div \left(\frac{5}{21}\right) = \left(-\frac{3}{8}\right) \div \frac{5}{21} = -\frac{3}{8} \cdot \frac{21}{5} = -\frac{63}{40}$

**77** $\frac{12}{33} \div \frac{21}{16} \cdot \left(\frac{-33}{40}\right) = \frac{12}{33} \cdot \frac{16}{21} \cdot \left(\frac{-33}{40}\right) = \frac{64}{231} \cdot \left(\frac{-33}{40}\right) = \frac{8 \cdot \cancel{8}}{7 \cdot \cancel{33}} \cdot \left(-\frac{\cancel{33}}{5 \cdot \cancel{8}}\right) = -\frac{8}{35}$

81  $\dfrac{3}{4}+\dfrac{3}{5}\div\dfrac{7}{6}-\dfrac{1}{2}=\dfrac{3}{4}+\dfrac{3}{5}\cdot\dfrac{6}{7}-\dfrac{1}{2}=\dfrac{3}{4}+\dfrac{18}{35}-\dfrac{1}{2}=\dfrac{3\cdot 35+18\cdot 4-1\cdot 70}{140}=\dfrac{105+72-70}{140}=\dfrac{107}{140}$

85  $\dfrac{3}{4}-\dfrac{3}{5}=\dfrac{3\cdot 5-3\cdot 4}{4\cdot 5}=\dfrac{15-12}{20}=\dfrac{3}{20}$ hour

## 1.6 Calculators and Approximations

### SEMIPROGRAMMED PROBLEMS

In problems 1–3, use a calculator to change each number to a decimal.

0.10714286      1 $\dfrac{3}{28}$ _____

4.28205128      2 $4\dfrac{11}{39}$ _____

5.9160798       3 $\sqrt{35}$ _____

In problems 4 and 5, round off the given numbers to the nearest tenth.

0.3          4 0.34 _____

2.6          5 2.58 _____

In problems 6 and 7, round off the given numbers to the nearest hundredth.

4.38         6 4.3752 _____

0.57         7 0.5649 _____

In problems 8 and 9, round off the given numbers to three significant digits.

39000        8 38976 _____

0.000982       9 0.0009824 _____

In problems 10 and 11, use a calculator to change each square root to a decimal, and then round off the results to four significant digits.

2.828        10 $\sqrt{8}=$ _____

6.083        11 $\sqrt{37}=$ _____

In problems 12 and 13, rewrite each number in scientific notation.

          12 5,780

5.78, $^3$          5,780 = _____ × 10 ——

          13 152,000,000

1.52, $^8$          152,000,000 = _____ × 10 ——

In problems 14 and 15, rewrite each number in expanded form.

38,700        14 $3.87 \times 10^4=$ _____

7,310,000       15 $7.31 \times 10^6=$ _____

**SOLUTIONS TO SELECTED ODD PROBLEMS** Section 1.6, text pages 44–45

Using an eight-digit calculator, we have for problems 1 and 5:

1  $\frac{4}{29} = 4 \div 29 = 0.13793103$
5  $\sqrt{26} = 5.0990195$
9  0.27 rounds off to 0.3 since 7 is greater than 5.
13  7.998 rounds off to 8.0 since 9 is greater than 5.
17  24.052 rounds off to 24.1 since 5 is the first digit dropped.
21  14.3649 rounds off to 14.36 since 4 is less than 5.
25  16.507 rounds off to 16.51 since 7 is greater than 5.
29  1.732 rounds off to 1.73 since 2 is less than 5.
33  368.1 rounds off to 368 since 1 is less than 5.
37  27.98 rounds off to 28.0 since 8 is greater than 5.
41  $\frac{1}{7} = 0.14285714$, which rounds off to 0.143 since 8 is greater than 5.
45  $\frac{11}{6} = 1.8333333$, which rounds off to 1.83 since 3 is less than 5.
49  $\sqrt{3} = 1.7320508$, which rounds off to 1.732 since 0 is less than 5.
53  $\sqrt{23} = 4.7958315$, which rounds off to 4.796 since 8 is greater than 5.
57  $\sqrt{111} = 10.535654$, which rounds off to 10.54 since the first digit dropped is 5.
61  $384,000 = 3.84 \times 10^5$ since the decimal point is moved 5 places to the left.
65  $2.1 \times 10^2 = 210$, by moving the decimal point 2 places to the right.
69  $3.12 \times 10^7 = 31,200,000$, by moving the decimal point 7 places to the right.
73  $5,980,000,000,000,000,000,000,000 = 5.98 \times 10^{24}$, since the decimal point is moved 24 places to the left.

## 1.7 The Language of Algebra

### SEMIPROGRAMMED PROBLEMS

In problems 1 and 2, rewrite each expression using exponential notation.

$x^4 y^2$

$2t^2 s^3 r$

1  $x \cdot x \cdot x \cdot x \cdot y \cdot y =$ _____

2  $2 \cdot t \cdot t \cdot s \cdot s \cdot s \cdot r =$ _____

In problem 3, use the formula $F = \frac{9}{5}C + 32$ to find the temperature $F$ in Fahrenheit when the temperature $C$ in Celsius is the given number.

3  $C = 30°$

$F = \frac{9}{5}C + 32$

30

$F = \frac{9}{5}(\underline{\phantom{xx}}) + 32$

54

$F = \underline{\phantom{xx}} + 32$

86°

$F = \underline{\phantom{xx}}$

In problem 4, use the formula $A = lw$ to find the area $A$ of the rectangle with the given length $l$ and width $w$.

4  $l = 8$ feet, $w = 5.3$ feet

$A = lw$

5.3

$A = (8)(\underline{\phantom{xx}})$

42.4 square feet

$A = \underline{\phantom{xx}}$

CHAPTER 1 NUMBERS AND THEIR PROPERTIES

In problem 5, use the formula $P = 2l + 2w$ to find the perimeter $P$ of the rectangle with the given length $l$ and width $w$.

    **5**   $l = 14$ inches, $w = 8$ inches

          $P = 2l + 2w$

14, 8                $P = 2(\underline{\phantom{xx}}) + 2(\underline{\phantom{xx}})$

28, 16               $P = \underline{\phantom{xx}} + \underline{\phantom{xx}}$

44 inches           $P = \underline{\phantom{xxxxxx}}$

In problem 6, use the formula $d = rt$ to find the distance $d$ an object travels when moving at a uniform rate $r$ during time $t$.

    **6**   $r = 50$ feet per minute, $t = 12$ minutes

          $d = rt$

50, 12               $d = (\underline{\phantom{xx}})(\underline{\phantom{xx}})$

600 feet            $d = \underline{\phantom{xxxxxx}}$

In problem 7, use the formula $A = \pi r^2$ to find the area $A$ of a circle with given radius $r$ (use $\pi = 3.14$).

    **7**   $r = 6$ inches

          $A = \pi r^2$

6                    $A = 3.14(\underline{\phantom{xx}})^2$

113.04 square inches    $A = \underline{\phantom{xxxxxx}}$

In problem 8, use the formula $A = s^2$ to find the area $A$ of the square with the given side $s$.

    **8**   $s = 7$ inches

          $A = s^2$

7                  $A = (\underline{\phantom{xx}})^2$

49 square inches       $A = \underline{\phantom{xxxxxx}}$

In problem 9, use the formula $V = s^3$ to find the volume $V$ of the cube with the given side $s$.

    **9**   $s = 4$ feet

          $V = s^3$

4                  $V = (\underline{\phantom{xx}})^3$

64 cubic feet          $V = \underline{\phantom{xxxxxx}}$

In problem 10, use the formula $V = \frac{4}{3}\pi r^3$ to find the volume $V$ of the sphere with given radius $r$ (use $\pi = 3.14$).

    **10**   $r = 6$ centimeters

           $V = \frac{4}{3}\pi r^3$

6                   $V = \frac{4}{3}(3.14)(\underline{\phantom{xx}})^3$

904.32 cubic centimeters    $V = \underline{\phantom{xxxxxx}}$

## SOLUTIONS TO SELECTED ODD PROBLEMS   Section 1.7, text pages 50–53

**1**   $x(y - 3)$; $x = 2$, $y = 5$

    $2(5 - 3) = 2(2) = 4$

5    $\dfrac{5}{y^2} + 8x; \; x = -2, \; y = -10$

$\dfrac{5}{(-10)^2} + 8(-2) = \dfrac{5}{100} - 16 = \dfrac{1}{20} - 16 = \dfrac{1 - 320}{20} = -\dfrac{319}{20}$

9    $5 \cdot 5 \cdot 5 = 5^3$        13    $(-t)(-t)(-t)(-t)(-t) = (-t)^5$     17    $3 \cdot 3 \cdot x \cdot x \cdot x \cdot x \cdot y \cdot y = 3^2 x^4 y^2$

21    $y^5 = y \cdot y \cdot y \cdot y \cdot y$        25    $(-x)^4 y^3 = (-x) \cdot (-x) \cdot (-x) \cdot (-x) \cdot y \cdot y \cdot y$

29    $F = \tfrac{9}{5}C + 32$        33    $F = \tfrac{9}{5}C + 32$
      $F = \tfrac{9}{5}(0) + 32$             $F = \tfrac{9}{5}(20) + 32$
      $F = 32°$                  $F = 68°$

37    (a)   $A = s^2$              (b)   $P = 4s$
           $A = 4^2$                 $P = 4(4)$
           $A = 16$ square feet        $P = 16$ feet

41    (a)   $A = \pi r^2$            (b)   $C = 2\pi r$
           $A = 3.14(10)^2$           $C = 2(3.14)(10)$
           $A = 314$ square inches    $C = 62.8$ inches

45    (a)   $A = \pi r^2$            (b)   $C = 2\pi r$
           $A = 3.14(6.2)^2$          $C = 2(3.14)(6.2)$
           $A = 120.7$ square meters   $C = 38.94$ meters

49    (a)   $A = lw$               (b)   $P = 2l + 2w$
           $A = (8)(3)$              $P = 2(8) + 2(3)$
           $A = 24$ square meters     $P = 22$ meters

53    (a)   $V = s^3$              (b)   $A = 6s^2$
           $V = (3)^3$               $A = 6(3)^2$
           $V = 27$ cubic inches      $A = 54$ square inches

57    (a)   $V = s^3$              (b)   $A = 6s^2$
           $V = (10)^3$              $A = 6(10)^2$
           $V = 1,000$ cubic centimeters    $A = 600$ square centimeters

61    (a)   $V = \dfrac{4\pi r^3}{3}$          (b)   $A = 4\pi r^2$
                                             $A = 4(3.14)(4.1)^2$
           $V = \dfrac{4(3.14)(4.1)^3}{3}$       $A = 211.13$ square inches

           $V = 288.55$ cubic inches

65    $d = rt$          69    $d = rt$          73    $S = P(1 + rt)$
      $d = (50)(3)$            $d = (600)(13)$          $S = 25{,}125[1 + (0.0925)(7)]$
      $d = 150$ miles         $d = 7{,}800$ centimeters    $S = 25{,}125(1.6475)$
                                                                     $S = \$41{,}393.44$

## CUMULATIVE REVIEW PROBLEM SET    Chapter 1

1    Translate the following statement into symbols: Three more than twice a number $x$ is less than or equal to 17.

2    Identify the base and exponent of $3^4$. Then write $3^4$ in expanded form and find its value.

3    Use the rule for the order of operations to find the value of $4[2(5 - 3) + 7] + 13$.

4    Use set notation to describe the set $B$ of the counting numbers greater than 3 but less than 10.

5    Express $-\tfrac{9}{5}$ as a decimal.        6    Express 2.38 as a quotient of integers.

In problems 7–10, state the property that justifies each statement.

7    $3(7) = 7(3)$                               8    $(-11) + 11 = 0$

9    $3(x + y) = 3x + 3y$                    10    If $5y = 0$, then $y = 0$

In problems 11–14, perform the indicated operations.

**11**  $(-18) + (+15)$    **12**  $(+23) - (-17)$    **13**  $(-2)(+3)(-5)$    **14**  $(-24) \div (-6)$

**15**  Evaluate $5 + [(-2) + 5]^3 + 3^2 \cdot (-1)$

In problems 16–19, perform the indicated operations and simplify.

**16**  $\dfrac{13}{24} + \dfrac{7}{18}$    **17**  $\dfrac{9}{49} - \dfrac{2}{49}$    **18**  $\dfrac{7}{18} \cdot \left(\dfrac{-24}{35}\right)$    **19**  $\dfrac{21}{40} \div \left(\dfrac{9}{-25}\right)$

**20**  Round off 27.3749 to the nearest hundredth.

**21**  Rewrite 789,000 in scientific notation.

**22**  Rewrite $3.42 \times 10^5$ in expanded form.

**23**  Rewrite $x \cdot x \cdot y \cdot y \cdot y \cdot y$ using exponential notation.

**24**  Use the formula $P = 2l + 2w$ to find the perimeter of a rectangle with a length of 10 inches and a width of 7 inches.

**25**  Use the formula $A = 6s^2$ to find the surface area of a cube whose side is 2.5 meters.

## Answers

**1**  $2x + 3 \leq 17$    **2**  Base = 3; exponent = 4; $3^4 = 3 \cdot 3 \cdot 3 \cdot 3 = 81$    **3**  57    **4**  $B = \{4, 5, 6, 7, 8, 9\}$
**5**  $-1.8$    **6**  $\frac{119}{50}$    **7**  Commutative property for multiplication    **8**  Inverse property for addition
**9**  Distributive property    **10**  Zero-factor property    **11**  $-3$    **12**  40    **13**  30    **14**  4
**15**  23    **16**  $\frac{67}{72}$    **17**  $\frac{1}{7}$    **18**  $-\frac{4}{15}$    **19**  $-\frac{35}{24}$    **20**  27.37    **21**  $7.89 \times 10^5$
**22**  342,000    **23**  $x^2 y^4$    **24**  34 inches    **35**  37.5 square meters

# CHAPTER 1    TESTS

## Chapter Test A

**1**  Translate the following statement into symbols: Four times the difference of $y$ and 3 is 15.

**2**  Identify the base and exponent in the expression $2^5$. Write $2^5$ in expanded form and find its value.

**3**  Use the rule for the order of operations to evaluate $6 + 2(3 - 1)^2 - 12 \div 3$.

**4**  Use set notation to describe the set $A$ of the counting number greater than 5 but less than 13.

**5**  Express each rational number as a decimal.
  (a)  $2\frac{1}{4}$    (b)  $-\frac{1}{3}$

**6**  Express each decimal in the form $\dfrac{a}{b}$, where $b \neq 0$.
  (a)  0.6    (b)  $-3.27$

**7**  Justify each statement by giving the appropriate property. Assume that all letters represent real numbers.
  (a)  $a + (x + 3) = (a + x) + 3$    (b)  $ay = ya$    (c)  $x(y + z) = xy + xz$
  (d)  $3 + (-3) = 0$    (e)  If $x + z = y + z$, then $x = y$.    (f)  $x(yz) = (xy)z$
  (g)  If $3x = 3y$, then $x = y$.    (h)  $x \cdot \dfrac{1}{x} = 1$, for $x \neq 0$    (i)  $a + 0 = a$

**8**  Perform the indicated operations.
  (a)  $17 + (-4)$    (b)  $(-41) + 9$    (c)  $102 - (-63)$    (d)  $(-4) \cdot (-23)$
  (e)  $(8) \cdot (-7)$    (f)  $(-128) \div (-4)$    (g)  $(-0.133) \div (0.19)$

**9** Perform the indicated operations and simplify.

(a) $\dfrac{17}{24} + \dfrac{5}{24}$ (b) $\dfrac{11}{20} - \dfrac{5}{12}$ (c) $\left(\dfrac{-12}{25}\right) \cdot \dfrac{15}{16}$ (d) $\dfrac{13}{30} \div \dfrac{26}{45}$

**10** Round off the given number as indicated.

(a) 3.149 to the nearest tenth (b) 0.507 to the nearest hundredth (c) 2.30584 four significant digits

**11** Rewrite 41,700 in scientific notation. **12** Rewrite $5.12 \times 10^3$ in expanded form.

**13** Rewrite each expression using exponential notation.

(a) $3 \cdot 3 \cdot x \cdot y \cdot y \cdot y$ (b) $(-3)(-2)(x)(x)(-x)(y)(y)$

**14** Use the formulas $A = lw$ and $P = 2l + 2w$ to find the area and perimeter of the rectangle whose length is 15 feet and width is 12 feet.

---

## Solutions

**1** $4(y - 3) = 15$  **2** Base = 2, exponent = 5; $2^5 = 2 \cdot 2 \cdot 2 \cdot 2 \cdot 2 = 32$

**3** $6 + 2(3 - 1)^2 - 12 \div 3 = 6 + 2(2)^2 - 12 \div 3$
$= 6 + 2(4) - 12 \div 3$
$= 6 + 8 - 4$
$= 10$

**4** $A = \{6, 7, 8, 9, 10, 11, 12\}$

**5** (a) $2\tfrac{1}{4} = 2.25$ (b) $-\tfrac{1}{3} = -0.333\cdots = -0.\overline{3}$

**6** (a) $0.6 = \tfrac{6}{10} = \tfrac{3}{5}$ (b) $-3.27 = -\tfrac{327}{100}$

**7** (a) Associative property of addition
(b) Commutative property of multiplication
(c) Distributive property
(d) Additive inverse
(e) Cancellation property with respect to addition
(f) Associative property of multiplication
(g) Cancellation property with respect to multiplication
(h) Multiplicative inverse
(i) Additive identity

**8** (a) 13 (b) $-32$ (c) 165 (d) 92
(e) $-56$ (f) 32 (g) $-0.7$

**9** (a) $\dfrac{17}{24} + \dfrac{5}{24} = \dfrac{17 + 5}{24} = \dfrac{22}{24} = \dfrac{11}{12}$

(b) $\dfrac{11}{20} - \dfrac{5}{12} = \dfrac{33}{60} - \dfrac{25}{60} = \dfrac{33 - 25}{60} = \dfrac{8}{60} = \dfrac{2}{15}$

(c) $\left(\dfrac{-12}{25}\right) \cdot \dfrac{15}{16} = \dfrac{(-3) \cdot \cancel{4}}{5 \cdot \cancel{5}} \cdot \dfrac{3 \cdot \cancel{5}}{\cancel{4} \cdot 4} = \dfrac{(-3) \cdot 3}{5 \cdot 4} = \dfrac{-9}{20} = -\dfrac{9}{20}$

(d) $\dfrac{13}{30} \div \dfrac{26}{45} = \dfrac{13}{30} \cdot \dfrac{45}{26} = \dfrac{\cancel{13}}{2 \cdot \cancel{15}} \cdot \dfrac{3 \cdot \cancel{15}}{2 \cdot \cancel{13}} = \dfrac{3}{2 \cdot 2} = \dfrac{3}{4}$

**10** (a) 3.149 rounds off to 3.1 since the first digit dropped is 4, which is less than 5.
(b) 0.507 rounds off to 0.51 since the first digit dropped is 7, which is greater than 5.
(c) 2.30584 rounds off to 2.306 since the first digit dropped is 8, which is greater than 5.

16  CHAPTER 1  NUMBERS AND THEIR PROPERTIES

11  $41,700 = 4.17 \times 10^4$, since the decimal point was moved 4 places to the left.

12  $5.12 \times 10^3 = 5,120$, since the exponent 3 indicates a movement of the decimal point 3 places to the right.

13  (a)  $3^2xy^3$  (b)  $-6x^3y^2$

14  $A = lw = 15 \times 12 = 180$ square feet; $P = 2l + 2w = 30 + 24 = 54$ feet

## Chapter Test B

*Multiple Choice:* Select the *one* correct answer for each of the following questions.

1  The statement, "Twice the sum of $x$ and 5 is less than 11," translates into _____.
   (a)  $2x + 5 < 11$   (b)  $2(x + 5) < 11$   (c)  $2x + 5 > 11$   (d)  $2(x + 5) \leq 11$

2  The value of of $8 + 3 \cdot 2^2 - 6 \div 2$ is _____.
   (a)  19   (b)  41   (c)  7   (d)  17

3  [number line graph with points at $-\frac{5}{2}, -1, 2, \frac{7}{2}$]  is the graph of:
   (a)  $\{-\frac{5}{2}, -1, 0, 2, \frac{7}{2}\}$   (b)  $\{-\frac{5}{2}, -1, 2, \frac{7}{2}\}$
   (c)  $\{-\frac{5}{2}, -1, 2, \frac{7}{2}, \ldots\}$   (d)  $\{-\frac{3}{2}, -1, 2, \frac{3}{2}\}$

4  $5\frac{1}{8}$ expressed as a decimal is _____.
   (a)  5.13   (b)  5.1   (c)  5.8   (d)  5.125

5  3.42 expressed as the quotient of two integers is _____.
   (a)  $\frac{171}{50}$   (b)  $\frac{342}{1,000}$   (c)  $\frac{173}{100}$   (d)  $\frac{342}{99}$

6  The property that justifies the following statement $(x + y) \cdot 3 = 3 \cdot (x + y)$ is the _____.
   (a)  associative   (b)  commutative   (c)  identity   (d)  distributive

7  If $5 \cdot z = 0$, then $z =$ _____.
   (a)  1   (b)  0   (c)  any real number   (d)  none of these

8  If $a$ and $b$ are real numbers, then $(-a)(-b) =$ _____.
   (a)  $-ab$   (b)  $a(-b)$   (c)  $ab$   (d)  none of these

9  If $x$ is a real number, and $x \neq 0$, then $x \cdot \frac{1}{x} =$ _____.
   (a)  1   (b)  0   (c)  $x$   (d)  none of these

10  $(-7) \cdot (2) \cdot (-5)$ is _____.
    (a)  $-70$   (b)  70   (c)  $-28$   (d)  28

11  $|-7| + |3|$ is _____.
    (a)  $-4$   (b)  10   (c)  $-10$   (d)  4

12  $7 - (-11)$ is _____.
    (a)  $-4$   (b)  4   (c)  18   (d)  $-18$

13  $(-42) \div (-6)$ is _____.
    (a)  $-7$   (b)  $-6$   (c)  6   (d)  7

14  $\frac{3}{8} + \frac{5}{12}$ is _____.
    (a)  $\frac{19}{24}$   (b)  $\frac{2}{5}$   (c)  $\frac{1}{3}$   (d)  $\frac{13}{24}$

15 $\frac{4}{11} \div \frac{5}{22} \cdot \frac{10}{3}$ is _____.
 (a) $\frac{12}{25}$ (b) $\frac{25}{12}$ (c) $\frac{3}{16}$ (d) $\frac{16}{3}$

16 4.1387 rounded off to the nearest hundredth is _____.
 (a) 414 (b) 4.14 (c) 4.139 (d) 4.1

17 35.027629 rounded off to four significant digits is _____.
 (a) 35.03 (b) 35.02763 (c) 35.0276 (d) 35.02

18 $3 \cdot 3 \cdot t \cdot t \cdot u \cdot u \cdot u \cdot u$ can be expressed as _____.
 (a) $9t^2 + u^4$ (b) $9t^2 u^4$ (c) $9t^2 + 9u^4$ (d) $9(tu)^6$

19 3,790 is written in scientific notation as _____.
 (a) $37.9 \times 10^2$ (b) $0.379 \times 10^4$ (c) $3.79 \times 10^3$ (d) $379 \times 10$

20 $-5x^2 y^3$ can be expressed as _____.
 (a) $-5 \cdot x \cdot x + y \cdot y \cdot y$
 (b) $-5 \cdot x \cdot x - 5 \cdot y \cdot y \cdot y$
 (c) $5 \cdot (-x) \cdot (-x) \cdot y \cdot y \cdot y$
 (d) $-5 \cdot x \cdot x \cdot y \cdot y \cdot y$

21 If $P = 2l + 2w$, then for $l = 12$ inches and $w = 6$ inches, $P = $ _____.
 (a) 36 square inches (b) 30 square inches
 (c) 36 inches (d) 30 inches

22 If $A = \pi r^2$, then for $r = 3$ feet, $A = $ _____.
 (a) $9\pi$ square feet (b) $9\pi$ feet (c) $6\pi$ square feet (d) $6\pi$ feet

## Answers

| 1 b | 2 d | 3 b | 4 d | 5 a | 6 b | 7 b | 8 c | 9 a | 10 b |
| 11 b | 12 c | 13 d | 14 a | 15 d | 16 b | 17 a | 18 b | 19 c |
| 20 d | 21 c | 22 a |

# 2 The Algebra of Polynomials

In this chapter we present different types of problems designed to reinforce the student's understanding of the properties of real numbers as they are applied to algebra. After completing the appropriate sections, the student should be able to:

1. Apply the properties of positive integral exponents.
2. Determine the degree and numerical coefficients of polynomials.
3. Evaluate polynomials.
4. Add and subtract polynomials.
5. Multiply and divide polynomials.
6. Factor polynomials.

## 2.1 Polynomials

### SEMIPROGRAMMED PROBLEMS

In problems 1–3, identify the polynomial as a monomial, binomial, or trinomial.

binomial      1   $3x^2y + 2b$ is a _____.
monomial     2   $6xyz^2$ is a _____.
trinomial      3   $3x^2 + 2ab + c^3$ is a _____.

In problems 4 and 5, find the degree of each polynomial.

2      4   $x^2 - 3x + 6$ _____
7      5   $3xy^2 + 2x^4y^3$ _____

In problems 6 and 7, find the numerical coefficients of each polynomial.

3, 5, 7      6   $3x^2 + 5x + 7$ _____
−2, 4, −5, 1      7   $-2x^3 + 4x^2 - 5x + 1$ _____

In problems 8–10, evaluate each polynomial for the indicated values of the variables.

8   $3x^2 + 2$ at $x = 1$

3, 5      $3(1)^2 + 2 =$ _____ $+ 2 =$ _____

9   $x^2 + x - 4$ at $x = 2$

4, 2      $(2)^2 + (2) - 4 =$ _____ $+ 2 - 4 =$ _____

10   $6a - 35a^2 + 11$ at $a = -2$

−12, −140, −141      $6(-2) - 35(-2)^2 + 11 =$ _____ $+$ _____ $+ 11 =$ _____

In problems 11 and 12, evaluate each polynomial of more than one variable for $x = 1$, $y = 2$, $z = -1$.

**11** $x^2 + 2xy + z^3$
$(1)^2 + 2(1)(2) + (-1)^3 =$ _____

4

**12** $3x^4 - 4xyz - z^3$
$3(1)^4 - 4(1)(2)(-1) - (-1)^3 =$ _____

12

## SOLUTIONS TO SELECTED ODD PROBLEMS  Section 2.1, text pages 62-63

**1** Binomial: degree = 1; 3, $-2$

**5** Trinomial: degree = 2; 1, $-5$, 6

**9** Trinomial: degree = 4; $-1$, $-1$, 13

**13** The degree of each term is 2, so that the degree of the polynomial is 2.

**17** The degree of the terms are found by adding exponents. Thus, $3 + 1 = 4$; $1 + 5 = 6$; and $2 + 3 = 5$. The degree of the polynomial is the largest sum, 6.

**21** $4x - 1$, for $x = 2$
$4(2) - 1 = 8 - 1 = 7$

**25** $y^3 - 2y^2 - y + 1$, for $y = -1$
$(-1)^3 - 2(-1)^2 - (-1) + 1 = -1 - 2 + 1 + 1 = -1$

**29** $x^5 - x^4 + x^3 - x^2 + x - 2$ for $x = 2$
$2^5 - 2^4 + 2^3 - 2^2 + 2 - 2 = 32 - 16 + 8 - 4 + 2 - 2 = 20$

**33** $x^3y^2 - 2yz^2 + xy^2$ for $x = 1$, $y = -1$, $z = 2$
$(1)^3(-1)^2 - 2(-1)(2)^2 + (1)(-1)^2$
$= (1)(1) - 2(-1)(4) + (1)(1) = 1 + 8 + 1 = 10$

**37** No, because of the division by $x$.

**41** (a) $h = -16t^2 + 64t$
$= -16(2)^2 + 64(2) = (-16)(4) + 128$
$= -64 + 128 = 64$ feet

(b) $h = -16t^2 + 64t$
$= -16(3)^2 + 64(3) = -16(9) + 192$
$= -144 + 192 = 48$ feet

## 2.2 Addition and Subtraction of Polynomials

### SEMIPROGRAMMED PROBLEMS

In problems 1-4, perform the indicated additions of the polynomials.

**1** $7x^2 + 2x^2$
$7x^2 + 2x^2 = (7 + 2)x^2 =$ _____

$9x^2$

**2** $(3x + 2) + (5x + 6)$
$(3x + 2) + (5x + 6) = (3 + 5)x + (2 + 6)$
$=$ _____

$8x + 8$

**3** $(x^2 + 3x + 5) + (2x^2 + x + 3)$
$(x^2 + 3x + 5) + (2x^2 + x + 3)$
$= (1 + 2)x^2 + (3 + 1)x + (5 + 3)$
$=$ _____

$3x^2 + 4x + 8$

**4** $(2x^2 - xy + y^2) + (3x^2 + 2xy - 3y^2)$
$(2x^2 - xy + y^2) + (3x^2 + 2xy - 3y^2)$
$= (2 + 3)x^2 + (-1 + 2)xy + [1 + (-3)]y^2$
$=$ _____

$5x^2 + xy - 2y^2$

In problems 5–8, perform the indicated subtractions of the polynomials.

5  $12y^2 - 7y^2$
   $12y^2 - 7y^2 = (12 - 7)y^2 = $ _____

$5y^2$

6  $(3x + 6) - (2x + 1)$
   $(3x + 6) - (2x + 1) = (3 - 2)x + (6 - 1) = $ _____

$x + 5$

7  $(5x^2 - 7) - (2x^2 + 3)$
   $(5x^2 - 7) - (2x^2 + 3) = (5 - 2)x^2 + (-7 - 3)$
   $= $ _____

$3x^2 - 10$

8  $(5x^2 + 3xy - 4y^2) - (2x^2 - 5xy + 6y^2)$
   $(5x^2 + 3xy - 4y^2) - (2x^2 - 5xy + 6y^2)$
   $= (5 - 2)x^2 + [3 - (-5)]xy + (-4 - 6)y^2$
   $= $ _____

$3x^2 + 8xy - 10y^2$

In problems 9 and 10, perform the indicated operations and combine like terms.

9  $(2x + 3) + (4x + 5) - (3x + 2)$
   $(2x + 3) + (4x + 5) - (3x + 2)$
   $= (2 + 4 - 3)x + (3 + 5 - 2)$
   $= $ _____

$3x + 6$

10  $(x^2 + 2x + 1) + (3x^2 - 4x + 6) - (2x^2 + x + 2)$
    $(x^2 + 2x + 1) + (3x^2 - 4x + 6) - (2x^2 + x + 2)$
    $= (1 + 3 - 2)x^2 + (2 - 4 - 1)x + (1 + 6 - 2)$
    $= $ _____

$2x^2 - 3x + 5$

## SOLUTIONS TO SELECTED ODD PROBLEMS  Section 2.2, text pages 68–69

1  $5x^2 + 7x^2 = (5 + 7)x^2 = 12x^2$

5  $-3t^2 + 7t^2 = (-3 + 7)t^2 = 4t^2$

9  $(2z^2 + 3z + 1) + (5z^2 + 2z + 4)$
   $= (2 + 5)z^2 + (3 + 2)z + (1 + 4) = 7z^2 + 5z + 5$

13  $(5c^2 - 3c^3 - c + 2c^4) + (4c^3 + 3c^4 - c^2 + 2c)$
    $= (2 + 3)c^4 + (-3 + 4)c^3 + (5 - 1)c^2 + (-1 + 2)c$
    $= 5c^4 + c^3 + 4c^2 + c$

17  $3x^2 - x^2 = 3x^2 + (-x^2) = [3 + (-1)]x^2 = 2x^2$

21  $(3t^3 + 2) - (-t^3 + 4) = (3t^3 + 2) + [-(-t^3 + 4)]$
    $= 3t^3 + 2 + t^3 - 4$
    $= (3 + 1)t^3 + (2 - 4)$
    $= 4t^3 - 2$

25  $(3s^4 - 4s^3 + 6s^2 + s - 1) - (4 - s + 2s^2 - 3s^3 - s^4)$
    $= (3s^4 - 4s^3 + 6s^2 + s - 1) + [-(4 - s + 2s^2 - 3s^3 - s^4)]$
    $= 3s^4 - 4s^3 + 6s^2 + s - 1 - 4 + s - 2s^2 + 3s^3 + s^4$
    $= (3 + 1)s^4 + (-4 + 3)s^3 + (6 - 2)s^2 + (1 + 1)s + (-1 - 4)$
    $= 4s^4 - s^3 + 4s^2 + 2s - 5$

29  $xy + 3xy + 8xy = (1 + 3 + 8)xy = 12xy$

33  $5mn^2 + (-3mn^2) + 2m^2n$
    $= [5 + (-3)]mn^2 + 2m^2n = 2mn^2 + 2m^2n$

37  $(7ts - 4) + (-3ts - 2)$
    $= 7ts - 4 - 3ts - 2$
    $= (7 - 3)ts + (-4 - 2)$
    $= 4ts - 6$

41  $(3w^2z + 4wz - 7wz^2) - (-2w^2z + 3wz^2 + wz)$
    $= (3w^2z + 4wz - 7wz^2) + [-(-2w^2z + 3wz^2 + wz)]$
    $= 3w^2z + 4wz - 7wz^2 + 2w^2z - 3wz^2 - wz$
    $= (3 + 2)w^2z + (4 - 1)wz + (-7 - 3)wz^2$
    $= 5w^2z + 3wz - 10wz^2$

45  $(t^3 - 2t^2 + 3t + 1) + (2t^3 + t^2 - 2t + 2) + (-t^3 + 3t^2 - 2t - 1)$
    $= t^3 - 2t^2 + 3t + 1 + 2t^3 + t^2 - 2t + 2 - t^3 + 3t^2 - 2t - 1$
    $= (1 + 2 - 1)t^3 + (-2 + 1 + 3)t^2 + (3 - 2 - 2)t + (1 + 2 - 1)$
    $= 2t^3 + 2t^2 - t + 2$

49  $(7u^3v^2 - 3u^2v + 2w) - (4w + u^2v - u^3v^2) - (-2u^3v^2 + 5u^2v + 3w)$
    $= (7u^3v^2 - 3u^2v + 2w) + [-(4w + u^2v - u^3v^2)] + [-(-2u^3v^2 + 5u^2v + 3w)]$
    $= 7u^3v^2 - 3u^2v + 2w - 4w - u^2v + u^3v^2 + 2u^3v^2 - 5u^2v - 3w$
    $= (7 + 1 + 2)u^3v^2 + (-3 - 1 - 5)u^2v + (2 - 4 - 3)w$
    $= 10u^3v^2 - 9u^2v - 5w$

53  $10x + 25(x - 6)$
    $= 10x + 25x - 150$
    $= 35x - 150$

## 2.3 Properties of Positive Integral Exponents

### SEMIPROGRAMMED PROBLEMS

In problems 1 and 2, use the property $a^m \cdot a^n = a^{m+n}$ to simplify each expression.

$7, x^9$      1  $x^2 \cdot x^7 = x^{2+\underline{\phantom{xx}}} = \underline{\phantom{xxx}}$

$1, x^7$      2  $x^4 \cdot x \cdot x^2 = x^{4+\underline{\phantom{xx}}+2} = \underline{\phantom{xxx}}$

In problems 3 and 4, use the property $(a^m)^n = a^{mn}$ to remove the parentheses and brackets in each expression.

$5, x^{15}$      3  $(x^5)^3 = x^{(\underline{\phantom{xx}}) \cdot 3} = \underline{\phantom{xxx}}$

$6, x^6$      4  $[(-x)^3]^2 = (-x)^{\underline{\phantom{xx}}} = \underline{\phantom{xxx}}$

In problems 5 and 6, use the property $(ab)^n = a^n b^n$ to remove the parentheses in each expression.

$m^5$      5  $(mn)^5 = \underline{\phantom{xxx}} \cdot n^5$

$2, 4x^2$      6  $(2x)^2 = 2^{\underline{\phantom{xx}}} x^2 = \underline{\phantom{xxx}}$

In problems 7 and 8, use the property $\left(\dfrac{a}{b}\right)^n = \dfrac{a^n}{b^n}$ to remove the parentheses in each expression.

$y^5$      7  $\left(\dfrac{x}{y}\right)^5 = \dfrac{x^5}{\underline{\phantom{xxx}}}$

$3, \dfrac{x^3}{27}$      8  $\left(\dfrac{x}{3}\right)^3 = \dfrac{x^3}{3^{\underline{\phantom{xx}}}} = \underline{\phantom{xxx}}$

## CHAPTER 2 THE ALGEBRA OF POLYNOMIALS

In problems 9 and 10, use the property that for $a \neq 0$ and $m > n$, $\dfrac{a^m}{a^n} = a^{m-n}$.

$x^2$  

9 $\dfrac{x^4}{x^2} = x^{4-2} = $ _____

$3, -27$  

10 $\dfrac{(-3)^7}{(-3)^4} = (-3)^{7-4} = (-3)$ ____ = _____

In problems 11 and 12, use the property that for $a \neq 0$ and $m < n$, $\dfrac{a^m}{a^n} = \dfrac{1}{a^{n-m}}$.

$\dfrac{1}{x^3}$

11 $\dfrac{x^4}{x^7} = \dfrac{1}{x^{7-4}} = $ _____

$5, -\dfrac{1}{x^5}$

12 $\dfrac{(-x)^3}{(-x)^8} = \dfrac{1}{(-x)^{\phantom{5}}} = $ _____

In problems 13 and 14, use the property that for $a \neq 0$ and $m = n$, $\dfrac{a^m}{a^n} = 1$.

1

13 $\dfrac{x^{10}}{x^{10}} = $ _____

1

14 $\dfrac{y^8}{y^8} = $ _____

In problems 15–17, use the properties of exponents above to simplify the expressions.

$x^{12}y^{18}$

15 $(x^2y^3)^6 = (x^2)^6(y^3)^6 = $ _____

16 $\left(\dfrac{3ab^3}{2a^4c}\right)^4 = \dfrac{(3ab^3)^4}{(2a^4c)^4} = \dfrac{3^4a^4(b^3)^4}{2^4(a^4)^4c^4}$

$81b^{12}$

$\phantom{16} = \dfrac{81a^4b^{12}}{16a^{16}c^4} = \dfrac{\phantom{81b^{12}}}{16a^{12}c^4}$

$10^4$

17 $(3 \times 10^3)(2.1 \times 10^4) = (3 \cdot 2.1) \cdot (10^3 \cdot $ ____ $)$

$10^7$

$\phantom{17} = 6.3 \times $ _____

## SOLUTIONS TO SELECTED ODD PROBLEMS   Section 2.3, text pages 76–77

1   $3^2 \cdot 3^3 = 3^{2+3} = 3^5 = 243$     5   $-x^5 \cdot x^7 = -x^{5+7} = -x^{12}$

9   $(-v)^3 \cdot (-v)^5 = (-v)^{3+5} = (-v)^8 = v^8$     13   $[(-2)^3]^2 = (-2)^{3 \cdot 2} = (-2)^6 = 64$

17   $(t^2)^{11} = t^{2 \cdot 11} = t^{22}$     21   $(2x)^4 = 2^4 \cdot x^4 = 16x^4$     25   $(xyz)^7 = x^7y^7z^7$

29   $[-3(-x)y]^3 = (-3)^3(-x)^3y^3$     33   $\left(\dfrac{-2}{3}\right)^3 = \dfrac{(-2)^3}{3^3} = \dfrac{-8}{27} = -\dfrac{8}{27}$
$\phantom{29} = (-27)(-x^3)y^3 = 27x^3y^3$

37   $\left(\dfrac{a}{-b}\right)^5 = \dfrac{a^5}{(-b)^5} = \dfrac{a^5}{-b^5} = -\dfrac{a^5}{b^5}$     41   $\dfrac{3^5}{3^2} = 3^{5-2} = 3^3 = 27$

45   $\dfrac{-x^8}{-x^3} = \dfrac{x^8}{x^3} = x^{8-3} = x^5$     49   $\dfrac{w^{13}}{w^{10}} = w^{13-10} = w^3$     53   $\dfrac{u^4v^7}{uv^3} = u^{4-1} \cdot v^{7-3} = u^3v^4$

57   $\left(\dfrac{3a^5b^3}{2a^2b^6}\right)^3 = \left(\dfrac{3a^{5-2}}{2b^{6-3}}\right)^3$     61   $(2 \times 10^5)(4.8 \times 10^9) = (2)(4.8)(10^5 \times 10^9)$
$\phantom{57} = \left(\dfrac{3a^3}{2b^3}\right)^3 = \dfrac{3^3(a^3)^3}{2^3(b^3)^3} = \dfrac{27a^9}{8b^9}$     $\phantom{61} = 9.6 \times 10^{5+9}$
$\phantom{61 \quad} = 9.6 \times 10^{14}$

65    $\dfrac{7.8 \times 10^9}{3.9 \times 10^5} = \left(\dfrac{7.8}{3.9}\right)\left(\dfrac{10^9}{10^5}\right) = 2 \times 10^{9-5} = 2 \times 10^4$

69    $(-1)^2 = 1, (-1)^4 = 1, (-1)^6 = 1, \ldots$   In general, $(-1)^n = 1$ where $n$ is an even positive integer.

## 2.4 Multiplication of Polynomials

### SEMIPROGRAMMED PROBLEMS

In problems 1–8, find the products.

1    $(-3x^2)(5x)$
$(-3x^2)(5x) = (-3 \cdot 5)(x^2 \cdot x) = $ _____

$-15x^3$

2    $(4x^2y)(3x^3y^2)$
$(4x^2y)(3x^3y^2) = (4 \cdot 3)(x^2 \cdot x^3)(y \cdot y^2) = $ _____

$12x^5y^3$

3    $-2x^2(x^2 - 3)$
$-2x^2(x^2 - 3) = (-2x^2)(x^2) + (-2x^2)(-3)$
$= $ _____

$-2x^4 + 6x^2$

4    $5xy(x^2 + 2xy - y^2)$
$5xy(x^2 + 2xy - y^2) = (5xy)(x^2) + (5xy)(2xy) + (5xy)(-y^2)$
$= $ _____

$5x^3y + 10x^2y^2 - 5xy^3$

5    $(2x - 1)(x^2 + x + 1)$
$(2x - 1)(x^2 + x + 1) = (2x - 1)x^2 + (2x - 1)x + (2x - 1)1$
$= $ _____ $+ 2x^2 - x + 2x - 1$
$= $ _____

$2x^3 - x^2$
$2x^3 + x^2 + x - 1$

6    $(x + y)(x^2 + xy + y^2)$
$(x + y)(x^2 + xy + y^2)$
$= (x + y)x^2 + (x + y)xy + $ _____
$= x^3 + x^2y + x^2y + xy^2 + $ _____
$= $ _____

$(x + y)y^2$
$xy^2 + y^3$
$x^3 + 2x^2y + 2xy^2 + y^3$

7    $(x - 5)(2x - 7)$
$(x - 5)(2x - 7) = 2x^2 - 7x - $ \_\_\_\_ $x + 35$
$= 2x^2 - $ \_\_\_\_ $x + 35$

10
17

8    $(x + 2y)(3x - 4y)$
$(x + 2y)(3x - 4y) = 3x^2 - 4$\_\_\_\_ $+ 6xy - 8y^2$
$= 3x^2 + $ \_\_\_\_$xy - 8y^2$

$xy$
2

In problems 9–15, use Special Products to find each product.

9    $(4y + 9z)^2$
$(4y + 9z)^2 = ($\_\_\_$)^2 + 2($\_\_\_$)($\_\_\_$) + ($\_\_\_$)^2$
$= $ _____

$4y, 4y, 9z, 9z$
$16y^2 + 72yz + 81z^2$

10    $(2x - 7y)^2$
$(2x - 7y)^2 = ($\_\_\_$)^2 - 2($\_\_\_$)($\_\_\_$) + ($\_\_\_$)^2$
$= $ _____

$2x, 2x, 7y, 7y$
$4x^2 - 28xy + 49y^2$

## CHAPTER 2 THE ALGEBRA OF POLYNOMIALS

$3x, 4$
$9x^2 - 16$

$x, 3, x, 3, 3$
$x^3 + 9x^2 + 27x + 27$

$x, 2y, x, 2y, 2y$
$x^3 - 6x^2y + 12xy^2 - 8y^3$

$4, a, 64 + a^3$

$3m, 1$
$27m^3 - 1$

11   $(3x - 4)(3x + 4)$
$(3x - 4)(3x + 4) = (\underline{\quad})^2 - (\underline{\quad})^2$
$= \underline{\qquad}$

12   $(x + 3)^3$
$(x + 3)^3 = x^3 + 3(\underline{\quad})^2(\underline{\quad}) + 3(\underline{\quad})(\underline{\quad})^2 + (\underline{\quad})^3$
$= \underline{\qquad}$

13   $(x - 2y)^3$
$(x - 2y)^3$
$= x^3 - 3(\underline{\quad})^2(\underline{\quad}) + 3(\underline{\quad})(\underline{\quad})^2 - (\underline{\quad})^3$
$= \underline{\qquad}$

14   $(4 + a)(16 - 4a + a^2)$
$(4 + a)(16 - 4a + a^2) = (\underline{\quad})^3 + (\underline{\quad})^3 = \underline{\qquad}$

15   $(3m - 1)(9m^2 + 3m + 1) = (\underline{\quad})^3 - (\underline{\quad})^3$
$= \underline{\qquad}$

## SOLUTIONS TO SELECTED ODD PROBLEMS   Section 2.4, text pages 87–88

1   $(2x^2)(3x^4) = (2 \cdot 3)(x^2 \cdot x^4) = 6x^6$

5   $(7u^2v^3)(-4u^3v^4) = [7 \cdot (-4)](u^2 \cdot u^3)(v^3 \cdot v^4) = -28u^5v^7$

9   $(2ab)(3a^2c)(-4b^2c^3) = [2 \cdot 3 \cdot (-4)](a \cdot a^2)(b \cdot b^2)(c \cdot c^3) = -24a^3b^3c^4$

13   $t^2(t + 2) = (t^2)(t) + 2(t^2) = t^3 + 2t^2$

17   $(-2xy^2)(2x^2 - 3xy + 5y^2)$
$= (-2xy^2)[2x^2 + (-3xy) + 5y^2]$
$= (-2xy^2)(2x^2) + (-2xy^2)(-3xy) + (-2xy^2)(5y^2)$
$= -4x^3y^2 + 6x^2y^3 - 10xy^4$

21   $\begin{array}{r} x + y - 1 \\ x + y \\ \hline x^2 + xy - x \\ xy + y^2 - y \\ \hline x^2 + 2xy - x + y^2 - y \end{array}$

25   $\begin{array}{r} m^3 + 2m^2 - 3m + 4 \\ m^2 + 3 \\ \hline m^5 + 2m^4 - 3m^3 + 4m^2 \\ + 3m^3 + 6m^2 - 9m + 12 \\ \hline m^5 + 2m^4 \phantom{- 3m^3} + 10m^2 - 9m + 12 \end{array}$

29   $\begin{array}{r} x^3 - 3x^2y + 3xy^2 - y^3 \\ x^2 + 2xy + y^2 \\ \hline x^5 - 3x^4y + 3x^3y^2 - x^2y^3 \\ 2x^4y - 6x^3y^2 + 6x^2y^3 - 2xy^4 \\ x^3y^2 - 3x^2y^3 + 3xy^4 - y^5 \\ \hline x^5 - x^4y - 2x^3y^2 + 2x^2y^3 + xy^4 - y^5 \end{array}$

33   $(u - 4)(u - 5) = u^2 - 9u + 20$
    First term:    $(u)(u) = u^2$
    Middle term: $(-5)(u) + (-4)(u) = -9u$
    Last term:    $(-4)(-5) = 20$

37   $(y + 3)(y - 6) = y^2 - 3y - 18$
    First term:    $(y)(y) = y^2$
    Middle term: $(-6)(y) + (3)(y) = -3y$
    Last term:    $(3)(-6) = -18$

41   $(5w + 4)(w - 1) = 5w^2 - w - 4$
    First term:    $(5w)(w) = 5w^2$
    Middle term: $(5w)(-1) + 4(w) = -w$
    Last term:    $(4)(-1) = -4$

45   $(6m - 5n)(4m + 3n) = 24m^2 - 2mn - 15n^2$
    First term:    $(6m)(4m) = 24m^2$
    Middle term: $(6m)(3n) + (-5n)(4m) = -2mn$
    Last term:    $(-5n)(3n) = -15n^2$

49   $(10v - 7)(5v - 8) = 50v^2 - 115v + 56$
    First term:    $(10v)(5v) = 50v^2$
    Middle term: $(10v)(-8) + (-7)(5v) = -115v$
    Last term:    $(-7)(-8) = 56$

53   $(2s + t)^2 = (2s)^2 + 2(2s)(t) + (t)^2$
$= 4s^2 + 4st + t^2$

57   $(3x - 5)^2 = (3x)^2 - 2(3x)(5) + (5)^2$
$= 9x^2 - 30x + 25$

61   $(x + y)(x - y) = x^2 - y^2$

65   $(2m + 9)(2m - 9) = (2m)^2 - (9)^2 = 4m^2 - 81$

69  $(5u + 6v)(5u - 6v) = (5u)^2 - (6v)^2 = 25u^2 - 36v^2$

73  Verify by the vertical scheme:
$$\begin{array}{r} a^2 - ab + b^2 \\ a + b \\ \hline a^3 - a^2b + ab^2 \\ a^2b - ab^2 + b^3 \\ \hline a^3 \qquad\qquad + b^3 \end{array}$$

77  Using Special Product (vi), we have:
$(x + 1)(x^2 - x + 1) = x^3 + 1^3 = x^3 + 1$

81  Using Special Product (vii), we have:
$(u - 3v)(u^2 + 3uv + 9v^2) = u^3 - (3v)^3 = u^3 - 27v^3$

85  $(2x + 3y)(4x^2 - 6xy + 9y^2) = (2x)^3 + (3y)^3 = 8x^3 + 27y^3$

89  $(u^2 - v^2)(u^4 + u^2v^2 + v^4) = (u^2)^3 - (v^2)^3 = u^6 - v^6$

## 2.5 Factoring Polynomials

### SEMIPROGRAMMED PROBLEMS

In problems 1–3, factor each polynomial by factoring out the common factor.

3

$x + 3y - 1$

$5 + 4xy + x^2y^2$

1  $5x + 15$
$5x + 15 = 5 \cdot x + 5 \cdot 3 = 5(x + \underline{\qquad})$

2  $x^3 + 3x^2y - x^2$
$x^3 + 3x^2y - x^2 = x^2 \cdot x + x^2 \cdot 3y - x^2 \cdot 1$
$\qquad = x^2(\underline{\qquad})$

3  $5xy + 4x^2y^2 + x^3y^3$
$5xy + 4x^2y^2 + x^3y^3 = xy \cdot 5 + xy \cdot 4xy + xy \cdot x^2y^2$
$\qquad = xy(\underline{\qquad})$

In problems 4–6, factor out the common binomial factor in each expression.

$2m + 5$

$z - y$

$m + 1$

4  $(2m + 5)x + (2m + 5)y$
$(2m + 5)x + (2m + 5)y = (\underline{\qquad})(x + y)$

5  $(x + y)z - (x + y)y$
$(x + y)z - (x + y)y = (x + y)(\underline{\qquad})$

6  $(x - 3y)m + x - 3y$
$(x - 3y)m + x - 3y = (x - 3y)(\underline{\qquad})$

In problems 7–10, factor each expression by grouping.

$ax - ay$
$a(x - y)$
$a + b$

$1 - x$
$-1$
$(m - 1)(x - 1)$

7  $ax - ay + bx - by$
$ax - ay + bx - by = (\underline{\qquad}) + (bx - by)$
$\qquad = \underline{\qquad} + b(x - y)$
$\qquad = (\underline{\qquad})(x - y)$

8  $mx + 1 - m - x$
$mx + 1 - m - x = (mx - m) + (\underline{\qquad})$
$\qquad = m(x - 1) + (\underline{\qquad})(x - 1)$
$\qquad = \underline{\qquad}$

$5z^2 - xz^2$
$z^2(5 - x)$
$(y^2 + z^2)(5 - x)$

9  $5y^2 - xz^2 - y^2x + 5z^2$
$5y^2 - xz^2 - y^2x + 5z^2 = (5y^2 - y^2x) + (\underline{\phantom{xxxx}})$
$= y^2(5 - x) + \underline{\phantom{xxxx}}$
$= \underline{\phantom{xxxx}}$

10  $3mx - n + 3nx - p + 3px - m$
$3mx - n + 3nx - p + 3px - m$

$3nx - n, 3px - p$
$n(3x - 1), p(3x - 1)$
$(3x - 1)(m + n + p)$

$= (3mx - m) + (\underline{\phantom{xxxx}}) + (\underline{\phantom{xxxx}})$
$= m(3x - 1) + \underline{\phantom{xxxx}} + \underline{\phantom{xxxx}}$
$= \underline{\phantom{xxxx}}$

## SOLUTIONS TO SELECTED ODD PROBLEMS   Section 2.5, text pages 92–93

1  $x^2 - x$:   GCF $= x$
$= (x)(x) - x(1) = x(x - 1)$

5  $4x^2 + 7xy$:   GCF $= x$
$= (x)(4x) + (x)(7y) = x(4x + 7y)$

9  $6p^2q + 24pq^2$:   GCF $= 6pq$
$= (6pq)(p) + (6pq)(4q) = 6pq(p + 4q)$

13  $12x^3y - 48x^2y^2$:   GCF $= 12x^2y$
$= (12x^2y)(x) - (12x^2y)(4y) = 12x^2y(x - 4y)$

17  $x^3y^2 + x^2y^3 + 2xy^4$:   GCF $= xy^2$
$= (xy^2)(x^2) + (xy^2)(xy) + (xy^2)(2y^2) = xy^2(x^2 + xy + 2y^2)$

21  $3x(2a + b) + 5y(2a + b) = (3x + 5y)(2a + b)$

25  $m(x - y) + (y - x)$
$= m(x - y) - (x - y)$, since $y - x = -(x - y)$
$= (m - 1)(x - y)$

29  $y(xy + 2)^3 - 5x(xy + 2)^2 + 7(xy + 2)$
$= (xy + 2)[y(xy + 2)^2 - 5x(xy + 2) + 7]$

33  $x^5 + 3x^4 + x + 3 = (x^5 + 3x^4) + (x + 3)$
$= x^4(x + 3) + (x + 3)$
$= (x^4 + 1)(x + 3)$

37  $ab^2 - b^2c - ad + cd = (ab^2 - b^2c) + (-ad + cd)$
$= b^2(a - c) - d(a - c)$
$= (b^2 - d)(a - c)$

41  $x^2 - ax + bx - ab = (x^2 - ax) + (bx - ab)$
$= x(x - a) + b(x - a)$
$= (x + b)(x - a)$

45  $2x^3 + x^2y - x^2 + 2xy + y^2 - y = (2x^3 + x^2y - x^2) + (2xy + y^2 - y)$
$= x^2(2x + y - 1) + y(2x + y - 1)$
$= (x^2 + y)(2x + y - 1)$

## 2.6  Factoring by Recognizing Special Products

### SEMIPROGRAMMED PROBLEMS

In problems 1–3, use the difference of two squares to factor each expression completely.

$2xy, 3m$
$2xy - 3m, 2xy + 3m$
$3ab^2, 4$
$3ab^2 - 4, 3ab^2 + 4$

$2(a - b), 7c$
$2(a - b) - 7c, 2(a - b) + 7c$

1  $4x^2y^2 - 9m^2 = (\underline{\phantom{xxx}})^2 - (\underline{\phantom{xx}})^2$
$= (\underline{\phantom{xxxxx}})(\underline{\phantom{xxxxx}})$

2  $9a^2b^4 - 16 = (\underline{\phantom{xxx}})^2 - (\underline{\phantom{xx}})^2$
$= (\underline{\phantom{xxxxx}})(\underline{\phantom{xxxxx}})$

3  $4(a - b)^2 - 49c^2$
$= [\underline{\phantom{xxxx}}]^2 - (\underline{\phantom{xx}})^2$
$= [\underline{\phantom{xxxxx}}][\underline{\phantom{xxxxx}}]$

## 2.6 SOLUTIONS TO SELECTED ODD PROBLEMS

In problems 4–7, use the sum or difference of two cubes to factor each expression completely.

$b$

$-, +$

$+, +$

$3b^2$

$12a^2b^2$

$2y$

$-(x-5)(2y)$

$(x-5)^2 - 2y(x-5) + 4y^2$

4.  $125a^3 + b^3 = (5a)^3 + (\underline{\phantom{b}})^3$
    $= (5a+b)(25a^2 \underline{\phantom{-}} 5ab \underline{\phantom{+}} b^2)$
5.  $27a^3 - b^3 = (3a)^3 - (b)^3$
    $= (3a-b)(9a^2 \underline{\phantom{+}} 3ab \underline{\phantom{+}} b^2)$
6.  $64a^6 + 27b^6 = (4a^2)^3 + (\underline{\phantom{xx}})^3$
    $= (4a^2 + 3b^2)(16a^4 - \underline{\phantom{xxxxx}} + 9b^4)$
7.  $(x-5)^3 + 8y^3 = (x-5)^3 + (\underline{\phantom{xx}})^3$
    $= [(x-5) + 2y][(x-5)^2 \underline{\phantom{xxxxx}} + 4y^2]$
    $= (x-5+2y)[\underline{\phantom{xxxxxxxxxxxxxx}}]$

In problems 8–10, use common factors, grouping, or special products to factor the expressions.

$x^2 - y^2$

$(x-y)(x+y)$

$x^3 - 8$

$x^2 + 2x + 4$

$z^2 + 10z + 25$

$z + 5$

$z + 5, x + 2y$

$(x + 2y - z - 5) \cdot (x + 2y + z + 5)$

8.  $x^5y^3 - x^3y^5 = x^3y^3(\underline{\phantom{xxxx}})$
    $= x^3y^3 \underline{\phantom{xxxxxx}}$
9.  $4x^4y^2 - 32xy^2 = 4xy^2(\underline{\phantom{xxxx}})$
    $= 4xy^2(x-2)(\underline{\phantom{xxxx}})$
10. $x^2 - z^2 + 4xy - 10z + 4y^2 - 25$
    $= (x^2 + 4xy + 4y^2) - (\underline{\phantom{xxxxx}})$
    $= (x+2y)^2 - (\underline{\phantom{xx}})^2$
    $= [(x+2y) - (\underline{\phantom{xx}})][(\underline{\phantom{xx}}) + (z+5)]$
    $= \underline{\phantom{xxxxxxxxxxxxxxxx}}$

**SOLUTIONS TO SELECTED ODD PROBLEMS**  Section 2.6, text pages 97–98

1.  $x^2 - 4 = x^2 - 2^2 = (x-2)(x+2)$
5.  $36 - 25t^2 = 6^2 - (5t)^2 = (6-5t)(6+5t)$
9.  $a^2b^2 - c^2 = (ab)^2 - c^2 = (ab-c)(ab+c)$
13. $81x^4 - 1 = (9x^2)^2 - 1^2 = (9x^2 - 1)(9x^2 + 1) = (3x-1)(3x+1)(9x^2+1)$
17. $(x+y)^2 - (a-b)^2 = [(x+y) - (a-b)][(x+y) + (a-b)]$
    $= (x+y-a+b)(x+y+a-b)$
21. $x^3 + 1 = x^3 + 1^3 = (x+1)(x^2 - x \cdot 1 + 1^2)$
    $= (x+1)(x^2 - x + 1)$
25. $27w^3 + z^3 = (3w)^3 + z^3 = (3w+z)[(3w)^2 - (3w)(z) + (z)^2]$
    $= (3w+z)(9w^2 - 3wz + z^2)$
29. $w^3 - 8y^3z^3 = w^3 - (2yz)^3 = (w-2yz)[(w)^2 + (w)(2yz) + (2yz)^2]$
    $= (w-2yz)(w^2 + 2wyz + 4y^2z^2)$
33. $(y+1)^3 + (w+2)^3 = [(y+1) + (w+2)][(y+1)^2 - (y+1)(w+2) + (w+2)^2]$
    $= (y+w+3)[(y+1)^2 - (y+1)(w+2) + (w+2)^2]$
37. $x^9 - 1 = (x^3)^3 - 1^3 = (x^3 - 1)[(x^3)^2 + (x^3)(1) + (1)^2]$
    $= (x-1)(x^2 + x + 1)(x^6 + x^3 + 1)$
41. $64y - 4y^3 = 4y(16 - y^2) = 4y(4-y)(4+y)$

45    $7x^7y + 7xy^7 = 7xy(x^6 + y^6) = 7xy[(x^2)^3 + (y^2)^3]$
$= 7xy(x^2 + y^2)[(x^2)^2 - (x^2)(y^2) + (y^2)^2]$
$= 7xy(x^2 + y^2)(x^4 - x^2y^2 + y^4)$

49    $2u^7 - 128u = 2u(u^6 - 64) = 2u(u^3 - 8)(u^3 + 8)$
$= 2u(u - 2)(u^2 + 2u + 4)(u + 2)(u^2 - 2u + 4)$

53    $9u^2 - 42uv + 49v^2 = (3u)^2 - 2(3u)(7v) + (7v)^2 = (3u - 7v)^2$

57    $w^2 - y^2 + 2w - 2yz + 1 - z^2 = (w^2 + 2w + 1) + (-y^2 - 2yz - z^2)$
$= (w^2 + 2w + 1) - (y^2 + 2yz + z^2)$
$= (w + 1)^2 - (y + z)^2$
$= [(w + 1) - (y + z)][(w + 1) + (y + z)]$
$= (w + 1 - y - z)(w + 1 + y + z)$

61    $4m^4 + n^4 = (4m^4 + 4m^2n^2 + n^4) - 4m^2n^2$
$= (2m^2 + n^2)^2 - (2mn)^2$
$= (2m^2 + n^2 - 2mn)(2m^2 + n^2 + 2mn)$

## 2.7   Factoring Trinomials of the Form $ax^2 + bx + c$

### SEMIPROGRAMMED PROBLEMS

In problems 1–8, factor each trinomial completely.

| | |
|---|---|
| 6, 7 | **1**   $x^2 + 13x + 42 = x^2 + (6 + 7)x + 7 \cdot 6$ <br> $= (x + \underline{\ \ \ })(x + \underline{\ \ \ })$ |
| $x + 6$ | **2**   $x^2 + 10x + 24 = (x + 4)(\underline{\ \ \ \ \ \ \ \ })$ |
| $b + 5$ | **3**   $b^2 - 6b - 55 = (b - 11)(\underline{\ \ \ \ \ \ \ \ })$ |
| 8, 9 | **4**   $a^2 + 17a + 72 = (a + \underline{\ \ \ })(a + \underline{\ \ \ })$ |
| $5y, 3y$ | **5**   $15y^2 - 8y + 1 = (\underline{\ \ \ } - 1)(\underline{\ \ \ } - 1)$ |
| $3x + 2y$ | **6**   $6x^2 - 5xy - 6y^2 = (2x - 3y)(\underline{\ \ \ \ \ \ \ \ })$ |
| $2m - 3n$ | **7**   $6m^2 - 13mn + 6n^2 = (3m - 2n)(\underline{\ \ \ \ \ \ \ \ })$ |
| $2x - 5$ | **8**   $10x^2 - 11x - 35 = (5x + 7)(\underline{\ \ \ \ \ \ \ \ })$ |

In problems 9–11, use common factors and the factoring of trinomials to factor each polynomial completely.

| | |
|---|---|
| $x^2 - 5x + 6$ | **9**   $x^3y - 5x^2y + 6xy = xy(\underline{\ \ \ \ \ \ \ \ })$ |
| $x - 2, x - 3$ | $= xy(\underline{\ \ \ \ })(\underline{\ \ \ \ })$ |
| $y^2 + 7y + 12$ | **10**   $aby^2 + 7aby + 12ab = ab(\underline{\ \ \ \ \ \ \ \ })$ |
| $y + 3, y + 4$ | $= ab(\underline{\ \ \ \ })(\underline{\ \ \ \ })$ |
| $mn^2x, y^2 - 2y - 15$ | **11**   $mn^2xy^2 - 2mn^2xy - 15mn^2x = (\underline{\ \ \ \ })(\underline{\ \ \ \ })$ |
| $mn^2x(y - 5)(y + 3)$ | $= \underline{\ \ \ \ \ \ \ \ \ \ \ \ \ \ \ \ }$ |

### SOLUTIONS TO SELECTED ODD PROBLEMS    Section 2.7, text page 105

**1**   $x^2 + 4x + 3$
     First term: $(x)(x)$; Last term: $3 \cdot 1$
     $x^2 + 4x + 3 = (x + 3)(x + 1)$

**5**   $y^2 + 15y + 36$
     First term: $(y)(y)$; Last term: $36 \cdot 1, 18 \cdot 2, 12 \cdot 3, 9 \cdot 4, 6 \cdot 6$
     $y^2 + 15y + 36 = (y + 12)(y + 3)$

9    $u^2 - 16u + 63$
       First term: $(u)(u)$; Last term: $63 \cdot 1, 21 \cdot 3, 9 \cdot 7$
       $u^2 - 16u + 63 = (u - 9)(u - 7)$

13    $x^2 - 7x - 18$
       First term: $(x)(x)$; Last term: $18 \cdot 1, 9 \cdot 2, 6 \cdot 3$
       $x^2 - 7x - 18 = (x - 9)(x + 2)$

17    $12 - x^2 - 4x = 12 - 4x - x^2$
       First term: $12 \cdot 1, 6 \cdot 2, 4 \cdot 3$; Last term: $(x)(x)$
       $12 - 4x - x^2 = (6 + x)(2 - x)$

21    $2w^2 + 7w + 3$
       First term: $(2w)(w)$; Last term: $3 \cdot 1$
       $2w^2 + 7w + 3 = (2w + 1)(w + 3)$

25    $5y^2 - 11y + 2$
       First term: $(5y)(y)$; Last term: $2 \cdot 1$
       $5y^2 - 11y + 2 = (5y - 1)(y - 2)$

29    $6x^2 + 13x + 6$
       First term: $(6x)(x), (3x)(2x)$; Last term: $6 \cdot 1, 3 \cdot 2$
       $6x^2 + 13x + 6 = (3x + 2)(2x + 3)$

33    $12v^2 + 17v - 5$
       First term: $(12v)(v), (6v)(2v), (4v)(3v)$; Last term: $5 \cdot 1$
       $12v^2 + 17v - 5 = (4v - 1)(3v + 5)$

37    $12 - 2w^2 - 5w = 12 - 5w - 2w^2$
       First term: $12 \cdot 1, 6 \cdot 2, 3 \cdot 4$; Last term: $(2w)(w)$
       $12 - 5w - 2w^2 = (3 - 2w)(4 + w)$

41    $5x^3 - 55x^2 + 140x$
       $= 5x(x^2 - 11x + 28)$    [Factoring out the GCF, $5x$]
       $= 5x(x - 4)(x - 7)$    [Factoring the trinomial]

45    $x^2y^2 + 10xy^2 + 21y^2$
       $= y^2(x^2 + 10x + 21)$    [Factoring out the GCF, $y^2$]
       $= y^2(x + 7)(x + 3)$    [Factoring the trinomial]

49    $wx^2y - 9wxy + 14wy$
       $= wy(x^2 - 9x + 14)$    [Factoring out the GCF, $wy$]
       $= wy(x - 7)(x - 2)$    [Factoring the trinomial]

## 2.8   Division of Polynomials

### SEMIPROGRAMMED PROBLEMS

In problems 1–4, divide as indicated and simplify.

$4x^3$

     1    $8x^5 \div 2x^2 = \dfrac{8x^5}{2x^2} =$ _____

$3x^2y^3$

     2    $12x^7y^4 \div 4x^5y = \dfrac{12x^7y^4}{4x^5y} =$ _____

     3    $(x^3 + 5x^2 + x) \div x = \dfrac{x^3 + 5x^2 + x}{x}$

           $= \dfrac{x^3}{x} + \dfrac{5x^2}{x} + \dfrac{x}{x}$

$x^2 + 5x + 1$

           $=$ _____

# CHAPTER 2 THE ALGEBRA OF POLYNOMIALS

$\dfrac{15x^4y}{5x^2y^2}$

$2x + \dfrac{3x^2}{y}$

6x

$3x^2$

$9x$

15

$12x^2$

$32x$

$15x$

$90x$

$4x^2$

$-14x^2$

0

---

**4** $(10x^3y^2 + 15x^4y) \div 5x^2y^2 = \dfrac{10x^3y^2 + 15x^4y}{5x^2y^2}$

$\qquad\qquad\qquad\qquad = \dfrac{10x^3y^2}{5x^2y^2} + \underline{\qquad}$

$\qquad\qquad\qquad\qquad = \underline{\qquad}$

In problems 5–10, divide each polynomial.

**5** $2x^2 + 9x + 9$ by $x + 3$

$$\begin{array}{r} 2x + 3 \\ x+3 \overline{\smash{)}\,2x^2 + 9x + 9} \\ \underline{2x^2 + \phantom{xxx}} \\ 3x + 9 \\ \underline{3x + 9} \\ 0 \end{array}$$

**6** $2x^3 - 9x^2 + 19x + 15$ by $2x - 3$

$$\begin{array}{r} x^2 - 3x + 5 \\ 2x-3 \overline{\smash{)}\,2x^3 - 9x^2 + 19x + 15} \\ \underline{2x^3 - \phantom{xxx}} \\ -6x^2 + 19x \\ \underline{-6x^2 + \phantom{xxx}} \\ 10x + 15 \\ \underline{10x - \phantom{xxx}} \\ 30 \end{array}$$

**7** $3x^3 - 4x^2 - 36x + 16$ by $x - 4$

$$\begin{array}{r} 3x^2 + 8x - 4 \\ x-4 \overline{\smash{)}\,3x^3 - 4x^2 - 36x + 16} \\ \underline{3x^3 - \phantom{xxx}} \\ 8x^2 - 36x \\ \underline{8x^2 - \phantom{xxx}} \\ -4x + 16 \\ \underline{-4x + 16} \\ 0 \end{array}$$

**8** $21x^4 - 42x^3 - 45x^2 + 102x - 24$ by $3x - 6$

$$\begin{array}{r} 7x^3 - \underline{\phantom{xxx}} + 4 \\ 3x-6 \overline{\smash{)}\,21x^4 - 42x^3 - 45x^2 + 102x - 24} \\ \underline{21x^4 - 42x^3} \\ -45x^2 + 102x \\ \underline{-45x^2 + \phantom{xxx}} \\ 12x - 24 \\ \underline{12x - 24} \\ 0 \end{array}$$

**9** $8x^3 + 6x^2 - 29x + 15$ by $2x + 5$

$$\begin{array}{r} \underline{\phantom{xxx}} - 7x + 3 \\ 2x+5 \overline{\smash{)}\,8x^3 + 6x^2 - 29x + 15} \\ \underline{8x^3 + 20x^2} \\ \underline{\phantom{xxxx}} - 29x \\ -14x^2 - 35x \\ 6x + 15 \\ 6x + 15 \\ \underline{\phantom{xxx}} \end{array}$$

$x^3$

$x^2y^2$

$xy^3 - y^4$
0

10  $x^4 - y^4$ by $x - y$

$$\begin{array}{r} \phantom{x-y)x^4+0x^3y+0x^2y^2+}+x^2y+xy^2+y^3 \\ x-y\overline{)x^4+0x^3y+0x^2y^2+0xy^3-y^4} \\ \underline{x^4-\phantom{0}x^3y} \\ x^3y \\ \underline{x^3y-\phantom{0}x^2y^2} \\ x^2y^2-\phantom{0}xy^3 \\ \underline{xy^3-y^4} \end{array}$$

## SOLUTIONS TO SELECTED ODD PROBLEMS   Section 2.8, text page 112

1  $\dfrac{6x^5}{2x^2} = 3x^3$

5  $\dfrac{14u^3vw^4}{7u^5vw^2} = \dfrac{2w^2}{u^2}$

9  $\dfrac{4x^2y^3 - 16xy^3 + 4xy}{2xy} = \dfrac{4x^2y^3}{2xy} - \dfrac{16xy^3}{2xy} + \dfrac{4xy}{2xy}$

$\phantom{9\ \dfrac{4x^2y^3 - 16xy^3 + 4xy}{2xy}} = 2xy^2 - 8y^2 + 2$

13  
$$\begin{array}{r} x-2 \\ x-5\overline{)x^2-7x+10} \\ \text{subtract}\longrightarrow \underline{x^2-5x} \\ -2x+10 \\ \text{subtract}\longrightarrow \underline{-2x+10} \\ 0 \end{array}$$

Check: $(x-5)(x-2) + 0 = x^2 - 7x + 10$

17  
$$\begin{array}{r} w^2+w-4 \\ w+2\overline{)w^3+3w^2-2w-5} \\ \text{subtract}\longrightarrow \underline{w^3+2w^2} \\ w^2-2w \\ \text{subtract}\longrightarrow \underline{w^2+2w} \\ -4w-5 \\ \text{subtract}\longrightarrow \underline{-4w-8} \\ 3 = R \end{array}$$

Check: $(w^2+w-4)(w+2) + 3 = w^3 + 3w^2 - 2w - 5$

21  
$$\begin{array}{r} x^2+3x+4 \\ x^2+2x-1\overline{)x^4+5x^3+\phantom{0}9x^2+5x-4} \\ \text{subtract}\longrightarrow \underline{x^4+2x^3-\phantom{0}x^2} \\ 3x^3+10x^2+5x \\ \text{substract}\longrightarrow \underline{3x^3+\phantom{0}6x^2-3x} \\ 4x^2+8x-4 \\ \text{subtract}\longrightarrow \underline{4x^2+8x-4} \\ 0 \end{array}$$

Check: $(x^2+3x+4)(x^2+2x-1) + 0 = x^4 + 5x^3 + 9x^2 + 5x - 4$

25  
$$\begin{array}{r} x+2 \\ x^4+x^3-2x^2+6x-12\overline{)x^5+3x^4+0x^3+2x^2+\phantom{0}0x-24} \\ \text{subtract}\longrightarrow \underline{x^5+\phantom{0}x^4-2x^3+6x^2-12x} \\ 2x^4+2x^3-4x^2+12x-24 \\ \text{subtract}\longrightarrow \underline{2x^4+2x^3-4x^2+12x-24} \\ 0 \end{array}$$

Check: $(x+2)(x^4+x^3-2x^2+6x-12) + 0 = x^5 + 3x^4 + 2x^2 - 24$

29
$$\require{enclose}\begin{array}{r}2m^2 - mn + 2n^2\phantom{)}\\m+n\enclose{longdiv}{2m^3 + m^2n + mn^2 + 4n^3}\end{array}$$

subtract ⟶ $2m^3 + 2m^2n$
$\phantom{xxxxxxxx}-m^2n + mn^2$
subtract ⟶ $-m^2n - mn^2$
$\phantom{xxxxxxxxxx}2mn^2 + 4n^3$
subtract ⟶ $2mn^2 + 2n^3$
$\phantom{xxxxxxxxxxxxxx}2n^3 = R$

Check: $(2m^2 - mn + 2n^2)(m+n) + 2n^3 = 2m^3 + m^2n + mn^2 + 4n^3$

33
$$\require{enclose}\begin{array}{r}x^2 + xy + y^2\phantom{)}\\x-y\enclose{longdiv}{x^3 + 0x^2y + 0xy^2 - y^3}\end{array}$$

subtract ⟶ $x^3 - x^2y$
$\phantom{xxxxxxx}x^2y + 0xy^2$
subtract ⟶ $x^2y - xy^2$
$\phantom{xxxxxxxxx}xy^2 - y^3$
subtract ⟶ $xy^2 - y^3$
$\phantom{xxxxxxxxxxx}0$

Check: $(x^2 + xy + y^2)(x-y) + 0 = x^3 - y^3$

## CUMULATIVE REVIEW PROBLEM SET  Chapters 1–2

1. Evaluate the expression $7^2 - 12 \div 3 \times 5 + 3(6+2)$.
2. Express $\frac{7}{3}$ as a decimal.
3. Name the property that justifies the statement $x(x+4) = x^2 + 4x$.
4. Evaluate $(-1) \cdot (+2) \cdot (-1) \cdot (-3) \cdot (-1)$.
5. Perform the operations and simplify the result. $\frac{2}{3} \cdot \frac{15}{12} + \frac{7}{24}$
6. Round off 2.01356 to four significant digits.
7. Rewrite 38,900 in scientific notation.
8. Find the value of $\frac{x-y}{5}$ for $x = 15$ and $y = -5$.
9. Rewrite $x^4y^3$ in expanded form.
10. Evaluate the formula $P = 2l + 2w$ for $l = 4.3$ meters and $w = 2.7$ meters.
11. Find the sum of $(3x^3 + 7x^2 - 8x + 5) + (x^3 - 3x^2 - 2x + 2)$.
12. Find the product of $(3u - v)(u^2 + uv - v^2)$.
13. Use a special product to find $(3x - y)(3x + y)$.
14. Use a special product to find $(2x + y)(4x^2 - 2xy + y^2)$.
15. Factor $18x^3y - 27x^2y^7 + 45x^4y^2$ by finding a common factor.
16. Use the method of grouping to factor $2x^2 - 3y + xy - 6x$.
17. Use a special product to factor $16 - 81y^4$.
18. Factor $28x^3y - 7xy^3$.
19. Factor $6w^2 - 29wz + 35z^2$.
20. Divide $x^3 - xy^2 + 6y^3$ by $x + 2y$.

---

*Answers*

1  53   2  $2.333\cdots = 2.\overline{3}$   3  Distributive property   4  6   5  $\frac{9}{8}$   6  2.014
7  $3.89 \times 10^4$   8  4   9  $x \cdot x \cdot x \cdot x \cdot y \cdot y \cdot y$   10  $P = 14$ meters   11  $4x^3 + 4x^2 - 10x + 7$

12  $3u^3 + 2u^2v - 4uv^2 + v^3$    13  $9x^2 - y^2$    14  $8x^3 + y^3$    15  $9x^2y(2x - 3y^6 + 5x^2y)$
16  $(2x + y)(x - 3)$    17  $(2 - 3y)(2 + 3y)(4 + 9y^2)$    18  $7xy(2x - y)(2x + y)$
19  $(3w - 7z)(2w - 5z)$    20  $x^2 - 2xy + 3y^2$

## CHAPTER 2  TESTS

### Chapter Test A

1  Perform the indicated operation and simplify the result.

   (a)  $x^3 \cdot x^4$     (b)  $\dfrac{x^7}{x^3}$     (c)  $(x^3)^4$     (d)  $(2x^2y^3)^4$     (e)  $\left(\dfrac{x^2}{y^3}\right)^4$

2  Determine the degree and coefficients of each polynomial.

   (a)  $x^5 - 3x^2 + 6$     (b)  $2x^4 + x^3 - x^2 + 1$     (c)  $x^2 - \tfrac{1}{2}xy^2 + \tfrac{2}{3}x^2y^3$

3  Evaluate each polynomial for the indicated value.

   (a)  $x^3 - 3x^2 + 3x - 1$, $x = 1$     (b)  $x^5 + 2x^4 - 8x^3 + 4x^2 + 2x - 8$, $x = 2$
   (c)  $-x^4 + 5x^3 + 10x - 12$, $x = -3$     (d)  $3x^2y - 4xy^3$ for $x = -1$, $y = 2$

4  Perform the indicated operations.

   (a)  $(x^2 - 3x + 6) + (2x^2 + 5x - 12)$
   (b)  $(3x^3 + 4x^2 - 5x + 7) + (-2x^3 + x^2 - 3)$
   (c)  $(x^4 - 3x^3 + 4x^2 - x + 2) - (2x^3 - x - 3)$
   (d)  $(2x^5 + 3x^3 - 7x - 1) - (-x^4 + x^2 + 3) + (x^5 + 2x^4 - 3x + 5)$

5  Determine the product.

   (a)  $(2x + 3)(x - 1)$     (b)  $(4x + 3)(4x - 3)$     (c)  $(x - 1)(x^2 + x + 1)$
   (d)  $(2x + 1)(4x^2 - 2x + 1)$     (e)  $(x - y)^3$     (f)  $(3x - y + 1)(3x + y - 1)$

6  Factor each expression completely.

   (a)  $xy^2 - 2x^2y - x^3y^3$     (b)  $4x^2 - 9y^2$     (c)  $x^2 + 5x + 6$
   (d)  $2x^2 + x - 1$     (e)  $8x^3 + 1$     (f)  $x^3 - 27y^3$
   (g)  $mx^2 - my^2 + mx - my$

7  Divide each term as indicated.

   (a)  $3x^2y^3 - 2x^3y^4$ by $x^2y$     (b)  $x^4 + 2x^3 - 3x^2 + x - 3$ by $x - 1$
   (c)  $x^4 + 2x^3 + 3x^2 + 2x + 1$ by $x^2 + x + 1$

### Solutions

1  (a)  $x^3 \cdot x^4 = x^{3+4} = x^7$     (b)  $\dfrac{x^7}{x^3} = x^{7-3} = x^4$     (c)  $(x^3)^4 = x^{3(4)} = x^{12}$

   (d)  $(2x^2y^3)^4 = 2^4(x^2)^4(y^3)^4 = 16x^8y^{12}$     (e)  $\left(\dfrac{x^2}{y^3}\right)^4 = \dfrac{(x^2)^4}{(y^3)^4} = \dfrac{x^8}{y^{12}}$

2  (a)  Degree = 5; 1, −3, 6     (b)  Degree = 4; 2, 1, −1, 1     (c)  Degree = 5; 1, $-\tfrac{1}{2}$, $\tfrac{2}{3}$

3  (a)  for $x = 1$; $x^3 - 3x^2 + 3x - 1 = 1^3 - 3(1)^2 + 3(1) - 1$
          $= 1 - 3(1) + 3(1) - 1 = 0$

   (b)  for $x = 2$; $x^5 + 2x^4 - 8x^3 + 4x^2 + 2x - 8$
          $= 2^5 + 2(2)^4 - 8(2)^3 + 4(2)^2 + 2(2) - 8$
          $= 32 + 32 - 64 + 16 + 4 - 8 = 12$

(c) for $x = -3$; $-x^4 + 5x^3 + 10x - 12$
$$= -(-3)^4 + 5(-3)^3 + 10(-3) - 12$$
$$= -(81) + 5(-27) + 10(-3) - 12$$
$$= -81 - 135 - 30 - 12 = -258$$

(d) for $x = -1$ and $y = 2$; $3x^2y - 4xy^2$
$$= 3(-1)^2(2) - 4(-1)(2)^3$$
$$= 3(1)(2) - 4(-1)(8)$$
$$= 6 + 32 = 38$$

**4** (a)
$$\begin{array}{r} x^2 - 3x + 6 \\ (+)\, 2x^2 + 5x - 12 \\ \hline 3x^2 + 2x - 6 \end{array}$$

(b)
$$\begin{array}{r} 3x^3 + 4x^2 - 5x + 7 \\ (+) -2x^3 + x^2 \quad\quad - 3 \\ \hline x^3 + 5x^2 - 5x + 4 \end{array}$$

(c)
$$\begin{array}{r} x^4 - 3x^3 + 4x^2 - x + 2 \\ (-) \quad\quad 2x^3 \quad\quad - x - 3 \\ \hline x^4 - 5x^3 + 4x^2 \quad\quad + 5 \end{array}$$

(d)
$$\begin{array}{r} 2x^5 \quad\quad + 3x^3 \quad\quad - 7x - 1 \\ (-) \quad - x^4 \quad\quad + x^2 \quad\quad + 3 \\ \hline 2x^5 + x^4 + 3x^3 - x^2 - 7x - 4 \end{array}$$

$$\begin{array}{r} 2x^5 + x^4 + 3x^3 - x^2 - 7x - 4 \\ (+) \quad x^5 + 2x^4 \quad\quad\quad - 3x + 5 \\ \hline 3x^5 + 3x^4 + 3x^3 - x^2 - 10x + 1 \end{array}$$

**5** (a) $(2x + 3)(x - 1) = 2x^2 - 2x + 3x - 3 = 2x^2 + x - 3$

(b) $(4x + 3)(4x - 3) = (4x)^2 - (3)^2 = 16x^2 - 9$

(c) $(x - 1)(x^2 + x + 1) = (x)^3 - (1)^3 = x^3 - 1$

(d) $(2x + 1)(4x^2 - 2x + 1) = (2x)^3 + (1)^3 = 8x^3 + 1$

(e) $(x - y)^3 = (x)^3 - 3(x)^2(y) + 3(x)(y)^2 - (y)^3 = x^3 - 3x^2y + 3xy^2 - y^3$

(f) $(3x - y + 1)(3x + y - 1) = [3x - (y - 1)][3x + (y - 1)] = (3x)^2 - (y - 1)^2$
$$= 9x^2 - (y^2 - 2y + 1) = 9x^2 - y^2 + 2y - 1$$

**6** (a) $xy^2 - 2x^2y - x^3y^3 = (xy)(y) - (xy)(2x) - (xy)(x^2y^2)$
$$= xy(y - 2x - x^2y^2)$$

(b) $4x^2 - 9y^2 = (2x)^2 - (3y)^2 = (2x - 3y)(2x + 3y)$

(c) $x^2 + 5x + 6 = (x + 2)(x + 3)$

(d) $2x^2 + x - 1 = (2x - 1)(x + 1)$

(e) $8x^3 + 1 = (2x)^3 + (1)^3 = (2x + 1)[(2x)^2 - (2x)(1) + (1)^2]$
$$= (2x + 1)(4x^2 - 2x + 1)$$

(f) $x^3 - 27y^3 = (x)^3 - (3y)^3 = (x - 3y)[(x)^2 + (x)(3y) + (3y)^2]$
$$= (x - 3y)(x^2 + 3xy + 9y^2)$$

(g) $mx^2 - my^2 + mx - my = (mx^2 - my^2) + (mx - my)$
$$= m(x^2 - y^2) + m(x - y)$$
$$= m(x - y)(x + y) + m(x - y)$$
$$= m(x - y)[(x + y) + 1]$$
$$= m(x - y)(x + y + 1)$$

**7** (a) $(3x^2y^3 - 2x^3y^4) \div x^2y = \dfrac{3x^2y^3}{x^2y} - \dfrac{2x^3y^4}{x^2y} = 3y^2 - 2xy^3$

(b)
$$\begin{array}{r} x^3 + 3x^2 \quad\quad + 1 \\ x - 1 \overline{\smash{)}\, x^4 + 2x^3 - 3x^2 + x - 3} \\ \underline{x^4 - \phantom{2}x^3} \\ 3x^3 - 3x^2 \\ \underline{3x^3 - 3x^2} \\ 0 + x \\ x - 1 \\ \hline -2 = R \end{array}$$

(c)
$$\begin{array}{r} x^2 + x + 1 \\ x^2 + x + 1 \overline{\smash{)}\, x^4 + 2x^3 + 3x^2 + 2x + 1} \\ \underline{x^4 + x^3 + x^2} \\ x^3 + 2x^2 + 2x \\ \underline{x^3 + x^2 + x} \\ x^2 + x + 1 \\ \underline{x^2 + x + 1} \\ 0 \end{array}$$

## Chapter Test B

*Multiple Choice:* Select the *one* correct answer for each of the following questions.

1. The expression $5^4$ is equal to _____.
   - (a) $5 \cdot 5 \cdot 5$
   - (b) 625
   - (c) 20
   - (d) none of these

2. The expression $2^7 \cdot 2^{10}$ is equal to _____.
   - (a) $2^{70}$
   - (b) $2^3$
   - (c) 140
   - (d) $2^{17}$

3. The expression $(x^5 y^2)^4$ is equal to _____.
   - (a) $x^{20} y^8$
   - (b) $x^9 y^6$
   - (c) $x^{19} y^6$
   - (d) $x^7 y^6$

4. The expression $(\frac{4}{7})^3$ is equal to _____.
   - (a) $\frac{4^3}{7^4}$
   - (b) $\frac{12}{21}$
   - (c) $\frac{64}{343}$
   - (d) none of these

5. The expression $\frac{a^5}{a^2}$ is equal to _____.
   - (a) $a^3$
   - (b) $a^7$
   - (c) $a^0$
   - (d) 1

6. The expression $\left(\frac{x^2 y^4 z}{x y^2 z^2}\right)^3$ is equal to _____.
   - (a) $\frac{x^3 y^6}{z^3}$
   - (b) $\frac{x}{z}$
   - (c) $\frac{x^3}{z^6}$
   - (d) $\frac{x^6}{z^4}$

7. The expression $x^2 + 5x + 3$ is called a _____.
   - (a) binomial
   - (b) trinomial
   - (c) monomial
   - (d) none of these

8. The degree of the polynomial expression $-3x^5 - 5x + 1$ is equal to _____.
   - (a) 4
   - (b) 3
   - (c) 5
   - (d) no degree

9. The coefficients of the polynomial expression $2x^3 + 5x^2 - 3x + 4$ are _____.
   - (a) 2, 5, −3, 4
   - (b) 2, 0, −3, 4
   - (c) 2, 5, 3, 4
   - (d) 2, 0, 0, 5, −3, 1

10. Evaluating the polynomial expression $5x^2 + 3x - 2$ when $x = 2$, we get _____.
    - (a) 18
    - (b) 26
    - (c) 6
    - (d) 24

11. The sum of $(3a^2 + 2b)$ and $(5a^2 - 7b)$ is equal to _____.
    - (a) $8a^2 + 2b$
    - (b) $8a^2 - 3b$
    - (c) $8a^2 - 5b$
    - (d) $-8a^2 - 2b$

12. The sum of $(-5x^2 + 6x + 3)$ and $(5x^2 - 6x - 3)$ is equal to _____.
    - (a) 0
    - (b) $10x^2 - 6x - 3$
    - (c) $-10x^2 - 12x + 3$
    - (d) $5x^2 - 3$

# CHAPTER 2 THE ALGEBRA OF POLYNOMIALS

13. Subtracting $(2x^2 + y - 3)$ from $(6x^2 + 2)$ we get _____.
    (a) $4x^2 + y - 5$
    (b) $4x^2 - y + 5$
    (c) $4x^2 + 5$
    (d) $-4x^2 - y + 5$

14. The product of $(x - 5)(2x - 7)$ is equal to _____.
    (a) $2x^2 + 17x + 35$
    (b) $2x^2 - 17x - 35$
    (c) $2x^2 - 17x + 35$
    (d) $-2x^2 - 17x + 35$

15. The product of $(2x + 3)^2$ is equal to _____.
    (a) $4x^2 - 12x + 9$
    (b) $4x^2 - 12x - 9$
    (c) $4x^2 + 12x - 9$
    (d) $4x^2 + 12x + 9$

16. The product of $(1 + 10x)(1 - 10x + 100x^2)$ is equal to _____.
    (a) $1 - 1000x^3$
    (b) $-1 - 1000x^3$
    (c) $1 + 1000x^3$
    (d) none of these

17. The product of $(5 - 4x)^3$ is equal to _____.
    (a) $125 - 300x + 240x^2 - 64x^3$
    (b) $125 + 300x - 240x^2 + 64x^3$
    (c) $125 + 300x + 240x^2 - 64x^3$
    (d) $-125 - 300x + 240x^2 - 64x^3$

18. Factoring the polynomial expression $5x^2 - 125x$ we get _____.
    (a) $5x(x - 25)$
    (b) $-5x(x - 25)$
    (c) $5x(x + 25)$
    (d) $-5x(-x - 25)$

19. Factoring $81 - x^4$ we get _____.
    (a) $(9 - x)(9 + x)$
    (b) $(9 + x^2)(9 + x^2)$
    (c) $(3 - x)(3 + x)(9 + x^2)$
    (d) $(9 - x^2)(9 - x^2)$

20. Factoring $6 + 11x - 35x^2$ we get _____.
    (a) $(2 - 7x)(3 - 5x)$
    (b) $(-2 + 7x)(-3 - 5x)$
    (c) $(2 + 7x)(3 + 5x)$
    (d) $(2 + 7x)(3 - 5x)$

21. Dividing $x^3 + 2x^2$ by $x^2$ we obtain _____.
    (a) $x^2 + 2$
    (b) $x + 2$
    (c) $x^2 + 2x$
    (d) $x^3 + 2$

22. Dividing $x^4 + x^3 - 2x^2 + 3x - 1$ by $x^2 - x + 1$ we obtain _____.
    (a) $x^2 + 2x - 1$
    (b) $x^2 - 2x + 1$
    (c) $x^2 + 1$
    (d) $x^2 - 1$

## Answers

| 1 b | 2 d | 3 a | 4 c | 5 a | 6 a | 7 b | 8 c | 9 a | 10 d |
|---|---|---|---|---|---|---|---|---|---|
| 11 c | 12 a | 13 b | 14 c | 15 d | 16 c | 17 a | 18 a | 19 c | |
| 20 d | 21 b | 22 a | | | | | | | |

# 3 The Algebra of Fractions

In this chapter we extend the rules for operations with rational numbers to include rational expressions. After completing the appropriate sections, the student should be able to:

1. Reduce a fraction to lowest terms.
2. Multiply and divide fractions.
3. Add and subtract fractions.
4. Simplify complex fractions.

## 3.1 Rational Expressions

### SEMIPROGRAMMED PROBLEMS

In problems 1 and 2, determine all values for which the rational expressions are not defined.

1 $\dfrac{x-1}{x-2}$

2     The denominator is zero for $x = $ _____.

2 $\dfrac{3x}{(x-1)(2x+1)}$

1, $-\dfrac{1}{2}$     The denominator is zero for $x = $ _____ or $x = $ _____.

In problems 3 and 4, indicate which pairs of fractions are equivalent.

3 $\dfrac{4x}{5}$ and $\dfrac{5x}{9}$

$\ne, \ne$     Since $(4x)(9)$ _____ $(5)(5x)$, then $\dfrac{4x}{5}$ _____ $\dfrac{5x}{9}$.

4 $\dfrac{5(x+3)}{10(x+3)}$ and $\dfrac{1}{2}$

$=, =$     Since $5(x+3)(2)$ _____ $10(x+3)$, then $\dfrac{5(x+3)}{10(x+3)}$ _____ $\dfrac{1}{2}$.

In problems 5–7, reduce each fraction to lowest terms.

5 $\dfrac{x^2 - x}{x^2 - 1}$

$x - 1, \dfrac{x}{x+1}$     $\dfrac{x^2 - x}{x^2 - 1} = \dfrac{x(\underline{\phantom{xx}})}{(x-1)(x+1)} = $ _____

6 $\dfrac{x^3 - 4x}{x^2 - 3x + 2}$

$\dfrac{x(x+2)}{x-1}$     $\dfrac{x^3 - 4x}{x^2 - 3x + 2} = \dfrac{x(x-2)(x+2)}{(x-1)(x-2)} = $ _____

**38** CHAPTER 3 THE ALGEBRA OF FRACTIONS

$b - 3, \dfrac{b-3}{b+8}$

7 $\dfrac{b^2 + 4b - 21}{b^2 + 15b + 56}$

$\dfrac{b^2 + 4b - 21}{b^2 + 15b + 56} = \dfrac{(b+7)(\underline{\qquad})}{(b+7)(b+8)} = \underline{\qquad}$

In problems 8–10, find the misssing expression that will make the two fractions equivalent.

8 $\dfrac{x-y}{5} = \dfrac{3(x-y)^2}{?}$

$5(3)(x-y), 15(x-y)$

$15(x-y)$

$\dfrac{x-y}{5} = \dfrac{(x-y)(3)(x-y)}{\underline{\qquad}} = \dfrac{3(x-y)^2}{\underline{\qquad}},$

so the unknown expression is $\underline{\qquad}$.

9 $\dfrac{x+1}{x-1} = \dfrac{?}{x^2 - 3x + 2}$

$(x-2), x^2 - x - 2$

$x^2 - x - 2$

$\dfrac{x+1}{x-1} = \dfrac{(x+1)\underline{\qquad}}{(x-1)(x-2)} = \dfrac{\underline{\qquad}}{x^2 - 3x + 2}$

so the unknown expression is $\underline{\qquad}$.

10 $\dfrac{x-3y}{2x-3y} = \dfrac{?}{2x^2 - 5xy + 3y^2}$

$x-y, x^2 - 4xy + 3y^2$

$x^2 - 4xy + 3y^2$

$\dfrac{x-3y}{2x-3y} = \dfrac{(x-3y)(\underline{\qquad})}{(2x-3y)(x-y)} = \dfrac{\underline{\qquad}}{2x^2 - 5xy + 3y^2},$

so the unknown expression is $\underline{\qquad}$.

## SOLUTIONS TO SELECTED ODD PROBLEMS   Section 3.1, text pages 124–125

1  $\dfrac{7x}{x-3}$ is undefined when $x - 3 = 0$, so $x = 3$.

5  $\dfrac{2x - 4}{(x-6)(x+7)}$ is undefined when $(x-6)(x+7) = 0$, so $x = 6$ or $x = -7$.

9  $\dfrac{v^2 + 2v - 3}{(v+8)(v+9)(v-4)}$ is undefined when $(v+8)(v+9)(v-4) = 0$, so $v = -8, v = -9$, or $v = 4$.

13  Not equivalent, because $7 \cdot 10 \neq 9 \cdot 8$, or $70 \neq 72$.

17  Not equivalent, because $(v+2)(3v-12) \neq 3(v^2 - 4)$, or $3v^2 - 6v - 24 \neq 3v^2 - 12$.

21  $\dfrac{15}{18} = \dfrac{5 \cdot \cancel{3}}{6 \cdot \cancel{3}} = \dfrac{5}{6}$

25  $\dfrac{m^2 + m}{m^2 - m} = \dfrac{\cancel{m}(m+1)}{\cancel{m}(m-1)} = \dfrac{m+1}{m-1}$

29  $\dfrac{4x^2 - 9}{6x^2 - 9x} = \dfrac{\cancel{(2x-3)}(2x+3)}{3x\cancel{(2x-3)}} = \dfrac{2x+3}{3x}$

33  $\dfrac{3x^2 + 7x + 4}{3x^2 - 5x - 12} = \dfrac{\cancel{(3x+4)}(x+1)}{\cancel{(3x+4)}(x-3)} = \dfrac{x+1}{x-3}$

37  $\dfrac{3x^3 - 3xy^2}{3xy^2 + 3x^2y - 6x^3} = \dfrac{\cancel{3x}(x-y)(x+y)}{\cancel{3x}(y+2x)(y-x)}$

$= \dfrac{(x-y)(x+y)}{(y+2x)(y-x)} = \dfrac{-\cancel{(y-x)}(x+y)}{(y+2x)\cancel{(y-x)}} = -\dfrac{x+y}{2x+y}$

41  $\dfrac{9}{16} = \dfrac{9}{16} \cdot \dfrac{4}{4} = \dfrac{36}{64}$

45  $\dfrac{3}{u+v} = \dfrac{3}{u+v} \cdot \dfrac{5(u+v)}{5(u+v)} = \dfrac{15u + 15v}{5(u+v)^2}$

49  $\dfrac{2t}{2t-1} = \dfrac{2t}{2t-1} \cdot \dfrac{2t+1}{2t+1} = \dfrac{2t(2t+1)}{4t^2 - 1} = \dfrac{4t^2 + 2t}{4t^2 - 1}$

53  $\dfrac{x+2}{x-2} = \dfrac{x+2}{x-2} \cdot \dfrac{2x-3}{2x-3} = \dfrac{(x+2)(2x-3)}{(x-2)(2x-3)} = \dfrac{2x^2+x-6}{2x^2-7x+6}$

57  $\dfrac{-7}{y-x} = \dfrac{7}{-(y-x)} = \dfrac{7}{x-y}$  (by Rule i)

61  $\dfrac{3-x}{y-x} = \dfrac{-(3-x)}{-(y-x)} = \dfrac{x-3}{x-y}$  (by Rule iii)

## 3.2 Multiplication and Division of Fractions

SEMIPROGRAMMED PROBLEMS

### Multiplication of Fractions

In problems 1–5, find the products and reduce them to lowest terms.

| | |
|---|---|
| | 1  $\dfrac{8x^2}{5wz} \cdot \dfrac{15w^2}{14x^3}$ |
| $15w^2$ | $\dfrac{8x^2}{5wz} \cdot \dfrac{15w^2}{14x^3} = \dfrac{8x^2 \cdot \underline{\phantom{xxx}}}{5wz \cdot 14x^3}$ |
| $7xz, \dfrac{12w}{7xz}$ | $= \dfrac{10x^2w \cdot 12w}{10x^2w \cdot \underline{\phantom{xxx}}} = \underline{\phantom{xxx}}$ |
| | 2  $\dfrac{24}{3x-6} \cdot \dfrac{x^2-4}{x+2}$ |
| | $\dfrac{24}{3x-6} \cdot \dfrac{x^2-4}{x+2} = \dfrac{24(x^2-4)}{(3x-6)(x+2)}$ |
| $x+2, 8$ | $= \dfrac{3(8)(x-2)(\underline{\phantom{xxx}})}{3(x-2)(x+2)} = \underline{\phantom{xxx}}$ |
| | 3  $\dfrac{x^2-144}{x+4} \cdot \dfrac{x^2-16}{x-12}$ |
| | $\dfrac{x^2-144}{x+4} \cdot \dfrac{x^2-16}{x-12}$ |
| $x^2-16$ | $= \dfrac{(x^2-144)(\underline{\phantom{xxx}})}{(x+4)(x-12)}$ |
| $(x-4)(x+4)$ | $= \dfrac{(x-12)(x+12)\underline{\phantom{xxx}}}{(x+4)(x-12)}$ |
| $(x+12)(x-4)$ | $= \underline{\phantom{xxx}}$ |
| | 4  $\dfrac{x^2-10x+25}{x^2-100} \cdot \dfrac{x^2+12x+20}{x^2-7x+10}$ |
| | $\dfrac{x^2-10x+25}{x^2-100} \cdot \dfrac{x^2+12x+20}{x^2-7x+10}$ |
| $x^2+12x+20$ | $= \dfrac{(x^2-10x+25)(\underline{\phantom{xxx}})}{(x^2-100)(x^2-7x+10)}$ |
| $(x-2)(x-5)$ | $= \dfrac{(x-5)(x-5)(x+10)(x+2)}{(x-10)(x+10)\underline{\phantom{xxx}}}$ |
| $\dfrac{(x-5)(x+2)}{(x-10)(x-2)}$ | $= \underline{\phantom{xxx}}$ |

## 40 CHAPTER 3 THE ALGEBRA OF FRACTIONS

$$5 \quad \frac{xy - y^2}{x^2 + xy} \cdot \frac{x^3 + y^3}{x^3 - y}$$

$x^2 - xy + y^2$

$$\frac{xy - y^2}{x^2 + xy} \cdot \frac{x^3 + y^3}{x^3 - y^3}$$

$x^2 - xy + y^2$

$$= \frac{y(x-y)}{x(x+y)} \cdot \frac{(x+y)(\underline{\qquad})}{(x-y)(x^2+xy+y^2)}$$

$$\frac{y(x^2 - xy + y^2)}{x(x^2 + xy + y^2)}$$

$$= \frac{y(x-y)(x+y)(\underline{\qquad})}{x(x+y)(x-y)(x^2+xy+y^2)}$$

$$= \underline{\qquad}$$

## Division of Fractions

In problems 6–10, find the quotients and reduce them to lowest terms.

$$6 \quad \frac{3x^2}{5y} \div \frac{6y^2}{2x^3}$$

$2x^3, \dfrac{x^5}{5y^3}$

$$\frac{3x^2}{5y} \div \frac{6y^2}{2x^3} = \frac{3x^2}{5y} \cdot \frac{\underline{\qquad}}{6y^2} = \underline{\qquad}$$

$$7 \quad \frac{x^2 - 2x + 1}{4x^2 - 1} \div \frac{4x - 4}{6x + 3}$$

$6x + 3$

$$\frac{x^2 - 2x + 1}{4x^2 - 1} \div \frac{4x - 4}{6x + 3} = \frac{x^2 - 2x + 1}{4x^2 - 1} \cdot \frac{\underline{\qquad}}{4x - 4}$$

$2x + 1$

$$= \frac{(x-1)^2(3)(\underline{\qquad})}{(2x-1)(2x+1)(4)(x-1)}$$

$\dfrac{3(x-1)}{4(2x-1)}$

$$= \underline{\qquad}$$

$$8 \quad \frac{9a^2 - 4}{a^2 + 4a - 12} \div \frac{12a + 8}{a^2 + a - 30}$$

$$\frac{9a^2 - 4}{a^2 + 4a - 12} \div \frac{12a + 8}{a^2 + a - 30}$$

$a^2 + a - 30$

$$= \frac{9a^2 - 4}{a^2 + 4a - 12} \cdot \frac{\underline{\qquad}}{12a + 8}$$

$a^2 + a - 30$

$$= \frac{(9a^2 - 4)(\underline{\qquad})}{(a^2 + 4a - 12)(12a + 8)}$$

$a + 6$

$$= \frac{(3a-2)(3a+2) \cdot (a-5)(\underline{\qquad})}{(a+6)(a-2) \cdot 4(3a+2)}$$

$\dfrac{(3a-2)(a-5)}{4(a-2)}$

$$= \underline{\qquad}$$

$$9 \quad \frac{x^2 + 5x + 6}{2x^2 + x - 1} \div \frac{x^2 - 9}{(x+1)^2}$$

$(x+1)^2$

$$\frac{x^2 + 5x + 6}{2x^2 + x - 1} \div \frac{x^2 - 9}{(x+1)^2} = \frac{x^2 + 5x + 6}{2x^2 + x - 1} \cdot \frac{\underline{\qquad}}{x^2 - 9}$$

$(x+1)^2$

$$= \frac{(x^2 + 5x + 6)\underline{\qquad}}{(2x^2 + x - 1)(x^2 - 9)}$$

$2x - 1$

$$= \frac{(x+2)(x+3)(x+1)(x+1)}{(\underline{\qquad})(x+1)(x-3)(x+3)}$$

$\dfrac{(x+2)(x+1)}{(2x-1)(x-3)}$

$$= \underline{\qquad}$$

## 3.2 SOLUTIONS TO SELECTED ODD PROBLEMS

10 $\dfrac{4x^2 - 36}{x^2 + 4x + 3} \div \dfrac{x^2 + x - 12}{72 - 6x - 6x^2}$

$72 - 6x - 6x^2$

$\dfrac{4x^2 - 36}{x^2 + 4x + 3} \div \dfrac{x^2 + x - 12}{72 - 6x - 6x^2}$

$x + 4$

$= \dfrac{4x^2 - 36}{x^2 + 4x + 3} \cdot \dfrac{\underline{\phantom{XXXX}}}{x^2 + x - 12}$

$\dfrac{-24(x-3)}{x+1}$

$= \dfrac{4(x+3)(x-3)(\underline{\phantom{XX}})(-6)(x-3)}{[(x+3)(x+1)][(x-3)(x+4)]}$

$= \underline{\phantom{XXXXXX}}$

In problems 11 and 12, perform the indicated operations and reduce the answer to lowest terms.

11 $\left[\dfrac{2x^2 + x - 1}{3x^4 y^2} \cdot \dfrac{2x^6 y^3}{x^2 + 2x + 1}\right] \div \dfrac{4x^2 - 1}{x^2 - 1}$

$x^2 - 1$

$\left[\dfrac{2x^2 + x - 1}{3x^4 y^2} \cdot \dfrac{2x^6 y^3}{x^2 + 2x + 1}\right] \div \dfrac{4x^2 - 1}{x^2 - 1}$

$x + 1$

$= \dfrac{2x^2 + x - 1}{3x^4 y^2} \cdot \dfrac{2x^6 y^3}{x^2 + 2x + 1} \cdot \dfrac{\underline{\phantom{XXXX}}}{4x^2 - 1}$

$\dfrac{2x^2 y(x-1)}{3(2x+1)}$

$= \dfrac{(2x-1)(x+1)(2x^6 y^3)(x-1)(\underline{\phantom{XX}})}{(3x^4 y^2)(x+1)^2(2x-1)(2x+1)}$

$= \underline{\phantom{XXXXXX}}$

12 $\dfrac{x^3 - 16x}{x^2 - 3x - 10} \div \left[\dfrac{x^3 + x^2 - 12x}{x^2 + 5x + 6} \cdot \dfrac{x^2 - 4}{x^2 - 9}\right]$

$x^3 + x^2 - 12x, x^2 - 4$

$\dfrac{x^3 - 16x}{x^2 - 3x - 10} \div \left[\dfrac{x^3 + x^2 - 12x}{x^2 + 5x + 6} \cdot \dfrac{x^2 - 4}{x^2 - 9}\right]$

$x + 3$

$= \dfrac{x^3 - 16x}{x^2 - 3x - 10} \cdot \dfrac{x^2 + 5x + 6}{\underline{\phantom{XXXX}}} \cdot \dfrac{x^2 - 9}{\underline{\phantom{XXXX}}}$

$\dfrac{(x-4)(x+3)^2}{(x-5)(x+2)(x-2)}$

$= \dfrac{x(x-4)(x+4)(x+3)(x+2)(x-3)(\underline{\phantom{XX}})}{(x-5)(x+2)x(x+4)(x-3)(x-2)(x+2)}$

$= \underline{\phantom{XXXXXX}}$

**SOLUTIONS TO SELECTED ODD PROBLEMS**   Section 3.2, text pages 130–131

1   $\dfrac{12}{13} \cdot \dfrac{39x}{60} = \dfrac{\cancel{12}^{1}}{\cancel{13}_{1}} \cdot \dfrac{\cancel{39x}^{3}}{\cancel{60}_{5}} = \dfrac{3x}{5}$

5   $\dfrac{\cancel{5x^2 y}}{\cancel{3t^2 s}} \cdot \dfrac{\cancel{6t s}^{2}}{\cancel{10x^2}_{\cancel{2}}} = \dfrac{y}{t}$

9   $\dfrac{3x+6}{5x+5} \cdot \dfrac{10x+10}{x^2 - 6x - 16} = \dfrac{3\cancel{(x+2)}}{\cancel{5}(x+1)} \cdot \dfrac{\cancel{10}^{2}(x+1)}{\cancel{(x+2)}(x-8)} = \dfrac{6}{x-8}$

13   $\dfrac{v^2 - 1}{v-3} \cdot \dfrac{3v^2 - 8v - 3}{v^2 - 10v + 9} = \dfrac{\cancel{(v-1)}(v+1)}{\cancel{(v-3)}} \cdot \dfrac{(3v+1)\cancel{(v-3)}}{(v-9)\cancel{(v-1)}} = \dfrac{(v+1)(3v+1)}{v-9}$

17   $\dfrac{3y^2 - y - 2}{3y^2 + y - 2} \cdot \dfrac{3y^2 - 5y + 2}{3y^2 + 5y + 2} = \dfrac{\cancel{(3y+2)}(y-1)}{\cancel{(3y-2)}(y+1)} \cdot \dfrac{\cancel{(3y-2)}(y-1)}{\cancel{(3y+2)}(y+1)} = \dfrac{(y-1)^2}{(y+1)^2}$

21   $\dfrac{x^3 - y^3}{3x + 3y} \cdot \dfrac{6x^2 + 12xy + 6y^2}{2x^2 + 2xy + 2y^2} = \dfrac{(x-y)\cancel{(x^2 + xy + y^2)}}{\cancel{3(x+y)}} \cdot \dfrac{\cancel{6}^{\cancel{2}}(x+y)(x+y)}{\cancel{2}(x^2 + xy + y^2)} = x^2 - y^2$

42  CHAPTER 3  THE ALGEBRA OF FRACTIONS

25  $\dfrac{4x^3}{y} \div \dfrac{2x}{3y^2} = \dfrac{\cancel{4x^3}^{2x^2}}{y} \cdot \dfrac{3y^{\cancel{2}\,y}}{\cancel{2x}} = 6x^2y$

29  $\dfrac{12 - 6y}{7y - 21} \div \dfrac{2y - 4}{y^2 - 9} = \dfrac{12 - 6y}{7y - 21} \cdot \dfrac{y^2 - 9}{2y - 4} = \dfrac{\cancel{6(2-y)}^{3}}{7\cancel{(y-3)}} \cdot \dfrac{\cancel{(y-3)}(y+3)}{\cancel{2(y-2)}^{-1}} = -\dfrac{3y + 9}{7}$

33  $\dfrac{a^2 - 4a + 4}{a + 2} \div \dfrac{a^2 - 4}{3a + 6} = \dfrac{a^2 - 4a + 4}{a + 2} \cdot \dfrac{3a + 6}{a^2 - 4} = \dfrac{(a - 2)(a - 2)}{a + 2} \cdot \dfrac{3(a + 2)}{\cancel{(a - 2)}(a + 2)} = \dfrac{3(a - 2)}{a + 2}$

37  $\dfrac{2x^2 - 5x - 3}{3x^2 - 5x - 2} \div \dfrac{2x^2 + 11x + 5}{x^2 + 3x - 10} = \dfrac{2x^2 - 5x - 3}{3x^2 - 5x - 2} \cdot \dfrac{x^2 + 3x - 10}{2x^2 + 11x + 5} = \dfrac{\cancel{(2x + 1)}(x - 3)}{(3x + 1)\cancel{(x - 2)}} \cdot \dfrac{(x + 5)\cancel{(x - 2)}}{\cancel{(2x + 1)}\cancel{(x + 5)}} = \dfrac{x - 3}{3x + 1}$

41  $\dfrac{8x^3 - 27}{9x^2 - 3x + 1} \div \dfrac{4x^2 + 6x + 9}{27x^3 + 1} = \dfrac{8x^3 - 27}{9x^2 - 3x + 1} \cdot \dfrac{27x^3 + 1}{4x^2 + 6x + 9}$

$= \dfrac{(2x - 3)\cancel{(4x^2 + 6x + 9)}}{\cancel{9x^2 - 3x + 1}} \cdot \dfrac{(3x + 1)\cancel{(9x^2 - 3x + 1)}}{\cancel{4x^2 + 6x + 9}} = (2x - 3)(3x + 1)$

45  $\left[\dfrac{b - 1}{21 - 4a - a^2} \cdot \dfrac{b - 2}{b - b^3}\right] \div \dfrac{2 - b}{a^2 + 6a - 7}$

First simplify inside brackets:

$\dfrac{\cancel{b - 1}^{-1}}{(7 + a)(3 - a)} \cdot \dfrac{b - 2}{b\cancel{(1 - b)}(1 + b)} = \dfrac{2 - b}{b(7 + a)(3 - a)(1 + b)}$

Then,

$\left[\dfrac{2 - b}{b(7 + a)(3 - a)(1 + b)}\right] \div \dfrac{2 - b}{a^2 + 6a - 7} = \dfrac{\cancel{2 - b}}{b\cancel{(7 + a)}(3 - a)(1 + b)} \cdot \dfrac{\cancel{(a + 7)}(a - 1)}{\cancel{2 - b}} = \dfrac{a - 1}{(3 - a)(b + b^2)}$

49  $\dfrac{2y^2 + 5y - 3}{6y^2 - 5y - 6} \div \left[\dfrac{2y^2 + 9y - 5}{12y^2 - y - 6} \cdot \dfrac{4y^2 + 9y - 9}{2y^2 + y - 6}\right]$

First simplify inside brackets:

$\dfrac{(2y - 1)(y + 5)}{(3y + 2)\cancel{(4y - 3)}} \cdot \dfrac{\cancel{(4y - 3)}(y + 3)}{(2y - 3)(y + 2)} = \dfrac{(2y - 1)(y + 5)(y + 3)}{(3y + 2)(2y - 3)(y + 2)}$

Then,

$\dfrac{2y^2 + 5y - 3}{6y^2 - 5y - 6} \div \dfrac{(2y - 1)(y + 5)(y + 3)}{(3y + 2)(2y - 3)(y + 2)} = \dfrac{\cancel{(2y - 1)}\cancel{(y + 3)}}{\cancel{(3y + 2)}\cancel{(2y - 3)}} \cdot \dfrac{\cancel{(3y + 2)}\cancel{(2y - 3)}(y + 2)}{\cancel{(2y - 1)}(y + 5)\cancel{(y + 3)}} = \dfrac{y + 2}{y + 5}$

## 3.3 Addition and Subtraction of Fractions

### SEMIPROGRAMMED PROBLEMS

In problems 1–10, perform the indicated operations and simplify the results.

7, 12, $\dfrac{3}{x}$

1  $\dfrac{5}{4x} + \dfrac{7}{4x}$

$\dfrac{5}{4x} + \dfrac{7}{4x} = \dfrac{5 + \underline{\quad\quad}}{4x} = \dfrac{\underline{\quad\quad}}{4x} = \underline{\quad\quad}$

$2x + 1$

2  $\dfrac{4x + 3}{4y} - \dfrac{2x + 1}{4y}$

$\dfrac{4x + 3}{4y} - \dfrac{2x + 1}{4y} = \dfrac{4x + 3 - (\underline{\quad\quad})}{4y}$

$2x + 2$, $\dfrac{x + 1}{2y}$

$= \dfrac{\underline{\quad\quad}}{4y} = \underline{\quad\quad}$

## 3.3 SEMIPROGRAMMED PROBLEMS

$-2x + 1, \dfrac{x+5}{y^2 - 2y + 7}$

3. $\dfrac{3x+4}{y^2-2y+7} + \dfrac{-2x+1}{y^2-2y+7}$

$\dfrac{3x+4}{y^2-2y+7} + \dfrac{-2x+1}{y^2-2y+7}$

$= \dfrac{3x+4+(\underline{\qquad})}{y^2-2y+7} = \underline{\qquad}$

4. $\dfrac{6}{5x} + \dfrac{1}{2y}$

$5x, 12y + 5x$

$\dfrac{6}{5x} + \dfrac{1}{2y} = \dfrac{6(2y)+1(\underline{\qquad})}{(5x)(2y)} = \dfrac{\underline{\qquad}}{10xy}$

5. $\dfrac{4}{x-2} + \dfrac{5}{x}$

$x - 2, 9x - 10$

$\dfrac{4}{x-2} + \dfrac{5}{x} = \dfrac{4x + 5(\underline{\qquad})}{(x-2)x} = \dfrac{\underline{\qquad}}{x^2 - 2x}$

6. $\dfrac{7}{x+3} - \dfrac{5}{x}$

$x + 3, 2x - 15$

$\dfrac{7}{x+3} - \dfrac{5}{x} = \dfrac{7x - 5(\underline{\qquad})}{(x+3)x} = \dfrac{\underline{\qquad}}{x^2 + 3x}$

7. $\dfrac{5}{14x^2y} - \dfrac{3}{21xy^2}$

$42x^2y^2$

The LCD is _____, so that

$6x$

$\dfrac{5}{14x^2y} - \dfrac{3}{21xy^2} = \dfrac{15y}{42x^2y^2} - \dfrac{\underline{\qquad}}{42x^2y^2}$

$15y - 6x$

$= \dfrac{\underline{\qquad}}{42x^2y^2}.$

8. $\dfrac{3}{x^2-16} + \dfrac{1}{x+4}$

$x^2 - 16$

The LCD is _____, so that

$x - 4$

$\dfrac{3}{x^2-16} + \dfrac{1}{x+4} = \dfrac{3}{x^2-16} + \dfrac{\underline{\qquad}}{x^2-16}$

$x - 1$

$= \dfrac{\underline{\qquad}}{x^2-16}.$

9. $\dfrac{2x}{x^2-4xy+4y^2} - \dfrac{y}{x^2-4y^2}$

$(x-2y)^2(x+2y)$

The LCD is _____, so that

$\dfrac{2x}{x^2-4xy+4y^2} - \dfrac{y}{x^2-4y^2}$

$x + 2y$

$= \dfrac{2x}{(x-2y)(x-2y)} - \dfrac{y}{(x-2y)(\underline{\qquad})}$

$x - 2y$

$= \dfrac{2x(x+2y) - y(\underline{\qquad})}{(x-2y)^2(x+2y)}$

$2x^2 + 4xy$

$= \dfrac{(\underline{\qquad}) - xy + 2y^2}{(x-2y)^2(x+2y)}$

$2x^2 + 3xy + 2y^2$

$= \dfrac{\underline{\qquad}}{(x-2y)^2(x+2y)}$

44  CHAPTER 3  THE ALGEBRA OF FRACTIONS

$(x + 5)(x + 1)(x - 1)$

$(x + 5)(x - 1)$

$x - 1, x + 1, x + 5$

$x^2 - x + 2x + 2 + 3x + 15$

$x^2 + 4x + 17$

10  $\dfrac{x}{x^2 + 6x + 5} + \dfrac{2}{x^2 + 4x - 5} + \dfrac{3}{x^2 - 1}$

The LCD is _____, so that

$\dfrac{x}{x^2 + 6x + 5} + \dfrac{2}{x^2 + 4x - 5} + \dfrac{3}{x^2 - 1}$

$= \dfrac{x}{(x + 5)(x + 1)} + \dfrac{2}{\rule{2cm}{0.4pt}} + \dfrac{3}{(x - 1)(x + 1)}$

$= \dfrac{x(\rule{1.5cm}{0.4pt}) + 2(\rule{1.5cm}{0.4pt}) + 3(\rule{1.5cm}{0.4pt})}{(x + 5)(x + 1)(x - 1)}$

$= \dfrac{\rule{3cm}{0.4pt}}{(x + 5)(x + 1)(x - 1)}$

$= \dfrac{\rule{3cm}{0.4pt}}{(x + 5)(x + 1)(x - 1)}.$

## SOLUTIONS TO SELECTED ODD PROBLEMS  Section 3.3, text pages 139–141

1  $\dfrac{5}{8x} + \dfrac{1}{8x} = \dfrac{5 + 1}{8x} = \dfrac{6}{8x} = \dfrac{3}{4x}$

5  $\dfrac{7}{12} - \dfrac{3}{12} = \dfrac{7 - 3}{12} = \dfrac{4}{12} = \dfrac{1}{3}$

9  $\dfrac{2x}{x^2 - 4} - \dfrac{4}{x^2 - 4} = \dfrac{2x - 4}{x^2 - 4} = \dfrac{2(x - 2)}{(x - 2)(x + 2)} = \dfrac{2}{x + 2}$

13  $\dfrac{3m}{m^2 + m - 2} + \dfrac{6}{m^2 + m - 2} = \dfrac{3m + 6}{m^2 + m - 2} = \dfrac{3(m + 2)}{(m - 1)(m + 2)} = \dfrac{3}{m - 1}$

17  $\dfrac{15}{8u^2 - 6u - 5} - \dfrac{12u}{8u^2 - 6u - 5} = \dfrac{15 - 12u}{8u^2 - 6u - 5} = \dfrac{\overset{-1}{3(5 - 4u)}}{(4u - 5)(2u + 1)} = \dfrac{-3}{2u + 1} = -\dfrac{3}{2u + 1}$

21  $\dfrac{5}{8x} + \dfrac{3}{7x} = \dfrac{5(7x) + 3(8x)}{(8x)(7x)} = \dfrac{35x + 24x}{(8x)(7x)} = \dfrac{59x}{(8x)(7x)} = \dfrac{59}{56x}$

25  $\dfrac{2x}{3} + \dfrac{9x}{10} = \dfrac{2x(10) + (9x)(3)}{3 \cdot 10} = \dfrac{20x + 27x}{30} = \dfrac{47x}{30}$

29  $\dfrac{9}{y - 5} + \dfrac{6}{y - 3} = \dfrac{9(y - 3) + 6(y - 5)}{(y - 5)(y - 3)} = \dfrac{9y - 27 + 6y - 30}{(y - 5)(y - 3)} = \dfrac{15y - 57}{(y - 5)(y - 3)}$

33  $\dfrac{x}{3x + 2} + \dfrac{1}{x - 4} = \dfrac{x(x - 4) + 1(3x + 2)}{(3x + 2)(x - 4)} = \dfrac{x^2 - 4x + 3x + 2}{(3x + 2)(x - 4)} = \dfrac{x^2 - x + 2}{(3x + 2)(x - 4)}$

37  $\dfrac{12}{5m^3n} - \dfrac{4}{3m^2n^5}$

The LCD is $15m^3n^5$, so

$\dfrac{12}{5m^3n} - \dfrac{4}{3m^2n^5} = \dfrac{12(3n^4)}{(5m^3n)(3n^4)} - \dfrac{4(5m)}{(3m^2n^5)(5m)} = \dfrac{36n^4 - 20m}{15m^3n^5}.$

41  $\dfrac{c}{c^2 - 9} - \dfrac{c - 1}{c^2 - 5c + 6} = \dfrac{c}{(c - 3)(c + 3)} - \dfrac{c - 1}{(c - 2)(c - 3)}$

The LCD is $(c - 3)(c + 3)(c - 2)$, so

$\dfrac{c}{c^2 - 9} - \dfrac{c - 1}{c^2 - 5c + 6} = \dfrac{c(c - 2)}{(c - 3)(c + 3)(c - 2)} - \dfrac{(c - 1)(c + 3)}{(c - 3)(c + 3)(c - 2)}$

$$= \frac{c^2 - 2c - [c^2 + 2c - 3]}{(c-3)(c+3)(c-2)} = \frac{c^2 - 2c - c^2 - 2c + 3}{(c-3)(c+3)(c-2)}$$

$$= \frac{3 - 4c}{(c-3)(c+3)(c-2)}.$$

**45** $\quad \dfrac{x-2}{x^2 + 10x + 16} + \dfrac{x+1}{x^2 + 9x + 14} = \dfrac{x-2}{(x+8)(x+2)} + \dfrac{x+1}{(x+2)(x+7)}$

The LCD is $(x + 8)(x + 2)(x + 7)$, so

$$\frac{x-2}{x^2 + 10x + 16} + \frac{x+1}{x^2 + 9x + 14} = \frac{(x-2)(x+7)}{(x+8)(x+2)(x+7)} + \frac{(x+1)(x+8)}{(x+8)(x+2)(x+7)}$$

$$= \frac{x^2 + 5x - 14 + x^2 + 9x + 8}{(x+8)(x+2)(x+7)}$$

$$= \frac{2x^2 + 14x - 6}{(x+8)(x+2)(x+7)}.$$

**49** $\quad \dfrac{4}{3t - 2} + \dfrac{1}{2 - 3t} = \dfrac{4}{3t - 2} - \dfrac{1}{3t - 2} = \dfrac{4 - 1}{3t - 2} = \dfrac{3}{3t - 2}$

**53** $\quad \dfrac{v}{v+2} - \dfrac{v}{v-2} - \dfrac{v^2}{v^2 - 4} = \dfrac{v}{v+2} - \dfrac{v}{v-2} - \dfrac{v^2}{(v+2)(v-2)}$

The LCD is $(v + 2)(v - 2)$, so

$$\frac{v}{v+2} - \frac{v}{v-2} - \frac{v^2}{v^2 - 4} = \frac{v(v-2)}{(v+2)(v-2)} - \frac{v(v+2)}{(v-2)(v+2)} - \frac{v^2}{(v-2)(v+2)}$$

$$= \frac{v^2 - 2v - v^2 - 2v - v^2}{(v+2)(v-2)} = \frac{-v^2 - 4v}{(v+2)(v-2)} = \frac{v^2 + 4v}{4 - v^2}.$$

**57** $\quad \dfrac{1}{x^2 - x - 6} + \dfrac{2}{x^2 - 6x + 9} - \dfrac{1}{x^2 + 4x + 4} = \dfrac{1}{(x-3)(x+2)} + \dfrac{2}{(x-3)^2} - \dfrac{1}{(x+2)^2}$

The LCD is $(x - 3)^2(x + 2)^2$, so

$$\frac{1}{x^2 - x - 6} + \frac{2}{x^2 - 6x + 9} - \frac{1}{x^2 + 4x + 4} = \frac{(x-3)(x+2)}{(x-3)^2(x+2)^2} + \frac{2(x+2)^2}{(x-3)^2(x+2)^2} - \frac{(x-3)^2}{(x-3)^2(x+2)^2}$$

$$= \frac{x^2 - x - 6}{(x-3)^2(x+2)^2} + \frac{2x^2 + 8x + 8}{(x-3)^2(x+2)^2} - \frac{x^2 - 6x + 9}{(x-3)^2(x+2)^2}$$

$$= \frac{x^2 - x - 6 + 2x^2 + 8x + 8 - x^2 + 6x - 9}{(x-3)^2(x+2)^2}$$

$$= \frac{2x^2 + 13x - 7}{(x-3)^2(x+2)^2}.$$

**61** $\quad x + 3\left(\dfrac{1}{x}\right) = x + \dfrac{3}{x} = \dfrac{x^2}{x} + \dfrac{3}{x} = \dfrac{x^2 + 3}{x}$

**65** $\quad F = \dfrac{1}{p} + \dfrac{1}{q}; \; p = x + 10; \; q = x - 2$

$$F = \frac{1}{x + 10} + \frac{1}{x - 2}$$

The LCD is $(x + 10)(x - 2)$, so

$$F = \frac{1}{x + 10} + \frac{1}{x - 2} = \frac{x - 2}{(x + 10)(x - 2)} + \frac{x + 10}{(x + 10)(x - 2)} = \frac{2x + 8}{(x + 10)(x - 2)}.$$

## 3.4 Complex Fractions

**SEMIPROGRAMMED PROBLEMS**

In problems 1–10, express each complex fraction as a simple fraction in lowest terms.

1. $\dfrac{\frac{3}{4}}{\frac{6}{5}}$

$\dfrac{15}{24}, \dfrac{5}{8}$

$\dfrac{\frac{3}{4}}{\frac{6}{5}} = \dfrac{\frac{3}{4}(20)}{\frac{6}{5}(20)} = \underline{\qquad} = \underline{\qquad}$

2. $\dfrac{\frac{3x}{5y}}{\frac{5x}{7y}}$

$25x, \dfrac{21}{25}$

$\dfrac{\frac{3x}{5y}}{\frac{5x}{7y}} = \dfrac{\left(\frac{3x}{5y}\right)(35y)}{\left(\frac{5x}{7y}\right)(35y)} = \dfrac{21x}{\underline{\qquad}} = \underline{\qquad}$

3. $\dfrac{\frac{3}{xy^2}}{\frac{2}{x^2y}}$

$3x$

$\dfrac{\frac{3}{xy^2}}{\frac{2}{x^2y}} = \dfrac{\left(\frac{3}{xy^2}\right)(x^2y^2)}{\left(\frac{2}{x^2y}\right)(x^2y^2)} = \dfrac{\underline{\qquad}}{2y}$

4. $\dfrac{x + \frac{1}{y}}{y + \frac{1}{x}}$

$xy^2 + y$

$\dfrac{x + \frac{1}{y}}{y + \frac{1}{x}} = \dfrac{\left(x + \frac{1}{y}\right)(xy)}{\left(y + \frac{1}{x}\right)(xy)} = \dfrac{x^2y + x}{\underline{\qquad}}$

$xy + 1, \dfrac{x}{y}$

$= \dfrac{x(xy + 1)}{y(\underline{\qquad})} = \underline{\qquad}$

5. $\dfrac{1 - \frac{3y}{x+y}}{1 - \frac{y}{x-y}}$

$(x-y)(x+y)$

$\dfrac{1 - \frac{3y}{x+y}}{1 - \frac{y}{x-y}} = \dfrac{\left(1 - \frac{3y}{x+y}\right)\underline{\qquad}}{\left(1 - \frac{y}{x-y}\right)(x-y)(x+y)}$

$x - y$

$= \dfrac{(x-y)(x+y) - 3y(\underline{\qquad})}{(x-y)(x+y) - y(x+y)}$

3.4 SEMIPROGRAMMED PROBLEMS 47

$x^2 - 3xy + 2y^2$

$x - 2y, \ x - y$

$$= \frac{x^2 - y^2 - 3xy + 3y^2}{x^2 - y^2 - xy - y^2}$$

$$= \frac{\rule{2cm}{0.4pt}}{x^2 - xy - 2y^2}$$

$$= \frac{(x-y)(\underline{\qquad})}{(x+y)(x-2y)} = \frac{\rule{1cm}{0.4pt}}{x+y}$$

6  $\dfrac{\dfrac{4}{xy^3}}{\dfrac{3}{x^2y^2}}$

$x^2y^2, \ \dfrac{4x}{3y}$

$$\frac{\dfrac{4}{xy^3}}{\dfrac{3}{x^2y^2}} = \frac{4}{xy^3} \div \frac{3}{x^2y^2}$$

$$= \frac{4}{xy^3} \cdot \frac{\rule{1cm}{0.4pt}}{3} = \underline{\qquad}$$

7  $\dfrac{\dfrac{1}{x} - \dfrac{1}{y}}{\dfrac{1}{x} + \dfrac{1}{y}}$

$xy$

$\dfrac{y-x}{y+x}$

$$\frac{\dfrac{1}{x} - \dfrac{1}{y}}{\dfrac{1}{x} + \dfrac{1}{y}} = \frac{\dfrac{y-x}{xy}}{\dfrac{y+x}{xy}} = \frac{y-x}{xy} \cdot \frac{\rule{1cm}{0.4pt}}{y+x}$$

$$= \underline{\qquad}$$

8  $\dfrac{a^2 - \dfrac{1}{a}}{a + \dfrac{1}{a} + 1}$

$a^3 - 1$

$a^2 + a + 1$

$a - 1$

$$\frac{a^2 - \dfrac{1}{a}}{a + \dfrac{1}{a} + 1} = \frac{\dfrac{\rule{1cm}{0.4pt}}{a}}{\dfrac{a^2 + 1 + a}{a}}$$

$$= \frac{(a-1)(a^2+a+1)a}{a(\underline{\qquad})}$$

$$= \underline{\qquad}$$

9  $\dfrac{x - \dfrac{1}{x}}{1 + \dfrac{2}{x} + \dfrac{1}{x^2}}$

$x^2 + 2x + 1$

$x(x+1)^2$

$\dfrac{x(x-1)}{x+1}$

$$\frac{x - \dfrac{1}{x}}{1 + \dfrac{2}{x} + \dfrac{1}{x^2}} = \frac{\dfrac{x^2-1}{x}}{\dfrac{\rule{1cm}{0.4pt}}{x^2}}$$

$$= \frac{(x-1)(x+1)x^2}{\rule{2cm}{0.4pt}}$$

$$= \underline{\qquad}$$

## CHAPTER 3 THE ALGEBRA OF FRACTIONS

$2x + y$

$2x + y$

$\dfrac{-y}{2x+y}$

10 $\dfrac{\dfrac{1}{x+y} - \dfrac{1}{x}}{\dfrac{1}{x+y} + \dfrac{1}{x}}$

$\dfrac{\dfrac{1}{x+y} - \dfrac{1}{x}}{\dfrac{1}{x+y} + \dfrac{1}{x}} = \dfrac{\dfrac{x-x-y}{x(x+y)}}{\dfrac{\phantom{xxxx}}{x(x+y)}}$

$= \dfrac{-y}{x(x+y)} \cdot \dfrac{x(x+y)}{\phantom{xxxx}}$

$= \underline{\phantom{xxxx}}$

## SOLUTIONS TO SELECTED ODD PROBLEMS  Section 3.4, text pages 144–145

**1** $\dfrac{\dfrac{2}{3}}{\dfrac{4}{5}}$

The LCD is 15, so

$\dfrac{\dfrac{2}{3} \cdot 15}{\dfrac{4}{5} \cdot 15} = \dfrac{\overset{1}{\cancel{2}} \cdot 5}{\cancel{4} \cdot 3} = \dfrac{5}{6}.$

**5** $\dfrac{3 + \dfrac{1}{c}}{2 - \dfrac{3}{c}}$

The LCD is $c$, so

$\dfrac{\left(3 + \dfrac{1}{c}\right)c}{\left(2 - \dfrac{3}{c}\right)c} = \dfrac{3c + 1}{2c - 3}.$

**9** $\dfrac{\dfrac{2}{1+v}}{3 + \dfrac{1}{1+v}}$

The LCD is $1 + v$, so

$\dfrac{\left(\dfrac{2}{1+v}\right)(1+v)}{\left(3 + \dfrac{1}{1+v}\right)(1+v)} = \dfrac{2}{3 + 3v + 1} = \dfrac{2}{3v + 4}.$

**13** $\dfrac{\dfrac{x}{y} - \dfrac{y}{x}}{\dfrac{x}{y} + \dfrac{y}{x}}$

The LCD is $xy$, so

$= \dfrac{\left(\dfrac{x}{y} - \dfrac{y}{x}\right)xy}{\left(\dfrac{x}{y} + \dfrac{y}{x}\right)xy} = \dfrac{\dfrac{x \cdot x\cancel{y}}{\cancel{y}} - \dfrac{y \cdot \cancel{x}y}{\cancel{x}}}{\dfrac{x \cdot x\cancel{y}}{\cancel{y}} + \dfrac{y \cdot \cancel{x}y}{\cancel{x}}} = \dfrac{x^2 - y^2}{x^2 + y^2}.$

**17** $\dfrac{\dfrac{3}{1-x} + \dfrac{x}{x-1}}{\dfrac{1}{1-x}} = \dfrac{\dfrac{-3}{x-1} + \dfrac{x}{x-1}}{\dfrac{-1}{x-1}}$

The LCD is $x - 1$, so

$\dfrac{\left(\dfrac{-3}{x-1} + \dfrac{x}{x-1}\right)(x-1)}{\dfrac{-1}{x-1} \cdot (x-1)} = \dfrac{-3 + x}{-1} = 3 - x.$

**21** $\dfrac{\dfrac{x-1}{x+1} - \dfrac{x+1}{x-1}}{\dfrac{x-1}{x+1} + \dfrac{x+1}{x-1}}$

The LCD is $(x + 1)(x - 1)$, so

$$\frac{\left(\dfrac{x-1}{x+1} - \dfrac{x+1}{x-1}\right)(x+1)(x-1)}{\left(\dfrac{x-1}{x+1} + \dfrac{x+1}{x-1}\right)(x+1)(x-1)} = \frac{\dfrac{(x-1)(x+1)(x-1)}{x+1} - \dfrac{(x+1)(x+1)(x-1)}{x-1}}{\dfrac{(x-1)(x+1)(x-1)}{x+1} + \dfrac{(x+1)(x+1)(x-1)}{x-1}}$$

$$= \frac{x^2 - 2x + 1 - x^2 - 2x - 1}{x^2 - 2x + 1 + x^2 + 2x + 1} = \frac{-4x}{2x^2 + 2} = \frac{-2(2x)}{2(x^2 + 1)} = \frac{-2x}{x^2 + 1}.$$

**25** $R = \dfrac{1}{\dfrac{1}{R_1} + \dfrac{1}{R_2}}$

The LCD is $R_1 R_2$, so

$$R = \frac{1(R_1 R_2)}{\left(\dfrac{1}{R_1} + \dfrac{1}{R_2}\right) R_1 R_2} = \frac{R_1 R_2}{\dfrac{R_1 R_2}{R_1} + \dfrac{R_1 R_2}{R_2}} = \frac{R_1 R_2}{R_1 + R_2}.$$

Substituting for $R_1$ and $R_2$ we get

$$R = \frac{(10)(15)}{10 + 15} = \frac{150}{25} = 6.$$

## CUMULATIVE REVIEW PROBLEM SET   Chapters 1–3

1. Translate the following statement into symbols: The product of 3 and the sum of a number $x$ and 4 is greater than 5 less than the number $x$.

2. Evaluate the expression $3^2 + (5 - 2)^3 \div 3^2 + 4(7 - 3)$.

3. Evaluate $-|-11|$.

4. Rewrite 57,800,000 in scientific notation.

5. Evaluate $F = \frac{9}{5}C + 32$ for $C = 20°$.

6. Evaluate $3x^2 - 2xy - y^3$ for $x = 2$, $y = -3$.

7. Simplify $\dfrac{(-3xy^2)^2}{(3x^2 y^4)^3}$.

8. Use a special product to find the product of $(3x + y)(9x^2 - 3xy + y^2)$.

9. Factor $18x^3 y - 3x^2 y^2 - 3xy^3$.

10. Divide $w^4 - 3w^3 + 7w - 11$ by $w + 1$.

11. Determine all values of $z$ for which the fraction $\dfrac{3z}{z^2 - z - 6}$ is not defined.

12. Determine if the fractions $\dfrac{3}{x - 2}$ and $\dfrac{3x + 6}{x^2 - 4}$ are equivalent.

13. Reduce $\dfrac{4x^2 - y^2}{2x^2 + 7xy + 3y^2}$ to lowest terms.

14. Find the missing numerator that will make the fractions equivalent.
$\dfrac{4w}{w + 4} = \dfrac{?}{w^3 + 64}.$

15. Find the following product and simplify the result.
$\dfrac{a^2 b^2 - 9}{x^2 - x - 2} \cdot \dfrac{x^3 - x^2 - 2x}{a^2 b^2 - 6ab + 9}$

16. Find the following quotient and simplify the result.
$\dfrac{(x + 5)^2}{x^2 - 4} \div \dfrac{x^2 - 25}{(x + 2)^2}$

**17** Find the following sum.

$$\frac{1}{x^2 - 9} + \frac{2}{x^2 - 6x + 9}$$

**18** Find the following difference.

$$\frac{x}{x^2y - y^3} - \frac{y}{x^2 + xy}$$

**19** Perform the indicated operations and simplify.

$$\frac{1}{y - 2} - \frac{1}{y - 1} + \frac{y}{y^2 + 2y - 3}$$

**20** Change to a simple fraction.

$$\frac{u + \dfrac{1}{v}}{v + \dfrac{1}{u}}$$

## Answers

**1** $3(x + 4) > x - 5$   **2** 28   **3** $-11$   **4** $5.78 \times 10^7$   **5** $F = 68°$   **6** 51   **7** $\dfrac{1}{3x^4y^8}$

**8** $27x^2 + y^3$ [using the special product $(a + b)(a^2 - ab + b^2) = a^3 + b^3$]   **9** $3xy(3x + y)(2x - y)$

**10** $w^3 - 4w^2 + 4w + 3, R = -14$   **11** $-2, 3$   **12** Equivalent   **13** $\dfrac{2x - y}{x + 3y}$

**14** $4w^3 - 16w^2 + 64w$   **15** $\dfrac{x(ab + 3)}{ab - 3}$   **16** $\dfrac{(x + 5)(x + 2)}{(x - 5)(x - 2)}$   **17** $\dfrac{3x + 3}{(x - 3)^2(x + 3)}$

**18** $\dfrac{x^2 - xy^2 + y^3}{xy(x^2 - y^2)}$   **19** $\dfrac{y^2 - y + 3}{(y - 1)(y - 2)(y + 3)}$   **20** $\dfrac{u}{v}$

## CHAPTER 3   TESTS

### Chapter Test A

**1** Specify the real value, if any, of the variables for which each expression is undefined.

   (a) $\dfrac{x}{x - 1}$   (b) $\dfrac{x - y}{2x - 1}$   (c) $\dfrac{x}{(x - 1)(x + 2)}$   (d) $\dfrac{x - y}{(x + 5)(3x + 7)}$

**2** Indicate which pairs of fractions are equivalent.

   (a) $\dfrac{3}{7}$ and $\dfrac{8}{11}$   (b) $\dfrac{x}{4}$ and $\dfrac{6x^2}{24x}$   (c) $\dfrac{5x^2 + 10}{15x^2 - 5}$ and $\dfrac{x^2 + 2}{3x^2 - 1}$

**3** Reduce the fraction to lowest terms.

   (a) $\dfrac{y - x}{x^2 - y^2}$   (b) $\dfrac{y^2 - xy - 2x^2}{12x^2 - 3y^2}$   (c) $\dfrac{a^2x + b^2y - b^2x - a^2y}{ax - ay^2 - by^2 + bx^2}$

**4** Perform the indicated operation and reduce the result to lowest terms.

   (a) $\dfrac{24}{3x - 6} \cdot \dfrac{x^2 - 4}{x + 2}$   (b) $\dfrac{2x - 3}{x^2 - 1} \cdot \dfrac{2x^2 + x - 3}{4x^2 - 9}$

   (c) $\dfrac{a^2bc}{abc^2} \div \dfrac{ab^2c}{ac}$   (d) $\dfrac{9x^2 - 4}{x^2 + 4x - 12} \div \dfrac{12x + 8}{x^2 + x - 30}$

   (e) $\dfrac{xy}{1} \cdot \dfrac{y^2 - 4xy}{y - x} \div \dfrac{16x^2y^2 - y^4}{4x^2 - 3xy - y^2}$

**5** Perform the indicated operation and simplify.

(a) $\dfrac{3}{x-1} - \dfrac{3}{x} - \dfrac{2}{x^2} - \dfrac{1}{x^3}$

(b) $\dfrac{7}{x+4} + \dfrac{5}{x-3}$

(c) $\dfrac{2x+1}{24x} - \dfrac{3x^2 - 2x}{36x^2}$

(d) $\dfrac{2x}{x^2 - 4xy + 4y^2} - \dfrac{y}{x^2 - 4y^2}$

**6** Simplify each complex fraction.

(a) $\dfrac{\dfrac{2}{x} + \dfrac{3}{y}}{\dfrac{4}{x} - \dfrac{5}{y}}$

(b) $\dfrac{2 + \dfrac{1}{x^2 - 4}}{\dfrac{3}{x-2} - \dfrac{1}{x+2}}$

---

## Solutions

**1** (a) $x - 1 = 0$
$x = 1$

(b) $2x - 1 = 0$
$2x = 1$
$x = \tfrac{1}{2}$

(c) $(x-1)(x+2) = 0$
$x - 1 = 0$ or $x + 2 = 0$
$x = 1$ or $x = -2$

(d) $(x+5)(3x+7) = 0$
$x + 5 = 0$ or $3x + 7 = 0$
$x = -5$ or $3x = -7$
$x = -\tfrac{7}{3}$

**2** (a) $\dfrac{3}{7} \stackrel{?}{=} \dfrac{8}{11}$
$3(11) \stackrel{?}{=} 8(7)$
$33 \neq 56$
Not equivalent

(b) $\dfrac{x}{4} \stackrel{?}{=} \dfrac{6x^2}{24x}$
$x(24x) \stackrel{?}{=} 4(6x^2)$
$24x^2 = 24x^2$
Equivalent

(c) $\dfrac{5x^2 + 10}{15x^2 - 5} \stackrel{?}{=} \dfrac{x^2 + 2}{3x^2 - 1}$
$(5x^2 + 10)(3x^2 - 1) \stackrel{?}{=} (15x^2 - 5)(x^2 + 2)$
$15x^4 + 25x^2 - 10 = 15x^4 + 25x^2 - 10$
Equivalent

**3** (a) $\dfrac{y - x}{x^2 - y^2} = \dfrac{-1(x-y)}{(x-y)(x+y)} = \dfrac{-1}{x+y} = -\dfrac{1}{x+y}$

(b) $\dfrac{y^2 - xy - 2x^2}{12x^2 - 3y^2} = \dfrac{(y - 2x)(y + x)}{3(4x^2 - y^2)} = \dfrac{-1(2x - y)(y + x)}{3(2x - y)(2x + y)} = -\dfrac{x + y}{3(2x + y)}$

(c) $\dfrac{a^2 x + b^2 y - b^2 x - a^2 y}{ax^2 - ay^2 - by^2 + bx^2} = \dfrac{(a^2 - b^2)x - (a^2 - b^2)y}{a(x^2 - y^2) + b(x^2 - y^2)} = \dfrac{(a^2 - b^2)(x - y)}{(a + b)(x^2 - y^2)} = \dfrac{(a - b)(a + b)(x - y)}{(a + b)(x - y)(x + y)} = \dfrac{a - b}{x + y}$

**4** (a) $\dfrac{24}{3x - 6} \cdot \dfrac{x^2 - 4}{x + 2} = \dfrac{24(x-2)(x+2)}{3(x-2)(x+2)} = \dfrac{8}{1} = 8$

(b) $\dfrac{2x - 3}{x^2 - 1} \cdot \dfrac{2x^2 + x - 3}{4x^2 - 9} = \dfrac{(2x - 3)(2x + 3)(x - 1)}{(x - 1)(x + 1)(2x - 3)(2x + 3)} = \dfrac{1}{x + 1}$

(c) $\dfrac{a^2 bc}{abc^2} \div \dfrac{ab^2 c}{ac} = \dfrac{a^2 bc}{abc^2} \cdot \dfrac{ac}{ab^2 c} = \dfrac{a^3 bc^2}{a^2 b^3 c^3} = \dfrac{a}{b^2 c}$

(d) $\dfrac{9x^2 - 4}{x^2 + 4x - 12} \div \dfrac{12x + 8}{x^2 + x - 30} = \dfrac{9x^2 - 4}{x^2 + 4x - 12} \cdot \dfrac{x^2 + x - 30}{12x + 8}$
$= \dfrac{(3x - 2)(3x + 2) \cdot (x + 6)(x - 5)}{(x + 6)(x - 2) \cdot 4(3x + 2)} = \dfrac{(3x - 2)(x - 5)}{4(x - 2)}$

(e) $\dfrac{xy}{1} \cdot \dfrac{y^2 - 4xy}{y - x} \div \dfrac{16x^2y^2 - y^4}{4x^2 - 3xy - y^2} = \dfrac{xy}{1} \cdot \dfrac{y^2 - 4xy}{y - x} \cdot \dfrac{4x^2 - 3xy - y^2}{16x^2y^2 - y^4}$

$= \dfrac{xy \cdot y(y - 4x)(4x + y)(x - y)}{(y - x)y^2(16x^2 - y^2)}$

$= \dfrac{-xy^2\cancel{(4x - y)}\cancel{(4x + y)}\cancel{(x - y)}}{-y^2\cancel{(x - y)}\cancel{(4x - y)}\cancel{(4x + y)}}$

$= x$

5 (a) $\dfrac{3}{x - 1} - \dfrac{3}{x} - \dfrac{2}{x^2} - \dfrac{1}{x^3} = \dfrac{3 \cdot x^3}{x^3(x - 1)} - \dfrac{3 \cdot x^2(x - 1)}{x^3(x - 1)} - \dfrac{2 \cdot x(x - 1)}{x^3(x - 1)} - \dfrac{1 \cdot (x - 1)}{x^3(x - 1)}$

$= \dfrac{3x^3 - 3x^3 + 3x^2 - 2x^2 + 2x - x + 1}{x^3(x - 1)}$

$= \dfrac{x^2 + x + 1}{x^3(x - 1)}$

(b) $\dfrac{7}{x + 4} + \dfrac{5}{x - 3} = \dfrac{7(x - 3) + 5(x + 4)}{(x + 4)(x - 3)} = \dfrac{12x - 1}{(x + 4)(x - 3)}$

(c) $\dfrac{2x + 1}{24x} - \dfrac{3x^2 - 2x}{36x^2} = \dfrac{(2x + 1)3x}{72x^2} - \dfrac{(3x^2 - 2x)2}{72x^2} = \dfrac{(6x^2 + 3x) - (6x^2 - 4x)}{72x^2} = \dfrac{7x}{72x^2} = \dfrac{7}{72x}$

(d) $\dfrac{2x}{x^2 - 4xy + 4y^2} - \dfrac{y}{x^2 - 4y^2} = \dfrac{2x}{(x - 2y)^2} - \dfrac{y}{(x - 2y)(x + 2y)}$

$= \dfrac{2x(x + 2y)}{(x - 2y)^2(x + 2y)} - \dfrac{y(x - 2y)}{(x - 2y)^2(x + 2y)}$

$= \dfrac{(2x^2 + 4xy) - (xy - 2y^2)}{(x - 2y)^2(x + 2y)}$

$= \dfrac{2x^2 + 3xy + 2y^2}{(x - 2y)^2(x + 2y)}$

6 (a) $\dfrac{\dfrac{2}{x} + \dfrac{3}{y}}{\dfrac{4}{x} - \dfrac{5}{y}} = \dfrac{\left(\dfrac{2}{x} + \dfrac{3}{y}\right)xy}{\left(\dfrac{4}{x} - \dfrac{5}{y}\right)xy} = \dfrac{2y + 3x}{4y - 5x}$

(b) $\dfrac{2 + \dfrac{1}{x^2 - 4}}{\dfrac{3}{x - 2} - \dfrac{1}{x + 2}} = \dfrac{\left(2 + \dfrac{1}{(x - 2)(x + 2)}\right)(x - 2)(x + 2)}{\left(\dfrac{3}{x - 2} - \dfrac{1}{x + 2}\right)(x - 2)(x + 2)}$

$= \dfrac{2(x - 2)(x + 2) + 1}{3(x + 2) - 1(x - 2)}$

$= \dfrac{2(x^2 - 4) + 1}{3x + 6 - x + 2}$

$= \dfrac{2x^2 - 7}{2x + 8}$

## Chapter Test B

*Multiple Choice:* Select the *one* correct answer for each of the following questions.

1 The expression $\dfrac{36}{44}$ is reduced to _____.

   (a) $\dfrac{12}{23}$       (b) $\dfrac{9}{21}$       (c) $\dfrac{18}{44}$       (d) $\dfrac{9}{11}$

2 The expression $-\dfrac{-3x+3y}{5x-5y}$ is reduced to _____.

(a) $\dfrac{3}{5}$ (b) $\dfrac{x-y}{5}$ (c) $\dfrac{x-y}{x+y}$ (d) $\dfrac{-x+y}{x-y}$

3 The expression $\dfrac{5(x-1)^2}{15(x-1)}$ is reduced to _____.

(a) $\dfrac{x-1}{5}$ (b) $\dfrac{5x-1}{15}$ (c) $\dfrac{x-1}{3}$ (d) $\dfrac{x^2-2x+1}{15(x-1)}$

4 The expression $\dfrac{x^2-10x+24}{x-6}$ is reduced to _____.

(a) $\dfrac{x-4}{x-6}$ (b) $x-4$ (c) $\dfrac{x^2+24}{-6}$ (d) $x^2-9x+18$

5 Performing $\dfrac{5}{4x}+\dfrac{7}{4x}$ yields _____.

(a) $\dfrac{12}{x}$ (b) $\dfrac{2}{x}$ (c) $\dfrac{3}{4x}$ (d) $\dfrac{3}{x}$

6 Performing $\dfrac{2}{x+3}+\dfrac{3}{x-5}$ yields _____.

(a) $\dfrac{5x+1}{(x+3)(x-5)}$ (b) $\dfrac{1-5x}{(x-3)(x+5)}$ (c) $\dfrac{5x-1}{(x+3)(x-5)}$ (d) $\dfrac{5x-1}{x^2+2x-15}$

7 Performing $\dfrac{2x}{x^2-4xy+4y^2}-\dfrac{y}{x^2-4y^2}$ yields _____.

(a) $\dfrac{2x^2-3xy+2y^2}{(x+2y)^2(x-2y)}$ (b) $\dfrac{2x^2+3xy+2y^2}{(x-2y)^2(x+2y)}$ (c) $\dfrac{2x^2-3xy-2y^2}{(x+2y)^2(x-2y)}$ (d) $\dfrac{-2x^2+3xy-2y^2}{(x-2y)^2(x+2y)}$

8 Performing $\dfrac{1}{x^4-1}-\dfrac{1}{x^2+1}+\dfrac{2}{x^2-1}$ yields _____.

(a) $\dfrac{x^2+4}{x^4-1}$ (b) $\dfrac{x-2}{x^2+1}$ (c) $\dfrac{x^2-4}{x^4+1}$ (d) $\dfrac{x^2+4}{x^4+1}$

9 Multiplying $\dfrac{8x^2}{5wz}$ by $\dfrac{15w^2}{14x^3}$ yields _____.

(a) $\dfrac{12w}{7xz}$ (b) $\dfrac{15w}{7xz}$ (c) $\dfrac{8w}{7xz}$ (d) $\dfrac{8x}{5wz}$

10 Multiplying $\dfrac{x^2-144}{x+4}$ by $\dfrac{x^2-16}{x-12}$ yields _____.

(a) $\dfrac{x+12}{x-4}$ (b) $\dfrac{x-12}{x+4}$ (c) $(x+12)(x-4)$ (d) $\dfrac{x^2-16}{x-12}$

11 Multiplying $\dfrac{x^2-4y^2}{(x+y)^2}$ by $\dfrac{x^2-y^2}{x^2-4xy+4y^2}$ yields _____.

(a) $\dfrac{(x-2y)(x+y)}{x^2+xy-2y^2}$ (b) $\dfrac{(-x-2y)(x+y)}{x^2-xy-2y^2}$ (c) $\dfrac{x^2-xy-2y^2}{(x+y)(x-2y)}$ (d) $\dfrac{(x+2y)(x-y)}{(x+y)(x-2y)}$

12 Multiplying $\dfrac{2a+b}{a^2-2ab}$ by $\dfrac{a^3-2a^2b}{4a^2-b^2}$ yields _____.

(a) $\dfrac{a}{-2b+a}$ (b) $\dfrac{a}{2a-b}$ (c) $\dfrac{a}{-2a+b}$ (d) $\dfrac{a}{a+2b}$

## CHAPTER 3 THE ALGEBRA OF FRACTIONS

13. Multiplying $\dfrac{x^2 - 10x + 25}{x^2 - 100}$ by $\dfrac{x^2 + 12x + 20}{x^2 - 7x + 10}$ yields _____.

    (a) $\dfrac{(x - 5)(x + 2)}{(x - 10)(x - 2)}$    (b) $\dfrac{(x - 5)(x - 2)}{(x - 10)(x - 2)}$    (c) $\dfrac{(x - 5)(x + 2)}{(-x + 10)(x - 2)}$    (d) $\dfrac{(x - 5)(x + 2)}{(x + 10)(x - 2)}$

14. Dividing $\dfrac{9x^2}{11}$ by $\dfrac{63x^2}{44}$ yields _____.

    (a) $\dfrac{9x^2}{44}$    (b) $\dfrac{63}{11}$    (c) $\dfrac{4}{7}$    (d) $\dfrac{63}{44}$

15. Dividing $\dfrac{4x}{y}$ by $\dfrac{3x}{5y}$ yields _____.

    (a) $\dfrac{4}{5}$    (b) $\dfrac{4x}{5y}$    (c) $\dfrac{5}{4}$    (d) $\dfrac{20}{3}$

16. Dividing $\dfrac{5x + 10y}{x - y}$ by $\dfrac{x^2 + 2xy}{x^2 - y^2}$ yields _____.

    (a) $\dfrac{5x + 5y}{x}$    (b) $\dfrac{5x - 5y}{x}$    (c) $\dfrac{5x + 5y}{y}$    (d) $\dfrac{x + 5y}{x}$

17. The complex fraction $\dfrac{\tfrac{3}{4}}{\tfrac{6}{5}}$ is equivalent to _____.

    (a) $\dfrac{6}{5}$    (b) $\dfrac{5}{8}$    (c) $\dfrac{4}{5}$    (d) $\dfrac{5}{3}$

18. The complex fraction $\dfrac{\tfrac{1}{x} - \tfrac{1}{y}}{\tfrac{1}{x} + \tfrac{1}{y}}$ is equivalent to _____.

    (a) $\dfrac{y + x}{y - x}$    (b) $\dfrac{y - x}{y + x}$    (c) $\dfrac{-y + x}{y - x}$    (d) $\dfrac{-y - x}{y + x}$

19. The LCD of the fractions $\tfrac{5}{12}$, $\tfrac{1}{10}$, and $\tfrac{7}{45}$ is _____.

    (a) 180    (b) 160    (c) 90    (d) 450

20. The LCD of the fractions $\dfrac{5x}{7y^3z^2}$, $\dfrac{9y^2}{14x^3z^3}$, and $\dfrac{11z}{21y^5x^6}$ is _____.

    (a) $42x^5y^5z^3$    (b) $42x^6y^4z^3$    (c) $42x^6y^5z^3$    (d) $42x^6y^6z^4$

---

**Answers**

1 d    2 a    3 c    4 b    5 d    6 c    7 b    8 a    9 a    10 c
11 d    12 b    13 a    14 c    15 d    16 a    17 b    18 b    19 a
20 c

# 4 Linear Equations and Inequalities

In this chapter we use the algebraic tools developed in Chapters 1–3 to solve linear equations and inequalities. After completing the appropriate sections, the student should be able to:

1. Solve linear equations.
2. Solve fractional equations.
3. Solve literal equations and formulas for specified variables or literal numbers.
4. Translate verbal expressions into algebraic expressions.
5. Solve word problems using linear equations or inequalities.
6. Solve linear inequalities.
7. Solve absolute-value equations and inequalities.

## 4.1 Equations

**SEMIPROGRAMMED PROBLEMS**

In problems 1–9, solve each first-degree equation.

|  |  |
|---|---|
| 2 <br> 6 | **1**    $x + 2 = 8$ <br> $x + 2 - 2 = 8 - \underline{\phantom{xx}}$ <br> $x = \underline{\phantom{xx}}$ |
| 3 <br> 7 | **2**    $y - 3 = 4$ <br> $y - 3 + 3 = 4 + \underline{\phantom{xx}}$ <br> $y = \underline{\phantom{xx}}$ |
| 3 <br> $-5$ | **3**    $3z = -15$ <br> $\dfrac{3z}{3} = \dfrac{-15}{\underline{\phantom{xx}}}$ <br> $z = \underline{\phantom{xx}}$ |
| 1 <br> 15 <br> 5 | **4**    $3x - 1 = 14$ <br> $3x - 1 + 1 = 14 + \underline{\phantom{xx}}$ <br> $3x = \underline{\phantom{xx}}$ <br> $x = \underline{\phantom{xx}}$ |
| 3, 3 <br> 16 <br> 4 | **5**    $4x + 3 = 19$ <br> $4x + 3 - \underline{\phantom{xx}} = 19 - \underline{\phantom{xx}}$ <br> $4x = \underline{\phantom{xx}}$ <br> $x = \underline{\phantom{xx}}$ |

55

# CHAPTER 4 LINEAR EQUATIONS AND INEQUALITIES

|  |  |
|---|---|
| 5x <br> 30 <br> 10 | **6**   $8x + 14 = 5x + 44$ <br> $\phantom{xx} 8x - \underline{\phantom{xx}} = 44 - 14$ <br> $\phantom{xxxxxx} 3x = \underline{\phantom{xx}}$ <br> $\phantom{xxxxxxxx} x = \underline{\phantom{xx}}$ |
| $-5x$ <br> 15 <br> 3 | **7**   $7x - 5 + 3x = 5x + 10$ <br> $\phantom{xx} 7x + 3x + \underline{\phantom{xxx}} = 10 + 5$ <br> $\phantom{xxxxxxxx} 5x = \underline{\phantom{xx}}$ <br> $\phantom{xxxxxxxxx} x = \underline{\phantom{xx}}$ |
| 2 <br> 12 <br> 12 <br> 2 | **8**   $8x = 10 + 2(x + 1)$ <br> $\phantom{xx} 8x = 10 + 2x + \underline{\phantom{xx}}$ <br> $8x - 2x = \underline{\phantom{xx}}$ <br> $\phantom{xx} 6x = \underline{\phantom{xx}}$ <br> $\phantom{xxx} x = \underline{\phantom{xx}}$ |
| $4x + 20$ <br> $4x - 27$ <br> $-6$ <br> $-6$ <br> $-2$ | **9**   $7(x - 3) = 4(x + 5) - 47$ <br> $7x - 21 = \underline{\phantom{xxxx}} - 47$ <br> $7x - 21 = \underline{\phantom{xxxx}}$ <br> $7x - 4x = \underline{\phantom{xx}}$ <br> $\phantom{xx} 3x = \underline{\phantom{xx}}$ <br> $\phantom{xxx} x = \underline{\phantom{xx}}$ |

In problems 10–12, express each repeating decimal as a quotient of integers.

|  |  |
|---|---|
| $7.\overline{7}$ <br> $7.\overline{7}$ <br> $7$ <br> $\frac{7}{9}$ | **10**   $0.\overline{7}$ <br> Let $x = 0.\overline{7}$, then $10x = \underline{\phantom{xx}}$ so that <br> $10x - x = \underline{\phantom{xx}} - 0.\overline{7}$ <br> or   $9x = \underline{\phantom{xx}}$ <br> $\phantom{xx} x = \underline{\phantom{xx}}$. |
| $23.\overline{23}$ <br> $23.\overline{23}$ <br> $23$ <br> $\frac{23}{99}$ | **11**   $0.\overline{23}$ <br> Let $x = 0.\overline{23}$, then $100x = \underline{\phantom{xxxx}}$ so that <br> $100x - x = \underline{\phantom{xxxx}} - 0.\overline{23}$ <br> or   $99x = \underline{\phantom{xx}}$ <br> $\phantom{xx} x = \underline{\phantom{xx}}$. |
| $3471.\overline{471}$ <br> $3471.\overline{471}$ <br> $3468$ <br> $3468, \frac{1156}{333}$ | **12**   $3.\overline{471}$ <br> Let $x = 3.\overline{471}$, then $1000x = \underline{\phantom{xxxxx}}$ so that <br> $1000x - x = \underline{\phantom{xxxxx}} - 3.\overline{471}$ <br> or   $999x = \underline{\phantom{xx}}$ <br> $x = \dfrac{\overline{\phantom{xxxx}}}{999} = \underline{\phantom{xx}}$. |

## SOLUTIONS TO SELECTED ODD PROBLEMS   Section 4.1, text page 156

**1**   $x + 3 = 10$ <br>
$\phantom{xx} x + 3 - 3 = 10 - 3$ <br>
$\phantom{xxxxxx} x = 7$

**5**   $w + 11 = 17$ <br>
$w + 11 - 11 = 17 - 11$ <br>
$\phantom{xxxxx} w = 6$

**9**   $-4u = 12$ <br>
$\dfrac{-4u}{-4} = \dfrac{12}{-4}$ <br>
$\phantom{xx} u = -3$

## 4.1 SOLUTIONS TO SELECTED ODD PROBLEMS

**13**
$15t = 75$
$\dfrac{15t}{15} = \dfrac{75}{15}$
$t = 5$

**17**
$-12b = 8$
$\dfrac{-12b}{-12} = \dfrac{8}{-12}$
$b = -\dfrac{2}{3}$

**21**
$6y + 7 = 31$
$6y + 7 - 7 = 31 - 7$
$6y = 24$
$\dfrac{6y}{6} = \dfrac{24}{6}$
$y = 4$

**25**
$3t - 5 = 20$
$3t - 5 + 5 = 20 + 5$
$3t = 25$
$\dfrac{3t}{3} = \dfrac{25}{3}$
$t = \dfrac{25}{3}$

**29**
$6x - 8 = 7 - x$
$6x - 8 + x = 7 - x + x$
$7x - 8 = 7$
$7x - 8 + 8 = 7 + 8$
$7x = 15$
$\dfrac{7x}{7} = \dfrac{15}{7}$
$x = \dfrac{15}{7}$

**33**
$5 - 9z = -8z + 3$
$5 - 9z + 8z = -8z + 3 + 8z$
$5 - z = 3$
$5 - z - 5 = 3 - 5$
$-z = -2$
$\dfrac{-z}{-1} = \dfrac{-2}{-1}$
$z = 2$

**37**
$12t + 1 = 25 - 12t$
$12t + 1 + 12t = 25 - 12t + 12t$
$24t + 1 = 25$
$24t + 1 - 1 = 25 - 1$
$24t = 24$
$\dfrac{24t}{24} = \dfrac{24}{24}$
$t = 1$

**41**
$1 - 2(5 - 2y) = 26 - 3y$
$1 - 10 + 4y = 26 - 3y$
$-9 + 4y = 26 - 3y$
$-9 + 4y + 3y = 26 - 3y + 3y$
$-9 + 7y = 26$
$-9 + 7y + 9 = 26 + 9$
$7y = 35$
$\dfrac{7y}{7} = \dfrac{35}{7}$
$y = 5$

**45**
$6(c - 10) + 3(2c - 7) = -45$
$6c - 60 + 6c - 21 = -45$
$12c - 81 = -45$
$12c - 81 + 81 = -45 + 81$
$12c = 36$
$\dfrac{12c}{12} = \dfrac{36}{12}$
$c = 3$

**49**
$34 - 3y = 8(7 - y) + 23$
$34 - 3y = 56 - 8y + 23$
$34 - 3y = 79 - 8y$
$34 - 3y + 8y = 79 - 8y + 8y$
$34 + 5y = 79$
$34 + 5y - 34 = 79 - 34$
$5y = 45$
$\dfrac{5y}{5} = \dfrac{45}{5}$
$y = 9$

**53** Let $x = 1.\overline{3}$. Then $10x = 13.\overline{3}$.
$10x - x = 13.\overline{3} - 1.\overline{3}$
$9x = 12$
$x = \dfrac{12}{9} = \dfrac{4}{3}$

**57** Let $x = -3.\overline{128}$. Then $1000x = -3128.\overline{128}$.
$1000x - x = -3128.\overline{128} - (-3.\overline{128})$
$999x = -3125$
$x = -\dfrac{3125}{999}$

**61**
$0.1347y - 6.738 = 0.2814y - 1.813$
$0.1347y - 6.738 - 0.2814y = 0.2814y - 1.813 - 0.2814y$
$-0.1467y - 6.738 = -1.813$
$-0.1467y - 6.738 + 6.738 = -1.813 + 6.738$
$-0.1467y = 4.925$
$\dfrac{-0.1467y}{-0.1467} = \dfrac{4.925}{-0.1467}$
$y = -33.572$

# 4.2 Equations Involving Fractions

## SEMIPROGRAMMED PROBLEMS

In problems 1–8, solve each equation.

**1** $\dfrac{x}{2} + \dfrac{x}{3} = 10$

6          $6\left(\dfrac{x}{2} + \dfrac{x}{3}\right) = (\underline{\quad})10$

60          $3x + 2x = \underline{\quad}$

60          $5x = \underline{\quad}$

12          $x = \underline{\quad}$

**2** $\dfrac{x-10}{8} + \dfrac{13}{4} = \dfrac{4x+6}{3}$

24        $24\left(\dfrac{x-10}{8} + \dfrac{13}{4}\right) = (\underline{\quad})\left(\dfrac{4x+6}{3}\right)$

8        $3(x-10) + 6(13) = (\underline{\quad})(4x+6)$

$32x + 48$      $3x - 30 + 78 = \underline{\qquad}$

$-29, 0$      $(\underline{\quad})x = \underline{\quad}$

0        $x = \underline{\quad}$

**3** $\left(\dfrac{6}{x} + \dfrac{x-3}{2x}\right) = 2$

$2x$      $2x\left(\dfrac{6}{x} + \dfrac{x-3}{2x}\right) = (\underline{\quad})2$

$4x$      $12 + x - 3 = \underline{\quad}$

12      $x - 4x = 3 - \underline{\quad}$

$-9$      $-3x = \underline{\quad}$

3      $x = \underline{\quad}$

**4** $\dfrac{2}{3x} + \dfrac{1}{6x} = \dfrac{1}{4}$

$12x, 12x$    $(\underline{\quad})\left(\dfrac{2}{3x} + \dfrac{1}{6x}\right) = (\underline{\quad})\dfrac{1}{4}$

$8 + 2$      $\underline{\qquad} = 3x$

10       $\underline{\qquad} = 3x$

$\dfrac{10}{3}$      $\underline{\qquad} = x$

**5** $\dfrac{8}{x-3} = 2$

$x - 3$      $(x-3)\dfrac{8}{x-3} = (\underline{\quad})2$

6      $8 = 2x - \underline{\quad}$

14      $\underline{\quad} = 2x$

7      $x = \underline{\quad}$

**6** $\dfrac{4}{x-8} = \dfrac{3}{x-9}$

$(x-8)(x-9)$

$x-8$

$24$

$3x$

$12$

$(x-2)(x+2)$

$x-2$

$x^2-4$

$-4$

$-4$

$-\frac{1}{2}$

$(x-2)(x+4)$

$x-2$

$x^2+2x-8$

$20$

$-36$

$12$

$(x-8)(x-9)\left(\dfrac{4}{x-8}\right) = \underline{\hspace{1cm}} \left(\dfrac{3}{x-9}\right)$

$\qquad 4(x-9) = 3(\underline{\hspace{1cm}})$

$\qquad 4x - 36 = 3x - \underline{\hspace{1cm}}$

$\qquad 4x - \underline{\hspace{1cm}} = -24 + 36$

$\qquad x = \underline{\hspace{1cm}}$

7  $\dfrac{x+2}{x-2} - \dfrac{x-2}{x+2} = 1 - \dfrac{x^2}{x^2-4}$

$(x-2)(x+2)\left(\dfrac{x+2}{x-2} - \dfrac{x-2}{x+2}\right) = \underline{\hspace{1cm}}\left(1 - \dfrac{x^2}{x^2-4}\right)$

$\qquad (x+2)^2 - (\underline{\hspace{1cm}})^2 = (x-2)(x+2) - x^2$

$\qquad x^2 + 4x + 4 - (x^2 - 4x + 4) = (\underline{\hspace{1cm}}) - x^2$

$\qquad x^2 + 4x + 4 - x^2 + 4x - 4 = \underline{\hspace{1cm}}$

$\qquad 8x = \underline{\hspace{1cm}}$

$\qquad x = \underline{\hspace{1cm}}$

8  $\dfrac{5}{x-2} + \dfrac{2x}{x+4} = 2$

$(x-2)(x+4)\left(\dfrac{5}{x-2} + \dfrac{2x}{x+4}\right) = \underline{\hspace{1cm}} \, 2$

$\qquad 5(x+4) + 2x(\underline{\hspace{1cm}}) = 2(x-2)(x+4)$

$\qquad 5x + 20 + 2x^2 - 4x = 2(\underline{\hspace{1cm}})$

$\qquad 2x^2 - 2x^2 + 5x - 4x - 4x = -16 - \underline{\hspace{1cm}}$

$\qquad -3x = \underline{\hspace{1cm}}$

$\qquad x = \underline{\hspace{1cm}}$

## SOLUTIONS TO SELECTED ODD PROBLEMS  Section 4.2, text pages 161–162

1  $\dfrac{2}{3} - \dfrac{5x}{3} = \dfrac{17}{3}$

The LCD is 3, so

$3\left(\dfrac{2}{3} - \dfrac{5x}{3}\right) = 3\left(\dfrac{17}{3}\right)$

$3\left(\dfrac{2}{3}\right) - 3\left(\dfrac{5x}{3}\right) = 17$

$\qquad 2 - 5x = 17$

$\qquad -5x = 15$

$\qquad x = -3$

5  $\dfrac{t}{6} - \dfrac{t}{7} = \dfrac{1}{42}$

The LCD is 42, so

$42\left(\dfrac{t}{6} - \dfrac{t}{7}\right) = 42\left(\dfrac{1}{42}\right)$

$42\left(\dfrac{t}{6}\right) - 42\left(\dfrac{t}{7}\right) = 1$

$\qquad 7t - 6t = 1$

$\qquad t = 1$

9  $\dfrac{u-1}{2} + \dfrac{u}{7} = \dfrac{11}{14}$

The LCD is 14, so

$14\left(\dfrac{u-1}{2} + \dfrac{u}{7}\right) = 14\left(\dfrac{11}{14}\right)$

$14\left(\dfrac{u-1}{2}\right) + 14\left(\dfrac{u}{7}\right) = 11$

$\qquad 7(u-1) + 2u = 11$

$\qquad 7u - 7 + 2u = 11$

$\qquad 9u - 7 = 11$

$\qquad 9u = 18$

$\qquad u = 2$

13  $\dfrac{5u}{4} - 1 = \dfrac{3u}{4} + \dfrac{1}{2}$

The LCD is 4, so

$4\left(\dfrac{5u}{4} - 1\right) = 4\left(\dfrac{3u}{4} + \dfrac{1}{2}\right)$

$4\left(\dfrac{5u}{4}\right) - 4(1) = 4\left(\dfrac{3u}{4}\right) + 4\left(\dfrac{1}{2}\right)$

$\qquad 5u - 4 = 3u + 2$

$\qquad 2u - 4 = 2$

$\qquad 2u = 6$

$\qquad u = 3$

**17** $\frac{1}{3}(3x - 2) + \frac{1}{2}(x - 3) = \frac{5}{6}$

The LCD is 6, so

$$6\left[\frac{1}{3}(3x - 2) + \frac{1}{2}(x - 3)\right] = 6\left(\frac{5}{6}\right)$$

$$6\left[\frac{1}{3}(3x - 2)\right] + 6\left[\frac{1}{2}(x - 3)\right] = 5$$

$$2(3x - 2) + 3(x - 3) = 5$$
$$6x - 4 + 3x - 9 = 5$$
$$9x - 13 = 5$$
$$9x = 18$$
$$x = 2$$

**21** $\frac{z - 2}{3} - \frac{z - 3}{5} = \frac{13}{15}$

The LCD is 15, so

$$15\left(\frac{z - 2}{3} - \frac{z - 3}{5}\right) = 15\left(\frac{13}{15}\right)$$

$$15\left(\frac{z - 2}{3}\right) - 15\left(\frac{z - 3}{5}\right) = 13$$

$$5(z - 2) - 3(z - 3) = 13$$
$$5z - 10 - 3z + 9 = 13$$
$$2z - 1 = 13$$
$$2z = 14$$
$$z = 7$$

**25** $\frac{1}{y} + \frac{2}{y} = 3 - \frac{3}{y}$

The LCD is $y$, so

$$y\left(\frac{1}{y} + \frac{2}{y}\right) = y\left(3 - \frac{3}{y}\right)$$

$$y\left(\frac{1}{y}\right) + y\left(\frac{2}{y}\right) = y(3) - y\left(\frac{3}{y}\right)$$

$$1 + 2 = 3y - 3$$
$$3 = 3y - 3$$
$$6 = 3y$$
$$2 = y$$

*Check:* substitute 2 for $y$ in original equation.

$$\frac{1}{2} + \frac{2}{2} = 3 - \frac{3}{2}$$

$$1\frac{1}{2} = 1\frac{1}{2}$$

Therefore, 2 is the solution.

**29** $\frac{x}{x - 1} - \frac{3}{x + 1} = 1$

The LCD is $(x - 1)(x + 1)$, so

$$(x - 1)(x + 1)\left(\frac{x}{x - 1} - \frac{3}{x + 1}\right) = (x - 1)(x + 1)(1)$$

$$(x - 1)(x + 1)\left(\frac{x}{x - 1}\right) - (x - 1)(x + 1)\left(\frac{3}{x + 1}\right) = x^2 - 1$$

$$x(x + 1) - 3(x - 1) = x^2 - 1$$
$$x^2 + x - 3x + 3 = x^2 - 1$$
$$-2x + 3 = -1$$
$$-2x = -4$$
$$x = 2$$

*Check:* substitute 2 for $x$ in original equation.

$$\frac{2}{2 - 1} - \frac{3}{2 + 1} = 1$$
$$2 - 1 = 1$$

Therefore, 2 is the solution.

**33** $\dfrac{-4}{3u} = \dfrac{3}{3u+1} + \dfrac{2}{u}$

The LCD is $3u(3u+1)$, so

$$3u(3u+1)\left(\dfrac{-4}{3u}\right) = 3u(3u+1)\left(\dfrac{3}{3u+1} + \dfrac{2}{u}\right)$$

$$-4(3u+1) = (3u)(3u+1)\left(\dfrac{3}{3u+1}\right) + 3u(3u+1)\left(\dfrac{2}{u}\right)$$

$$-12u - 4 = (3u)3 + 6(3u+1)$$
$$-12u - 4 = 9u + 18u + 6$$
$$-12u - 4 = 27u + 6$$
$$-39u - 4 = 6$$
$$-39u = 10$$
$$u = -\tfrac{10}{39}$$

*Check:* substitute $-\tfrac{10}{39}$ for $u$ in original equation.

$$\dfrac{-4}{3(-\tfrac{10}{39})} = \dfrac{3}{3(-\tfrac{10}{39})+1} + \dfrac{2}{-\tfrac{10}{39}}$$

$$\dfrac{4}{\tfrac{10}{13}} = \dfrac{3}{\tfrac{3}{13}} + \left(-\dfrac{2}{\tfrac{10}{39}}\right)$$

$$\dfrac{26}{5} = 13 - \dfrac{39}{5}$$

$$\dfrac{26}{5} = \dfrac{26}{5}$$

Therefore, $-\tfrac{10}{39}$ is the solution.

**37** $\dfrac{4}{y-2} = \dfrac{5y}{y^2-4} - \dfrac{y+3}{y^2-2y}$

The LCD is $y(y-2)(y+2)$, so

$$y(y-2)(y+2)\left(\dfrac{4}{y-2}\right) = y(y-2)(y+2)\left(\dfrac{5y}{y^2-4} - \dfrac{y+3}{y^2-2y}\right)$$

$$4y(y+2) = y(y-2)(y+2)\left(\dfrac{5y}{y^2-4}\right) - y(y-2)(y+2)\left(\dfrac{y+3}{y^2-2y}\right)$$

$$4y^2 + 8y = 5y^2 - (y+2)(y+3)$$
$$4y^2 + 8y = 5y^2 - (y^2 + 5y + 6)$$
$$4y^2 + 8y = 5y^2 - y^2 - 5y - 6$$
$$4y^2 + 8y = 4y^2 - 5y - 6$$
$$13y = -6$$
$$y = -\tfrac{6}{13}$$

*Check:* substitute $-\tfrac{6}{13}$ for $y$ in the original equation.

$$\dfrac{4}{-\tfrac{6}{13}-2} = \dfrac{5(-\tfrac{6}{13})}{(-\tfrac{6}{13})^2-4} - \dfrac{(-\tfrac{6}{13}+3)}{(-\tfrac{6}{13})^2-2(-\tfrac{6}{13})}$$

$$\dfrac{4}{-\tfrac{32}{13}} = \dfrac{-\tfrac{30}{13}}{-\tfrac{640}{169}} - \dfrac{\tfrac{33}{13}}{\tfrac{192}{169}}$$

$$-\dfrac{13}{8} = \dfrac{39}{64} - \dfrac{143}{64}$$

$$-\dfrac{13}{8} = -\dfrac{13}{8}$$

Therefore, $-\tfrac{6}{13}$ is the solution.

41  $\dfrac{1}{t(t-1)} - \dfrac{1}{t} = \dfrac{1}{t-1}$

The LCD is $t(t-1)$, so

$$t(t-1)\left[\dfrac{1}{t(t-1)} - \dfrac{1}{t}\right] = t(t-1)\left[\dfrac{1}{t-1}\right]$$

$$t(t-1)\left[\dfrac{1}{t(t-1)}\right] - t(t-1)\left[\dfrac{1}{t}\right] = t$$

$$1 - (t-1) = t$$
$$1 - t + 1 = t$$
$$2 = 2t$$
$$1 = t$$

Check: For $t = 1$, the denominator $t - 1 = 1 - 1 = 0$. Since division by 0 is undefined, 1 is not a solution. Hence, there is no solution for this equation.

## 4.3 Literal Equations and Formulas

### SEMIPROGRAMMED PROBLEMS

In problems 1–6, solve each equation for $x$.

**1**  $cx + d = a$

$cx = a - \underline{\phantom{d}}$     $d$

$x = \underline{\phantom{\dfrac{a-d}{c}}}$     $\dfrac{a-d}{c}$

**2**  $ax + 2 = bx + 5$

$ax - \underline{\phantom{bx}} = 5 - \underline{\phantom{2}}$     $bx, 2$

$(\underline{\phantom{a-b}})x = \underline{\phantom{3}}$     $a - b, 3$

$x = \underline{\phantom{\dfrac{3}{a-b}}}$     $\dfrac{3}{a-b}$

**3**  $a(x - b) = c + b(x + a)$

$ax - ab = c + \underline{\phantom{bx+ab}}$     $bx + ab$

$ax - \underline{\phantom{bx}} = c + 2(\underline{\phantom{ab}})$     $bx, ab$

$(\underline{\phantom{a-b}})x = \underline{\phantom{c+2ab}}$     $a - b, c + 2ab$

$x = \underline{\phantom{\dfrac{c+2ab}{a-b}}}$     $\dfrac{c + 2ab}{a - b}$

**4**  $\dfrac{x}{a} - \dfrac{1}{c} = \dfrac{a}{b}$

$abc\left(\dfrac{x}{a} - \dfrac{1}{c}\right) = (\underline{\phantom{abc}})\dfrac{a}{b}$     $abc$

$bcx - \underline{\phantom{ab}} = a^2c$     $ab$

$bcx = a^2c + \underline{\phantom{ab}}$     $ab$

$x = \dfrac{\overline{\phantom{a^2c+ab}}}{bc}$     $a^2c + ab$

**5**  $\dfrac{8x + 2a}{a} + \dfrac{6x + a}{2a} = 2x + 3$

### 4.3 SEMIPROGRAMMED PROBLEMS

|  |  |
|---|---|
| $2a$ | $2a\left(\dfrac{8x + 2a}{a} + \dfrac{6x + a}{2a}\right) = \underline{\phantom{xx}}(2x + 3)$ |
| $6x + a$ | $16x + 4a + (\underline{\phantom{xx}}) = 4ax + 6a$ |
| $a$ | $16x + 6x - 4ax = 6a - 4a - \underline{\phantom{xx}}$ |
| $22x$ | $\underline{\phantom{xx}} - 4ax = a$ |
| $22 - 4a$ | $x(\underline{\phantom{xx}}) = a$ |
| $22 - 4a$ | $x = \dfrac{a}{\underline{\phantom{xx}}}$ |
|  | 6   $\dfrac{a+b}{cx-d} - \dfrac{a-b}{cx+d} = 0$ |
|  | $\dfrac{a+b}{cx-d} = \dfrac{\underline{\phantom{xx}}}{cx+d}$ |
| $a - b$ |  |
| $cx + d$ | $(cx-d)(cx+d)\left(\dfrac{a+b}{cx-d}\right) = (cx-d)(\underline{\phantom{xx}})\left(\dfrac{a-b}{cx+d}\right)$ |
| $(cx - d)(a - b)$ | $(cx+d)(a+b) = \underline{\phantom{xx}}$ |
| $bd$ | $cxa + cxb + ad + bd = cxa - cxb - ad + \underline{\phantom{xx}}$ |
| $-2ad$ | $cxa - cxa + cxb + cxb = \underline{\phantom{xx}}$ |
|  | $2cxb = -2ad$ |
| $\dfrac{-ad}{bc}$ | $x = \underline{\phantom{xx}}$ |

In problems 7–12, solve the formulas for the indicated unknown.

7   $E = IR$, for $I$

|  |  |
|---|---|
| $\dfrac{1}{R}$ | $\left(\underline{\phantom{xx}}\right)E = \left(\dfrac{1}{R}\right)IR$ |
| $\dfrac{E}{R}$ | $\underline{\phantom{xx}} = I$ |

8   $y = mx + b$, for $x$

|  |  |
|---|---|
| $b, b$ | $y - \underline{\phantom{xx}} = mx + b - \underline{\phantom{xx}}$ |
| $y - b$ | $\underline{\phantom{xx}} = mx$ |
| $\dfrac{y-b}{m}$ | $\underline{\phantom{xx}} = x$ |

9   $S = 2\pi r^2 + 2\pi rh$, for $h$

|  |  |
|---|---|
| $2\pi r^2$ | $S - \underline{\phantom{xx}} = 2\pi rh$ |
| $\dfrac{1}{2\pi r}, \dfrac{1}{2\pi r}$ | $\left(\underline{\phantom{xx}}\right)(S - 2\pi r^2) = \left(\underline{\phantom{xx}}\right)2\pi rh$ |
| $\dfrac{S - 2\pi r^2}{2\pi r}$ | $\underline{\phantom{xx}} = h$ |

10   $s = \dfrac{a - rl}{1 - r}$, for $l$

|  |  |
|---|---|
| $1 - r$ | $(\underline{\phantom{xx}})s = a - rl$ |
| $s - sr$ | $rl = a - (\underline{\phantom{xx}})$ |
| $a - s + sr$ | $l = \dfrac{\underline{\phantom{xx}}}{r}$ |

## CHAPTER 4  LINEAR EQUATIONS AND INEQUALITIES

| | |
|---|---|
| $ax$ | **11** $\dfrac{E}{a} = \dfrac{R+x}{x}$, for $x$ |
| $ax$ | $(\underline{\quad})\dfrac{E}{a} = ax\left(\dfrac{R+x}{x}\right)$ |
| $aR$ | $Ex = aR + \underline{\quad}$ |
| $E - a$ | $Ex - ax = \underline{\quad}$ |
| $\dfrac{aR}{E-a}$ | $(\underline{\qquad})x = aR$ |
| | $x = \underline{\qquad}$ |
| | **12** $V = \dfrac{\pi h^2}{3}(3x - h)$, for $x$ |
| $\pi h^3$ | $3V = \pi h^2(3x) - \underline{\quad}$ |
| $3V$ | $\underline{\quad} = 3\pi h^2 x - \pi h^3$ |
| $\pi h^3$ | $3V + \underline{\quad} = 3\pi h^2 x$ |
| $3V + \pi h^3$ | $x = \dfrac{\overline{\qquad}}{3\pi h^2}$ |

## SOLUTIONS TO SELECTED ODD PROBLEMS   Section 4.3, text pages 166–167

**1**  $6x + 7c = 37c$, for $x$.
$6x + 7c - 7c = 37c - 7c$
$6x = 30c$
$x = \dfrac{30c}{6}$
$x = 5c$

**5**  $12z - 4b = 6z - 7b$, for $z$.
$12z = 6z - 3b$
$6z = -3b$
$z = \dfrac{-3b}{6}$
$z = -\dfrac{b}{2}$

**9**  $4x - 3a - (10x + 7a) = 0$, for $x$.
$4x - 3a - 10x - 7a = 0$
$-6x - 10a = 0$
$-6x = 10a$
$x = \dfrac{10a}{-6}$
$x = -\dfrac{5a}{3}$

**13**  $5(4r - 3c) - 2(7r - 9c) = 0$, for $r$.
$20r - 15c - 14r + 18c = 0$
$6r + 3c = 0$
$6r = -3c$
$r = \dfrac{-3c}{6}$
$r = -\dfrac{c}{2}$

**17**  $3(a - 2b) + 4(b + a) = 5$, for $a$.
$3a - 6b + 4b + 4a = 5$
$7a - 2b = 5$
$7a = 5 + 2b$
$a = \dfrac{5 + 2b}{7}$

**21**  $\dfrac{b - x}{3} = \dfrac{2a - b}{4} - \dfrac{3x}{5}$, for $x$.
$60\left(\dfrac{b-x}{3}\right) = 60\left(\dfrac{2a-b}{4}\right) - 60\left(\dfrac{3x}{5}\right)$
$20(b - x) = 15(2a - b) - 12(3x)$
$20b - 20x = 30a - 15b - 36x$
$-20x = 30a - 35b - 36x$
$16x = 30a - 35b$
$x = \dfrac{30a - 35b}{16}$

25 $\quad \dfrac{2t+a}{4} - \dfrac{6t+3a}{7} = \dfrac{15a}{28}$, for $t$.

$$28\left(\dfrac{2t+a}{4}\right) - 28\left(\dfrac{6t+3a}{7}\right) = 28\left(\dfrac{15a}{28}\right)$$
$$7(2t+a) - 4(6t+3a) = 15a$$
$$14t + 7a - 24t - 12a = 15a$$
$$-10t - 5a = 15a$$
$$-10t = 20a$$
$$t = \dfrac{20a}{-10}$$
$$t = -2a$$

29 $\quad \dfrac{3}{x} - \dfrac{4}{b} = \dfrac{5}{3b}$, for $x$.

$$3bx\left(\dfrac{3}{x} - \dfrac{4}{b}\right) = 3bx\left(\dfrac{5}{3b}\right)$$
$$3bx\left(\dfrac{3}{x}\right) - 3bx\left(\dfrac{4}{b}\right) = 5x$$
$$9b - 12x = 5x$$
$$-12x = 5x - 9b$$
$$-17x = -9b$$
$$x = \dfrac{-9b}{-17}$$
$$x = \dfrac{9b}{17}$$

33 $\quad \dfrac{1}{y} + \dfrac{2}{y+a} = \dfrac{3}{y-a}$, for $y$.

$$y(y+a)(y-a)\left(\dfrac{1}{y} + \dfrac{2}{y+a}\right) = y(y+a)(y-a)\left(\dfrac{3}{y-a}\right)$$
$$y(y+a)(y-a)\left(\dfrac{1}{y}\right) + y(y+a)(y-a)\left(\dfrac{2}{y+a}\right) = 3y(y+a)$$
$$(y+a)(y-a) + 2y(y-a) = 3y(y+a)$$
$$y^2 - a^2 + 2y^2 - 2ay = 3y^2 + 3ay$$
$$3y^2 - a^2 - 2ay = 3y^2 + 3ay$$
$$-a^2 - 2ay = 3ay$$
$$-2ay = 3ay + a^2$$
$$-5ay = a^2$$
$$y = \dfrac{a^2}{-5a}$$
$$y = -\dfrac{a}{5}$$

37 $\quad V = \ell wh$, for $\ell$.

$$\dfrac{V}{wh} = \dfrac{\ell wh}{wh}$$
$$\dfrac{V}{wh} = \ell$$

41 $\quad V = gt$, for $t$.

$$\dfrac{V}{g} = \dfrac{gt}{g}$$
$$\dfrac{V}{g} = t$$

45 $\quad F = mx + b$, for $m$.

$$F - b = mx$$
$$\dfrac{F-b}{x} = m$$

49 $\quad S = \dfrac{n}{2}(a + \ell)$, for $a$.

$$2S = n(a + \ell)$$
$$2S = an + n\ell$$
$$2S - n\ell = an$$
$$\dfrac{2S - n\ell}{n} = a$$

53 $\quad L = a + (m-1)d$, for $d$.

$$L - a = (m-1)d$$
$$\dfrac{L - a}{m - 1} = d$$

## 4.4 Translating Verbal Expressions into Algebraic Expressions

### SEMIPROGRAMMED PROBLEMS

In problems 1–10, translate each statement into algebraic form.

1 What number is 2 more than 5 times another number represented by $x$? 5 times the number $x$ is _____. Therefore, 2 more than $5x$ is _____.

$5x$

$5x + 2$

# 66 CHAPTER 4 LINEAR EQUATIONS AND INEQUALITIES

$2y$
$2y - 7$


$2$
$z + 2, z + 4$
$3z + 6$



$8x, 8x - 5$



$3d$
$10d$

$3d, 15d$
$10d, 15d, 25d$

$0.5$
$0.5w + 12$

$0.15$
$0.15x, 1.15x$


$0.3z$
$0.3z, 0.7z$

$2x$
$2x + 3$
$2x + 3, 2x^2 + 3x$


$0.05$
$0.05x, 1.05x$

---

2  If the first number is $y$, what number is 7 less than twice the first? Twice the first number is ____. Therefore, 7 less than $2y$ is _____.

3  What is the sum of three consecutive odd integers, if $z$ represents the smallest of these integers? Since $z$ represents the smallest odd integer, and consecutive odd integers differ by ____, the next consecutive odd integers are _____ and _____. Therefore, their sum is _____.

4  A man is 8 times as old as his son is now. If $x$ represents his son's age now, how old was the man 5 years ago? The man's age now is ____. Therefore, his age 5 years ago was _____.

5  Represent the number of cents in a coin collection containing nickels and dimes, if $d$ represents the number of dimes and there are 3 times as many nickels as dimes. Since there are 3 times as many nickels as dimes, then ____ represents the number of nickels. Each dime is worth 10 cents, so that the value of the dimes is ____ cents. Each nickel is worth 5 cents, so the value of the nickels is $5(\_\_\_) = \_\_\_$ cents. Thus, the value of the coin collection is ____ + ____ = ____ cents.

6  What number is 12 more than 50% of a number represented by $w$? 50% of $w$ is $(\_\_\_)w$, so that the desired number is _____.

7  What number exceeds the number represented by $x$ by 15% of $x$? 15% of $x = (\_\_\_)x$. Therefore, the number that exceeds $x$ by 15% of $x$ is $x + \_\_\_ = \_\_\_$.

8  A television set originally priced at $z$ dollars is discounted by 30% for a sale. What is the sale price of the television? The amount of discount is 30% of $z = \_\_\_$. Therefore, the sale price is $z - \_\_\_ = \_\_\_$ dollars.

9  The length of a rectangle is 3 more than twice its width. If its width is $x$ feet, what is its area? Twice its width is ____. Therefore, its length = _____. The formula for the area of a rectangle is $A = \ell w$, so that the desired area $= (\_\_\_)x = \_\_\_$.

10  John invested $y$ dollars in a bank that pays 5% annual simple interest. How much money did John accumulate after 1 year? The interest for 1 year = 5% of $x = (\_\_\_)x$. Therefore, John accumulated $x + \_\_\_ = \_\_\_$ dollars after 1 year.

## SOLUTIONS TO SELECTED ODD PROBLEMS  Section 4.4, text pages 169–171

1  $(x + 5) + 1 = x + 6$

5  5 times the first number $x$ is $5x$. Therefore, 8 less than 5 times the first is $5x - 8$.

9  Since $x =$ the smallest odd integer, then $x + 2 =$ the next consecutive odd integer so that $x^2 + (x + 2)^2$ represents the sum of their squares.

13 $(11x + 7) + 9x = 20x + 7$

17 At present, $x$ = age of youngest daughter; $x + 2$ = age of second daughter; $x + 4$ = age of third daughter; $x + 6$ = age of fourth daughter. In 3 years, their ages will be $x + 3$; $x + 2 + 3$; $x + 4 + 3$; $x + 6 + 3$. Therefore, the sum of their ages in 3 years is $(x + 3) + (x + 5) + (x + 7) + (x + 9) = 4x + 24$.

21 $5(3n) + 10(4d) + 25(5q) = (15n + 40d + 125q)$ cents

25 30% of $x = 0.3x$. Therefore, 7 less than 30% of $x$ is $0.3x - 7$.

29 80% of $t = 0.8t$. Therefore, 8 more than 80% of $t$ is $0.8t + 8$.

33 Using $I = prt$, we have
$5x + 2 = p(0.12)(4)$
$$\frac{5x + 2}{0.48} = p$$

37 $x$ = length ($\ell$)
$x - 8$ = width ($w$)
$A = \ell w$
$A = x(x - 8)$

## 4.5 Applications of Linear Equations—Word Problems

### SEMIPROGRAMMED PROBLEMS

In problems 1–15, set up an equation and solve for the unknown value.

### Number Problems

6
$6x - 7 = 17$, 4
4

$3x + 2x$
$3x + 2x + 15 = 45$
6, 6

### Age Problems

$x + 5, x + 7$
$x + 12$
$x + 7 + x + 12,$
8, 8
13

$x + 28, x + 15$

15 years
$x + 28 + 15$, 13
13, 41

1 If a number is multiplied by 6 and then 7 is subtracted from this product, the result is 17. Find the number. Let $x$ = the number. Then $6x$ = the number after multiplying by ____. The equation is _____. Solving the equation, $x$ = ____. Therefore, the number is ____.

2 Three times a number is added to twice the number, and when 15 is added to the sum, the result is 45. Find the number. Let $x$ = the number. Three times the number plus twice the number is _____. The equation is _____. Solving the equation, $x$ = ____. Therefore, the number is ____.

3 Raúl is 5 years older than José, and 7 years from now the sum of their ages will be 35 years. How old is each now? Let $x$ years = José's age now. Then _____ years is Raúl's age now. Thus _____ represents José's age in 7 years, and _____ represents Raúl's age in 7 years. The equation is _____ = 35. Solving the equation, $x$ = ____. Therefore, ____ years is José's age now, and ____ years is Raúl's age now.

4 John is 28 years younger than his father. In 15 years his father's age will be twice John's age. How old is each now? Let $x$ years = John's age now. Then _____ years is his father's age now, and _____ years is John's age 15 years from now. $(x + 28 + 15)$ years is his father's age _____ from now. The equation is _____ = $2(x + 15)$. Solving the equation, $x$ = ____. Therefore, ____ years is John's age now, and ____ years is his father's age now.

## Money Value Problems

**5** Joe and Tom sold 17 boxes of candy for $10.00. They sold some for 50 cents and some for 75 cents. How many did they sell at each price? Let $x$ = the number of boxes of candy sold at 75 cents. Then _____ boxes of candy sold at 50 cents. $75x$ is the price of _____ boxes of candy, and _____ is the price of $17 - x$ boxes of candy. The equation is _____. Solving the equation, $x =$ _____. Therefore, _____ boxes of candy were sold at 75 cents, and _____ were sold at 50 cents.

$17 - x$
$x, 50(17 - x)$
$75x + 50(17 - x) = 1{,}000$
$6, 6$
$11$

**6** A purse contains twice as many dimes as nickels, and the total value of the nickels and dimes is $7.50. How many of each coin are there in the purse? Let $x$ = the number of nickels, so that _____ represents the number of dimes. Then _____ represents the value of the nickels, and _____ represents the value of the dimes. The equation is _____ = 7.50. Solving the equation, $x =$ _____. Therefore, there are _____ nickels and _____ dimes.

$2x$
$0.05x$
$0.10(2x)$
$0.05x + 0.2x, 30$
$30, 60$

## Finance and Investment Problems

**7** A television set is on sale at 30% off. If the savings is $179.70, find the original price of the television set. Let $x$ = the original price. The equation is formed as follows:

$30\% x =$ _____
$0.3x =$ _____
$3x =$ _____.

Solving the equation, $x =$ _____. Therefore, the original price of the television set was _____.

179.70
179.70
1797
599
$599

**8** A man invested $30,000.00, part of it at 8% interest and the remainder at 9%. The interest for 1 year is $2,600.00. How much did he invest at 8% and how much at 9%? Let $x$ = the amount invested at 8%. Then $(30{,}000 - x)$ is the amount invested at 9%. The equation is formed as follows:

$8\%x + 9\%(30{,}000 - x) =$ _____
$\frac{8}{100}x + \frac{9}{100}(30{,}000 - x) =$ _____
$8x + 9(30{,}000 - x) = 260{,}000$

Solving the equation, $x =$ _____. Therefore, _____ is invested at 8% and _____ is invested at 9%.

2,600
2,600

10,000, $10,000
$20,000

## Geometric Problems

**9** The length of a rectangle is 5 greater than its width, and its perimeter is 38. Find its length and its width. Let $x$ units = the width of the rectangle. Then _____ units is its length, and its perimeter is _____ units. The equation is $2x + 2(x + 5) =$ _____.

$x + 5$
$2x + 2(x + 5), 38$

## 4.5 SEMIPROGRAMMED PROBLEMS

| | |
|---|---|
| 7, 7 <br> 12 | Solving the equation, $x =$ \_\_\_\_\_. Therefore, the width is \_\_\_\_\_ units, and the length is \_\_\_\_\_ units. |
| | **10** The sides of two squares differ by 4 inches and their areas differ by 80 square inches. Find the lengths of the sides of the squares. Let $x$ units = the length of the side of the smaller square, so that |
| $x + 4$ | _____ units is the length of the side of the larger square. The |
| $x + 4, 8x + 16$ | equation is (_____)$^2 - x^2 = 80$, or _____ $= 80$. Solving the |
| 8, 12 | equation, $x =$ \_\_\_\_\_, so $x + 4 =$ \_\_\_\_\_. Therefore, the lengths of the |
| 8 inches and 12 inches | sides of the square are _____. |

### Motion Problems

| | |
|---|---|
| | **11** Two automobiles, traveling 40 miles per hour and 60 miles per hour, start from the same place. If the faster automobile leaves 4 hours after the slower one, how long will it take it to overtake the slower automobile? Let $x =$ the number of hours the faster automobile |
| $x + 4$ | traveled. Then _____ represents the number of hours the slower automobile traveled. Using the formula $d = rt$ and the fact that both automobiles traveled the same distance, the equation is formed as follows: |
| | $60x = 40(_____)$ |
| $x + 4$ | $60x = _____$. |
| $40x + 160$ | Solving this equation, $x =$ \_\_\_\_\_. Therefore, it took \_\_\_\_\_ hours for |
| 8, 8 | the faster automobile to overtake the slower one. |
| | **12** An automobile maintained a certain average speed for 2 hours and then increased its speed by 10 miles per hour for 3 hours. If the automobile traveled 230 miles during the 5 hours, how fast was the automobile traveling during the first 2 hours? Let $x =$ the speed of |
| $x + 10$ | the automobile during the first 2 hours. Then _____ represents its speed during the last 3 hours. Using the formula $d = rt$, the equation is formed as follows: |
| | $2x + 3(_____) = 230$. |
| $x + 10$ | Solving this equation, $x =$ \_\_\_\_\_. Therefore, the automobile was |
| 40 | traveling \_\_\_\_\_ miles per hour during the first 2 hours. |
| 40 | |

### Mixture Problems

| | |
|---|---|
| | **13** A druggist has a 10% solution and an 18% solution. How much of each must he use to obtain 40 ounces of a 12% solution. Let $x =$ the |
| $40 - x$ | amount of the 10% solution used. Then _____ represents the amount of the 18% solution needed. The equation is formed as follows: |
| $40 - x$ | $10\% x + 18\%(_____) = 12\%(40)$ |
| $40 - x$ | $0.10x + 0.18(_____) = 0.12(40)$ |
| $40 - x$ | $10x + 18(_____) = 12(40)$ |
| $720 - 18x$ | $10x + _____ = 480$. |

# CHAPTER 4 LINEAR EQUATIONS AND INEQUALITIES

30, 30
10

|  |  |
|---|---|
|  | Solving this equation, $x =$ _____. Therefore, _____ ounces of the 10% solution and _____ ounces of the 18% solution are required. |
| 50 − x | **14** A merchant has walnuts that sell for $1.50 per pound and peanuts that sell for $1.00 per pound. How many pounds of each must be used to make 50 pounds of a mixture worth $1.30 per pound? Let $x =$ the number of pounds of walnuts used. Then _____ represents the number of pounds of peanuts used. The equation is formed as follows: |
| 50 − x | $1.50x + 1.00(\underline{\hspace{1cm}}) = 1.30(50)$ |
| 50 − x | $1.50x + \underline{\hspace{1cm}} = 65.00.$ |
| 30, 30 | Solving this equation, $x =$ _____. Therefore, _____ pounds of walnuts |
| 20 | and _____ pounds of peanuts are required. |

## Work Problems

|  |  |
|---|---|
|  | **15** A large truck can haul gravel needed for a certain concrete job in 24 hours and a small truck can haul the gravel in 56 hours. How long should it take if the two trucks work together? Let $x =$ the number of hours it will take both trucks to haul the gravel. The large |
| $\frac{1}{24}$ of the work | truck will do _____ in 1 hour, and the small truck |
| $\frac{1}{56}$ of the work | will do _____ in 1 hour. Then both trucks will |
| $\frac{1}{x}$ of the work | do _____ in 1 hour. The equation is |
| $\frac{1}{24} + \frac{1}{56} = \frac{1}{x}$, $16\frac{4}{5}$ | _____. Solving the equation, $x =$ _____. Therefore, it |
| $16\frac{4}{5}$ hours | will take _____ for both trucks to do the work. |

## SOLUTIONS TO SELECTED ODD PROBLEMS   Section 4.5, text pages 183–186

**1**  Let $x =$ the number.
Then $2x + 3 = 57$
$2x = 54$
$x = 27$
Check: $2(27) + 3 = 54 + 3 = 57$

**5**  Let $x =$ the number.
Then $\frac{1}{4}x = \frac{1}{6}x + 3$
$3x = 2x + 36$
$x = 36$
Check: $\frac{1}{4}(36) = \frac{1}{6}(36) + 3$
$9 = 6 + 3$
$9 = 9$

**9**  Let $x =$ brother's age now; $x + 3 =$ Raul's age now; $x + 4 =$ brother's age in 4 years; $x + 7 =$ Raul's age in 4 years.
Then $(x + 4) + (x + 7) = 33$
$2x + 11 = 33$
$2x = 22$
$x = 11$    (brother's age now)
$x + 3 = 14$    (Raul's age now)
Check: $11 + 4 = 15$, $14 + 4 = 18$ and $15 + 18 = 33.$

**13**  Let $q =$ quarters; $q + 4 =$ dimes; $3(q + 4) =$ nickels.
Then $0.25q + 0.10(q + 4) + 0.05[3(q + 4)] = 6.50$
$25q + 10(q + 4) + 5[3(q + 4)] = 650$
$25q + 10q + 40 + 15q + 60 = 650$
$50q + 100 = 650$
$50q = 550$
$q = 11$

Answer is 11 quarters, 15 dimes, 45 nickels.
*Check:* $0.25(11) + 0.10(15) + 0.05(45) = 6.50$
$$2.75 + 1.50 + 2.25 = 6.50$$
$$6.50 = 6.50$$

17  Let $d$ = number of dimes; $d - 20$ = number of quarters.
Then $0.10(d) + 0.25(d - 20) = 12.50$
$$10d + 25(d - 20) = 1250$$
$$10d + 25d - 500 = 1250$$
$$35d = 1750$$
$$d = 50$$
Answer is 50 dimes, 30 quarters.
*Check:* $50(0.10) + 30(0.25) = 12.50$
$$12.50 = 12.50$$

21  Let $x$ = original salary (in dollars).
Then $x + 0.11x = 222$
$$1.11x = 222$$
$$x = 200$$
*Check:* $200 + 0.11(200) = 222$
$$200 + 22 = 222$$

25  Let $n$ = bill before tax and tip
Then $n + 0.04n + 0.15n = 19.04$
$$1.19n = 19.04$$
$$n = 16$$
*Check:* $16 + 0.04(16) + (0.15)(16) = 19.04$
$$16 + 0.64 + 2.40 = 19.04$$
$$19.04 = 19.04$$

29  Let $n$ = amount at 13% (in dollars); $18,000 - n$ = amount at 14%.
Then $0.13n + 0.14(18,000 - n) = 2,395$
$$0.13n + 2,520 - 0.14n = 2,395$$
$$-0.01n = -125$$
$$n = 12,500$$
*Check:* $(0.13)(12500) + 0.14(18000 - 12500) = 2395$
$$1625 + 770 = 2395$$
$$2395 = 2395$$

33  Let $x$ = amount at 7% (in dollars).
Then $0.07x + 8000(0.14) = 0.115(x + 8000)$
$$0.07x + 1120 = 0.115x + 920$$
$$-0.045x = -200$$
$$x = 4,444.44$$
*Check:* $0.07(4,444.44) + 8,000(0.14) = 0.115(4,444.44 + 8,000)$
$$311.11 + 1,120 = 0.115(12,444.44)$$
$$1,431.11 = 1,431.11$$

37  $w$ = width = $x$
$l$ = length = $4x$

$2l + 2w = P$ (Formula for perimeter of a rectangle)
$$2(x) + 2(4x) = 150$$
$$2x + 8x = 150$$
$$10x = 150$$
$$x = 15$$
Length = 60 meters, width = 15 meters.
*Check:* $P = 2l + 2w$
$$150 = 2(60) + 2(15)$$
$$150 = 120 + 30$$
$$150 = 150$$

**41.**

|  | Rate (in miles per hour) | Distance (in miles) | Time (in hours) = $\dfrac{\text{Distance}}{\text{Rate}}$ |
|---|---|---|---|
| Jog | $r + 5$ | 12 | $\dfrac{12}{r+5}$ |
| Walk | $r$ | 8 | $\dfrac{8}{r}$ |

$$\frac{12}{r+5} = \frac{8}{r}$$

$12r = 8(r + 5)$
$12r = 8r + 40$
$4r = 40$
$r = 10$ miles per hour

Time to walk 8 miles $= \dfrac{8}{r} = \dfrac{8}{10} = \dfrac{4}{5}$ hour = 48 minutes.

Check: Jog time $= \frac{12}{15} = \frac{4}{5}$ and walk time $= \frac{8}{10} = \frac{4}{5}$

**45.**

|  | Rate (in miles per hour) | Distance (in miles) | Time (in hours) = $\dfrac{\text{Distance}}{\text{Rate}}$ |
|---|---|---|---|
| Jog | $r$ | 15 | $\dfrac{15}{r}$ |
| Bike | $2r$ | 15 | $\dfrac{15}{2r}$ |

$$\frac{15}{r} + \frac{15}{2r} = 3$$

$30 + 15 = 6r$
$45 = 6r$
$\dfrac{45}{6} = r$

Jogging rate is $7\frac{1}{2}$ miles per hour, biking rate is $2(7\frac{1}{2}) = 15$ miles per hour.

Check: $\dfrac{15}{7\frac{1}{2}} + \dfrac{15}{15} = 3$

**49.**

|  | Gallons of Gasohol | Concentration of Alcohol | Gallons of Alcohol |
|---|---|---|---|
| First tank | $x$ | 0.09 | $0.09x$ |
| Second tank | $300{,}000 - x$ | 0.12 | $0.12(300{,}000 - x)$ |
| Mixture | $300{,}000$ | 0.10 | $0.10(300{,}000)$ |

$0.09x + 0.12(300{,}000 - x) = 0.10(300{,}000)$
$0.09x + 36{,}000 - 0.12x = 30{,}000$
$-0.03x = -6{,}000$
$x = 200{,}000$

Answer is 200,000 gallons of 9% solution, 100,000 gallons of 12% solution.
Check: $0.09(200{,}000) + 0.12(100{,}000) = 0.10(300{,}000)$
$18{,}000 + 12{,}000 = 30{,}000$

**53** Let $x$ = time for B to pollute the air alone.

| Smokestack | Time | Rate | Amount of Pollution in 20 Hours |
|---|---|---|---|
| B | $x$ | $\dfrac{1}{x}$ | $20\left(\dfrac{1}{x}\right) = \dfrac{20}{x}$ |
| A |  | $1.25\left(\dfrac{1}{x}\right) = \dfrac{5}{4x}$ | $20\left(\dfrac{5}{4x}\right) = \dfrac{25}{x}$ |

Equation is $\dfrac{20}{x} + \dfrac{25}{x} = 1$

$$\dfrac{45}{x} = 1$$

$$x = 45$$

Check: $\dfrac{20}{45} + \dfrac{25}{45} = \dfrac{45}{45} = 1$

Therefore, it would take smokestack B 45 hours to create as much pollution on its own as both smokestacks could in 20 hours.

**57** Let $x$ = number of hours to mow the lawn together.

|  | Time (in hours) | Rate | Part Done in $x$ Hours |
|---|---|---|---|
| John | $1\dfrac{1}{3} = \dfrac{4}{3}$ | $\dfrac{1}{\frac{4}{3}} = \dfrac{3}{4}$ | $\dfrac{3}{4}x = \dfrac{3x}{4}$ |
| Tom | $2$ | $\dfrac{1}{2}$ | $\dfrac{1}{2}x = \dfrac{x}{2}$ |

Equation is $\dfrac{3x}{4} + \dfrac{x}{2} = 1$

$$3x + 2x = 4$$
$$5x = 4$$
$$x = \dfrac{4}{5}$$

Check: $\dfrac{3}{4}\left(\dfrac{4}{5}\right) + \dfrac{1}{2}\left(\dfrac{4}{5}\right) = \dfrac{3}{5} + \dfrac{2}{5} = 1$

Therefore, it takes $\dfrac{4}{5}$ hour to mow the lawn together.

## 4.6 Inequalities

### SEMIPROGRAMMED PROBLEMS

In problems 1–4, sketch the graph of each solution set.

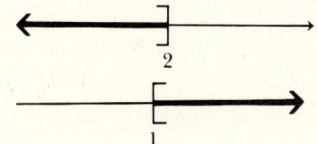

**1** $\{x \mid x \leq 2\}$

**2** $\{x \mid x \geq 1\}$

# CHAPTER 4 LINEAR EQUATIONS AND INEQUALITIES

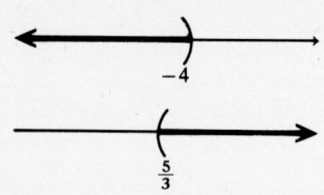

3  $\{x \mid x < -4\}$ _____

4  $\{x \mid x > \frac{5}{3}\}$ _____

In problems 5–7, graph the solution set of each compound inequality.

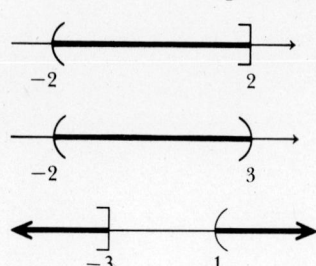

5  $\{x \mid x > -2\} \cap \{x \mid x \leq 2\}$ _____

6  $\{x \mid x > -2\} \cap \{x \mid x < 3\}$ _____

7  $\{x \mid x \leq -3\} \cup \{x \mid x > 1\}$ _____

In problems 8–10, insert the symbols $>$ or $<$ in the blank so that each resulting statement is true.

$<$          8  Since $5 < 8$, then $5 - 20$ ____ $8 - 20$.

$>$          9  If $x > y$, then $4x$ ____ $4y$.

$>$         10  If $u < v$, then $-3u$ ____ $-3v$.

In problems 11–18, state the property of the inequality that justifies each statement.

addition         11  $3 < 7$ so that $3 + x < 7 + x$ _____.

multiplication   12  $x < y$ so that $4x < 4y$ _____.

transitive       13  $p < q$ and $q < m$ so that $p < m$ _____.

division         14  $a < b$ so that $\dfrac{a}{-2} > \dfrac{b}{-2}$ _____.

multiplication   15  $2 < 6$ and $y < 0$ so that $2y > 6y$ _____.

transitive       16  $13 > 4$ and $4 > x$ so that $13 > x$ _____.

multiplication   17  $-\dfrac{y}{6} \leq 5$ so that $y \geq -30$ _____.

addition         18  $x - 8 \geq 2$ so that $x \geq 10$ _____.

## SOLUTIONS TO SELECTED ODD PROBLEMS   Section 4.6, text pages 195–196

37  $<$, because of multiplications property (i) of inequalities

41  <, because of transitive property of inequalities
45  <, because of multiplication property (i) of inequalities
49  addition property of inequalities
53  multiplication property (ii) of inequalities
57  multiplication property (ii) of inequalities

## 4.7 Linear Inequalities

### SEMIPROGRAMMED PROBLEMS

In problems 1–9, solve each inequality and sketch the graph of its solution set.

1    $3x - 5 < 7$

5               $3x < 7 + \underline{\phantom{xx}}$

12            $3x < \underline{\phantom{xx}}$

4             $x < \underline{\phantom{xx}}$

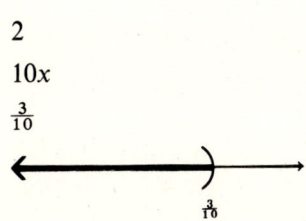

The graph is _____.

2    $3x - 2 < -7x + 1$

2            $3x < -7x + 1 + (\underline{\phantom{xx}})$

$10x$          $\underline{\phantom{xx}} < 3$

$\frac{3}{10}$         $x < \underline{\phantom{xx}}$

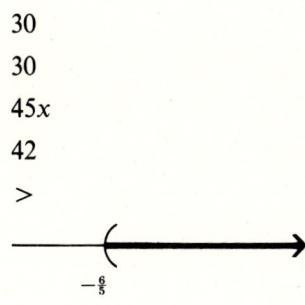

The graph is _____.

3    $\frac{1}{3}x - \frac{2}{5} < \frac{3}{2}x + 1$

30       $30(\frac{1}{3}x - \frac{2}{5}) < \underline{\phantom{xx}}(\frac{3}{2}x + 1)$

30       $10x - 12 < 45x + \underline{\phantom{xx}}$

$45x$     $10x - \underline{\phantom{xx}} < 30 + 12$

42       $-35x < \underline{\phantom{xx}}$

>           $x \underline{\phantom{xx}} -\frac{6}{5}$

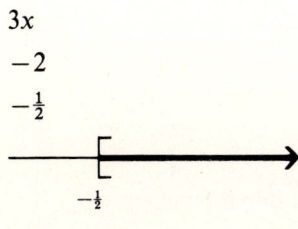

The graph is _____.

4    $x + 6 \geq 4 - 3x$

$3x$        $x + \underline{\phantom{xx}} \geq 4 - 6$

$-2$       $4x \geq \underline{\phantom{xx}}$

$-\frac{1}{2}$      $x \geq \underline{\phantom{xx}}$

The graph is _____.

5    $\frac{2}{3}(2x - 1) - \frac{1}{2}(3x + 2) \leq 3$

6       $6[\frac{2}{3}(2x - 1) - \frac{1}{2}(3x + 2)] \leq \underline{\phantom{xx}}(3)$

18     $4(2x - 1) - 3(3x + 2) \leq \underline{\phantom{xx}}$

6      $8x - 4 - 9x - \underline{\phantom{xx}} \leq 18$

# CHAPTER 4 LINEAR EQUATIONS AND INEQUALITIES

−x  
28  
≥  

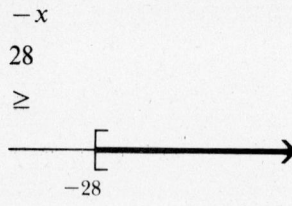

5  
3x − 7  
7  
12  
≤  

−1 + 4  
−2x  
>  

5  
2x + 3  
1 − 3  
−2  
x  

1  
−4, 6  
−2, 3  

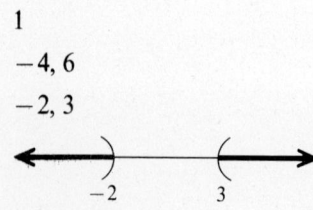

4,875, $51\frac{30}{95}$  
52

_____ − 4 − 6 ≤ 18  
−x ≤ _____  
x _____ − 28  
The graph is _____.

6    $\dfrac{3x - 7}{5} \geq 2x + 1$

$5\left(\dfrac{3x - 7}{5}\right) \geq ($_____$)(2x + 1)$

_____ ≥ 10x + 5  
3x − 10x ≥ 5 + _____  
−7x ≥ _____  
x _____ $-\dfrac{12}{7}$  
The graph is _____.

7    −3 < 1 − 2x < 4  
−1 − 3 < −1 + 1 − 2x < _____  
−4 < _____ < 3  
2 > x _____ $-\dfrac{3}{2}$  
The graph is _____.

8    $-0.2 \leq \dfrac{2x + 3}{5} \leq 0.2$

$5(-0.2) \leq 5\left(\dfrac{2x + 3}{5}\right) \leq ($_____$)(0.2)$

−1 ≤ _____ ≤ 1  
−1 − 3 ≤ 2x ≤ _____  
−4 ≤ 2x ≤ _____  
−2 ≤ _____ ≤ −1  
The graph is _____.

9    2x − 1 < −5    or    2x − 1 > 5  
2x − 1 + 1 < −5 + 1   or   2x − 1 + 1 > 5 + _____  
2x < _____    or    2x > _____  
x < _____    or    x > _____  
The graph is _____.

10   Mr. Jones bought a car for $5,875 by making a down payment of $1,000 and agreeing to monthly payments of $95. In how many months will he pay the balance due? Let x = the number of months required to pay off the balance. Then  
95x ≥ _____ or x ≥ _____.  
Therefore, it will take _____ full months to pay the balance due.

## SOLUTIONS TO SELECTED ODD PROBLEMS   Section 4.7, text pages 203–205

**1**
$$x - 2 < 4$$
$$x - 2 + 2 < 4 + 2$$
$$x < 6$$

**5**
$$4 + x \leq -1$$
$$-4 + 4 + x \leq -1 - 4$$
$$x \leq -5$$

**9**
$$-2x < -3$$
$$\frac{-2x}{-2} > \frac{-3}{-2}$$
$$x > \frac{3}{2}$$

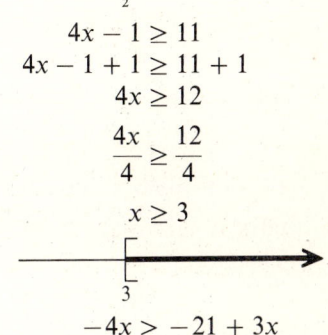

**13**
$$7x \leq -2$$
$$\frac{7x}{7} \leq \frac{-2}{7}$$
$$x \leq \frac{-2}{7}$$

**17**
$$-5x < -1$$
$$\frac{-5x}{-5} > \frac{-1}{-5}$$
$$x > \frac{1}{5}$$

**21**
$$4x - 1 \geq 11$$
$$4x - 1 + 1 \geq 11 + 1$$
$$4x \geq 12$$
$$\frac{4x}{4} \geq \frac{12}{4}$$
$$x \geq 3$$

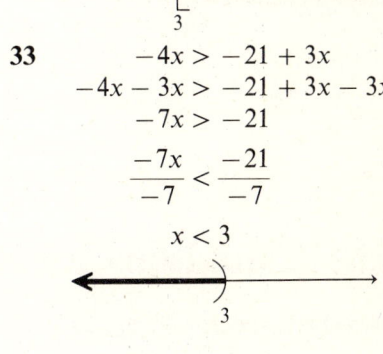

**25**
$$-5x + 2 > 12$$
$$-5x + 2 - 2 > 12 - 2$$
$$-5x > 10$$
$$\frac{-5x}{-5} < \frac{10}{-5}$$
$$x < -2$$

**29**
$$5 - 3x \geq 7$$
$$-5 + 5 - 3x \geq -5 + 7$$
$$-3x \geq 2$$
$$\frac{-3x}{-3} \leq \frac{2}{-3}$$
$$x \leq -\frac{2}{3}$$

**33**
$$-4x > -21 + 3x$$
$$-4x - 3x > -21 + 3x - 3x$$
$$-7x > -21$$
$$\frac{-7x}{-7} < \frac{-21}{-7}$$
$$x < 3$$

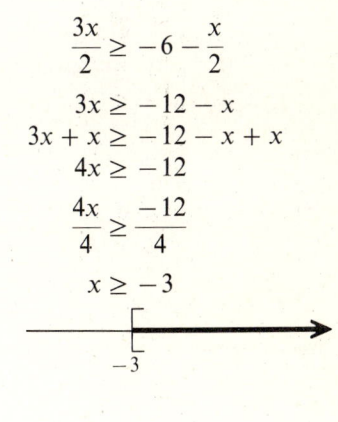

**37**
$$4(3 - x) \geq 2(x - 1)$$
$$12 - 4x \geq 2x - 2$$
$$-4x \geq 2x - 2 - 12$$
$$-4x \geq 2x - 14$$
$$-2x - 4x \geq -2x + 2x - 14$$
$$-6x \geq -14$$
$$\frac{-6x}{-6} \leq \frac{-14}{-6}$$
$$x \leq \frac{7}{3}$$

**41**
$$7(x - 3) \leq 4(x + 5) - 47$$
$$7x - 21 \leq 4x + 20 - 47$$
$$7x - 21 \leq 4x - 27$$
$$7x - 21 + 21 \leq 4x - 27 + 21$$
$$7x \leq 4x - 6$$
$$7x - 4x \leq 4x - 4x - 6$$
$$3x \leq -6$$
$$\frac{3x}{3} \leq \frac{-6}{3}$$
$$x \leq -2$$

**45**
$$\frac{3x}{2} \geq -6 - \frac{x}{2}$$
$$3x \geq -12 - x$$
$$3x + x \geq -12 - x + x$$
$$4x \geq -12$$
$$\frac{4x}{4} \geq \frac{-12}{4}$$
$$x \geq -3$$

**49**
$$\tfrac{1}{3}(2x + 3) \leq \tfrac{3}{4}x$$
$$4(2x + 3) \leq 9x$$
$$8x + 12 \leq 9x$$
$$-x + 12 \leq 0$$
$$-x \leq -12$$
$$x \geq 12$$

**53**
$$\frac{3x + 7}{7} - \frac{2x - 1}{3} \leq 1$$
$$3(3x + 7) - 7(2x - 1) \leq 21$$
$$9x + 21 - 14x + 7 \leq 21$$
$$-5x + 28 \leq 21$$
$$-5x + 28 - 28 \leq 21 - 28$$
$$-5x \leq -7$$
$$\frac{-5x}{-5} \geq \frac{-7}{-5}$$
$$x \geq \frac{7}{5}$$

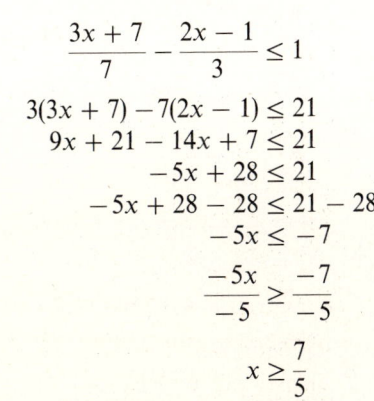

**57**
$$-5 \leq 3x - 1 \leq 5$$
$$-5 + 1 \leq 3x - 1 + 1 \leq 5 + 1$$
$$-4 \leq 3x \leq 6$$
$$-\frac{4}{3} \leq \frac{3x}{3} \leq \frac{6}{3}$$
$$-\frac{4}{3} \leq x \leq 2$$

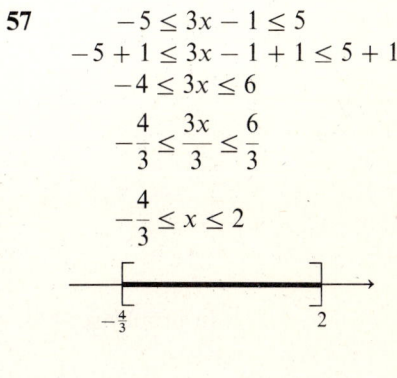

## CHAPTER 4  LINEAR EQUATIONS AND INEQUALITIES

**61**
$$-1 < 5 - x < 3$$
$$-1 - 5 < 5 - x - 5 < 3 - 5$$
$$-6 < -x < -2$$
$$6 > x > 2 \text{ or } 2 < x < 6$$

```
        (           )
————————2———————————6————————
```

**65**
$$2x - 1 < 3x + 7 \leq x + 9$$
$$2x - 1 < 3x + 7 \text{ and } 3x + 7 \leq x + 9$$
$$-x - 1 < 7 \quad \text{and } 2x + 7 \leq 9$$
$$-x < 8 \quad \text{and} \quad 2x \leq 2$$
$$x > -8 \quad \text{and} \quad x \leq 1$$
$$-8 < x \leq 1$$

```
        (                    ]
————————-8———————————————————1————————
```

**69**
$$6 - x > 5 \quad \text{or} \quad 6 - x < -5$$
$$-x > -1 \quad \text{or} \quad -x < -11$$
$$x < 1 \quad \text{or} \quad x > 11$$

```
                )           (
————————————————1———————————11————————
```

**73**
$$5x - 4 > 1 \quad \text{or} \quad 5x - 4 < -1$$
$$5x > 5 \quad \text{or} \quad 5x < 3$$
$$x > 1 \quad \text{or} \quad x < \tfrac{3}{5}$$

```
                )           (
————————————————3/5—————————1————————
```

**77** Five less than 4 times $x$ is expressed as $4x - 5$. So the inequality is $4x - 5 < 13$.

**81** $66 \leq F \leq 84$ and $F = \tfrac{9}{5}C + 32$.
Therefore, $66 \leq \tfrac{9}{5}C + 32 \leq 84$
$$66 - 32 \leq \tfrac{9}{5}C + 32 - 32 \leq 84 - 32$$
$$34 \leq \tfrac{9}{5}C \leq 52$$
$$\tfrac{5}{9}(34) \leq \tfrac{5}{9}(\tfrac{9}{5}C) \leq \tfrac{5}{9}(52)$$
$$\tfrac{170}{9} \leq C \leq \tfrac{260}{9}$$

## 4.8  Equations Involving Absolute Values

### SEMIPROGRAMMED PROBLEMS

In problems 1–4, use the formula $d = |a - b|$ to find the distance between the points whose coordinates are the given numbers.

|  | |
|---|---|
| 11 | **1**  5 and 11 |
| -6 | $d = \|5 - \underline{\quad}\|$ |
| 6 | $= \|\underline{\quad}\|$ |
|  | $= \underline{\quad}$ |
|  | **2**  -7 and 8 |
| 8 | $d = \|-7 - \underline{\quad}\|$ |
| -15 | $= \|\underline{\quad}\|$ |
| 15 | $= \underline{\quad}$ |
|  | **3**  5 and -3 |
| -3 | $d = \|5 - \underline{\quad}\|$ |
| 8 | $= \|\underline{\quad}\|$ |
| 8 | $= \underline{\quad}$ |
|  | **4**  -13 and -6 |
| -13 | $d = \|\underline{\quad} - (-6)\|$ |
| -7 | $= \|\underline{\quad}\|$ |
| 7 | $= \underline{\quad}$ |

In problems 5–13, solve each equation.

|  | |
|---|---|
|  | **5**  $\|x\| = 3$ |
| -3 | $x = 3 \text{ or } x = \underline{\quad}$ |

## 4.8 SOLUTIONS TO SELECTED ODD PROBLEMS

| | |
|---|---|
| 7 | **6** $\|x\| = 7$ |
| | $x = -7$ or $x = \underline{\phantom{xx}}$ |
| 15 | **7** $\|3x\| = 15$ |
| | $3x = -15$ or $3x = \underline{\phantom{xx}}$ |
| 5 | $x = -5$ or $x = \underline{\phantom{xx}}$ |
| | **8** $\|x\| = -3$ |
| | There is no value of $x$ to satisfy this equation, since the absolute value |
| nonnegative | is always $\underline{\phantom{xxxxxxxx}}$. Therefore, the equation has no |
| solution | $\underline{\phantom{xxxxxx}}$. |
| | **9** $\|x - 2\| = 3$ |
| | $x - 2 = 3$ or $x - 2 = \underline{\phantom{xx}}$ |
| $-3$ | |
| $-1$ | Solving each equation, $x = 5$ or $x = \underline{\phantom{xx}}$. |
| | **10** $\|x + \tfrac{1}{2}\| = \tfrac{1}{4}$ |
| | $x + \tfrac{1}{2} = \tfrac{1}{4}$ or $x + \tfrac{1}{2} = \underline{\phantom{xx}}$ |
| $-\tfrac{1}{4}$ | |
| $-\tfrac{3}{4}$ | Solving each equation, $x = -\tfrac{1}{4}$ or $x = \underline{\phantom{xx}}$. |
| | **11** $\|2x - 5\| = 9$ |
| | $2x - 5 = 9$ or $2x - 5 = \underline{\phantom{xx}}$ |
| $-9$ | Solving each equation for $x$, we have |
| $-4$ | $2x = 14$ or $2x = \underline{\phantom{xx}}$ |
| 7 | $x = \underline{\phantom{xx}}$ or $x = -2$. |
| | **12** $\|2 - 3x\| = 6$ |
| | $2 - 3x = 6$ or $2 - 3x = \underline{\phantom{xx}}$ |
| $-6$ | Solving each equation for $x$, we have |
| | $-3x = 4$ or $-3x = \underline{\phantom{xx}}$ |
| $-8$ | |
| $\tfrac{8}{3}$ | $x = -\tfrac{4}{3}$ or $x = \underline{\phantom{xx}}$. |
| | **13** $\|2x - 1\| = \|5 - x\|$ |
| | $2x - 1 = -(5 - x)$ or $2x - 1 = \underline{\phantom{xxxx}}$ |
| $5 - x$ | |
| $x - 5, x, 1$ | $2x - 1 = \underline{\phantom{xxxx}}$ or $2x + \underline{\phantom{xx}} = 5 + \underline{\phantom{xx}}$ |
| $x - 4, 3x$ | $2x = \underline{\phantom{xxxx}}$ $\underline{\phantom{xx}} = 6$ |
| $-4, 2$ | $x = \underline{\phantom{xx}}$ $x = \underline{\phantom{xx}}$ |

## SOLUTIONS TO SELECTED ODD PROBLEMS   Section 4.8, text page 209

**1** $d = |3 - 8| = |-5| = 5$

**5** $d = |-8 - (-1)| = |-7| = 7$

**9** $d = |3.5 - (-1.7)| = |5.2| = 5.2$

**13** $|x| = 2$
$x = 2$ or $x = -2$

**17** $|w| + 2 = 5$
$|w| = 3$
$w = 3$ or $w = -3$

**21** $|-7y| = 14$
$-7y = 14$ or $-7y = -14$
$y = -2$ or $y = 2$

**25** $|3z| - 1 = 14$
$|3z| = 15$
$3z = 15$ or $3z = -15$
$z = 5$ or $z = -5$

**29** $|2y| = -4$
No solution, by Property 1(ii), page 207

33. $|5 - q| = 6$
$5 - q = 6$ or $5 - q = -6$
$-q = 1$ or $-q = -11$
$q = -1$ or $q = 11$

37. $4|2 - 7x| = 16$
$|2 - 7x| = 4$
$2 - 7x = 4$ or $2 - 7x = -4$
$-7x = 2$ or $-7x = -6$
$x = -\frac{2}{7}$ or $x = \frac{6}{7}$

41. $|3t + 2| = 0$
$3t + 2 = 0$ [Property 1(iii), page 207]
$3t = -2$
$t = -\frac{2}{3}$

45. $|\frac{4}{5}x - 5| = 10$
$\frac{4}{5}x - 5 = -10$ or $\frac{4}{5}x - 5 = 10$
$\frac{4}{5}x = -5$ or $\frac{4}{5}x = 15$
$4x = -25$ or $4x = 75$
$x = -\frac{25}{4}$ or $x = \frac{75}{4}$

49. $|5x + 3| + 2 = 8$
$|5x + 3| = 6$
$5x + 3 = 6$ or $5x + 3 = -6$
$5x = 3$ or $5x = -9$
$x = \frac{3}{5}$ or $x = -\frac{9}{5}$

53. $|2x + 3(x - 1) + 1| = 4$
$|2x + 3x - 3 + 1| = 4$
$|5x - 2| = 4$
$5x - 2 = 4$ or $5x - 2 = -4$
$5x = 6$ or $5x = -2$
$x = \frac{6}{5}$ or $x = -\frac{2}{5}$

57. $|2x - 1| = |4x - 1|$
$2x - 1 = -(4x - 1)$ or $2x - 1 = 4x - 1$
$2x - 1 = -4x + 1$ or $2x = 4x$
$6x - 1 = 1$ or $-2x = 0$
$6x = 2$
$x = \frac{1}{3}$ or $x = 0$

61. $|4x + 1| = |4x - 1|$
$4x + 1 = 4x - 1$ or $4x + 1 = -(4x - 1)$
$1 = -1$    $4x + 1 = -4x + 1$
But $1 \neq -1$    $8x = 0$
$x = 0$
Therefore, the solution is 0.

65. $|x - 4| = |4 - x|$
$x - 4 = 4 - x$ or $x - 4 = -(4 - x)$
$2x = 8$ or $x - 4 = -4 + x$
$x = 4$ or $x - 4 = x - 4$, which is true for all real numbers.
Therefore, the solution set is the set of all real numbers.

69. $|2x - 7| = |7 - 2x|$ is true for all real numbers, since $|7 - 2x| = |-(2x - 7)| = |2x - 7|$. Also, $|7 - 2x| = 7 - 2x$ if $7 - 2x \geq 0$, or equivalently, if $x \leq \frac{7}{2}$. Since all the numbers in the given set except 4 satisfy this condition, then all the given numbers except 4 are solutions. (This result can be verified by substituting each number into the given equation.)

## 4.9 Inequalities Involving Absolute Values

### SEMIPROGRAMMED PROBLEMS

In problems 1–11, solve each inequality and sketch the graph of its solution set.

1. $|x| < 2$
   _____ $< x < 2$
   The solution set consists of all real numbers $x$ such that
   _____.
   The graph is _____.

$-2$

$-2 < x < 2$

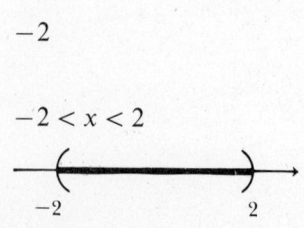

2. $|x| > 4$
   $x > 4$ or $x <$ _____
   The solution set consists of all real numbers $x$ such that
   _____.
   The graph is _____.

$-4$

$x > 4$ or $x < -4$

4.9 SEMIPROGRAMMED PROBLEMS    81

−5

−5 ≤ x ≤ 5

[graph: −5 to 5, closed brackets]

−7, 7

x ≤ −7 or x ≥ 7

[graph: arrows at −7 and 7, closed brackets outward]

|3x|
6
2

−2 < x < 2

[graph: −2 to 2, open parens]

−6
−4

−4 < x < 8

[graph: −4 to 8, open parens]

−2
−5, −1

x < −5 or x > −1

[graph: arrows at −5 and −1, open parens outward]

−3
2

−4 ≤ x ≤ 2

[graph: −4 to 2, closed brackets]

−5

---

3   $|x| \leq 5$

   _____ $\leq x \leq 5$

   The solution set consists of all real numbers $x$ such that
   _____.

   The graph is _____.

4   $|x| \geq 7$

   $x \leq$ _____ or $x \geq$ _____

   The solution set consists of all real numbers $x$ such that
   _____.

   The graph is _____.

5   $|-3x| < 6$

   $|-3x| =$ _____

   $-6 < 3x <$ _____

   $-2 < x <$ _____

   The solution set consists of all real numbers $x$ such that
   _____.

   The graph is _____.

6   $|x - 2| < 6$

   _____ $< x - 2 < 6$

   _____ $< x < 8$

   The solution set consists of all real numbers $x$ such that
   _____.

   The graph is _____.

7   $|x + 3| > 2$

   $x + 3 <$ _____ or $x + 3 > 2$

   $x <$ _____ or $x >$ _____

   The solution set consists of all real numbers $x$ such that
   _____.

   The graph is _____.

8   $|x + 1| \leq 3$

   _____ $\leq x + 1 \leq 3$

   $-4 \leq x \leq$ _____

   The solution set consists of all real numbers $x$ such that
   _____.

   The graph is _____.

9   $|2x - 7| \geq 5$

   $2x - 7 \geq 5$ or $2x - 7 \leq$ _____

12
1

$x \geq 6$ or $x \leq 1$

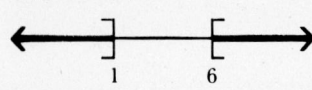

$-12$
$-16$
$-\frac{8}{3}$

$-\frac{8}{3} \leq x \leq \frac{16}{3}$

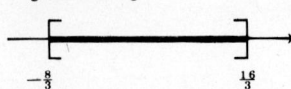

$3 - 2x$

4

$x \geq 4$ or $x \leq -1$

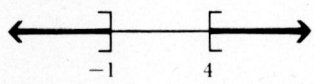

Solving for $x$, we have
$2x \geq$ _____ or $2x \leq 2$
$x \geq 6$ or $x \leq$ _____.
The solution set consists of all real numbers $x$ such that
_____.
The graph is _____.

10. $|4 - 3x| \leq 12$
_____ $\leq 4 - 3x \leq 12$
_____ $\leq -3x \leq 8$
_____ $\leq x \leq \frac{16}{3}$
The solution set consists of all real numbers $x$ such that
_____.
The graph is _____.

11. $|3 - 2x| \geq 5$
$3 - 2x \geq 5$ or _____ $\leq -5$
Solving for $x$, we have
$x \leq -1$ or $x \geq$ _____.
The solution set consists of all real numbers $x$ such that
_____.
The graph is _____.

## SOLUTIONS TO SELECTED ODD PROBLEMS   Section 4.9, text pages 214–215

1. $|x| < 1$
   $-1 < x < 1$

9. $|\frac{1}{3}x| \leq 2$
   $-2 \leq \frac{1}{3}x \leq 2$
   $-6 \leq x \leq 6$

17. $|x - 1| < 3$
    $-3 < x - 1 < 3$
    $-2 < x < 4$

25. $|3x - 4| < 7$
    $-7 < 3x - 4 < 7$
    $-3 < 3x < 11$
    $-1 < x < \frac{11}{3}$

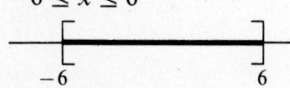

5. $|3x| < 15$
   $-15 < 3x < 15$
   $-5 < x < 5$

13. $|-6x| \geq 18$
    $-6x \geq 18$ or $-6x \leq -18$
    $x \leq -3$ or $x \geq 3$

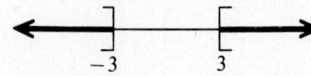

21. $|x - 4| > 5$
    $x - 4 > 5$ or $x - 4 < -5$
    $x > 9$ or $x < -1$

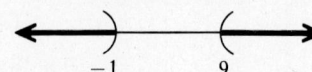

29. $|8x + 3| \geq 7$
    $8x + 3 \geq 7$ or $8x + 3 \leq -7$
    $8x \geq 4$ or $8x \leq -10$
    $x \geq \frac{1}{2}$ or $x \leq -\frac{5}{4}$

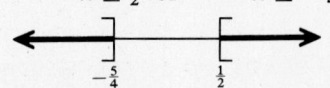

**33** $|4x - 1| \geq 11$
$4x - 1 \geq 11$ or $4x - 1 \leq -11$
$4x \geq 12$ or $4x \leq -10$
$x \geq 3$ or $x \leq -\frac{5}{2}$

**37** $\left|\frac{1}{3} - \frac{x}{6}\right| \geq \frac{1}{2}$
$\frac{1}{3} - \frac{x}{6} \geq \frac{1}{2}$ or $\frac{1}{3} - \frac{x}{6} \leq -\frac{1}{2}$
$2 - x \geq 3$ or $2 - x \leq -3$
$-x \geq 1$ or $-x \leq -5$
$x \leq -1$ or $x \geq 5$

**41** $\frac{1}{4}|2x - 1| \leq 3$
$-3 \leq \frac{1}{4}(2x - 1) \leq 3$
$-12 \leq 2x - 1 \leq 12$
$-11 \leq 2x \leq 13$
$-\frac{11}{2} \leq x \leq \frac{13}{2}$

**45** $|2x + 1| + 4 \leq 7$
$|2x + 1| \leq 3$
$-3 \leq 2x + 1 \leq 3$
$-4 \leq 2x \leq 2$
$-2 \leq x \leq 1$

**49** $|8 - 7x| + 9 > 1$
$|8 - 7x| > -8$
Answer is all real numbers because the absolute value of any number is $\geq 0$.

**53** $|2x + 3| > -3$
Answer is all real numbers because the absolute value of any number is $\geq 0$.

**57** $|3 - 5x| < 0$
Answer is no solution because the absolute value of any number is $\geq 0$.

## CUMULATIVE REVIEW PROBLEM SET   Chapters 1–4

1. Evaluate $-|-13|$.
2. Round off 3.1057 to three significant digits.
3. Rewrite $3.27 \times 10^8$ in expanded form.
4. Rewrite $7 \cdot x \cdot x \cdot y \cdot y \cdot y$ using exponential notation.
5. Evaluate $5x^2 + 2xy - 2y^3$ for $x = 3$ and $y = -2$.
6. Perform the addition and subtraction.
   $(3x^2 - 2x + 7) + (4x^2 + 9x - 3) - (2x^2 - 4x + 5)$.
7. Find each product.
   **(a)** $(3x - 7y)(3x + 7y)$  **(b)** $(4x + y)(16x^2 - 4xy + y^2)$  **(c)** $(5w + 2y)^2$
8. Factor each expression completely.
   **(a)** $81x^3 - 9xy^2$  **(b)** $2u^3v + 10u^2v^2 - 28uv^3$
9. Divide $x^4 + 2x^3 - 4x^2 + x - 8$ by $x + 2$.
10. Reduce to lowest terms.
    $$\frac{ax + ay - bx - by}{x^2 + 2xy + y^2}$$
11. Find the product and simplify the result.
    $$\frac{4t^2 - 64}{2t^2 - 8t} \cdot \frac{t - 4}{t + 4}$$
12. Find the quotient and simplify the result.
    $$\frac{x^3 - 1}{x^2 - 4y^2} \div \frac{x^2 + x + 1}{x + 2y}$$

# CHAPTER 4 LINEAR EQUATIONS AND INEQUALITIES

**13** Perform the indicated operations.

$$\frac{3}{x^2+3x+2} + \frac{2}{x^2-1} - \frac{1}{x^2+x-2}$$

**14** Simplify.

$$\frac{1+\frac{1}{n}}{n+2+\frac{1}{n}}$$

**15** Solve each equation.

(a) $3(2x-1) + 4 = 6(1 + 3x)$  (b) $\dfrac{4}{y-3} = \dfrac{12}{y+1}$  (c) $y - a = m(x - b)$; solve for $b$.

**16** Graph each solution set on a number line.

(a) $\{x \mid x \geq -2\}$  (b) $\{x \mid x < 4\} \cap \{x \mid x \geq -7\}$  (c) $\{x \mid x \leq -3\} \cup \{x \mid x \geq -1\}$

**17** Solve the inequality and graph its solution set.
$3(2 - 4x) \leq 5(x - 3)$

**18** Solve the absolute value equation.
$|x - 3| = 4$

**19** Solve each absolute value inequality and graph the solution set.

(a) $|2x - 1| < 3$  (b) $|3x + 1| \geq 5$

**20** A farmer used 740 feet of fencing to enclose a rectangular field. If the length of the field is 20 feet more than 4 times its width, find the length and width of the field.

---

## Answers

1 $-13$  2 $3.11$  3 $327{,}000{,}000$  4 $7x^2y^3$  5 $49$  6 $5x^2 + 11x - 1$
7(a) $9x^2 - 49y^2$  (b) $64x^3 + y^3$  (c) $25w^2 + 20wy + 4y^2$  8(a) $9x(3x - y)(3x + y)$
(b) $2uv(u - 2v)(u + 7v)$  9 $x^3 - 4x + 9$, $R = -26$  10 $\dfrac{a-b}{x+y}$  11 $\dfrac{2(t-4)}{t}$  12 $\dfrac{x-1}{x-2y}$
13 $\dfrac{4x}{(x+1)(x+2)(x-1)}$  14 $\dfrac{1}{n+1}$  15(a) $-\dfrac{5}{12}$  (b) $5$  (c) $\dfrac{mx - y + a}{m}$
16(a) 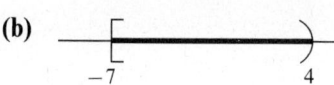 (b) 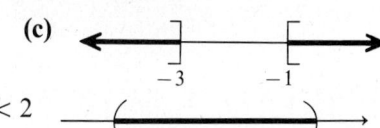 (c) 
17 $x \geq \dfrac{21}{17}$   18 $-1, 7$   19(a) $-1 < x < 2$
(b) $x \leq -2$ or $x \geq \dfrac{4}{3}$   20 length = 300 feet, width = 70 feet

---

# CHAPTER 4 TESTS

## Chapter Test A

**1** Solve each first-degree equation for $x$.

(a) $3x - 1 = 2x + 1$   (b) $x - 3(2 - x) = 2(x + 1) - 2$

(c) $2b + 3(x + a) = 5b + 6$   (d) $\dfrac{1}{2x} - \dfrac{3}{x} = \dfrac{-25}{22}$

**2** Solve each formula for the indicated variable.

(a) $V = bh$, for $h$   (b) $s = \frac{1}{2}(a + b + c)$ for $c$

**3** Graph each solution set on a number line.

(a) $\{x \mid x < -\frac{3}{2}\}$   (b) $\{x \mid x > -4\} \cap \{x \mid x \leq 0\}$   (c) $\{x \mid x \geq 2\} \cup \{x \mid x < -5\}$

**4** Solve each first-degree inequality for $x$.

(a) $2x - 1 < 3$   (b) $x + 1 > 3 - 2x$

(c) $4(2 - x) \leq -2(x + 1)$   (d) $3(1 - 2x) - 2(x + 1) < 0$

5   Find the solutions of each absolute-value equation.
    (a)  $|3x| = 6$
    (b)  $|2x - 1| = 5$

6   Find the solution set of each absolute-value inequality.
    (a)  $|x| < 11$
    (b)  $|x| \geq 8$
    (c)  $|3x - 4| \leq 5$
    (d)  $|2x - 3| > 4$

7   Translate the following into algebraic form: In a football game, one team scored 8 more points than the other. If $x$ represents the number of points scored by the winning team, how many points did the losing team score?

8   How many pounds of peanuts at $4.20 per pound should be mixed with 20 pounds of walnuts at $6.00 per pound to give a mixture worth $5.20 per pound?

9   Two cars depart at the same time from two cities that are 285 miles apart. The cars are traveling toward each other. Find the average speed of each car if they meet at the end of 3 hours and one car travels 15 miles per hour faster than the other.

---

## Solutions

1   (a)  $3x - 1 = 2x + 1$
         $3x = 2x + 2$
         $x = 2$

    (b)  $x - 3(2 - x) = 2(x + 1) - 2$
         $x - 6 + 3x = 2x + 2 - 2$
         $-6 + 4x = 2x$
         $2x = 6$
         $x = 3$

    (c)  $2b + 3(x + a) = 5b + 6$
         $2b + 3x + 3a = 5b + 6$
         $3x = 3b - 3a + 6$
         $x = \dfrac{3b - 3a + 6}{3}$
         $x = b - a + 2$

    (d)  $\dfrac{1}{2x} - \dfrac{3}{x} = \dfrac{-25}{22}$
         $22x\left(\dfrac{1}{2x} - \dfrac{3}{x}\right) = 22x\left(\dfrac{-25}{22}\right)$
         $11 - 66 = -25x$
         $-55 = -25x$
         $25x = 55$
         $x = \dfrac{55}{25} = \dfrac{11}{5}$

2   (a)  $V = bh$, for $h$
         $bh = V$
         $\dfrac{bh}{b} = \dfrac{V}{b}$
         $h = \dfrac{V}{b}$

    (b)  $s = \tfrac{1}{2}(a + b + c)$, for $c$
         $2s = 2 \cdot \tfrac{1}{2}(a + b + c)$
         $2s = a + b + c$
         $2s - a - b = c$

3   (a)  ←————————) $-\tfrac{3}{2}$
    (b)  (————] $-4$ to $0$
    (c)  ←————) $-5$  [————→ $2$

4   (a)  $2x - 1 < 3$
         $2x < 4$
         $x < 2$

    (b)  $x + 1 > 3 - 2x$
         $3x + 1 > 3$
         $3x > 2$
         $x > \tfrac{2}{3}$

    (c)  $4(2 - x) \leq -2(x + 1)$
         $8 - 4x \leq -2x - 2$
         $8 - 2x \leq -2$
         $-2x \leq -10$
         $x \geq 5$

    (d)  $3(1 - 2x) - 2(x + 1) < 0$
         $3 - 6x - 2x - 2 < 0$
         $-8x + 1 < 0$
         $-8x < -1$
         $x > \tfrac{1}{8}$

5   (a)  $|3x| = 6$
         $3x = -6$ or $3x = 6$
         $x = -2$ or $x = 2$

    (b)  $|2x - 1| = 5$
         $2x - 1 = -5$ or $2x - 1 = 5$
         $2x = -4$ or $2x = 6$
         $x = -2$ or $x = 3$

6   (a)    $|x| < 11$
             $-11 < x < 11$

  (b)    $|x| \geq 8$
             $x \leq -8$ or $x \geq 8$

  (c)    $|3x - 4| \leq 5$
             $-5 \leq 3x - 4 \leq 5$
             $-1 \leq 3x \leq 9$
             $-\frac{1}{3} \leq x \leq 3$

  (d)    $|2x - 3| > 4$
             $2x - 3 < -4$ or $2x - 3 > 4$
             $2x < -1$ or $\quad 2x > 7$
             $x < -\frac{1}{2}$ or $\quad x > \frac{7}{2}$

7    If $x$ represents the winning team's score, then the losing team's score is 8 points less, or $x - 8$.

8    Let $x =$ number of pounds of peanuts.
Then $4.20x + 6(20) = 5.20(x + 20)$
or      $42x + 60(20) = 52(x + 20)$
          $42x + 1200 = 52x + 1040$
                $160 = 10x$
                 $16 = x$
There will be 16 pounds of peanuts required.

9    Let $x =$ average speed of slower car; $x + 15 =$ average speed of faster car.
Then $3x =$ distance traveled by slower car; $3(x + 15) =$ distance traveled by faster car.
so that, $3x + 3(x + 15) = 285$
              $6x + 45 = 285$
                $6x = 240$
                 $x = 40$
             $x + 15 = 55$
Therefore, the slower car had an average speed of 40 miles per hour, and the faster car had an average speed of 55 miles per hour.

## Chapter Test B

*Multiple Choice:* Select the *one* correct answer for each of the following questions.

1    If $7x + 2 = 2$, then the solution is _____.

   (a)    2           (b)    0           (c)    $\frac{2}{7}$           (d)    $-\frac{2}{7}$

2    If $8x + 14 = 5x + 44$, then the solution is _____.

   (a)    14           (b)    0           (c)    10           (d)    44

3    If $\frac{7x}{4} + 5 = \frac{3x}{4} + 1$, then the solution is _____.

   (a)    $-4$           (b)    3           (c)    $-2$           (d)    $-3$

4    If $\frac{x}{2} + \frac{x}{3} = 10$, then the solution is _____.

   (a)    3           (b)    $-12$           (c)    2           (d)    12

5    If $\frac{x - 10}{3} + \frac{13}{4} = \frac{4x + 6}{3}$, then the solution is _____.

   (a)    3           (b)    $-3$           (c)    $-\frac{25}{12}$           (d)    12

6    If $\frac{6}{x} + \frac{x - 3}{2x} = 2$, then the solution is _____.

   (a)    2           (b)    $-3$           (c)    $-2$           (d)    3

7    If $\frac{x}{a} - \frac{1}{c} = \frac{a}{b}$, and $a$, $b$, and $c$ are constants, then the solution is _____.

   (a)    $a^2c - ab$       (b)    $-a^2c + ab$       (c)    $\frac{a^2c + ab}{bc}$       (d)    $\frac{a^2c - ab}{bc}$

8  If a number $x$ is multiplied by 6 and then 7 is subtracted from this product, the result is represented by _____.
   (a) $7 - 6x$  (b) $6(x - 7)$  (c) $6(7 - x)$  (d) $6x - 7$

9  Joanne told her daughter, "I am now 3 times as old as you, but in 10 years I will be twice as old as you will be." How old is Joanne? ____
   (a) 37  (b) 15  (c) 30  (d) 38

10  $0.\overline{53}$ can be expressed as ____.
   (a) $\frac{8}{15}$  (b) $\frac{53}{100}$  (c) $\frac{53}{99}$  (d) $5\frac{1}{3}$

11  If $A = \pi r^2 + 2\pi rh$, then $h = $ _____.
   (a) $\frac{A - r}{2}$  (b) $\frac{A - 2\pi r}{\pi r^2}$  (c) $\frac{A - 2}{r}$  (d) $\frac{A - \pi r^2}{2\pi r}$

12  If $3x - 5 < 7$, then the solution set is ____.
   (a) $x < -4$  (b) $x < 2$  (c) $x < 12$  (d) $x < 4$

13  If $\frac{1}{3}x - \frac{2}{5} < \frac{3}{2}x + 1$, then the solution set is _____.
   (a) $x < -\frac{6}{5}$  (b) $x > -\frac{6}{5}$  (c) $x < \frac{6}{5}$  (d) $x > \frac{6}{5}$

14  If $-3 < 1 - 2x < 4$, then the solution set is _____.
   (a) $-\frac{3}{2} < x < 2$  (b) $-\frac{3}{2} \leq x < 2$  (c) $-\frac{3}{2} \leq x \leq 2$  (d) none of these

15  If $\frac{3x - 7}{5} \geq 2x + 1$, then the solution set is _____.
   (a) $x \leq -\frac{12}{7}$  (b) $x \geq -\frac{12}{7}$  (c) $x \leq \frac{12}{7}$  (d) $x \geq \frac{12}{7}$

16  If $|3x| = 6$, then the solution set is _____.
   (a) $-2, 2$  (b) $-3, 3$  (c) $-1, 1$  (d) 3

17  If $|x + \frac{1}{2}| = \frac{1}{4}$, then the solutions are _____.
   (a) $\frac{3}{4}, \frac{5}{4}$  (b) $-\frac{3}{4}, -\frac{1}{4}$  (c) $-\frac{1}{4}, \frac{3}{4}$  (d) $-\frac{3}{4}, \frac{1}{4}$

18  If $|x| > 4$, then the solution set is _____.
   (a) $x > 4$ or $x < -4$  (b) $x > -4$ or $x < 4$  (c) $x > -4$ and $x < -1$  (d) $x = -4$ or $x \geq 4$

19  If $|4 - 3x| \leq 12$, then the solution set is _____.
   (a) $-12 \leq x \leq 12$  (b) $-4 \leq x \leq 4$  (c) $-13 \leq 4 - 3x \leq 12$  (d) $-\frac{8}{3} \leq x \leq \frac{16}{3}$

20  If $|2x - 7| \geq 5$, then the solution set is _____.
   (a) $x \geq -6$ or $x \leq -1$  (b) $x \geq 6$ or $x \leq 1$
   (c) $x \leq 6$ or $x \geq -1$  (d) $x < 6$ or $x > 1$

21  The graph of the solution set $\{x | x > 3\} \cap \{x | x \leq 7\}$ is ____.

   (a) 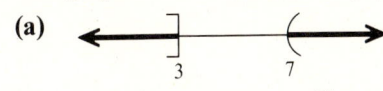  (b)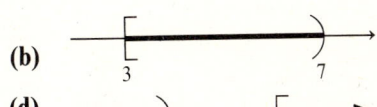

   (c) 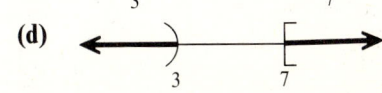  (d)

---

## Answers

1 b   2 c   3 a   4 d   5 c   6 d   7 c   8 d   9 c   10 c
11 d  12 d  13 b  14 a  15 a  16 a  17 b  18 a  19 d
20 b  21 c

# 5 Exponents, Radicals, and Complex Numbers

In this chapter we extend the use of exponents to include integral and rational exponents. Then we consider radicals and their properties. After completing this chapter, the student should be able to:

1. Interpret the meaning of $x^n$ for any negative and zero values of $n$.
2. Use the basic properties of rational exponents.
3. Indicate a particular root of a number using radicals.
4. Perform operations on radical expressions.
5. Perform operations on complex numbers.

## 5.1 Zero and Negative Integer Exponents

### SEMIPROGRAMMED PROBLEMS

In problems 1–8, rewrite each expression so that it contains only positive exponents and simplify the result.

1   1

0, 0

$\dfrac{1}{6^2}, \dfrac{1}{36}$

$\dfrac{1}{y^3}, \dfrac{x^2}{y^3}$

$\dfrac{1}{3^2} \cdot \dfrac{1}{y^2}, 9y^2$

$\dfrac{1}{xy}, y + x, y + x$

$\dfrac{1}{y}, y - x, \dfrac{y+x}{y-x}$

$y^2, y^2 + x^2$

---

1   $5^0 = $ \_\_\_\_

2   $\dfrac{x^0 - y^0}{x^0 + y^0} = \dfrac{\rule{1cm}{0.15mm}}{1 + 1} = $ \_\_\_\_

3   $6^{-2} = $ \_\_\_\_ $ = $ \_\_\_\_

4   $x^2 y^{-3} = x^2 \left( \rule{1cm}{0.15mm} \right) = $ \_\_\_\_

5   $\dfrac{3^{-2} x^0 y^{-2}}{z^{-4}} = \dfrac{\rule{1cm}{0.15mm}}{\dfrac{1}{z^4}} = \dfrac{\dfrac{1}{9y^2}}{\dfrac{1}{z^4}} = \dfrac{z^4}{\rule{1cm}{0.15mm}}$

6   $\dfrac{x^{-1} + y^{-1}}{(xy)^{-1}} = \dfrac{\dfrac{1}{x} + \dfrac{1}{y}}{\dfrac{1}{xy}} = \dfrac{\dfrac{\rule{1cm}{0.15mm}}{xy}}{\dfrac{1}{xy}} = $ \_\_\_\_

7   $\dfrac{x^{-1} + y^{-1}}{x^{-1} - y^{-1}} = \dfrac{\dfrac{1}{x} + \dfrac{1}{y}}{\dfrac{1}{x} - \dfrac{\rule{0.5cm}{0.15mm}}{\rule{0.5cm}{0.15mm}}} = \dfrac{\dfrac{y+x}{xy}}{\dfrac{\rule{1cm}{0.15mm}}{xy}} = $ \_\_\_\_

8   $x^{-2} + y^{-2} = \dfrac{1}{x^2} + \dfrac{1}{\rule{0.5cm}{0.15mm}} = \dfrac{\rule{1cm}{0.15mm}}{x^2 y^2}$

88

In problems 9–18, use the properties of exponents to rewrite each expression so that it contains only positive exponents, and simplify the result.

| | |
|---|---|
| $6^{-1}, \frac{1}{6}$ | **9** $\quad 6^{-3} \cdot 6^2 = 6^{-3+2} = \underline{\quad} = \underline{\quad}$ |
| $x^{-2+3}, x$ | **10** $\quad x^{-2} \cdot x^3 = \underline{\quad} = \underline{\quad}$ |
| $5^3, 125$ | **11** $\quad (5^{-3})^{-1} = 5^{(-3)(-1)} = \underline{\quad} = \underline{\quad}$ |
| $x^{-6}, \frac{1}{x^6}$ | **12** $\quad (x^{-3})^2 = \underline{\quad} = \underline{\quad}$ |
| $y^{-2}, \frac{1}{y^2}, \frac{25}{y^2}$ | **13** $\quad (5^{-1}y)^{-2} = 5^{(-1)(-2)}(\underline{\quad}) = 5^2\left(\underline{\quad}\right) = \underline{\quad}$ |
| $y^{-3}, \frac{x^3}{y^3}$ | **14** $\quad (x^{-1}y)^{-3} = x^3(\underline{\quad}) = \underline{\quad}$ |
| $a^2$ | **15** $\quad \left(\dfrac{a^{-1}}{b^{-1}}\right)^{-2} = \dfrac{\underline{\quad}}{b^2}$ |
| $x^2, 7x^2$ | **16** $\quad \left(\dfrac{x^{-2}}{7}\right)^{-1} = \dfrac{\underline{\quad}}{7^{-1}} = \underline{\quad}$ |
| $x^{-9}, \dfrac{1}{x^9}$ | **17** $\quad \dfrac{x^{-7}}{x^2} = x^{-7-2} = \underline{\quad} = \underline{\quad}$ |
| $x^{-2}, \dfrac{1}{x^2}$ | **18** $\quad \dfrac{x^3}{x^5} = x^{3-5} = \underline{\quad} = \underline{\quad}$ |

In problems 19 and 20, rewrite each expression without negative exponents and simplify.

| | |
|---|---|
| $(xy^{-5})^{-2}, x^{-2}, \dfrac{y^8}{x^8}$ | **19** $\quad \left(\dfrac{xy^{-5}}{x^{-3}y^{-1}}\right)^{-2} = \dfrac{\underline{\quad}}{(x^{-3}y^{-1})^{-2}} = \dfrac{(\underline{\quad})y^{10}}{x^6 y^2} = \underline{\quad}$ |
| $(2x^3)^{-3}, 2^{-3}x^{-9}$ | **20** $\quad \left(\dfrac{2x^3}{3x^{-1}y^2}\right)^{-3} = \dfrac{\underline{\quad}}{(3x^{-1}y^2)^{-3}} = \dfrac{\underline{\quad}}{3^{-3}x^3 y^{-6}}$ |
| $27y^6, \dfrac{27y^6}{8x^{12}}$ | $\qquad = \dfrac{\underline{\quad}}{8x^3 \cdot x^9} = \underline{\quad}$ |

## SOLUTIONS TO SELECTED ODD PROBLEMS  Section 5.1, text pages 228–230

**1** $\quad 4^0 = 1$

**5** $\quad (-3y^2)^0 = 1$

**9** $\quad 5^{-2} = \dfrac{1}{5^2} = \dfrac{1}{25}$

**13** $\quad (-10)^{-1} = \dfrac{1}{(-10)^1} = -\dfrac{1}{10}$

**17** $\quad p^{-5} = \dfrac{1}{p^5}$

**21** $\quad u^{-3}v^{-6} = \dfrac{1}{u^3 v^6}$

**25** $\quad \dfrac{10^0}{10^{-3}} = \dfrac{1}{10^{-3}} = 10^3 = 1000$

**29** $\quad 2^2 \cdot 3^2 \cdot 6^{-3} = \dfrac{2^2 \cdot 3^2}{6^3} = \dfrac{4 \cdot 9}{216} = \dfrac{1}{6}$

**33** $\quad \dfrac{2^{-2} + 5^{-2}}{10^{-3}} = \dfrac{\frac{1}{2^2} + \frac{1}{5^2}}{\frac{1}{10^3}} = \dfrac{\frac{1}{4} + \frac{1}{25}}{\frac{1}{1,000}} = \dfrac{\frac{29}{100}}{\frac{1}{1,000}} = 290$

**37** $\quad \dfrac{5^{-1} - 3^{-2}}{(45)^{-1}} = \dfrac{\frac{1}{5} - \frac{1}{3^2}}{\frac{1}{45}} = \dfrac{\frac{1}{5} - \frac{1}{9}}{\frac{1}{45}} = \dfrac{\frac{4}{45}}{\frac{1}{45}} = 4$

**41** $\quad \dfrac{1}{x^{-2}} - x^{-2} = x^2 - \dfrac{1}{x^2} = \dfrac{x^4 - 1}{x^2}$

**45** $\quad \dfrac{4 - x^{-2}}{2 + x^{-1}} = \dfrac{4 - \frac{1}{x^2}}{2 + \frac{1}{x}} = \dfrac{\frac{4x^2 - 1}{x^2}}{\frac{2x + 1}{x}} = \dfrac{4x^2 - 1}{x^2} \cdot \dfrac{x}{2x + 1} = \dfrac{2x - 1}{x}$

**49** $\quad 7^{-1} \cdot 7^3 = 7^2 = 49$

**53** $\quad x^{-3}x^{-2} = x^{-5} = \dfrac{1}{x^5}$

CHAPTER 5 EXPONENTS, RADICALS, AND COMPLEX NUMBERS

57  $[(-4)^2]^{-2} = (-4)^{-4} = \dfrac{1}{(-4)^4} = \dfrac{1}{256}$

61  $(5^{-1}p^4)^{-2} = 5^2 p^{-8} = \dfrac{5^2}{p^8} = \dfrac{25}{p^8}$

65  $\left(\dfrac{p^2}{q^2}\right)^{-2} = \dfrac{p^{-4}}{q^{-4}} = \dfrac{q^4}{p^4}$

69  $\dfrac{x^{-3}}{x^5} = x^{-3-5} = x^{-8} = \dfrac{1}{x^8}$

73  $\dfrac{(a+b)^{-2}}{(a+b)^{-3}} = (a+b)^{-2-(-3)} = a+b$

77  $\left(\dfrac{b^{-1}}{3a^{-1}}\right)\cdot\left(\dfrac{3a}{b}\right)^{-1} = \left(\dfrac{a}{3b}\right)\left(\dfrac{b}{3a}\right) = \dfrac{1}{9}$

81  $\left(\dfrac{a^2 c^{-2} b^{-1}}{(abc)^{-1}}\right)^{-3} = \left(\dfrac{a^2 \cdot (abc)}{c^2 b}\right)^{-3} = \left(\dfrac{a^3 bc}{c^2 b}\right)^{-3} = \left(\dfrac{a^3}{c}\right)^{-3} = \left(\dfrac{c}{a^3}\right)^3 = \dfrac{c^3}{a^9}$

85  $\left[\left(\dfrac{a^{-1}}{b}\right)^{-1}\cdot\left(\dfrac{b^{-1}}{a}\right)^{-1}\right]^{-2} = \left[\left(\dfrac{a}{b^{-1}}\right)\cdot\left(\dfrac{b}{a^{-1}}\right)\right]^{-2} = [(ab)\cdot(ab)]^{-2} = (a^2 b^2)^{-2} = a^{-4} b^{-4} = \dfrac{1}{a^4 b^4}$

89  $\dfrac{(1.12\times 10^{-3})\cdot(8.25\times 10^{-5})}{(2.35\times 10^{-6})^2} = \dfrac{9.24\times 10^{-8}}{5.5225\times 10^{-12}} = 1.67316\times 10^4$

93  (a)  $m = \dfrac{Pr}{12\left[1-\left(1+\dfrac{r}{12}\right)^{-12t}\right]} = \dfrac{(12{,}000)(0.12)}{12\left[1-\left(1+\dfrac{0.12}{12}\right)^{-12(4)}\right]} = 316.01$

Therefore, the monthly payment is $316.01.

(b)  $I = 12m\left[t - \dfrac{1-\left(1+\dfrac{r}{12}\right)^{-12t}}{r}\right] = 12(316.01)\left[4 - \dfrac{1-\left(1+\dfrac{0.12}{12}\right)^{-12(4)}}{0.12}\right] = 3{,}168.33$

Therefore, the total interest charge is $3,168.33.

97  (a)  $m = \dfrac{Pr}{12\left[1-\left(1+\dfrac{r}{12}\right)^{-12t}\right]}$

The downpayment is ($50,000)(0.20) = $10,000.
The remaining amount is $50,000 − $10,000 = $40,000.

$m = \dfrac{(40{,}000)(0.10)}{12\left[1-\left(1+\dfrac{0.10}{12}\right)^{-12(30)}\right]} = 351.03$

Therefore, the monthly payment is $351.03.

(b)  $P_k = P\left[\left(1+\dfrac{r}{12}\right)^{12t} - \left(1+\dfrac{r}{12}\right)^{k-1}\right] = (40{,}000)\left[\left(1+\dfrac{0.10}{12}\right)^{12(10)} - \left(1+\dfrac{0.10}{12}\right)^{10-1}\right] = 46{,}375.21$

Therefore, the amount to be paid is $50,000 − $46,375.21 = $3,624.79.

101  $d = r\cdot t$
$93{,}000{,}000 = (310{,}000)t$
$9.3\times 10^7 = (3.1\times 10^5)t$
$\dfrac{9.3\times 10^7}{3.1\times 10^5} = t$
$3\times 10^2 = t$

## 5.2 Roots and Rational Exponents

### SEMIPROGRAMMED PROBLEMS

In problems 1–3, find the principal root (if it is defined).

2, 2, 2
−4, −4
is not defined

1  $\sqrt[4]{16} =$ _____ since (_____)$^4$ = 16 and _____ > 0.

2  $\sqrt[3]{-64} =$ _____ since only (_____)$^3$ = −64.

3  $\sqrt[4]{-16} =$ _____.

## 5.2 SEMIPROGRAMMED PROBLEMS

In problems 4 and 5, rewrite each expression using rational exponents.

$17^{1/2}$

4  $\sqrt{17}$

$\sqrt{17} = \underline{\qquad}$

$(a+b)^3$, $(a+b)^{3/4}$

5  $\sqrt[4]{(a+b)^3}$

$\sqrt[4]{(a+b)^3} = (\underline{\qquad})^{1/4} = \underline{\qquad}$

In problems 6 and 7, rewrite each expression using radicals.

6  $6^{2/3}$

$6, \sqrt[3]{36}$

$6^{2/3} = \sqrt[3]{(\underline{\qquad})^2} = \underline{\qquad}$

7  $(2x)^{3/2}$

$2x, \sqrt{8x^3}$

$(2x)^{3/2} = \sqrt{(\underline{\qquad})^3} = \underline{\qquad}$

In problems 8–11, find the value of each expression.

8  $4^{3/2}$

$4, 2, 8$

$4^{3/2} = (\sqrt{\underline{\qquad}})^3 = (\underline{\qquad})^3 = \underline{\qquad}$

9  $(-8)^{5/3}$

$-8, -2, -32$

$(-8)^{5/3} = (\sqrt[3]{\underline{\qquad}})^5 = (\underline{\qquad})^5 = \underline{\qquad}$

10  $(-27)^{-4/3}$

$(-27)^{4/3}, \frac{1}{81}$

$(-27)^{-4/3} = \dfrac{1}{\underline{\qquad}} = \underline{\qquad}$

11  $(-4)^{-5/2}$

$(-4)^{5/2}$, undefined

$(-4)^{-5/2} = \dfrac{1}{\underline{\qquad}} = \underline{\qquad}$

In problems 12–16, simplify each expression. Assume that all variables are restricted to values for which all expressions are defined.

$x^{7/4}$

12  $x^{5/2} \cdot x^{-3/4} = x^{5/2 - 3/4} = \underline{\qquad}$

$3^3, 27$

13  $(3^{-1/5})^{-15} = 3^{(-1/5)(-15)} = \underline{\qquad} = \underline{\qquad}$

$3^{-3}x^9, \dfrac{x^9}{3^3}, \dfrac{x^9}{27}$

14  $(81x^{-12})^{-3/4} = (81)^{-3/4} \cdot (x^{-12})^{-3/4} = (3^4)^{-3/4} \cdot (x^{-12})^{-3/4}$

$= \underline{\qquad} = \underline{\qquad} = \underline{\qquad}$

$x^{1/8}$

15  $\dfrac{x^{-3/4}}{x^{-7/8}} = x^{-3/4 - (-7/8)} = \underline{\qquad}$

$\frac{1}{16}, \dfrac{x^6}{16z^6 y^2}$

16  $\left(\dfrac{-64x^{-9}y^3}{z^{-9}}\right)^{-2/3} = \dfrac{(-64)^{-2/3}(x^{-9})^{-2/3}(y^3)^{-2/3}}{(z^{-9})^{-2/3}}$

$= \dfrac{\underline{\qquad} x^6 y^{-2}}{z^6} = \underline{\qquad}$

In problems 17 and 18, perform each multiplication. Write each answer with positive exponents.

17  $x^{2/3}(2x^{4/3} - 3x^{1/3})$

$3x^{1/3}$

$= x^{2/3}(2x^{4/3}) - x^{2/3}(\underline{\qquad})$

$1/3, 2x^2 - 3x$

$= 2x^{2/3 + 4/3} - 3x^{2/3 + \underline{\qquad}} = \underline{\qquad}$

$x^{-1/4}, x^{-1/4}$

$x^{-1/2}, x^{-1/4}$

$x^{1/2}, x^{1/4}$

$1 + 4x^{1/4} + 4x^{1/2}$

18   $(x^{-1/4} + 2)^2 = (\underline{\phantom{xx}})^2 + 2(\underline{\phantom{xx}})(2) + (2)^2$

$= \underline{\phantom{xxxx}} + 4(\underline{\phantom{xx}}) + 4$

$= \dfrac{1}{\underline{\phantom{xx}}} + \dfrac{4}{\underline{\phantom{xx}}} + 4$

$= \dfrac{\overline{\phantom{xxxxxxxxxx}}}{x^{1/2}}$

## SOLUTIONS TO SELECTED ODD PROBLEMS   Section 5.2, text pages 239–241

1   $\sqrt{16} = 4$    5   $\sqrt[3]{-27} = -3$    9   $\sqrt{13} = 13^{1/2}$

13   $\sqrt[4]{a^3 b^2} = (a^3 b^2)^{1/4} = a^{3/4} b^{1/2}$    17   $3^{5/6} = \sqrt[6]{3^5} = \sqrt[6]{243}$

21   $(4xy)^{5/7} = \sqrt[7]{(4)^5 x^5 y^5} = \sqrt[7]{1{,}024 x^5 y^5}$    25   $-16^{-1/2} = -\dfrac{1}{16^{1/2}} = \dfrac{-1}{\sqrt{16}} = -\dfrac{1}{4}$

29   $\left(\dfrac{9}{16}\right)^{1/2} = \dfrac{9^{1/2}}{16^{1/2}} = \dfrac{\sqrt{9}}{\sqrt{16}} = \dfrac{3}{4}$    33   $(-8)^{4/3} = (\sqrt[3]{-8})^4 = (-2)^4 = 16$

37   $81^{3/4} = (\sqrt[4]{81})^3 = (3)^3 = 27$    41   $27^{8/12} = 27^{2/3} = (\sqrt[3]{27})^2 = (3)^2 = 9$

45   $9^{-3/2} = \dfrac{1}{9^{3/2}} = \dfrac{1}{(\sqrt{9})^3} = \dfrac{1}{(3)^3} = \dfrac{1}{27}$    49   $2^{1/3} 2^{2/3} = 2^{3/3} = 2$

53   $(5^{1/7})^{14} = 5^2 = 25$    57   $(8p^9)^{4/3} = 8^{4/3} \cdot p^{12} = (\sqrt[3]{8})^4 p^{12} = 2^4 p^{12} = 16 p^{12}$

61   $\left(\dfrac{125}{y^3}\right)^{-1/3} = \dfrac{125^{-1/3}}{y^{-1}} = \dfrac{y}{125^{1/3}} = \dfrac{y}{5}$    65   $\dfrac{x^{1/3}}{x^{-1/6}} = x^{1/3 - (-1/6)} = x^{1/2}$

69   $(6a^{7/2} b^{-3/2})^2 (4a^{-1/3} b^{-2/3})^3 = (6^2 a^7 b^{-3})(4^3 a^{-1} b^{-2})$

$= (36)(64) a^6 b^{-5}$

$= \dfrac{2304 a^6}{b^5}$

73   $\dfrac{(125 x^7 y^{-5})^{-1/3}}{(64 x^2 y^8)^{-1/6}} = \dfrac{125^{-1/3} x^{-7/3} y^{5/3}}{64^{-1/6} x^{-1/3} y^{-4/3}} = \dfrac{64^{1/6} y^3}{125^{1/3} x^2} = \dfrac{2 y^3}{5 x^2}$

77   $\left(\dfrac{81 p^{-12}}{q^{16}}\right)^{-1/4} \cdot \left(-\dfrac{p^{-2/3}}{q^{1/3}}\right)^3 = \left(\dfrac{81^{-1/4} p^3}{q^{-4}}\right)\left(-\dfrac{p^{-2}}{q}\right) = -\dfrac{81^{-1/4} p}{q^{-3}} = -\dfrac{pq^3}{81^{1/4}} = -\dfrac{pq^3}{3}$

81   $(2a^{1/2} - b^{-1/2})(2a^{1/2} + b^{-1/2}) = (2a^{1/2})^2 - (b^{-1/2})^2 = 4a - b^{-1} = 4a - \dfrac{1}{b} = \dfrac{4ab - 1}{b}$

85   $(x^{1/3} + y^{1/3})(x^{2/3} - x^{1/3} y^{1/3} + y^{2/3}) = (x^{1/3})^3 + (y^{1/3})^3 = x + y$

89   $4x^{-1/2} + x^{1/2} = \dfrac{4}{x^{1/2}} + x^{1/2} = \dfrac{4 + x}{x^{1/2}}$

93   $\dfrac{-x^2}{(x^2 + 1)^{1/2}} + (x^2 + 1)^{1/2} = \dfrac{-x^2 + (x^2 + 1)}{(x^2 + 1)^{1/2}} = \dfrac{1}{(x^2 + 1)^{1/2}}$

97   $A = \sqrt{s(s-a)(s-b)(s-c)}$

$s = \dfrac{a + b + c}{2} = \dfrac{42.3 + 28.7 + 37.1}{2} = 54.05$

$A = \sqrt{(54.05)(54.05 - 42.3)(54.05 - 28.7)(54.05 - 37.1)}$

$A = 522.4$ square meters

## 5.3 Radicals

### SEMIPROGRAMMED PROBLEMS

In problems 1–13, use the properties of radicals to simplify each expression. Assume that variables are restricted to values for which all expressions are defined.

| Answers | Problems |
|---|---|
| 25, 25, $5\sqrt{2}$ | 1. $\sqrt{50} = \sqrt{(\underline{\phantom{xx}})2} = \sqrt{\underline{\phantom{xx}}}\sqrt{2} = \underline{\phantom{xx}}$ |
| $4x^2$, $4x^2$, $2x\sqrt{2x}$ | 2. $\sqrt{8x^3} = \sqrt{(\underline{\phantom{xx}})2x} = \sqrt{\underline{\phantom{xx}}}\sqrt{2x} = \underline{\phantom{xx}}$ |
| $27x^3$, $27x^3$, $3x\sqrt[3]{2x^2}$ | 3. $\sqrt[3]{54x^5} = \sqrt[3]{(\underline{\phantom{xx}})2x^2} = \sqrt[3]{\underline{\phantom{xx}}}\sqrt[3]{2x^2} = \underline{\phantom{xx}}$ |
| $\sqrt{121}$, $\frac{9}{11}$ | 4. $\sqrt{\dfrac{81}{121}} = \dfrac{\sqrt{81}}{\underline{\phantom{xx}}} = \underline{\phantom{xx}}$ |
| $x^{10}$, $x^2$ | 5. $\sqrt[5]{\dfrac{x^{10}}{32}} = \dfrac{\sqrt[5]{\underline{\phantom{xx}}}}{\sqrt[5]{32}} = \dfrac{\underline{\phantom{xx}}}{2}$ |
| $y^7$ | 6. $\sqrt[5]{y^{35}} = (\sqrt[5]{y^5})^7 = \underline{\phantom{xx}}$ |
| $(-x)^3$, $-x^3$ | 7. $\sqrt[11]{-x^{33}} = (\sqrt[11]{-x^{11}})^3 = \underline{\phantom{xx}} = \underline{\phantom{xx}}$ |
| $x^{18}$, $\|x\|$ | 8. $\sqrt[3]{\sqrt[6]{x^{18}}} = \sqrt[18]{\underline{\phantom{xx}}} = \underline{\phantom{xx}}$ |
| $-x^2$ | 9. $\sqrt[7]{\sqrt[5]{-x^{70}}} = \sqrt[35]{-x^{70}} = \underline{\phantom{xx}}$ |
| $25x^2y^4$, $25x^2y^4$, $5xy^2\sqrt{3xy}$ | 10. $\sqrt{75x^3y^5} = \sqrt{(\underline{\phantom{xx}})3xy} = \sqrt{\underline{\phantom{xx}}}\sqrt{3xy} = \underline{\phantom{xx}}$ |
| $-8x^3y^6$, $-8x^3y^6$, $-2xy^2\sqrt[3]{2xy}$ | 11. $\sqrt[3]{-16x^4y^7} = \sqrt[3]{(\underline{\phantom{xx}})2xy} = \sqrt[3]{\underline{\phantom{xx}}}\sqrt[3]{2xy} = \underline{\phantom{xx}}$ |
| $72x^2y$, $36x^2$, $\dfrac{6x\sqrt{2y}}{\|z\|}$ | 12. $\sqrt{\dfrac{72x^2y}{z^2}} = \dfrac{\sqrt{\underline{\phantom{xx}}}}{\sqrt{z^2}} = \dfrac{\sqrt{\underline{\phantom{xx}}}\sqrt{2y}}{\|z\|} = \underline{\phantom{xx}}$ |
| $-54y^5$, $-27y^3$, $\dfrac{-3y\sqrt[3]{2y^2}}{2x}$ | 13. $\sqrt[3]{\dfrac{-54y^5}{8x^3}} = \dfrac{\sqrt[3]{\underline{\phantom{xx}}}}{\sqrt[3]{8x^3}} = \dfrac{\sqrt[3]{\underline{\phantom{xx}}}\sqrt[3]{2y^2}}{2x} = \underline{\phantom{xx}}$ |

In problems 14 and 15, express each pair of radical expressions as radicals having a common index; then simplify.

| | |
|---|---|
| $x^7$, $\sqrt[6]{x^6}$, $x\sqrt[6]{x}$ | 14. $\sqrt[3]{x^2} \cdot \sqrt{x} = \sqrt[3\times2]{x^{2\times2}} \cdot \sqrt[3\times2]{x^{1\times3}} = \sqrt[6]{x^4} \cdot \sqrt[6]{x^3}$ $= \sqrt[6]{\underline{\phantom{xx}}} = (\underline{\phantom{xx}})\sqrt[6]{x} = \underline{\phantom{xx}}$ |
| $x^{22}$ $x^2$, $x\sqrt[10]{x}$ | 15. $\sqrt[5]{x^3} \cdot \sqrt[4]{x^2} = \sqrt[5\times4]{x^{3\times4}} \cdot \sqrt[4\times5]{x^{2\times5}}$ $= \sqrt[20]{x^{12}} \cdot \sqrt[20]{x^{10}} = \sqrt[20]{\underline{\phantom{xx}}}$ $= x\sqrt[20]{\underline{\phantom{xx}}} = \underline{\phantom{xx}}$ |

### SOLUTIONS TO SELECTED ODD PROBLEMS  Section 5.3, text pages 248–249

1. $\sqrt{27} = \sqrt{9}\sqrt{3} = 3\sqrt{3}$

5. $\sqrt{288} = \sqrt{144} \cdot \sqrt{2} = 12\sqrt{2}$

9. $\sqrt[3]{162c^2} = \sqrt[3]{27} \cdot \sqrt[3]{6c^2} = 3\sqrt[3]{6c^2}$

13. $\sqrt[3]{8x^4} = \sqrt[3]{8} \cdot \sqrt[3]{x^3} \cdot \sqrt[3]{x} = 2x\sqrt[3]{x}$

17. $\sqrt{98p^3} = \sqrt{49p^2} \cdot \sqrt{2p} = 7p\sqrt{2p}$

21. $\sqrt[3]{\dfrac{-5}{8}} = \dfrac{\sqrt[3]{-5}}{\sqrt[3]{8}} = \dfrac{-\sqrt[3]{5}}{2}$

25 $\sqrt{\dfrac{3w^3}{4w^5}} = \dfrac{\sqrt{3}}{\sqrt{4w^2}} = \dfrac{\sqrt{3}}{2|w|}$

29 $\sqrt{\dfrac{3}{25x^2}} = \dfrac{\sqrt{3}}{\sqrt{25x^2}} = \dfrac{\sqrt{3}}{5|x|}$

33 $\sqrt{\dfrac{3}{9x^4}} = \dfrac{\sqrt{3}}{\sqrt{9x^4}} = \dfrac{\sqrt{3}}{3x^2}$

37 $\sqrt[3]{-y^9} = -\sqrt[3]{y^9} = -y^3$

41 $\sqrt{(a+b)^4} = \sqrt{[(a+b)^2]^2} = (a+b)^2$

45 $\sqrt[5]{-32y^{20}z^{10}} = \sqrt[5]{-32} \cdot \sqrt[5]{y^{20}} \cdot \sqrt[5]{z^{10}} = -2y^4z^2$

49 $(\sqrt[4]{5})^4 = 5$ [by Property (iii)]

53 $(\sqrt[11]{x})^{22} = x^{22/11} = x^2$

57 $\sqrt[3]{\sqrt[10]{m^{30}}} = \sqrt[30]{m^{30}} = m$

61 $\sqrt{18a^3b^7} \cdot \sqrt{2ab^3} = \sqrt{36a^4b^{10}} = 6a^2b^5$

65 $\sqrt[8]{x^{12}} \cdot \sqrt[8]{x^5y^{-8}} \cdot \sqrt[8]{x^2y^9} = \sqrt[8]{x^{19}y} = \sqrt[8]{x^{16}} \cdot \sqrt[8]{x^3y} = x^2\sqrt[8]{x^3y}$

69 $\sqrt[4]{\dfrac{a^4b^3}{a^3b}} \cdot \dfrac{\sqrt[4]{a^5b}}{\sqrt[4]{ab^{-1}}} = \dfrac{\sqrt[4]{a^9b^4}}{\sqrt[4]{a^4}} = \sqrt[4]{a^5b^4} = \sqrt[4]{a^4b^4} \cdot \sqrt[4]{a} = ab\sqrt[4]{a}$

73 $\sqrt[3]{x} \cdot \sqrt{x} = \sqrt[6]{x^2} \cdot \sqrt[6]{x^3} = \sqrt[6]{x^5}$

77 $\sqrt[5]{ab^2} \cdot \sqrt[4]{a^2b^3} = \sqrt[20]{a^4b^8} \cdot \sqrt[20]{a^{10}b^{15}} = \sqrt[20]{a^{14}b^{23}} = b\sqrt[20]{a^{14}b^3}$

81 $\sqrt{x^2 + 8x + 16} = \sqrt{(x+4)^2} = |x+4|$

85 $\sqrt{(4.3 - 2.4)^2 + (-5.7 - 4.2)^2} = \sqrt{(1.9)^2 + (-9.9)^2} = \sqrt{3.61 + 98.01} = \sqrt{101.62} = 10.08$

89 $f = \dfrac{1}{2\pi\sqrt{LC}}$

$= \dfrac{1}{2\pi\sqrt{(3.57 \times 10^{-8}) \cdot (121 \times 10^{-12})}}$

$= \dfrac{1}{2\pi\sqrt{431.97 \times 10^{-20}}}$

$= \dfrac{1}{2\pi\sqrt{431.9}(10^{-10})} = \dfrac{10^{10}}{130.58}$

$= 0.00765761 \times 10^{10} = 7.65761 \times 10^7$

## 5.4 Addition and Subtraction of Radical Expressions

### SEMIPROGRAMMED PROBLEMS

In problems 1–10, simplify by combining similar terms. Assume that all variables are restricted to values for which all radicals are defined.

| | |
|---|---|
| $5, 8\sqrt{2}$ | 1  $3\sqrt{2} + 5\sqrt{2} = (3 + \underline{\quad})\sqrt{2} = \underline{\quad\quad}$ |
| $2, 3\sqrt{3}$ | 2  $5\sqrt{3} - 2\sqrt{3} = (5 - \underline{\quad})\sqrt{3} = \underline{\quad\quad}$ |
| 9 | 3  $3\sqrt{50} + 5\sqrt{18} = 3\sqrt{25 \cdot 2} + 5\sqrt{(\underline{\quad})2}$ |
| 9 | $\quad\quad\quad\quad\quad\quad\quad = 3\sqrt{25}\sqrt{2} + 5\sqrt{\underline{\quad}}\sqrt{2}$ |
| 3 | $\quad\quad\quad\quad\quad\quad\quad = 3 \cdot 5\sqrt{2} + 5(\underline{\quad})\sqrt{2}$ |
| 15 | $\quad\quad\quad\quad\quad\quad\quad = 15\sqrt{2} + \underline{\quad}\sqrt{2}$ |
| 15 | $\quad\quad\quad\quad\quad\quad\quad = (15 + \underline{\quad})\sqrt{2}$ |
| $30\sqrt{2}$ | $\quad\quad\quad\quad\quad\quad\quad = \underline{\quad\quad}$ |
| $-1$ | 4  $-5\sqrt{xy} + 3\sqrt{xy} - \sqrt{xy} = (-5 + 3 + \underline{\quad})\sqrt{xy}$ |
| $-3\sqrt{xy}$ | $\quad\quad\quad\quad\quad\quad\quad\quad\quad\quad\quad = \underline{\quad\quad}$ |

| | |
|---|---|
| 17 | 5  $17\sqrt[3]{x^2} + 10\sqrt[3]{x^2} - 34\sqrt[3]{x^2} = (\underline{\quad} + 10 - 34)\sqrt[3]{x^2}$ |
| $-7\sqrt[3]{x^2}$ | $= \underline{\qquad}$ |
| $-1, 6\sqrt[4]{x}$ | 6  $3\sqrt[4]{x} + 4\sqrt[4]{x} - \sqrt[4]{x} = (3 + 4 + \underline{\quad})\sqrt[4]{x} = \underline{\quad}$ |
| 14 | 7  $28\sqrt[7]{3y} + 14\sqrt[7]{3y} + 8\sqrt[7]{3y} = (28 + \underline{\quad} + 8)\sqrt[7]{3y}$ |
| $50\sqrt[7]{3y}$ | $= \underline{\qquad}$ |
| $64 \cdot 5$ | 8  $\sqrt[3]{40} - \sqrt[3]{135} + \sqrt[3]{320} = \sqrt[3]{8 \cdot 5} - \sqrt[3]{27 \cdot 5} + \sqrt[3]{\underline{\quad}}$ |
| 64 | $= \sqrt[3]{8}\sqrt[3]{5} - \sqrt[3]{27}\sqrt[3]{5} + \sqrt[3]{\underline{\quad}}\sqrt[3]{5}$ |
| 4 | $= 2\sqrt[3]{5} - 3\sqrt[3]{5} + (\underline{\quad})\sqrt[3]{5}$ |
| 4 | $= (2 - 3 + \underline{\quad})\sqrt[3]{5}$ |
| $3\sqrt[3]{5}$ | $= \underline{\qquad}$ |
| $\sqrt{7 \cdot 16}$ | 9  $\sqrt{63} + 2\sqrt{112} - \sqrt{252} = \sqrt{7 \cdot 9} + 2(\underline{\quad}) - \sqrt{7 \cdot 36}$ |
| $\sqrt{7} \cdot \sqrt{16}$ | $= \sqrt{7} \cdot \sqrt{9} + 2(\underline{\quad}) - \sqrt{7} \cdot \sqrt{36}$ |
| $8, 5\sqrt{7}$ | $= 3\sqrt{7} + \underline{\quad}\sqrt{7} - 6\sqrt{7} = \underline{\quad}$ |
| $9x^4$ | 10  $\sqrt{4x^3} - \sqrt{9x^5} = \sqrt{4x^2 \cdot x} - \sqrt{(\underline{\quad})x}$ |
| $9x^4$ | $= \sqrt{4x^2}\sqrt{x} - \sqrt{\underline{\quad}}\sqrt{x}$ |
| $3x^2$ | $= 2x\sqrt{x} - (\underline{\quad})\sqrt{x}$ |
| $(2x - 3x^2)\sqrt{x}$ | $= \underline{\qquad}$ |

**SOLUTIONS TO SELECTED ODD PROBLEMS**  Section 5.4, text pages 252–253

1  $5\sqrt{7} + 3\sqrt{7} = (5 + 3)\sqrt{7} = 8\sqrt{7}$

5  $8\sqrt[3]{4} - 3\sqrt[3]{4} + 2\sqrt[3]{4} = (8 - 3 + 2)\sqrt[3]{4} = 7\sqrt[3]{4}$

9  $\sqrt{72} - 2\sqrt{8} + \sqrt{2} = \sqrt{36 \cdot 2} - 2\sqrt{4 \cdot 2} + \sqrt{2} = 6\sqrt{2} - 4\sqrt{2} + \sqrt{2} = (6 - 4 + 1)\sqrt{2} = 3\sqrt{2}$

13  $4\sqrt{20} - 2\sqrt{45} + \sqrt{80} = 4\sqrt{4 \cdot 5} - 2\sqrt{9 \cdot 5} + \sqrt{16 \cdot 5} = 8\sqrt{5} - 6\sqrt{5} + 4\sqrt{5} = (8 - 6 + 4)\sqrt{5} = 6\sqrt{5}$

17  $\sqrt{75x} - \sqrt{3x} - \sqrt{12x} = \sqrt{25(3x)} - \sqrt{3x} - \sqrt{4(3x)} = 5\sqrt{3x} - \sqrt{3x} - 2\sqrt{3x} = (5 - 1 - 2)\sqrt{3x} = 2\sqrt{3x}$

21  $\sqrt{18p} + \sqrt{50p} - \sqrt{2p} = \sqrt{9(2p)} + \sqrt{25(2p)} - \sqrt{2p} = 3\sqrt{2p} + 5\sqrt{2p} - \sqrt{2p} = (3 + 5 - 1)\sqrt{2p} = 7\sqrt{2p}$

25  $\dfrac{3\sqrt[4]{m^9}}{m} - 5\sqrt[4]{m^5} + \sqrt[8]{m^{10}} = \dfrac{3\sqrt[4]{m^8 \cdot m}}{m} - 5\sqrt[4]{m^4 \cdot m} + \sqrt[4 \cdot 2]{m^{5 \cdot 2}}$

$= \dfrac{3m^2\sqrt[4]{m}}{m} - 5m\sqrt[4]{m} + \sqrt[4]{m^5}$

$= 3m\sqrt[4]{m} - 5m\sqrt[4]{m} + m\sqrt[4]{m}$
$= (3m - 5m + m)\sqrt[4]{m}$
$= -m\sqrt[4]{m}$

29  $2y\sqrt{y} - 7\sqrt{y^3} + \dfrac{1}{7y}\sqrt{4y^3} = 2y\sqrt{y} - 7\sqrt{y^2 \cdot y} + \dfrac{\sqrt{4y^2 \cdot y}}{7y}$

$= 2y\sqrt{y} - 7y\sqrt{y} + \dfrac{2y\sqrt{y}}{7y}$

$= \left(2y - 7y + \dfrac{2}{7}\right)\sqrt{y}$

$= \left(-5y + \dfrac{2}{7}\right)\sqrt{y}$

33  $5p\sqrt[3]{p^4q} - 7q\sqrt[3]{-pq^4} + 5\sqrt[3]{p^7q} = 5p\sqrt[3]{p^3 \cdot pq} - 7q\sqrt[3]{-q^3 \cdot pq} + 5\sqrt[3]{p^6 \cdot pq}$
$= 5p \cdot p\sqrt[3]{pq} - 7q(-q)\sqrt[3]{pq} + 5p^2\sqrt[3]{pq}$
$= 5p^2\sqrt[3]{pq} + 7q^2\sqrt[3]{pq} + 5p^2\sqrt[3]{pq}$
$= (5p^2 + 7q^2 + 5p^2)\sqrt[3]{pq}$
$= (10p^2 + 7q^2)\sqrt[3]{pq}$

37  $7\sqrt[3]{m^5} + 3m\sqrt[3]{m^2} + 2m\sqrt[3]{8m^2} = 7\sqrt[3]{m^3 \cdot m^2} + 3m\sqrt[3]{m^2} + 2m\sqrt[3]{8 \cdot m^2}$
$= 7m\sqrt[3]{m^2} + 3m\sqrt[3]{m^2} + 4m\sqrt[3]{m^2}$
$= (7m + 3m + 4m)\sqrt[3]{m^2} = 14m\sqrt[3]{m^2}$

41  $\sqrt{9x^2(x-2y)} - \sqrt{36y^2(x-2y)} + 2\sqrt{(x-2y)^3} = 3|x|\sqrt{x-2y} - 6|y|\sqrt{x-2y} + 2(x-2y)\sqrt{x-2y}$
$= (3|x| - 6|y| + 2x - 4y)\sqrt{x-2y}$

## 5.5 Multiplication and Division of Radical Expressions

### SEMIPROGRAMMED PROBLEMS

In problems 1–8, find the products and express the results in simplest form. Assume that all variables are restricted to values for which all radicals are defined.

$\sqrt{2}, 2$

$\sqrt{x}$

$x^4, x^2$

$2\sqrt{2}$

$2\sqrt{2}$

$2$

$3\sqrt{2}$

$\sqrt{2}$

$6$

$5 + 2\sqrt{6}$

$2$

$1$

$-1 + \sqrt{3}$

$(\sqrt{y})^2$

$y$

$2\sqrt{x}$

$-2$

$\sqrt{x}$

$\sqrt{x}\sqrt{3}$

$\sqrt{3x}$

$1, 5\sqrt{3x}$

1  $\sqrt{2}(3 - \sqrt{2}) = 3\sqrt{2} - (\underline{\phantom{xx}})\sqrt{2} = 3\sqrt{2} - \underline{\phantom{xx}}$

2  $\sqrt{x}(\sqrt{x^3} - \sqrt{x}) = \sqrt{x}\sqrt{x^3} - (\underline{\phantom{xx}})\sqrt{x}$
$= \sqrt{\underline{\phantom{xx}}} - x = \underline{\phantom{xx}} - x$

3  $(\sqrt{2} + 1)(2\sqrt{2} + 1) = 2\sqrt{2}\sqrt{2} + \sqrt{2} + \underline{\phantom{xx}} + 1$
$= 4 + \sqrt{2} + \underline{\phantom{xx}} + 1$
$= 5 + (1 + \underline{\phantom{xx}})\sqrt{2}$
$= 5 + \underline{\phantom{xx}}$

4  $(\sqrt{3} + \sqrt{2})^2 = (\sqrt{3})^2 + 2\sqrt{3} \cdot \sqrt{2} + (\underline{\phantom{xx}})^2$
$= 3 + 2\sqrt{\underline{\phantom{xx}}} + 2$
$= \underline{\phantom{xx}}$

5  $(2 - \sqrt{3})(1 + \sqrt{3}) = 2 + (\underline{\phantom{xx}})\sqrt{3} - \sqrt{3} - \sqrt{3}\sqrt{3}$
$= 2 + (2 - \underline{\phantom{xx}})\sqrt{3} - 3$
$= \underline{\phantom{xx}}$

6  $(\sqrt{x} - \sqrt{y})(\sqrt{x} + \sqrt{y}) = (\sqrt{x})^2 - \underline{\phantom{xx}}$
$= x - \underline{\phantom{xx}}$

7  $(2 + \sqrt{x})(3 - \sqrt{x}) = 2 \cdot 3 - \underline{\phantom{xx}} + 3\sqrt{x} - \sqrt{x}\sqrt{x}$
$= 6 + (\underline{\phantom{xx}} + 3)\sqrt{x} - x$
$= 6 + \underline{\phantom{xx}} - x$

8  $(2\sqrt{3} + \sqrt{x})(\sqrt{3} + 2\sqrt{x})$
$= 2\sqrt{3}\sqrt{3} + 2\sqrt{3} \cdot 2\sqrt{x} + \underline{\phantom{xx}} + 2\sqrt{x}\sqrt{x}$
$= 2 \cdot 3 + 4\sqrt{3x} + \underline{\phantom{xx}} + 2x$
$= 6 + (4 + \underline{\phantom{xx}})\sqrt{3x} + 2x = 6 + \underline{\phantom{xx}} + 2x$

In problems 9–16, rationalize the denominator of the given fractions. Assume that all variables are restricted to values for which all radicals are defined.

$\sqrt{2}, \dfrac{3\sqrt{2}}{2}$

9  $\dfrac{3}{\sqrt{2}} = \dfrac{3(\underline{\phantom{xx}})}{\sqrt{2}\sqrt{2}} = \underline{\phantom{xx}}$

$\sqrt{3}, \sqrt{3}\sqrt{3}, \dfrac{\sqrt{3x}}{3}$

$\sqrt[3]{5}, \sqrt[3]{5}, \sqrt[3]{25}$

$\sqrt{2}+1$

$\sqrt{2}$

$2\sqrt{2}+2$

$2\sqrt{2}+2$

$\sqrt{5}+\sqrt{3}$

$4(\sqrt{5}+\sqrt{3})$

$2\sqrt{5}+2\sqrt{3}$

$2\sqrt{5}+2\sqrt{3}$

$\sqrt{x}-1$

$x, \dfrac{x-\sqrt{x}}{x-1}$

$\sqrt{3}-1$

$1$

$3-2\sqrt{3}+1, 2-\sqrt{3}$

$3-\sqrt{x}$

$3\sqrt{x}$

$5\sqrt{x}$

10. $\sqrt{\dfrac{x}{3}} = \dfrac{\sqrt{x}}{\underline{\phantom{xx}}} = \dfrac{\sqrt{x}\sqrt{3}}{\underline{\phantom{xx}}} = \underline{\phantom{xxx}}$

11. $\dfrac{7}{\sqrt[3]{5}} = \dfrac{7(\underline{\phantom{x}})(\underline{\phantom{x}})}{\sqrt[3]{5}\cdot\sqrt[3]{5}\cdot\sqrt[3]{5}} = \dfrac{7(\underline{\phantom{xx}})}{5}$

12. $\dfrac{2}{\sqrt{2}-1} = \dfrac{2(\underline{\phantom{xx}})}{(\sqrt{2}-1)(\sqrt{2}+1)}$

  $= \dfrac{2(\underline{\phantom{x}})+2}{(\sqrt{2})^2 - 1^2}$

  $= \dfrac{\underline{\phantom{xxxx}}}{2-1}$

  $= \underline{\phantom{xxxx}}$

13. $\dfrac{4}{\sqrt{5}-\sqrt{3}} = \dfrac{4(\underline{\phantom{xxx}})}{(\sqrt{5}-\sqrt{3})(\sqrt{5}+\sqrt{3})}$

  $= \dfrac{\underline{\phantom{xxxx}}}{(\sqrt{5})^2-(\sqrt{3})^2}$

  $= \dfrac{4\sqrt{5}+4\sqrt{3}}{5-3} = \dfrac{2(\underline{\phantom{xxx}})}{2}$

  $= \underline{\phantom{xxxx}}$

14. $\dfrac{\sqrt{x}}{\sqrt{x}+1} = \dfrac{\sqrt{x}(\underline{\phantom{xx}})}{(\sqrt{x}+1)(\sqrt{x}-1)}$

  $= \dfrac{\underline{\phantom{x}}-\sqrt{x}}{(\sqrt{x})^2-1^2} = \underline{\phantom{xxx}}$

15. $\dfrac{\sqrt{3}-1}{\sqrt{3}+1} = \dfrac{(\sqrt{3}-1)(\underline{\phantom{xx}})}{(\sqrt{3}+1)(\sqrt{3}-1)}$

  $= \dfrac{\sqrt{3}\sqrt{3}-\sqrt{3}-\sqrt{3}+\underline{\phantom{x}}}{3-1}$

  $= \dfrac{\underline{\phantom{xxxx}}}{2} = \underline{\phantom{xxx}}$

16. $\dfrac{2-\sqrt{x}}{3+\sqrt{x}} = \dfrac{(2-\sqrt{x})(\underline{\phantom{xx}})}{(3+\sqrt{x})(3-\sqrt{x})}$

  $= \dfrac{6-2\sqrt{x}-\underline{\phantom{xx}}+\sqrt{x}\sqrt{x}}{3^2-(\sqrt{x})^2}$

  $= \dfrac{6-\underline{\phantom{xx}}+x}{9-x}$

## SOLUTIONS TO SELECTED ODD PROBLEMS   Section 5.5, text pages 259–260

1. $\sqrt{12}\cdot\sqrt{3} = \sqrt{36} = 6$

5. $(2\sqrt{6})(3\sqrt{7}) = 6\sqrt{42}$

9. $(7\sqrt{3})(-11\sqrt{6}) = -77\sqrt{18} = -77\sqrt{9}\sqrt{2} = -77(3)\sqrt{2} = -231\sqrt{2}$

13. $\sqrt[3]{10}\cdot\sqrt[3]{75} = \sqrt[3]{750} = \sqrt[3]{125}\sqrt[3]{6} = 5\sqrt[3]{6}$

17. $\sqrt{x+1}\cdot\sqrt{x+3} = \sqrt{(x+1)(x+3)} = \sqrt{x^2+4x+3}$

21. $\sqrt{2}(2-\sqrt{2}) = 2\sqrt{2}-(\sqrt{2})^2 = 2\sqrt{2}-2$

25. $\sqrt{3}(\sqrt{5x}-\sqrt{10y}) = \sqrt{3}\sqrt{5x}-\sqrt{3}\cdot\sqrt{10y} = \sqrt{15x}-\sqrt{30y}$

29. $\sqrt{11}(2\sqrt{3}-4\sqrt{11}) = 2\sqrt{11}\sqrt{3}-4(\sqrt{11})^2 = 2\sqrt{33}-44$

33 $(\sqrt{3} - \sqrt{2})(2\sqrt{3} + \sqrt{2}) = 2\sqrt{3}\sqrt{3} - 2\sqrt{3}\sqrt{2} + \sqrt{2}\sqrt{3} - \sqrt{2}\sqrt{2}$
$= 6 - 2\sqrt{6} + \sqrt{6} - 2 = 4 - \sqrt{6}$

37 $(4\sqrt{2} - \sqrt{3})(5\sqrt{2} + 2\sqrt{3}) = 20\sqrt{2}\sqrt{2} - 5\sqrt{2}\sqrt{3} + 8\sqrt{2}\sqrt{3} - 2\sqrt{3}\sqrt{3}$
$= 40 - 5\sqrt{6} + 8\sqrt{6} - 6 = 34 + 3\sqrt{6}$

41 Using $(a - b)^2 = a^2 - 2ab + b^2$, $(\sqrt{5} - \sqrt{3})^2 = (\sqrt{5})^2 - 2\sqrt{5}\sqrt{3} + (\sqrt{3})^2 = 5 - 2\sqrt{15} + 3 = 8 - 2\sqrt{15}$

45 Using $(a - b)^2 = a^2 - 2ab + b^2$, $(2\sqrt{x} - \sqrt{y})^2 = (2\sqrt{x})^2 - 2(2)\sqrt{x}\sqrt{y} + (\sqrt{y})^2 = 4x - 4\sqrt{xy} + y$

49 Using $(a - b)(a + b) = a^2 - b^2$, $(2\sqrt{3} - 1)(2\sqrt{3} + 1) = (2\sqrt{3})^2 - (1)^2 = 12 - 1 = 11$

53 Using $(a - b)(a + b) = a^2 - b^2$, $(3\sqrt{x} - 11)(3\sqrt{x} + 11) = (3\sqrt{x})^2 - (11)^2 = 9x - 121$

57 $\dfrac{2}{\sqrt{3}} = \dfrac{2 \cdot \sqrt{3}}{\sqrt{3} \cdot \sqrt{3}} = \dfrac{2\sqrt{3}}{3}$

61 $\sqrt{\dfrac{1}{7y}} = \dfrac{1}{\sqrt{7y}} = \dfrac{1 \cdot \sqrt{7y}}{\sqrt{7y} \cdot \sqrt{7y}} = \dfrac{\sqrt{7y}}{7y}$

65 $\dfrac{5}{\sqrt[3]{9x}} = \dfrac{5 \cdot \sqrt[3]{3x^2}}{\sqrt[3]{9x} \cdot \sqrt[3]{3x^2}} = \dfrac{5\sqrt[3]{3x^2}}{\sqrt[3]{27x^3}} = \dfrac{5\sqrt[3]{3x^2}}{3x}$

69 $\dfrac{7}{\sqrt[6]{2}} = \dfrac{7(\sqrt[6]{2})^5}{\sqrt[6]{2}(\sqrt[6]{2})^5} = \dfrac{7\sqrt[6]{2^5}}{(\sqrt[6]{2})^6} = \dfrac{7\sqrt[6]{32}}{2}$

73 $\dfrac{3}{\sqrt{5} + \sqrt{2}} = \dfrac{3(\sqrt{5} - \sqrt{2})}{(\sqrt{5} + \sqrt{2})(\sqrt{5} - \sqrt{2})} = \dfrac{3(\sqrt{5} - \sqrt{2})}{(\sqrt{5})^2 - (\sqrt{2})^2} = \dfrac{3(\sqrt{5} - \sqrt{2})}{5 - 2} = \dfrac{3(\sqrt{5} - \sqrt{2})}{3} = \sqrt{5} - \sqrt{2}$

77 $\dfrac{\sqrt{2}}{1 + \sqrt{2}} = \dfrac{\sqrt{2}}{1 + \sqrt{2}} \cdot \dfrac{1 - \sqrt{2}}{1 - \sqrt{2}} = \dfrac{\sqrt{2}(1 - \sqrt{2})}{1 - 2} = \dfrac{\sqrt{2} - 2}{-1} = 2 - \sqrt{2}$

81 $\dfrac{\sqrt{y}}{3\sqrt{x} - 2\sqrt{y}} = \dfrac{\sqrt{y}}{(3\sqrt{x} - 2\sqrt{y})} \cdot \dfrac{(3\sqrt{x} + 2\sqrt{y})}{(3\sqrt{x} + 2\sqrt{y})} = \dfrac{\sqrt{y}(3\sqrt{x} + 2\sqrt{y})}{9x - 4y} = \dfrac{3\sqrt{xy} + 2y}{9x - 4y}$

85 $\dfrac{4\sqrt{2} + 3\sqrt{5}}{7\sqrt{5} - 3\sqrt{2}} = \dfrac{(4\sqrt{2} + 3\sqrt{5})}{(7\sqrt{5} - 3\sqrt{2})} \cdot \dfrac{(7\sqrt{5} + 3\sqrt{2})}{(7\sqrt{5} + 3\sqrt{2})} = \dfrac{28\sqrt{10} + 12(\sqrt{2})^2 + 21(\sqrt{5})^2 + 9\sqrt{10}}{(7\sqrt{5})^2 - (3\sqrt{2})^2}$
$= \dfrac{28\sqrt{10} + 24 + 105 + 9\sqrt{10}}{245 - 18} = \dfrac{37\sqrt{10} + 129}{227}$

89 Set $a = \sqrt[3]{5}$ $b = \sqrt[3]{2}$, then $a^3 - b^3 = (\sqrt[3]{5})^3 - (\sqrt[3]{2})^3 = 5 - 2 = 3$

93 Evaluating $3x^2 + 12x - 5$ for $x = \dfrac{-6 - \sqrt{51}}{3}$, we have

$3\left(\dfrac{-6 - \sqrt{51}}{3}\right)^2 + 12\left(\dfrac{-6 - \sqrt{51}}{3}\right) - 5 = 3\left(\dfrac{87 + 12\sqrt{51}}{9}\right) + 4(-6 - \sqrt{51}) - 5$
$= \dfrac{87 + 12\sqrt{51}}{3} - 24 - 4\sqrt{51} - 5$
$= 29 + 4\sqrt{51} - 24 - 4\sqrt{51} - 5 = 0$

97 $\sqrt{\dfrac{5}{3}} - \dfrac{15}{\sqrt{15}} + \dfrac{7\sqrt{15}}{3} = \dfrac{\sqrt{5}}{\sqrt{3}} - \dfrac{15}{\sqrt{15}} + \dfrac{7\sqrt{15}}{3} = \dfrac{\sqrt{5}\sqrt{3}}{\sqrt{3}\sqrt{3}} - \dfrac{15\sqrt{15}}{\sqrt{15}\sqrt{15}} + \dfrac{7}{3}\sqrt{15}$
$= \dfrac{\sqrt{15}}{3} - \dfrac{15\sqrt{15}}{15} + \dfrac{7}{3}\sqrt{15} = \dfrac{\sqrt{15}}{3} - \sqrt{15} + \dfrac{7}{3}\sqrt{15} = \dfrac{5}{3}\sqrt{15} = \dfrac{5\sqrt{15}}{3}$

## 5.6 Complex Numbers

### SEMIPROGRAMMED PROBLEMS

In problems 1 and 2, write each expression in the form $a + bi$, where $a$ and $b$ are real numbers.

$-1, -1, 7i$

1 $\sqrt{-49}$
$\sqrt{-49} = \sqrt{49(\underline{\phantom{xx}})} = \sqrt{49}\sqrt{\underline{\phantom{xx}}} = \underline{\phantom{xxx}}$

## 5.6 SEMIPROGRAMMED PROBLEMS

**2**  $3 - \sqrt{-16}$

−1

−1, 4i

$3 - \sqrt{-16} = 3 - \sqrt{16(\underline{\phantom{xx}})}$
$= 3 - \sqrt{16}\sqrt{\underline{\phantom{xx}}} = 3 - \underline{\phantom{xx}}$

In Problems 3–13, perform the indicated operations on the complex numbers.

7 − 4

22 + 3i

1 − 3i

6 − 4

20 + 2i

−5 + 4

−1 − i

4 − 7

−4 − 3i

−7 − 8

−15 − 9i

$i^2$

−1

27

3i

$9i^2$

−9, 25

3 − 2i, 2

13

49, 53

$\sqrt{3}$, 4

**3**  $(13 + 7i) + (9 - 4i) = (13 + 9) + (\underline{\phantom{xx}})i$
$= \underline{\phantom{xx}}$

**4**  $(1 + i) + (-4i) = 1 + (1 - 4)i = \underline{\phantom{xx}}$

**5**  $(12 + 6i) + (8 - 4i) = 20 + (\underline{\phantom{xx}})i$
$= \underline{\phantom{xx}}$

**6**  $(4 - 5i) - (5 - 4i) = (4 - 5) + (\underline{\phantom{xx}})i$
$= \underline{\phantom{xx}}$

**7**  $(3 + 4i) - (7 + 7i) = (3 - 7) + (\underline{\phantom{xx}})i$
$= \underline{\phantom{xx}}$

**8**  $(-7 - 6i) - (8 + 3i) = (\underline{\phantom{xx}}) + (-6 - 3)i$
$= \underline{\phantom{xx}}$

**9**  $(5 + 6i)(9 + 3i) = 45 + 15i + 54i + 18(\underline{\phantom{xx}})$
$= 45 + 69i + 18(\underline{\phantom{xx}})$
$= \underline{\phantom{xx}} + 69i$

**10**  $(4 - 3i)(4 + 3i) = 4^2 - (\underline{\phantom{xx}})^2$
$= 16 - (\underline{\phantom{xx}})$
$= 16 - (\underline{\phantom{xx}}) = \underline{\phantom{xx}}$

**11**  If $z = 3 + 2i$, then $\bar{z} = \underline{\phantom{xx}}$ and $z\bar{z} = (3)^2 + (\underline{\phantom{xx}})^2$
$= \underline{\phantom{xx}}$.

**12**  $(7 - 2i)(7 + 2i) = \underline{\phantom{xx}} + 4 = \underline{\phantom{xx}}$

**13**  $(\sqrt{3} + i)(\sqrt{3} - i) = (\underline{\phantom{xx}})^2 + 1 = \underline{\phantom{xx}}$

In problems 14 and 15, write each fraction in the form $a + bi$.

3 − 4i

9i

17i, −6 − 17i

$-\frac{6}{25}, -\frac{17}{25}$

2 − 2i

$2^2$

4

8, $\frac{5}{4}$, $\frac{1}{4}$

**14**  $\dfrac{2 - 3i}{3 + 4i} = \dfrac{(2 - 3i)(\underline{\phantom{xx}})}{(3 + 4i)(3 - 4i)}$

$= \dfrac{6 - 8i - (\underline{\phantom{xx}}) + 12i^2}{3^2 + 4^2}$

$= \dfrac{6 - (\underline{\phantom{xx}}) - 12}{9 + 16} = \dfrac{\underline{\phantom{xx}}}{25}$

$= \underline{\phantom{xx}} + \underline{\phantom{xx}} i$

**15**  $\dfrac{2 + 3i}{2 + 2i} = \dfrac{(2 + 3i)(2 - 2i)}{(2 + 2i)(\underline{\phantom{xx}})}$

$= \dfrac{4 - 4i + 6i - 6i^2}{2^2 + \underline{\phantom{xx}}}$

$= \dfrac{4 + 2i + 6}{4 + \underline{\phantom{xx}}}$

$= \dfrac{10 + 2i}{\underline{\phantom{xx}}} = \underline{\phantom{xx}} + \underline{\phantom{xx}} i$

# 100 CHAPTER 5 EXPONENTS, RADICALS, AND COMPLEX NUMBERS

In problems 16 and 17, write each expression as $i$, $-1$, $-i$, or 1.

$(i^4)^6$, 1, $-1$

**16** $i^{26}$

$i^{26} = i^{24+2} = i^{24} \cdot i^2 = (\underline{\phantom{xx}})i^2 = (\underline{\phantom{xx}})i^2 = \underline{\phantom{xx}}$

$i^3, (i^4)^{13}, 1, -i$

**17** $i^{55}$

$i^{55} = i^{52+3} = i^{52} \cdot (\underline{\phantom{xx}}) = (\underline{\phantom{xx}})i^3 = (\underline{\phantom{xx}})i^3 = \underline{\phantom{xx}}$

## SOLUTIONS TO SELECTED ODD PROBLEMS  Section 5.6, text pages 267–268

**1** $\sqrt{-81} = \sqrt{81}i = 9i$

**5** $-\sqrt{-x^4} = -\sqrt{x^4}i = -x^2 i$

**9** $5 + \sqrt{-81} = 5 + \sqrt{81}i = 5 + 9i$

**13** $\sqrt{-4} \cdot \sqrt{-9} = 2i \cdot 3i = 6i^2 = -6$

**17** $4x + 8i = 20 + 2yi$
$4x = 20$ and $8 = 2y$
$x = 5$ and $4 = y$

**21** $(-5 + 3i) + (5i - 1) = (-5 - 1) + (3 + 5)i = -6 + 8i$

**25** $(-2 - 3i) - (-3 - 2i) = -2 - 3i + 3 + 2i = (-2 + 3) + (-3 + 2)i = 1 - i$

**29** $(6 - 8i) - (5 + 3i) = 6 - 8i - 5 - 3i = (6 - 5) + (-8 - 3)i = 1 - 11i$

**33** $4i(4 - 5i) = 4i(4) - 4i(5i)$
$= 16i - 20i^2$
$= 16i - 20(-1) = 20 + 16i$

**37** $(2 - 4i)(3 + 2i) = 6 + 4i - 12i - 8i^2$
$= 6 - 8i - 8(-1) = 14 - 8i$

**41** $(-3 - 2i)(2 + 5i) = -6 - 15i - 4i - 10i^2$
$= -6 - 19i - 10(-1) = 4 - 19i$

**45** $(3 + 2i)^2 = 3^2 + 2(3)(2i) + (2i)^2$
$= 9 + 12i + 4i^2$
$= 9 + 12i + 4(-1) = 5 + 12i$

**49** $(5 - 4i)(5 + 4i) = 5^2 - (4i)^2$
$= 25 - 16i^2$
$= 25 - 16(-1) = 41$

**53** $(2 - 7i)(2 + 7i) = 2^2 - (7i)^2$
$= 4 - 49i^2$
$= 4 - 49(-1) = 53$

**57** **(a)** $\overline{2i} = -2i$   **(b)** $(2i)(-2i) = -4i^2 = -4(-1) = 4$

**61** $\dfrac{7}{5 - 4i} = \dfrac{7}{5 - 4i} \cdot \dfrac{5 + 4i}{5 + 4i} = \dfrac{7(5 + 4i)}{5^2 + 4^2} = \dfrac{35 + 28i}{41} = \dfrac{35}{41} + \dfrac{28}{41}i$

**65** $\dfrac{3 + 4i}{-2 + 5i} = \dfrac{3 + 4i}{-2 + 5i} \cdot \dfrac{-2 - 5i}{-2 - 5i} = \dfrac{-6 - 15i - 8i - 20i^2}{(-2)^2 + 5^2} = \dfrac{-6 - 23i - 20(-1)}{4 + 25} = \dfrac{14 - 23i}{29} = \dfrac{14}{29} - \dfrac{23}{29}i$

**69** $\dfrac{i}{-1 - i} = \dfrac{i}{-1 - i} \cdot \dfrac{-1 + i}{-1 + i} = \dfrac{i(-1 + i)}{(-1)^2 + 1^2} = \dfrac{-i + i^2}{2} = \dfrac{-1 - i}{2} = -\dfrac{1}{2} - \dfrac{1}{2}i$

**73** $\dfrac{3 - 2i}{5 - 2i} = \dfrac{3 - 2i}{5 - 2i} \cdot \dfrac{5 + 2i}{5 + 2i} = \dfrac{15 + 6i - 10i - 4i^2}{5^2 + 2^2} = \dfrac{15 - 4i - 4(-1)}{25 + 4} = \dfrac{19 - 4i}{29} = \dfrac{19}{29} - \dfrac{4}{29}i$

**77** $i^{29} = i^{28} \cdot i = (i^4)^7 \cdot i = 1^7 \cdot i = i$

**81** $i^{65} = i^{64} \cdot i = (i^4)^{16} \cdot i = 1^{16} \cdot i = i$

**85** $i^{-7} = \dfrac{1}{i^7} = \dfrac{1}{i^7} \cdot \dfrac{i}{i} = \dfrac{i}{i^8} = \dfrac{i}{(i^4)^2} = \dfrac{i}{1^2} = i$

**89** $2\left(\dfrac{-3 - \sqrt{7}i}{4}\right)^2 + 3\left(\dfrac{-3 - \sqrt{7}i}{4}\right) + 2 = 2\left(\dfrac{9 + 6\sqrt{7}i + 7i^2}{16}\right) + \dfrac{-9 - 3\sqrt{7}i}{4} + 2$

$= \dfrac{2 + 6\sqrt{7}i}{8} + \dfrac{-18 - 6\sqrt{7}i}{8} + \dfrac{16}{8} = \dfrac{0}{8} = 0$

## CUMULATIVE REVIEW PROBLEM SET  Chapters 1–5

**1** Find the value of $12 + 3 \cdot 2^4 \div 2^2 - 5$ by using the rule for the order of operations.

**2** Find the value of $\dfrac{24}{y - x}$ for $x = -2$ and $y = 4$.

**3** Factor completely.
   (a) $9x^3y - 81xy^3$
   (b) $2x^3yz^2 + 7x^2y^2z^2 - 15xy^3z^2$

**4** Perform each operation.
   (a) $(3x^2 - 2x + 7) + (x^2 + 4x - 11) - (2x^2 - 3x + 4)$
   (b) $(2u + v)(u^2 - uv + 3v^2)$
   (c) $(x^3 - x^2 + x + 14) \div (x + 2)$

**5** Perform the indicated operations and simplify the result.
   (a) $\dfrac{x^2 - x - 2}{x^2 - x - 6} \cdot \dfrac{2x + x^2}{x^2 - 2x}$
   (b) $\dfrac{u^2 + 5u + 6}{u^2 + u - 2} \div \dfrac{u^2 + 7u + 12}{u^2 + 3u - 4}$
   (c) $\dfrac{4}{2y^2 + y - 3} - \dfrac{1}{2y^2 + 5y + 3} + \dfrac{2}{y^2 - 1}$

**6** Simplify $\dfrac{\dfrac{1}{x} - \dfrac{1}{y}}{\dfrac{x}{y} - \dfrac{y}{x}}$.

**7** Solve each equation.
   (a) $5(x + 2) - 3(x - 1) = 4(x + 1) + 3$
   (b) $\dfrac{z + 4}{3} - \dfrac{z - 2}{5} = \dfrac{z + 2}{2}$
   (c) $|3(x - 2)| = 15$

**8** Solve each inequality.
   (a) $1 - 3x \le 16$
   (b) $|2x - 1| > 7$

**9** Graph each solution set on a number line.
   (a) $\{x \mid x > -2\} \cap \{x \mid x \le 5\}$
   (b) $\{x \mid x \le 3\} \cup \{x \mid x \ge 8\}$

**10** Maureen is 5 years younger than Tom, and the sum of their ages in 3 years is 63. What are their ages now?

**11** Rewrite each expression so that it contains only positive exponents and simplify.
   (a) $3x^{-2}$
   (b) $(2x^{-3})^{-4}$
   (c) $\left(\dfrac{x^{-4}y^{-2}}{x^{-5}y^{-1}}\right)^{-3}$
   (d) $\left(\dfrac{-m^{10}n^{15}}{32p^{25}}\right)^{-2/5}$

**12** Find the principal root.
   (a) $\sqrt{49}$
   (b) $\sqrt[3]{-216}$

**13** Use properties of radicals to simplify.
   (a) $\sqrt{50x^3y^7}$
   (b) $\sqrt{\dfrac{16u^3}{121w^4}}$

**14** Perform the indicated operations and simplify the result.
   (a) $\sqrt{45} + \sqrt{20}$
   (b) $(\sqrt{x} + \sqrt{y})(2\sqrt{x} - \sqrt{y})$

**15**

**16** Rationalize the denominator of each expression and simplify the result.
   (a) $\dfrac{4}{\sqrt{2}}$
   (b) $\dfrac{\sqrt{x} + 1}{\sqrt{x} - 1}$

**17** Write each expression in the form $a + bi$, where $a$ and $b$ are real numbers.
   (a) $2 - 3\sqrt{-49}$
   (b) $(3 - 4i) + (7 + 2i)$
   (c) $(-5 + 3i) - (-2 + 6i)$
   (d) $(3 + 4i)(2 - 5i)$

**18** Find the conjugate of $-3 + 4i$.

**19** Perform the division and express the answer in the form $a + bi$.
   $\dfrac{2 + i}{3 + 2i}$

**20** Simplify $i^{83}$.

## Answers

1. 19  2. 4  3(a) $9xy(x-3y)(x+3y)$  (b) $xyz^2(2x-3y)(x+5y)$  4(a) $2x^2+5x-8$
(b) $2u^3-u^2v+5uv^2+3v^3$  (c) $x^2-3x+7$  5(a) $\dfrac{x+1}{x-3}$  (b) 1  (c) $\dfrac{7y+11}{(2y+3)(y-1)(y+1)}$
6. $-\dfrac{1}{x+y}$  7(a) 3  (b) 2  (c) $-3, 7$  8(a) $x \geq -5$  (b) $x < -3$ or $x > 4$

9(a) ⟵─(─────]⟶
       $-2$    $5$

(b) ⟵───]    [───⟶
          $3$   $8$

10. 26, 31  11(a) $\dfrac{3}{x^2}$  (b) $\dfrac{x^{12}}{16}$  (c) $\dfrac{y^3}{x^3}$  (d) $\dfrac{4p^{10}}{m^4n^6}$

12(a) 7  (b) $-6$  13(a) $5xy^3\sqrt{2xy}$  (b) $\dfrac{4u\sqrt{u}}{11w^2}$  15(a) $5\sqrt{5}$  (b) $2x+\sqrt{xy}-y$

16(a) $2\sqrt{2}$  (b) $\dfrac{x+2\sqrt{x}+1}{x-1}$  17(a) $2-21i$  (b) $10-2i$  (c) $-3-3i$

(d) $26-7i$  18. $-3-4i$  19. $\dfrac{8}{13}-\dfrac{1}{13}i$  20. $-i$


## CHAPTER 5  TESTS

### Chapter Test A

1. Write each expression with positive exponents and simplify.
   (a) $x^{-2}y^{-3}$
   (b) $5x^{-6}$
   (c) $\left(\tfrac{3}{4}\right)^{-4}$
   (d) $(3x^{-1})^{-2}$
   (e) $\dfrac{1}{3^{-2}}+\dfrac{1}{3^0}$
   (f) $3(x+y)^{-1}$
   (g) $\dfrac{1}{x^{-2}+y^{-2}}$
   (h) $\dfrac{a^{-2}}{b^{-3}}$
   (i) $\left(\dfrac{x^3y^{-2}}{x^{-2}y}\right)^{-2}$

2. Simplify and write the answer in scientific notation.
   $$\dfrac{(8 \times 10^4) \cdot (3 \times 10^{-5})}{(2 \times 10^{-2}) \cdot (4 \times 10^4)}$$

3. Use the properties of rational exponents to simplify each expression. Assume that all bases are positive.
   (a) $5^{-1/2} \cdot 5^{5/2}$
   (b) $(3^{-1/3})^{-15}$
   (c) $(8x^3)^{1/3}$
   (d) $\left(\tfrac{9}{4}\right)^{-3/2}$
   (e) $\dfrac{64^{2/3}}{64^{-1/2}}$
   (f) $\left(\dfrac{8x^3}{27y^6}\right)^{-1/3}$

4. Use the properties of radicals to perform each operation and simplify.
   (a) $\sqrt{3} \cdot \sqrt{27}$
   (b) $\sqrt[3]{\dfrac{x^6}{8}}$
   (c) $\sqrt[5]{x^{10}}$
   (d) $\sqrt[3]{\sqrt{64}}$
   (e) $\sqrt[3]{-64x^8y^{10}}$
   (f) $\sqrt[4]{\dfrac{5y^5}{16x^8}}$

5. Perform the indicated operations and simplify the results.
   (a) $2\sqrt{50x}+50\sqrt{2x}$
   (b) $\sqrt[3]{250}-\sqrt[3]{16}+2\sqrt[3]{54}$
   (c) $(\sqrt{2}+1)(2\sqrt{2}+1)$
   (d) $(\sqrt{x}-3\sqrt{y})^2$
   (e) $(5\sqrt{x}-3)(5\sqrt{x}+3)$

**6** Rationalize each denominator.

(a) $\dfrac{5}{\sqrt{7}}$  (b) $\dfrac{\sqrt{3}-\sqrt{2}}{\sqrt{3}+\sqrt{2}}$  (c) $\dfrac{x}{\sqrt{x}+y}$

**7** Express each term in the form $a + bi$.

(a) $-2 + \sqrt{-9}$  (b) $(3 - 2i) + (-2 + 5i)$  (c) $(7 - 3i) - (5 + i)$

(d) $(-2 + 3i)(1 - 2i)$  (e) $\dfrac{1 + 2i}{2 + i}$

---

## Solutions

**1** (a) $x^{-2}y^{-3} = \dfrac{1}{x^2} \cdot \dfrac{1}{y^3}$  (b) $5x^{-6} = 5 \cdot \dfrac{1}{x^6}$  (c) $\left(\tfrac{3}{4}\right)^{-4} = \left(\tfrac{4}{3}\right)^4$

$= \dfrac{1}{x^2 y^3}$  $= \dfrac{5}{x^6}$  $= \dfrac{256}{81}$

(d) $(3x^{-1})^{-2} = 3^{-2}(x^{-1})^{-2}$  (e) $\dfrac{1}{3^{-2}} + \dfrac{1}{3^0} = 3^2 + \dfrac{1}{1}$  (f) $3(x+y)^{-1} = 3\left(\dfrac{1}{x+y}\right)$

$= \dfrac{1}{3^2} \cdot x^2$  $= 9 + 1$  $= \dfrac{3}{x+y}$

$= \dfrac{x^2}{9}$  $= 10$

(g) $\dfrac{1}{x^{-2}+y^{-2}} = \dfrac{1}{\dfrac{1}{x^2}+\dfrac{1}{y^2}}$  (h) $\dfrac{a^{-2}}{b^{-3}} = \dfrac{b^3}{a^2}$  (i) $\left(\dfrac{x^3 y^{-2}}{x^{-2} y}\right)^{-2} = (x^5 y^{-3})^{-2}$

$= \dfrac{1}{\dfrac{y^2+x^2}{x^2 y^2}}$  $= (x^5)^{-2}(y^{-3})^{-2}$

$= x^{-10} y^6$

$= \dfrac{x^2 y^2}{y^2 + x^2}$  $= \dfrac{y^6}{x^{10}}$

**2** $\dfrac{(8 \times 10^4)\cdot(3 \times 10^{-5})}{(2 \times 10^{-2})\cdot(4 \times 10^4)} = \dfrac{(8 \times 3)\cdot(10^4 \times 10^{-5})}{(2 \times 4)\cdot(10^{-2} \times 10^4)} = \dfrac{24 \times 10^{-1}}{8 \times 10^2} = 3 \times 10^{-3}$

**3** (a) $5^{-1/2} \cdot 5^{5/2} = 5^{-1/2+5/2}$  (b) $(3^{-1/3})^{-15} = 3^{(-1/3)(-15)}$  (c) $(8x^3)^{1/3} = 8^{1/3}(x^3)^{1/3}$

$= 5^2$  $= 3^5$  $= 2x$

$= 25$  $= 243$

(d) $\left(\dfrac{9}{4}\right)^{-3/2} = \left(\dfrac{4}{9}\right)^{3/2}$  (e) $\dfrac{64^{2/3}}{64^{-1/2}} = 64^{2/3-(-1/2)}$  (f) $\left(\dfrac{8x^3}{27y^6}\right)^{-1/3} = \left(\dfrac{27y^6}{8x^3}\right)^{1/3}$

$= \dfrac{4^{3/2}}{9^{3/2}}$  $= 64^{7/6}$  $= \dfrac{27^{1/3}(y^6)^{1/3}}{8^{1/3}(x^3)^{1/3}}$

$= \dfrac{2^3}{3^3}$  $= 2^7$  $= \dfrac{3y^2}{2x}$

$= \dfrac{8}{27}$  $= 128$

**4** (a) $\sqrt{3} \cdot \sqrt{27} = \sqrt{3(27)}$  (b) $\sqrt[3]{\dfrac{x^6}{8}} = \dfrac{\sqrt[3]{x^6}}{\sqrt[3]{8}}$  (c) $\sqrt[5]{x^{10}} = \sqrt[5]{(x^2)^5}$

$= \sqrt{81}$  $= x^2$

$= 9$  $= \dfrac{x^2}{2}$

**104** CHAPTER 5 EXPONENTS, RADICALS, AND COMPLEX NUMBERS

(d) $\sqrt[3]{\sqrt{64}} = \sqrt[3 \times 2]{64}$
$= \sqrt[6]{64}$
$= 2$

(e) $\sqrt[3]{-64x^8y^{10}}$
$= \sqrt[3]{(-64x^6y^9)(x^2y)}$
$= \sqrt[3]{(-4x^2y^3)^3}\sqrt[3]{x^2y}$
$= -4x^2y^3\sqrt[3]{x^2y}$

(f) $\sqrt[4]{\dfrac{5y^5}{16x^8}} = \dfrac{\sqrt[4]{5y^5}}{\sqrt[4]{16x^8}}$
$= \dfrac{\sqrt[4]{y^4(5y)}}{\sqrt[4]{(2x^2)^4}}$
$= \dfrac{y\sqrt[4]{5y}}{2x^2}$

**5** (a) $2\sqrt{50x} + 50\sqrt{2x} = 2\sqrt{25(2x)} + 50\sqrt{2x}$
$= 2\sqrt{25}\sqrt{2x} + 50\sqrt{2x}$
$= 10\sqrt{2x} + 50\sqrt{2x}$
$= 60\sqrt{2x}$

(b) $\sqrt[3]{250} - \sqrt[3]{16} + 2\sqrt[3]{54} = \sqrt[3]{125(2)} - \sqrt[3]{8(2)} + 2\sqrt[3]{27(2)}$
$= \sqrt[3]{125}\sqrt[3]{2} - \sqrt[3]{8}\sqrt[3]{2} + 2\sqrt[3]{27}\sqrt[3]{2}$
$= [5 - 2 + 2(3)]\sqrt[3]{2}$
$= 9\sqrt[3]{2}$

(c) $(\sqrt{2} + 1)(2\sqrt{2} + 1) = 2(\sqrt{2})^2 + \sqrt{2} + 2\sqrt{2} + 1$
$= 2(2) + 3\sqrt{2} + 1$
$= 5 + 3\sqrt{2}$

(d) $(\sqrt{x} - 3\sqrt{y})^2 = (\sqrt{x})^2 - 2(\sqrt{x})(3\sqrt{y}) + (3\sqrt{y})^2$
$= x - 6\sqrt{xy} + 9y$

(e) $(5\sqrt{x} - 3)(5\sqrt{x} + 3) = (5\sqrt{x})^2 - 3^2$
$= 25x - 9$

**6** (a) $\dfrac{5}{\sqrt{7}} = \dfrac{5 \cdot \sqrt{7}}{\sqrt{7} \cdot \sqrt{7}}$
$= \dfrac{5\sqrt{7}}{7}$

(b) $\dfrac{\sqrt{3} - \sqrt{2}}{\sqrt{3} + \sqrt{2}} = \dfrac{(\sqrt{3} - \sqrt{2})}{(\sqrt{3} + \sqrt{2})} \cdot \dfrac{(\sqrt{3} - \sqrt{2})}{(\sqrt{3} - \sqrt{2})}$
$= \dfrac{(\sqrt{3})^2 - 2\sqrt{3} \cdot \sqrt{2} + (\sqrt{2})^2}{(\sqrt{3})^2 - (\sqrt{2})^2}$
$= \dfrac{3 - 2\sqrt{6} + 2}{3 - 2}$
$= 5 - 2\sqrt{6}$

(c) $\dfrac{x}{\sqrt{x} + y} = \dfrac{x}{(\sqrt{x} + y)} \cdot \dfrac{(\sqrt{x} - y)}{(\sqrt{x} - y)} = \dfrac{x(\sqrt{x} - y)}{(\sqrt{x})^2 - y^2} = \dfrac{x\sqrt{x} - xy}{x - y^2}$

**7** (a) $-2 + \sqrt{-9} = -2 + \sqrt{9(-1)} = -2 + \sqrt{9}i = -2 + 3i$

(b) $(3 - 2i) + (-2 + 5i) = [3 + (-2)] + [(-2) + 5]i = 1 + 3i$

(c) $(7 - 3i) - (5 + i) = (7 - 5) + (-3 - 1)i = 2 - 4i$

(d) $(-2 + 3i)(1 - 2i) = -2 + 7i - 6i^2 = -2 + 7i - 6(-1) = 4 + 7i$

(e) $\dfrac{1 + 2i}{2 + i} = \dfrac{(1 + 2i)}{(2 + i)} \cdot \dfrac{(2 - i)}{(2 - i)} = \dfrac{2 + 3i - 2i^2}{2^2 + 1^2} = \dfrac{4 + 3i}{5} = \dfrac{4}{5} + \dfrac{3}{5}i$

## Chapter Test B

*Multiple Choice:* Select the *one* correct answer for each of the following questions.

**1** The expression $(\tfrac{3}{5})^0$ is equal to _____.

   (a) 0        (b) $\tfrac{3}{5}$        (c) $\tfrac{5}{3}$        (d) 1

**2** The expression $\dfrac{5^4}{5^4}$ is equal to _____.

   (a) 20        (b) 1        (c) 0        (d) $\tfrac{20}{625}$

3  The expression $(x^2 + 5)^0$ is equal to _____.
   (a) 1     (b) 0     (c) $x^2 + 5$     (d) $\dfrac{1}{x^2 + 5}$

4  The expression $\dfrac{(1.2 \times 10^{-4})(10 \times 10^2)}{6 \times 10^3}$ is equal to _____.
   (a) $2 \times 10$     (b) $2 \times 10^{-5}$     (c) $2 \times 10^5$     (d) $2 \times 10^{-11}$

5  The expression $5^{-2}$ is equal to _____.
   (a) $5^2$     (b) $\tfrac{1}{5}$     (c) $\tfrac{1}{25}$     (d) $-25$

6  The expression $(\tfrac{3}{4})^{-1}$ is equal to _____.
   (a) $\tfrac{3}{4}$     (b) $-\tfrac{3}{4}$     (c) $-\tfrac{4}{3}$     (d) $\tfrac{4}{3}$

7  The principal square root of 49, denoted by $\sqrt{49}$, is _____.
   (a) $-7$     (b) 7     (c) $-\sqrt{7}$     (d) $\sqrt{7}$

8  The principal cube root of $-64$, denoted by $\sqrt[3]{-64}$, is _____.
   (a) 8     (b) undefined     (c) $-4$     (d) 4

9  The value of $(27)^{4/3}$ is equal to _____.
   (a) 81     (b) 243     (c) 16     (d) 27

10  The value of $\left(\dfrac{16x^4}{y^8}\right)^{-3/4}$ is equal to _____.
    (a) $\dfrac{y^6}{-8x^3}$     (b) $\dfrac{y^6}{16x^3}$     (c) $\dfrac{y^6}{6x^3}$     (d) $\dfrac{y^6}{8x^3}$

11  Performing $x^{4/3} \cdot x^{2/5}$ yields _____.
    (a) $x^{26/15}$     (b) $x^{8/15}$     (c) $x^{10/6}$     (d) $x^{10/8}$

12  Performing $x^{-2/3} \cdot x^{5/6}$ yields _____.
    (a) $x^{5/3}$     (b) $x^{1/6}$     (c) $x^{-5/3}$     (d) $x^{-5/4}$

13  $(x^{4/3})^{-3/4} =$ _____.
    (a) $x^{-3}$     (b) $x$     (c) $x^4$     (d) $x^{-1}$

14  The product $\sqrt{18} \cdot \sqrt{2}$ is equal to _____.
    (a) $\sqrt{2}$     (b) $\sqrt{18}$     (c) $\sqrt{6}$     (d) 6

15  The quotient $\dfrac{\sqrt{3x}}{\sqrt{12}}$ is equal to _____.
    (a) $\dfrac{\sqrt{x}}{2}$     (b) $\dfrac{x}{2}$     (c) $\dfrac{x}{4}$     (d) $\dfrac{x}{12}$

16  $\sqrt{250z^3y^{10}}$ simplifies to _____.
    (a) $5zy^5\sqrt{2z}$     (b) $5z^2y^5\sqrt{10z}$     (c) $5z^2y\sqrt{10zy^9}$     (d) $5zy^5\sqrt{10z}$

17  $\sqrt{\sqrt[3]{x^{12}}} =$ _____.
    (a) $x^2$     (b) $x$     (c) $\sqrt[5]{x^{12}}$     (d) $\sqrt{x^9}$

18  $\sqrt[6]{8u^9} =$ _____.
    (a) $\sqrt[3]{4u^9}$     (b) $u\sqrt{2u}$     (c) $\sqrt{2u}$     (d) $\sqrt{8u^3}$

19  Performing $\sqrt{8} + \sqrt{18} + \sqrt{98}$ yields _____.
    (a) $12\sqrt{2}$     (b) $\sqrt{8} + \sqrt{18}$     (c) $3\sqrt{8} + 5\sqrt{2}$     (d) $7\sqrt{2}$

20  The product $(\sqrt{33} - \sqrt{6})(\sqrt{33} + \sqrt{6})$ is equal to _____.
    (a) 33     (b) $\sqrt{39}$     (c) 27     (d) $\sqrt{198}$

**21** Rationalizing the denominator of $\dfrac{3}{\sqrt{5}-\sqrt{3}}$ yields _____.

(a) $\dfrac{3\sqrt{5}-3\sqrt{3}}{2}$ 
(b) $\sqrt{5}+\sqrt{3}$ 
(c) $\dfrac{3\sqrt{5}+3\sqrt{3}}{2}$ 
(d) $\dfrac{3}{\sqrt{5}+\sqrt{3}}$

**22** Expressing $5+\sqrt{-25}$ in the form $a+bi$ yields _____.

(a) $5+i$ 
(b) $5-i$ 
(c) $5+5i$ 
(d) $5-5i$

**23** The product $(7+i)(3-2i)$ is expressed in the form $a+bi$ as _____.

(a) $21+i$ 
(b) $23-11i$ 
(c) $23+11i$ 
(d) $23-i$

**24** The conjugate of $3+2i$ is _____.

(a) $-3+2i$ 
(b) $-3-2i$ 
(c) $3-2i$ 
(d) $3+2i$

**25** Writing $\dfrac{3-2i}{3+2i}$ in the form $a+bi$ yields _____.

(a) $\tfrac{5}{13}-\tfrac{12}{13}i$ 
(b) $5-12i$ 
(c) $\tfrac{5}{13}+\tfrac{12}{13}i$ 
(d) $12-5i$

---

**Answers**

1 d  2 b  3 a  4 b  5 c  6 d  7 b  8 c  9 a  10 d
11 a  12 b  13 d  14 d  15 a  16 d  17 a  18 b  19 a
20 c  21 c  22 c  23 b  24 c  25 a

# 6 Nonlinear Equations and Inequalities

In this chapter we solve equations of the form $ax^2 + bx + c = 0$, $a \neq 0$. Such equations are called *quadratic equations* (second-degree equations). After completing the appropriate sections, the student should be able to:

1. Solve equations by the factoring method.
2. Solve equations by the completing-the-square method.
3. Solve equations by using the quadratic formula.
4. Solve word problems that involve applications of quadratic equations.
5. Solve equations quadratic in form.
6. Solve equations involving radicals.
7. Solve equations involving rational exponents.
8. Solve quadratic inequalities and rational inequalities.

## 6.1 Solving Quadratic Equations by Factoring

### SEMIPROGRAMMED PROBLEMS

In problems 1–10, solve each equation by the factoring method.

| | |
|---|---|
| | **1**   $x^2 - 5x + 4 = 0$ |
| $x - 4$ | $(x - 1)(\underline{\phantom{xxxx}}) = 0$ |
| $0$ | $x - 1 = 0 \quad$ or $\quad x - 4 = \underline{\phantom{xx}}$ |
| $1$ | $x = \underline{\phantom{xx}} \quad$ or $\quad x = 4$ |
| | **2**   $x^2 - 6x + 8 = 0$ |
| $x - 4$ | $(x - 2)(\underline{\phantom{xxxx}}) = 0$ |
| $x - 4$ | $x - 2 = 0 \quad$ or $\quad \underline{\phantom{xxxx}} = 0$ |
| $2, 4$ | $x = \underline{\phantom{xx}} \quad$ or $\quad x = \underline{\phantom{xx}}$ |
| | **3**   $49x^2 - 16 = 0$ |
| $7x + 4$ | $(7x - 4)(\underline{\phantom{xxxx}}) = 0$ |
| $7x + 4$ | $7x - 4 = 0 \quad$ or $\quad \underline{\phantom{xxxx}} = 0$ |
| $\frac{4}{7}, -\frac{4}{7}$ | $x = \underline{\phantom{xx}} \quad$ or $\quad x = \underline{\phantom{xxxx}}$ |
| | **4**   $x^2 + x = 72$ |
| $x^2 + x - 72$ | The equation in standard form is $\underline{\phantom{xxxxxx}} = 0$. |
| $x + 9, x - 8$ | $(\underline{\phantom{xxxx}})(\underline{\phantom{xxxx}}) = 0$ |
| $x - 8$ | $x + 9 = 0 \quad$ or $\quad \underline{\phantom{xxxx}} = 0$ |
| $-9, 8$ | $x = \underline{\phantom{xx}} \quad$ or $\quad x = \underline{\phantom{xx}}$ |

107

## CHAPTER 6 NONLINEAR EQUATIONS AND INEQUALITIES

| | |
|---|---|
| $5x + 7$ | **5**   $25x^2 + 70x + 49 = 0$ |
| $5x + 7$ | $(\underline{\qquad})^2 = 0$ |
| $-\frac{7}{5}, -\frac{7}{5}$ | $5x + 7 = 0 \quad$ or $\quad \underline{\qquad} = 0$ |
| | $x = \underline{\qquad} \quad$ or $\quad x = \underline{\qquad}$ |
| | **6**   $x^2 = 4x + 21$ |
| | The equation in standard form is |
| $x^2 - 4x - 21$ | $\underline{\qquad\qquad} = 0.$ |
| $x - 7$ | $(x + 3)(\underline{\qquad}) = 0$ |
| $x - 7$ | $x + 3 = 0 \quad$ or $\quad \underline{\qquad} = 0$ |
| $-3, 7$ | $x = \underline{\qquad} \quad$ or $\quad x = \underline{\qquad}$ |
| | **7**   $(x - 9)(x + 9) + x + 81 = 0$ |
| | $x^2 + \underline{\qquad} = 0$ |
| $x$ | $x(\underline{\qquad}) = 0$ |
| $x + 1$ | $x = \underline{\qquad} \quad$ or $\quad x + 1 = \underline{\qquad}$ |
| $0, 0$ | $x = \underline{\qquad} \quad$ or $\quad x = \underline{\qquad}$ |
| $0, -1$ | |
| | **8**   $(2x - 1)(x + 7) = 2x - 7$ |
| | $2x^2 + \underline{\qquad} - 7 = 2x - 7$ |
| $13x$ | $2x^2 + \underline{\qquad} = 0$ |
| $11x$ | $x(\underline{\qquad}) = 0$ |
| $2x + 11$ | $x = 0 \quad$ or $\quad \underline{\qquad} = 0$ |
| $2x + 11$ | $x = \underline{\qquad} \quad$ or $\quad x = \underline{\qquad}$ |
| $0, -\frac{11}{2}$ | |
| | **9**   $\dfrac{x}{2x + 1} + \dfrac{1}{x + 2} = 2$ |
| | The LCD of the fractions is $\underline{\qquad\qquad}$. |
| $(2x + 1)(x + 2)$ | $(2x + 1)(x + 2)\left(\dfrac{x}{2x + 1} + \dfrac{1}{x + 2}\right) = 2\underline{\qquad\qquad}$ |
| $(2x + 1)(x + 2)$ | |
| $2x + 1$ | $x(x + 2) + \underline{\qquad} = 2(2x + 1)(x + 2)$ |
| $4x^2 + 10x + 4$ | $x^2 + 2x + 2x + 1 = \underline{\qquad\qquad}$ |
| $0$ | $3x^2 + 6x + 3 = \underline{\qquad}$ |
| $0$ | $x^2 + 2x + 1 = \underline{\qquad}$ |
| $0$ | $(x + 1)(x + 1) = \underline{\qquad}$ |
| $0, 0$ | $x + 1 = \underline{\qquad} \quad$ or $\quad x + 1 = \underline{\qquad}$ |
| $-1$ | $x = -1 \quad$ or $\quad x = \underline{\qquad}$ |
| | **10**   $\dfrac{x - 3}{x + 3} = \dfrac{2}{x}$ |
| | The LCD is $\underline{\qquad}$. |
| $x(x + 3)$ | $x(x + 3)\left(\dfrac{x - 3}{x + 3}\right) = \underline{\qquad}\left(\dfrac{2}{x}\right)$ |
| $x(x + 3)$ | $x(x - 3) = \underline{\qquad}$ |
| $2(x + 3)$ | $x^2 - 3x = \underline{\qquad}$ |
| $2x + 6$ | $x^2 - 5x - 6 = \underline{\qquad}$ |
| $0$ | $(x - 6)(\underline{\qquad}) = 0$ |
| $x + 1$ | $x - 6 = 0 \quad$ or $\quad x + 1 = \underline{\qquad}$ |
| $0$ | $x = \underline{\qquad} \quad$ or $\quad x = \underline{\qquad}$ |
| $6, -1$ | |

**SOLUTIONS TO SELECTED ODD PROBLEMS**  Section 6.1, text pages 279–280

1.  $x^2 - 3x + 2 = 0$
    $(x - 2)(x - 1) = 0$
    $x - 2 = 0$ or $x - 1 = 0$
    $x = 2$ or $x = 1$

5.  $t^2 - t = 20$
    $t^2 - t - 20 = 0$
    $(t - 5)(t + 4) = 0$
    $t - 5 = 0$ or $t + 4 = 0$
    $t = 5$ or $t = -4$

9.  $u^2 - u = 12$
    $u^2 - u - 12 = 0$
    $(u - 4)(u + 3) = 0$
    $u - 4 = 0$ or $u + 3 = 0$
    $u = 4$ or $u = -3$

13. $10y^2 + y - 2 = 0$
    $(2y + 1)(5y - 2) = 0$
    $2y + 1 = 0$ or $5y - 2 = 0$
    $2y = -1$ or $5y = 2$
    $y = -\frac{1}{2}$ or $y = \frac{2}{5}$

17. $-10t^2 + 11t - 3 = 0$
    $10t^2 - 11t + 3 = 0$
    $(2t - 1)(5t - 3) = 0$
    $2t - 1 = 0$ or $5t - 3 = 0$
    $t = \frac{1}{2}$ or $t = \frac{3}{5}$

21. $6u^2 + 7u = 20$
    $6u^2 + 7u - 20 = 0$
    $(2u + 5)(3u - 4) = 0$
    $2u + 5 = 0$ or $3u - 4 = 0$
    $u = -\frac{5}{2}$ or $u = \frac{4}{3}$

25. $x^2 = 7x$
    $x^2 - 7x = 0$
    $x(x - 7) = 0$
    $x = 0$ or $x - 7 = 0$
    $x = 0$ or $x = 7$

29. $49z^2 - 14z + 1 = 0$
    $(7z - 1)(7z - 1) = 0$
    $7z - 1 = 0$ or $7z - 1 = 0$
    $z = \frac{1}{7}$

33. $z(3z + 11) = 20$
    $3z^2 + 11z = 20$
    $3z^2 + 11z - 20 = 0$
    $(3z - 4)(z + 5) = 0$
    $3z - 4 = 0$ or $z + 5 = 0$
    $z = \frac{4}{3}$ or $z = -5$

37. $r(2r - 19) = 33$
    $2r^2 - 19r = 33$
    $2r^2 - 19r - 33 = 0$
    $(2r + 3)(r - 11) = 0$
    $2r + 3 = 0$ or $r - 11 = 0$
    $r = -\frac{3}{2}$ or $r = 11$

41. $(2x + 1)(3x - 2) = 10$
    $6x^2 - x - 2 = 10$
    $6x^2 - x - 12 = 0$
    $(3x + 4)(2x - 3) = 0$
    $3x + 4 = 0$ or $2x - 3 = 0$
    $x = -\frac{4}{3}$ or $x = \frac{3}{2}$
    Check: For $x = -\frac{4}{3}$,
    $[2(-\frac{4}{3}) + 1][3(-\frac{4}{3}) - 2] = 10$
    $[-\frac{8}{3} + 1][-4 - 2] = 10$
    $(-\frac{5}{3})(-6) = 10$
    $10 = 10$
    For $x = \frac{3}{2}$,
    $[2(\frac{3}{2}) + 1][3(\frac{3}{2}) - 2] = 10$
    $[3 + 1][\frac{9}{2} - 2] = 10$
    $(4)(\frac{5}{2}) = 10$
    $10 = 10$

45. $\dfrac{15}{(x - 2)^2} + \dfrac{2}{x - 2} = 1$
    The LCD is $(x - 2)^2$, so
    $15 + 2(x - 2) = (x - 2)^2$
    $15 + 2x - 4 = x^2 - 4x + 4$
    $-x^2 + 6x + 7 = 0$
    $x^2 - 6x - 7 = 0$
    $(x - 7)(x + 1) = 0$
    $x - 7 = 0$ or $x + 1 = 0$
    $x = 7$ or $x = -1$

49. $x^2 - 2ax - 15a^2 = 0$, for $x$.
    $(x + 3a)(x - 5a) = 0$
    $x + 3a = 0$ or $x - 5a = 0$
    $x = -3a$ or $x = 5a$

53. $6m^2 + mb = 2b^2$, for $m$.
    $6m^2 + mb - 2b^2 = 0$
    $(3m + 2b)(2m - b) = 0$
    $3m + 2b = 0$ or $2m - b = 0$
    $m = \dfrac{-2b}{3}$ or $m = \dfrac{b}{2}$

## 6.2 Solving Quadratic Equations by Roots Extraction and Completing the Square

**SEMIPROGRAMMED PROBLEMS**

In problems 1–5, solve each equation by the roots extraction method.

4, 2

−2

1.  $x^2 = 4$
    $x = \pm\sqrt{\underline{\phantom{xxx}}} = \pm\underline{\phantom{xxx}}$
    so that
    $x = \underline{\phantom{xxx}}$ or $x = 2$

**110** CHAPTER 6 NONLINEAR EQUATIONS AND INEQUALITIES

25
25, 5

−5, 5

**2**   $3x^2 = 75$
$x^2 =$ _____
$x = \pm\sqrt{\_\_\_\_} = \pm\_\_\_\_$
so that
$x =$ _____ or $x =$ _____

11
$\frac{11}{2}$
$\frac{11}{2}, \frac{\sqrt{22}}{2}$

$-\frac{\sqrt{22}}{2}, \frac{\sqrt{22}}{2}$

**3**   $2x^2 - 11 = 0$
$2x^2 =$ _____
$x^2 =$ _____
$x = \pm\sqrt{\_\_\_\_} = \pm\_\_\_\_$
so that
$x =$ _____ or $x =$ _____

−4
$-\frac{4}{49}$
$-\frac{4}{49}, \frac{2}{7}i$

$-\frac{2i}{7}, \frac{2i}{7}$

**4**   $49x^2 + 4 = 0$
$49x^2 =$ _____
$x^2 =$ _____
$x = \pm\sqrt{\_\_\_\_} = \pm\_\_\_\_$
so that
$x =$ _____ or $x =$ _____

$\pm\sqrt{\frac{16}{9}}, \pm\frac{4}{3}$
$\frac{1}{3} \pm \frac{4}{3}$

$-1, \frac{5}{3}$

**5**   $(x - \frac{1}{3})^2 = \frac{16}{9}$
$x - \frac{1}{3} =$ _____ = _____
$x =$ _____
so that
$x = \frac{1}{3} - \frac{4}{3} =$ _____ or $x = \frac{1}{3} + \frac{4}{3} =$ _____

In problems 6–8, complete the quadratic square and express the result as the square of a binomial.

$\frac{3}{2}$
$x^2 + 3x + \frac{9}{4}$
$x + \frac{3}{2}$

**6**   $x^2 + 3x$
Add (_____)$^2$ to the expression $x^2 + 3x$.
The resulting perfect square is _____.
So $x^2 + 3x + \frac{9}{4} = ($ _____ $)^2$.

2
$x + 2$

**7**   $x^2 + 4x$
Add (_____)$^2$ to the expression $x^2 + 4x$.
The resulting perfect square is $x^2 + 4x + 4 = ($ _____ $)^2$.

$\frac{2}{5}$
$x - \frac{2}{5}$

**8**   $x^2 - \frac{4}{5}x$
Add (_____)$^2$ to the expression $x^2 - \frac{4}{5}x$.
The resulting perfect square is $x^2 - \frac{4}{5}x + \frac{4}{25} = ($ _____ $)^2$.

In problems 9–14, solve each equation by completing the square.

−1

**9**   $x^2 - 5x + 1 = 0$
$x^2 - 5x =$ _____

$\frac{25}{4}$

$\frac{25}{4}, \frac{25}{4}$

$\frac{21}{4}$

$\frac{21}{4}, \frac{\sqrt{21}}{2}$

$-\frac{\sqrt{21}}{2}$

$\frac{5}{2} - \frac{\sqrt{21}}{2}, \frac{5}{2} + \frac{\sqrt{21}}{2}$

$-5$

$4$

$4, 4$

$-1$

$-1, i$

$-i, i$

$-2 - i, -2 + i$

$2$

$\frac{1}{2}$

$-\frac{1}{2}$

$\frac{25}{16}$

$\frac{25}{16}, \frac{25}{16}$

$x - \frac{5}{4}$

$x - \frac{5}{4}$

$-\sqrt{\frac{17}{16}}, -\frac{\sqrt{17}}{4}, \sqrt{\frac{17}{16}}, \frac{\sqrt{17}}{4}$

$\frac{5}{4} - \frac{\sqrt{17}}{4}, \frac{5}{4} + \frac{\sqrt{17}}{4}$

$5$

$-\frac{4}{5}$

$\frac{9}{25}$

---

## 6.2 SEMIPROGRAMMED PROBLEMS

Completing the square by adding $[\frac{1}{2}(-5)]^2 = $ _____,

$x^2 - 5x + $ _____ $= -1 + $ _____

$\left(x - \frac{5}{2}\right)^2 = $ _____.

$x - \frac{5}{2} = \pm\sqrt{\text{\_\_\_\_}} = \pm\text{_____}$

so that

$x - \frac{5}{2} = $ _____ or $x - \frac{5}{2} = \frac{\sqrt{21}}{2}$

$x = $ _____ or $x = $ _____

**10** $x^2 + 4x + 5 = 0$

$x^2 + 4x = $ _____

Completing the square by adding $[\frac{1}{2}(4)]^2 = $ _____,

$x^2 + 4x + $ _____ $= -5 + $ _____

$(x + 2)^2 = $ _____.

$x + 2 = \pm\sqrt{\text{\_\_\_\_}} = \pm\text{_____}$

so that

$x + 2 = $ _____ or $x + 2 = $ _____

$x = $ _____ or $x = $ _____

**11** $2x^2 - 5x + 1 = 0$

Divide both sides of the equation by _____ to make the coefficient of the $x^2$ term 1.

$x^2 - \frac{5}{2}x + $ _____ $= 0$

$x^2 - \frac{5}{2}x = $ _____

adding $\left[\frac{1}{2}\left(-\frac{5}{2}\right)\right]^2 = $ _____ to both sides.

$x^2 - \frac{5}{2}x + $ _____ $= -\frac{1}{2} + $ _____

$\left(\text{_____}\right)^2 = \frac{17}{16}$

_____ $= \pm\sqrt{\frac{17}{16}}$

$x - \frac{5}{4} = $ _____ $= $ _____ or $x - \frac{5}{4} = $ _____ $= $ _____

$x = $ _____ or $x = $ _____

**12** $5x^2 - 6x - 4 = 0$

Divide both sides of the equation by _____ to make the coefficient of the $x^2$ term 1.

$x^2 - \frac{6}{5}x + \left(\text{_____}\right) = 0$

$x^2 - \frac{6}{5}x = \frac{4}{5}$, and $\left[\frac{1}{2}\left(-\frac{6}{5}\right)\right]^2 = $ _____

$\frac{9}{25}, \frac{9}{25}$

$x - \frac{3}{5}$

$x - \frac{3}{5}, \frac{\sqrt{29}}{5}$

$-\frac{\sqrt{29}}{5}, \frac{\sqrt{29}}{5}$

$\frac{3 - \sqrt{29}}{5}, \frac{3 + \sqrt{29}}{5}$

3

0

$\frac{2}{3}, \frac{2}{3}, \frac{1}{9}$

$\frac{1}{9}, \frac{1}{9}$

$\frac{7}{9}$

$\frac{7}{9}, \frac{\sqrt{7}}{3}$

$-\frac{\sqrt{7}}{3}, \frac{\sqrt{7}}{3}$

$-\frac{1}{3} + \frac{\sqrt{7}}{3}$

2

0

$-\frac{3}{2}, \frac{3}{2}, \frac{9}{16}$

$\frac{9}{16}, \frac{9}{16}$

$-\frac{15}{16}$

$-\frac{15}{16}, \frac{\sqrt{15}}{4} i$

$\frac{\sqrt{15}i}{4}$

$\frac{-3 - \sqrt{15}i}{4}, \frac{-3 + \sqrt{15}i}{4}$

$x^2 - \frac{6}{5}x + \underline{\phantom{xx}} = \frac{4}{5} + \underline{\phantom{xx}}$

$\left(\underline{\phantom{xxxx}}\right)^2 = \frac{29}{25}$

$\underline{\phantom{xx}} = \pm\sqrt{\frac{29}{25}} = \pm\underline{\phantom{xx}}$

$x - \frac{3}{5} = \underline{\phantom{xx}}$ or $x - \frac{3}{5} = \underline{\phantom{xx}}$

$x = \underline{\phantom{xx}}$ or $x = \underline{\phantom{xx}}$

13  $3x^2 + 2x - 2 = 0$

Divide both sides of the equation by ___ so that

$x^2 + \frac{2}{3}x - \frac{2}{3} = \underline{\phantom{xx}}$

$x^2 + \frac{2}{3}x = \underline{\phantom{xx}}$ and $\left[\frac{1}{2}\left(\underline{\phantom{xx}}\right)\right]^2 = \underline{\phantom{xx}}$

$x^2 + \frac{2}{3}x + \underline{\phantom{xx}} = \frac{2}{3} + \underline{\phantom{xx}}$

$\left(x + \frac{1}{3}\right)^2 = \underline{\phantom{xx}}$

$x + \frac{1}{3} = \pm\sqrt{\underline{\phantom{xx}}} = \pm\underline{\phantom{xx}}$

so that

$x + \frac{1}{3} = \underline{\phantom{xx}}$ or $x + \frac{1}{3} = \underline{\phantom{xx}}$

$x = -\frac{1}{3} - \frac{\sqrt{7}}{3}$ or $x = \underline{\phantom{xx}}$

14  $2x^2 + 3x + 3 = 0$

Divide both sides of the equation by ___ so that

$x^2 + \frac{3}{2}x + \frac{3}{2} = \underline{\phantom{xx}}$

$x^2 + \frac{3}{2}x = \underline{\phantom{xx}}$ and $\left[\frac{1}{2}\left(\underline{\phantom{xx}}\right)\right]^2 = \underline{\phantom{xx}}$

$x^2 + \frac{3}{2}x + \underline{\phantom{xx}} = -\frac{3}{2} + \underline{\phantom{xx}}$

$\left(x + \frac{3}{4}\right)^2 = \underline{\phantom{xx}}$

$x + \frac{3}{4} = \pm\sqrt{\underline{\phantom{xx}}} = \pm\underline{\phantom{xx}}$

so that

$x + \frac{3}{4} = -\frac{\sqrt{15}}{4}i$ or $x + \frac{3}{4} = \underline{\phantom{xx}}$

$x = \underline{\phantom{xx}}$ or $x = \underline{\phantom{xx}}$

## SOLUTIONS TO SELECTED ODD PROBLEMS  Section 6.2, text pages 289–290

**1**  $x^2 = 64$
$x = \pm\sqrt{64} = \pm 8$
$x = 8$ or $-8$

**5**  $4z^2 = 60$
$z^2 = 15$
$z = \pm\sqrt{15}$
$z = \sqrt{15}$ or $-\sqrt{15}$

**9**  $(2m + 1)^2 = 36$
$2m + 1 = \pm\sqrt{36}$
$2m + 1 = \pm 6$
$2m + 1 = -6$ or $2m + 1 = 6$
$2m = -7$ or $2m = 5$
$m = -\frac{7}{2}$ or $m = \frac{5}{2}$

**13**  $(4y - 3)^2 - 16 = 0$
$(4y - 3)^2 = 16$
$4y - 3 = \pm\sqrt{16} = \pm 4$
$4y - 3 = 4$ or $4y - 3 = -4$
$4y = 7$ or $4y = -1$
$y = \frac{7}{4}$ or $y = -\frac{1}{4}$

**17**  $(u - 6)^2 + 25 = 0$
$(u - 6)^2 = -25$
$u - 6 = \pm\sqrt{-25} = \pm 5i$
$u - 6 = 5i$ or $u - 6 = -5i$
$u = 6 + 5i$ or $u = 6 - 5i$

**21**  $x^2 + 6x + 9 = (x + 3)^2$

**25**  $m^2 + 20m + 100 = (m + 10)^2$

**29**  $c^2 + 17c + \frac{289}{4} = (c + \frac{17}{2})^2$

**33**  $y^2 - 12y - 17 = 0$
$y^2 - 12y = 17$
$y^2 - 12y + 36 = 17 + 36$
$(y - 6)^2 = 53$
$y - 6 = \pm\sqrt{53}$
$y - 6 = \sqrt{53}$ or $y - 6 = -\sqrt{53}$
$y = 6 + \sqrt{53}$ or $y = 6 - \sqrt{53}$

**37**  $5x^2 - 10x - 1 = 0$
$5x^2 - 10x = 1$
$x^2 - 2x = \frac{1}{5}$
$x^2 - 2x + 1 = \frac{1}{5} + 1$
$(x - 1)^2 = \frac{6}{5}$
$x - 1 = \pm\sqrt{\frac{6}{5}} = \pm\sqrt{\frac{6 \cdot 5}{5 \cdot 5}} = \pm\frac{\sqrt{30}}{5}$
$x - 1 = -\frac{\sqrt{30}}{5}$ or $x - 1 = +\frac{\sqrt{30}}{5}$
$x = 1 - \frac{\sqrt{30}}{5}$ or $x = 1 + \frac{\sqrt{30}}{5}$
$x = \frac{5 - \sqrt{30}}{5}$ or $x = \frac{5 + \sqrt{30}}{5}$

**41**  $5m^2 - 8m + 17 = 0$
$5m^2 - 8m = -17$
$m^2 - \frac{8}{5}m = \frac{-17}{5}$
$m^2 - \frac{8}{5}m + \frac{16}{25} = \frac{-17}{5} + \frac{16}{25}$
$\left(m - \frac{4}{5}\right)^2 = \frac{-69}{25}$
$m - \frac{4}{5} = \sqrt{\frac{-69}{25}}$ or $m - \frac{4}{5} = -\sqrt{\frac{-69}{25}}$
$m = \frac{4}{5} + \sqrt{\frac{-69}{25}}$ or $m = \frac{4}{5} - \sqrt{\frac{-69}{25}}$
$m = \frac{4}{5} + \frac{\sqrt{69}}{5}i$ or $m = \frac{4}{5} - \frac{\sqrt{69}}{5}i$

**45**  $3t^2 - 8t + 4 = 0$
$3t^2 - 8t = -4$
$t^2 - \frac{8}{3}t = -\frac{4}{3}$
$t^2 - \frac{8}{3}t + \frac{16}{9} = -\frac{4}{3} + \frac{16}{9}$
$(t - \frac{4}{3})^2 = \frac{4}{9}$
$t - \frac{4}{3} = \sqrt{\frac{4}{9}}$ or $t - \frac{4}{3} = -\sqrt{\frac{4}{9}}$
$t - \frac{4}{3} = \frac{2}{3}$ or $t - \frac{4}{3} = -\frac{2}{3}$
$t = 2$ or $t = \frac{2}{3}$

## CHAPTER 6 NONLINEAR EQUATIONS AND INEQUALITIES

**49**
$$16y^2 - 24y = -5$$
$$y^2 - \tfrac{3}{2}y = -\tfrac{5}{16}$$
$$y^2 - \tfrac{3}{2}y + \tfrac{9}{16} = -\tfrac{5}{16} + \tfrac{9}{16}$$
$$(y - \tfrac{3}{4})^2 = \tfrac{1}{4}$$
$$y - \tfrac{3}{4} = \sqrt{\tfrac{1}{4}} \text{ or } y - \tfrac{3}{4} = -\sqrt{\tfrac{1}{4}}$$
$$y - \tfrac{3}{4} = \tfrac{1}{2} \text{ or } y - \tfrac{3}{4} = -\tfrac{1}{2}$$
$$y = \tfrac{5}{4} \text{ or } y = \tfrac{1}{4}$$

**53**
$$-gt^2 + vt = s, \text{ for } t$$
$$t^2 - \frac{v}{g}t = -\frac{s}{g}, g > 0$$
$$t^2 - \frac{v}{g}t + \frac{v^2}{4g^2} = -\frac{s}{g} + \frac{v^2}{4g^2}$$
$$\left(t - \frac{v}{2g}\right)^2 = \frac{v^2 - 4gs}{4g^2}$$
$$t - \frac{v}{2g} = +\sqrt{\frac{v^2 - 4gs}{4g^2}} \text{ or } t - \frac{v}{2g} = -\sqrt{\frac{v^2 - 4gs}{4g^2}}$$
$$t = \frac{v + \sqrt{v^2 - 4gs}}{2g} \text{ or } t = \frac{v - \sqrt{v^2 - 4gs}}{2g}$$

**57**
$$2\left(-\tfrac{3}{4} + \tfrac{\sqrt{7}}{4}i\right)^2 + 3\left(-\tfrac{3}{4} + \tfrac{\sqrt{7}}{4}i\right) + 2 = 0$$
$$2\left(\tfrac{9}{16} - \tfrac{3\sqrt{7}}{8}i + \tfrac{7}{16}i^2\right) - \tfrac{9}{4} + \tfrac{3\sqrt{7}}{4}i + 2 = 0$$
$$2\left(\tfrac{1}{8} - \tfrac{3\sqrt{7}}{8}i\right) - \tfrac{9}{4} + \tfrac{3\sqrt{7}}{4}i + 2 = 0$$
$$\tfrac{1}{4} - \tfrac{3\sqrt{7}}{4}i - \tfrac{9}{4} + \tfrac{3\sqrt{7}}{4}i + 2 = 0$$
$$0 = 0$$

**61**
$$s^2 = 64h$$
$$s^2 = 64(0.75)$$
$$s^2 = 48$$
$$s = \sqrt{48} \text{ or } 4\sqrt{3} \text{ feet per second}$$

## 6.3 Using the Quadratic Formula to Solve Quadratic Equations

### SEMIPROGRAMMED PROBLEMS

$\dfrac{-b \pm \sqrt{b^2 - 4ac}}{2a}$

**1** The solutions of the quadratic equation $ax^2 + bx + c = 0$, $a \neq 0$, are expressed by the quadratic formula

$$x = \underline{\hspace{3cm}}.$$

In problems 2–8, solve each equation by using the quadratic formula.

**2** $x^2 + 6x + 5 = 0$

1, 6, 5     $a = \underline{\quad}, b = \underline{\quad}, c = \underline{\quad}$

6     $x = \dfrac{-\underline{\quad} \pm \sqrt{6^2 - 4(1)(5)}}{2(1)}$

16, 4     $= \dfrac{-6 \pm \sqrt{\underline{\quad}}}{2} = \dfrac{-6 \pm \underline{\quad}}{2}$

4     $x = \dfrac{-6 - \underline{\quad}}{2}$ or $x = \dfrac{-6 + 4}{2}$

$-5, -1$     $x = \underline{\quad}$ or $x = \underline{\quad}$

**3** $x^2 - 2x = 15$

$x^2 - 2x - 15$     The standard form of the equation is $\underline{\hspace{4cm}} = 0$.

$1, -2, -15$     $a = \underline{\quad}, b = \underline{\quad}, c = \underline{\quad}$

$-2$     $x = \dfrac{-(-2) \pm \sqrt{(\underline{\quad})^2 - 4(1)(-15)}}{2(1)}$

60, 8     $= \dfrac{2 \pm \sqrt{4 + \underline{\quad}}}{2} = \dfrac{2 \pm \underline{\quad}}{2}$

6.3 SEMIPROGRAMMED PROBLEMS   115

| | |
|---|---|
| 8, 8 | $x = \dfrac{2 - \underline{\phantom{xx}}}{2}$ or $x = \dfrac{2 + \underline{\phantom{xx}}}{2}$ |
| $-3, 5$ | $x = \underline{\phantom{xx}}$ or $x = \underline{\phantom{xx}}$ |

4  $4x^2 - 8x = -13$

$4x^2 - 8x + 13$    The standard form of the equation is $\underline{\phantom{xxxxxx}} = 0$.

$4, -8, 13$    $a = \underline{\phantom{xx}}, b = \underline{\phantom{xx}}, c = \underline{\phantom{xx}}$

$-8, -8, 4, 13$    $x = \dfrac{-(\underline{\phantom{x}}) \pm \sqrt{(\underline{\phantom{x}})^2 - 4(\underline{\phantom{x}})(\underline{\phantom{x}})}}{2(4)}$

$-144, 12i$    $= \dfrac{8 \pm \sqrt{\underline{\phantom{xxx}}}}{2(4)} = \dfrac{8 \pm \underline{\phantom{xx}}}{8}$

$12i, 12i$    $x = \dfrac{8 - \underline{\phantom{xx}}}{8}$ or $x = \dfrac{8 + \underline{\phantom{xx}}}{8}$

$\dfrac{2-3i}{2}, \dfrac{2+3i}{2}$    $x = \underline{\phantom{xxxx}}$ or $x = \underline{\phantom{xxxx}}$

5  $5x^2 + 13x = 6$

$5x^2 + 13x - 6$    The standard form of the equation is $\underline{\phantom{xxxxxx}} = 0$.

$5, 13, -6$    $a = \underline{\phantom{xx}}, b = \underline{\phantom{xx}}, c = \underline{\phantom{xx}}$

$13, 13, 5, -6$    $x = \dfrac{-\underline{\phantom{xx}} \pm \sqrt{(\underline{\phantom{x}})^2 - 4(\underline{\phantom{x}})(\underline{\phantom{x}})}}{2(5)}$

$289, 17$    $= \dfrac{-13 \pm \sqrt{\underline{\phantom{xxx}}}}{10} = \dfrac{-13 \pm \underline{\phantom{xx}}}{10}$

$-3, \tfrac{2}{5}$    $x = \underline{\phantom{xx}}$ or $x = \underline{\phantom{xx}}$

6  $\dfrac{x^2 - 3}{3} + \dfrac{x}{4} = 2$

$12$    $12\left(\dfrac{x^2 - 3}{3} + \dfrac{x}{4}\right) = (\underline{\phantom{xx}})(2)$

$3$    $4(x^2 - 3) + \underline{\phantom{xx}}x = 24$

$24$    $4x^2 - 12 + 3x = \underline{\phantom{xx}}$

$4x^2 + 3x - 36$    The standard form of the equation is $\underline{\phantom{xxxxxx}} = 0$.

$4, 3, -36$    $a = \underline{\phantom{xx}}, b = \underline{\phantom{xx}}, c = \underline{\phantom{xx}}$

$3, 3, 4, -36$    $x = \dfrac{-\underline{\phantom{xx}} \pm \sqrt{(\underline{\phantom{x}})^2 - 4(\underline{\phantom{x}})(\underline{\phantom{x}})}}{2(4)}$

$\dfrac{-3 \pm \sqrt{585}}{8}, 65$    $= \underline{\phantom{xxxxx}} = \dfrac{-3 \pm 3\sqrt{\underline{\phantom{xx}}}}{8}$

$\dfrac{-3 - 3\sqrt{65}}{8}, \dfrac{-3 + 3\sqrt{65}}{8}$    $x = \underline{\phantom{xxxxx}}$ or $x = \underline{\phantom{xxxxx}}$

7  $(2x + 1)(x + 1) = -2$

$2x^2 + 3x + 3$    The standard form of this equation is $\underline{\phantom{xxxxxx}} = 0$.

$2, 3, 3$    $a = \underline{\phantom{xx}}, b = \underline{\phantom{xx}}, c = \underline{\phantom{xx}}$

$3$    $x = \dfrac{-3 \pm \sqrt{9 - 4(2)(\underline{\phantom{xx}})}}{2(2)}$

$-15$    $= \dfrac{-3 \pm \sqrt{(\underline{\phantom{xxx}})}}{4} = \dfrac{-3 \pm \sqrt{15}i}{4}$

$\dfrac{-3 - \sqrt{15}i}{4}, \dfrac{-3 + \sqrt{15}i}{4}$    $x = \underline{\phantom{xxxxx}}$ or $x = \underline{\phantom{xxxxx}}$

# CHAPTER 6  NONLINEAR EQUATIONS AND INEQUALITIES

**8** $x^2 + x + k = kx^2$, $k \leq \frac{1}{2}$

$(1-k)x^2 + x + k$     The standard form of the equation is _____ = 0.

$1-k, 1, k$     $a =$ _____, $b =$ _____, $c =$ _____

$1, 1, k$     $x = \dfrac{-\underline{\phantom{xx}} \pm \sqrt{(\underline{\phantom{xx}})^2 - 4(1-k)(\underline{\phantom{xx}})}}{2(1-k)}$

$1 - 4k + 4k^2,\ \dfrac{-1 \pm (1-2k)}{2(1-k)}$     $= \dfrac{-1 \pm \sqrt{\underline{\phantom{xxxx}}}}{2(1-k)} =$ _____

$-1,\ \dfrac{k}{k-1}$     $x =$ _____ or $x =$ _____

In problem 9, fill in the missing information regarding the quadratic discriminant.

**9** If $ax^2 + bx + c = 0$, $a \neq 0$, we define the discriminant of the quadratic

$b^2 - 4ac$     equation as _____. If

real     **(i)** $b^2 - 4ac > 0$, there are two unequal _____ solutions;

one     **(ii)** $b^2 - 4ac = 0$, there is only _____ real solution, a double solution;

complex     **(iii)** $b^2 - 4ac < 0$, there are two unequal _____ solutions.

In problems 10–12, use the discriminant to determine the number and the kind of solutions to each quadratic equation.

**10** $3x^2 - 4x + 1 = 0$

$3, -4, 1$     $a =$ _____, $b =$ _____, $c =$ _____

$-4, 3, 1, 4$     $b^2 - 4ac = (\underline{\phantom{xx}})^2 - 4(\underline{\phantom{xx}})(\underline{\phantom{xx}}) =$ _____

real and unequal     Since $4 > 0$, the equation has two _____ solutions.

**11** $25x^2 + 20x + 4 = 0$

$25, 20, 4$     $a =$ _____, $b =$ _____, $c =$ _____

$20, 25, 4, 0$     $b^2 - 4ac = (\underline{\phantom{xx}})^2 - 4(\underline{\phantom{xx}})(\underline{\phantom{xx}}) =$ _____

one     Therefore, there is only _____ real solution (a double solution).

**12** $5x^2 + 2x + 2 = 0$

$5, 2, 2$     $a =$ _____, $b =$ _____, $c =$ _____

$2, 5, 2, -36$     $b^2 - 4ac = (\underline{\phantom{xx}})^2 - 4(\underline{\phantom{xx}})(\underline{\phantom{xx}}) =$ _____

complex     Therefore, there are two _____ solutions.

## SOLUTIONS TO SELECTED ODD PROBLEMS  Section 6.3, text page 295

**1** $x^2 - 5x + 4 = 0$
$a = 1, b = -5, c = 4$. Thus,

$x = \dfrac{-(-5) \pm \sqrt{(-5)^2 - 4(1)(4)}}{2(1)}$

$x = \dfrac{5 \pm \sqrt{9}}{2} = \dfrac{5 \pm 3}{2}$

$x = \dfrac{5 - 3}{2}$ or $x = \dfrac{5 + 3}{2}$

$x = 1$ or $x = 4$

**5** $2u^2 - 5u - 3 = 0$
$a = 2, b = -5, c = -3$. Thus,

$u = \dfrac{-(-5) \pm \sqrt{(-5)^2 - 4(2)(-3)}}{2(2)}$

$u = \dfrac{5 \pm \sqrt{49}}{4} = \dfrac{5 \pm 7}{4}$

$u = \dfrac{5 - 7}{4}$ or $u = \dfrac{5 + 7}{4}$

$u = -\dfrac{1}{2}$ or $u = 3$

## 6.3 SOLUTIONS TO SELECTED ODD PROBLEMS

**9**   $6t^2 - 8t - 3 = 0$
$a = 6, b = -8, c = -3.$ Thus,
$$t = \frac{-(-8) \pm \sqrt{(-8)^2 - 4(6)(-3)}}{2(6)}$$
$$t = \frac{8 \pm \sqrt{136}}{12} = \frac{8 \pm 2\sqrt{34}}{12} = \frac{4 \pm \sqrt{34}}{6}$$
$$t = \frac{4 - \sqrt{34}}{6} \text{ or } t = \frac{4 + \sqrt{34}}{6}$$

**13**   $5y^2 - 2y + 7 = 0$
$a = 5, b = -2, c = 7.$ Thus,
$$y = \frac{-(-2) \pm \sqrt{(-2)^2 - 4(5)(7)}}{2(5)}$$
$$y = \frac{2 \pm \sqrt{-136}}{10} = \frac{2 \pm 2\sqrt{34}i}{10}$$
$$y = \frac{1 \pm \sqrt{34}i}{5} = \frac{1}{5} \pm \frac{\sqrt{34}}{5}i$$
$$y = \frac{1}{5} - \frac{\sqrt{34}}{5}i \text{ or } y = \frac{1}{5} + \frac{\sqrt{34}}{5}i$$

**17**   $15y^2 + 2y - 8 = 0$
$a = 15, b = 2, c = -8.$ Thus,
$$y = \frac{-2 \pm \sqrt{2^2 - 4(15)(-8)}}{2(15)}$$
$$y = \frac{-2 \pm \sqrt{484}}{30} = \frac{-2 \pm 22}{30}$$
$$y = \frac{-2 - 22}{30} \text{ or } y = \frac{-2 + 22}{30}$$
$$y = -\frac{4}{5} \text{ or } y = \frac{2}{3}$$

**21**   $4y(y - 1) = 19$
$4y^2 - 4y - 19 = 0$
$a = 4, b = -4, c = -19.$ Thus,
$$y = \frac{-(-4) \pm \sqrt{(-4)^2 - 4(4)(-19)}}{2(4)}$$
$$y = \frac{4 \pm \sqrt{320}}{8} = \frac{4 \pm 8\sqrt{5}}{8} = \frac{1 \pm 2\sqrt{5}}{2}$$
$$y = \frac{1 - 2\sqrt{5}}{2} \text{ or } y = \frac{1 + 2\sqrt{5}}{2}$$

**25**   $(y - 1)(y - 5) = 9$
$y^2 - 6y + 5 = 9$
$y^2 - 6y - 4 = 0$
$a = 1, b = -6, c = -4.$ Thus, for $n \neq 0$
$$y = \frac{-(-6) \pm \sqrt{(-6)^2 - 4(1)(-4)}}{2(1)}$$
$$y = \frac{6 \pm \sqrt{52}}{2} = \frac{6 \pm 2\sqrt{13}}{2} = 3 \pm \sqrt{13}$$
$$y = 3 - \sqrt{13} \text{ or } y = 3 + \sqrt{13}$$

**29**   $1.47y^2 - 3.82y - 5.71 = 0$
$a = 1.47, b = -3.82, c = -5.71.$ Thus,
$$y = \frac{-(-3.82) \pm \sqrt{(-3.82)^2 - 4(1.47)(-5.71)}}{2(1.47)}$$
$$y = \frac{3.82 \pm \sqrt{48.16}}{2.94} = \frac{3.82 \pm 6.94}{2.94}$$
$$y = \frac{3.82 - 6.94}{2.94} \text{ or } y = \frac{3.82 + 6.94}{2.94}$$
$$y = -1.06 \text{ or } y = 3.66$$

**33**   $nx^2 + mnx - m^2 = 0,$ for $x.$
$a = n, b = mn, c = -m^2.$ Thus,
$$x = \frac{-mn \pm \sqrt{(mn)^2 - 4(n)(-m^2)}}{2n} = \frac{-mn \pm \sqrt{m^2n^2 + 4m^2n}}{2n}$$
$$x = \frac{-mn \pm \sqrt{m^2(n^2 + 4n)}}{2n} = \frac{-mn \pm m\sqrt{n^2 + 4n}}{2n}$$
$$x = \frac{-mn - m\sqrt{n^2 + 4n}}{2n} \text{ or } x = \frac{-mn + m\sqrt{n^2 + 4n}}{2n}$$

**37**   $x^2 + 6x - 7 = 0$
$a = 1, b = 6, c = -7.$ Thus,
$b^2 - 4ac = 6^2 - 4(1)(-7) = 36 + 28 = 64 > 0$
Therefore, there are two unequal real roots.

**41**   $t^2 + 3t + 5 = 0$
$a = 1, b = 3, c = 5.$ Thus,
$b^2 - 4ac = 3^2 - 4(1)(5) = 9 - 20 = -11 < 0$
Therefore, there are two unequal complex roots.

**45**   $9z^2 + 30z + 25 = 0$
$a = 9, b = 30, c = 25.$ Thus,
$b^2 - 4ac = 30^2 - 4(9)(25) = 900 - 900 = 0$
Therefore, there is one real root.

**49**   $2x^2 + 7x + 3k = 0$
$a = 2, b = 7, c = 3k.$ Thus,
$b^2 - 4ac < 0$ if $7^2 - 4(2)(3k) < 0$
$49 - 24k < 0$ or $k > \frac{49}{24}$

# 6.4 Applications of Quadratic Equations

## SEMIPROGRAMMED PROBLEMS

| | |
|---|---|
| $x + 1$ | **1** Find two consecutive positive integers whose product is 56. Let $x =$ the smaller positive integer. Then _____ = the next positive integer. |
| $x, x + 1$ | The product of the two integers is 56. Therefore, ____(_____) = 56 |
| $x^2 + x - 56$ | and _____ = 0, so that solving the equation by factoring, we have |
| 0 | $(x + 8)(x - 7) =$ _____ |
| 0, 0 | $x + 8 =$ _____ or $x - 7 =$ _____ |
| $-8, 7$ | $x =$ _____ or $x =$ _____. |
| positive | $-8$ is not a solution, since we need _____ integers. Therefore, |
| 7, 8 | the two positive integers are $x =$ _____ and $x + 1 =$ _____. |
| | **2** The area of a rectangle is 192 square inches. Its length is 4 inches more than its width. Find its dimensions. Let $x =$ the length of the rectangle in inches. Then _____ = the width of the rectangle |
| $x - 4$ | in inches. The area of a rectangle is given by the formula |
| length, width | area = _____ × _____, so that $192 = x(x - 4)$ or |
| $x^2 - 4x - 192$ | _____ = 0. Solving the equation for $x$, we have |
| $x - 16, x + 12$ | (_____)(_____) = 0. Then |
| $16, -12$ | $x =$ _____ or $x =$ _____. |
| | It does not make sense to say that the side of the rectangle is |
| $-12, 16$ | _____. Therefore, the length is _____ inches and the width is |
| 12 | _____ inches. |
| | *Check:* |
| | The length $x = 16$ and the width $x - 4 = 12$. |
| | Then $x(x - 4) = (16)(12) \stackrel{?}{=} 192$. Yes. |
| | **3** Adding a border to a rectangular rug 6 feet by 9 feet increases its area by 34 square feet. Find the width of the rug after adding the border. To clarify the problem, we shall draw a diagram and represent the given information on it. |
| $6 + x$ | 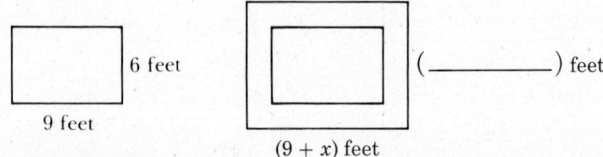 |
| | The area of the new rug is expressed by $(9 + x)(6 + x) =$ |
| 54, 54 | _____ + 34 or $54 + 15x + x^2 =$ _____ + 34, so that |
| $x^2 + 15x - 34$ | _____ = 0. Solving the equation, we have |
| $x - 2$ | $(x + 17)($ _____ $) = 0$. Then |
| $-17, 2$ | $x =$ _____ or $x =$ _____. |
| | Since it does not make sense to represent sides of the rug by |
| negative | _____ numbers, $-17$ is rejected as a solution to the problem. |
| 2, 8 feet | Hence, the width of the rug is $6 + x = 6 +$ _____ = _____. |

4 A rectangular plot of ground is 20 feet wide and 32 feet long. Its length and width are increased by the same amount, thereby doubling its area. By how much is each dimension increased? Draw a diagram to represent the given information.

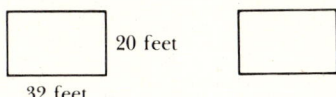

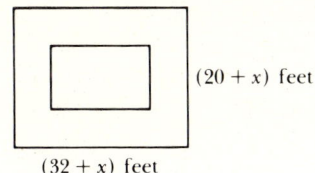

$(32 + x)(20 + x)$
32, 20, 1,280
$x^2 + 52x - 640$
$-26 + 2\sqrt{329}, -26 - 2\sqrt{329}$

Let $x$ = the number of feet its dimensions are increased. The area of the new rectangle is given by _____. Forming the equation, we have $(32 + x)(20 + x) = 2(\_\_\_)(\_\_\_) =$ _____ or _____ = 0. Solving for $x$, we obtain
$x =$ _____ or $x =$ _____.
$-26 - 2\sqrt{329}$ has no meaning in this problem and is therefore discarded. The solution is $x = -26 + 2\sqrt{329} =$
$-26 + 36.28 =$ _____.

10.28 feet

5 Bill and Al start from a certain crossroad at the same time, Bill going north at 4 miles per hour and Al going east at 3 miles per hour. In how many hours will they be 16 miles apart if the measurement is to be considered in a straight line across the country? Draw a diagram to represent the given information.

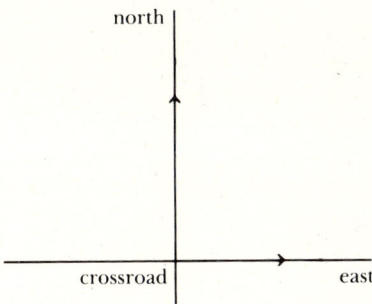

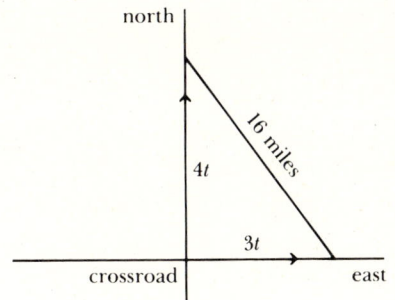

$4t$

$(4t)^2 + (3t)^2, 25t^2$
$25t^2 - 256$
$5t - 16, 5t + 16$
$\frac{16}{5}, -\frac{16}{5}$

Let $t$ = the required time in hours. After $t$ hours, Al will have traveled $3t$ miles going east and Bill will have traveled _____ miles going north. Therefore, the equation is
_____ = $(16)^2$ or _____ = 256.
_____ = 0
(_____)(_____) = 0
$t =$ _____ or $t =$ _____.
The solution $t = -\frac{16}{5}$ may be interpreted as meaning that if Bill and Al had been traveling along these roads before reaching the crossroad, they would have been 16 miles apart $\frac{16}{5}$ hours before meeting.

6 A boat goes 144 miles up a river and back again in 15 hours. Find the speed each way if the speed down the river is 8 miles greater than

$x + 8$

$\dfrac{144}{x}$

$\dfrac{144}{x + 8}$

15, 384

$5x^2 - 56x - 384$

$5x + 24, x - 16$

$-\dfrac{24}{5}, 16$

$-\dfrac{24}{5}$

16

the speed up the river. Let $x =$ the speed up the river in miles per hour. Then _____ = the speed down the river in miles per hour.

The time going up the river is _____ hours, and the time going down the river is _____ hours. Forming the equation, we have

$$\dfrac{144}{x} + \dfrac{144}{x + 8} = \underline{\qquad} \quad \text{or} \quad 5x^2 - 56x = \underline{\qquad}$$

$$\underline{\hspace{3cm}} = 0$$

$(\underline{\hspace{2cm}})(\underline{\hspace{2cm}}) = 0$

$x = \underline{\qquad}$ or $x = \underline{\qquad}$.

Since it makes no sense to have a negative speed, reject _____ as a solution to the problem. Therefore, the speed up the river is _____ miles per hour.

*Check:*

$x = 16$ in the equation above.

$$\dfrac{144}{16} + \dfrac{144}{16 + 8} \stackrel{?}{=} 15$$

$$9 + 6 \stackrel{?}{=} 15. \quad \text{Yes.}$$

## SOLUTIONS TO SELECTED ODD PROBLEMS  Section 6.4, text pages 301–304

**1** Let $n =$ first positive integer.
Then, $n + 1 =$ next positive integer.
Therefore, $n(n + 1) = 30$
$n^2 + n - 30 = 0$
$(n + 6)(n - 5) = 0$
$n + 6 = 0 \quad n - 5 = 0$
$n = -6 \quad n = 5$
We reject $n = -6$, so that the numbers are 5 and $5 + 1 = 6$.

**5** Let $x =$ David's age.
So that $x^2 = 5x + 6$
$x^2 - 5x - 6 = 0$
$(x + 1)(x - 6) = 0$
$x + 1 = 0 \quad x - 6 = 0$
$x = -1 \quad x = 6$
Therefore, David is 6 years old.

**9** Let $l =$ length and $w =$ width, so that by the perimeter formula,
$2l + 2w = 148$ or $l = 74 - w$.
By the area formula, $lw = 1{,}320$, so that
$(74 - w)w = 1{,}320$
$74w - w^2 = 1{,}320$
$w^2 - 74w + 1{,}320 = 0$
$(w - 30)(w - 44) = 0$
$w = 30 \quad \text{or} \quad w = 44$
So that,
$l = 74 - 30 \quad \text{or} \quad l = 74 - 44$
$l = 44 \qquad\qquad l = 30$.
Thus, the dimensions are 44 feet and 30 feet.

13  Let $x$ = width of walk.

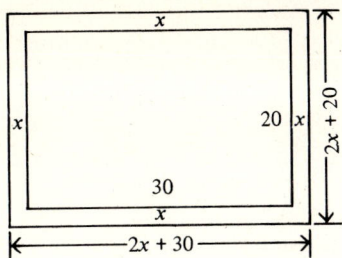

Area of outer rectangle is $(2x + 30)(2x + 20)$, which consists of area of lawn and area of flower border.
Therefore, $(2x + 30)(2x + 20) = 2(20)(30)$
$$4x^2 + 100x + 600 = 1200$$
$$4x^2 + 100x - 600 = 0$$
$$x^2 + 25x - 150 = 0$$
$$(x + 30)(x - 5) = 0$$
$x = -30$ or $x = 5$
Therefore, the width of the flower border is 5 feet.

17  Let $x$ = length of longer rafter.
Then, $x - 7$ = length of shorter rafter.

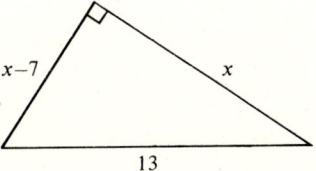

$$(x - 7)^2 + x^2 = 13^2$$
$$x^2 - 14x + 49 + x^2 = 169$$
$$2x^2 - 14x - 120 = 0$$
$$x^2 - 7x - 60 = 0$$
$$(x - 12)(x + 5) = 0$$
$x = 12$ or $x = -5$
Therefore, the rafters are 12 meters and $12 - 7 = 5$ meters.

21  $h = 128t - 16t^2$
For $h = 112$ feet, thus, $112 = 128t - 16t^2$
$$16t^2 - 128t + 112 = 0$$
$$t^2 - 8t + 7 = 0$$
$$(t - 1)(t - 7) = 0$$
$t = 1$ or $t = 7$
Therefore, $h = 112$ feet when $t = 1$ second or $t = 7$ seconds.

25  Let $x$ = Andy's average speed.

|       | Average Speed ($r$) | Distance ($d$) | Time $t$ $\left(t = \dfrac{d}{r}\right)$ |
|-------|---------------------|----------------|------------------------------------------|
| Andy  | $x$                 | 300            | $\dfrac{300}{x}$                         |
| Mario | $x - 20$            | 300            | $\dfrac{300}{x - 20}$                    |

Since 30 minutes = $\frac{1}{2}$ hour, we have
$$\frac{300}{x-20} - \frac{300}{x} = \frac{1}{2}.$$
$$300(x)(2) - 300(x-20)(2) = (x-20)x$$
$$600x - 600x + 12{,}000 = x^2 - 20x$$
$$x^2 - 20x - 12{,}000 = 0$$
$$(x - 120)(x + 100) = 0$$
$x = 120$ or $x = -100$
Therefore, Andy's average speed is 120 miles per hour.

**29** $R = 200x - \frac{1}{3}x^2$
For $R = 10{,}800$, thus, $10{,}800 = 200x - \frac{1}{3}x^2$
$$\frac{1}{3}x^2 - 200x + 10{,}800 = 0$$
$$x^2 - 600x + 32{,}400 = 0$$
$$(x - 540)(x - 60) = 0$$
$x = 540$ or $x = 60$
Therefore, the price per unit greater than $60 is $540.

## 6.5 Equations in Quadratic Form

### SEMIPROGRAMMED PROBLEMS

In problems 1–6, find the solutions of each equation.

| | |
|---|---|
| | **1** $x^4 - 15x^2 + 36 = 0$ |
| $x^2, x^4$ | Let $u =$ _____; then $u^2 =$ _____. Substitute $u$ into the equation |
| $u^2 - 15u + 36$ | to obtain _____ $= 0$. Then |
| $u - 3, u - 12$ | (_____)(_____) $= 0$. |
| 3, 12 | $u =$ _____ or $u =$ _____. |
| | Since $u = x^2$ and $u = 3$ or $u = 12$, we have |
| 3, 12 | $x^2 =$ _____ or $x^2 =$ _____ |
| $-\sqrt{3}, \sqrt{3}$ | $x =$ _____ or $x =$ _____ |
| $-2\sqrt{3}, 2\sqrt{3}$ | $x =$ _____ or $x =$ _____. |
| $-2\sqrt{3}, -\sqrt{3}, \sqrt{3}, 2\sqrt{3}$ | The solutions are _____. |
| | **2** $x^{-2} + x^{-1} - 6 = 0$ |
| $x^{-1}, x^{-2}$ | Let $u =$ _____; then $u^2 =$ _____. Substitute $u$ into the equation |
| $u^2 + u - 6, u - 2, u + 3$ | to obtain _____ $= 0$. Then (_____)(_____) $= 0$. |
| $2, -3$ | $u =$ _____ or $u =$ _____. |
| | Since $u = x^{-1}$ and $u = 2$ or $u = -3$, we have |
| $-3$ | $x^{-1} = 2$ or $x^{-1} =$ _____ |
| $2, -3$ | $\dfrac{1}{x} =$ _____ or $\dfrac{1}{x} =$ _____ |
| $\frac{1}{2}, -\frac{1}{3}$ | $x =$ _____ or $x =$ _____. |
| $-\frac{1}{3}, \frac{1}{2}$ | The solutions are _____. |
| | **3** $x^{-4} - 17x^{-2} + 16 = 0$ |
| $x^{-2}, x^{-4}$ | Let $u =$ _____; then $u^2 =$ _____. Substitute $u$ into the equation |
| $u$ | to obtain $u^2 - 17$_____ $+ 16 = 0$. Then |

$u - 1, u - 16$
1, 16

1, 16

1, 16

$1, \frac{1}{16}$
$1, -1$
$\frac{1}{4}, -\frac{1}{4}$
$1, -1, \frac{1}{4}, -\frac{1}{4}$

$u^2$
$u^2 - u - 6$
$u - 3, u + 2$
$3, -2$

$3, -2$
$x^2 + 2x - 3, x^2 + 2x + 2$

$x - 1, 2$

$-3, 1$
$-1 - i, -1 + i$
$-3, 1, -1 - i, -1 + i$

$\dfrac{1}{u}$

$u + \dfrac{2}{u} - 3, 3u$
$u - 1, u - 2$
1, 2

1, 2
$x^2 - x - 1, x^2 - 2x - 2$
$1 \pm \sqrt{3}$

$\dfrac{1 - \sqrt{5}}{2}, \dfrac{1 + \sqrt{5}}{2}, 1 - \sqrt{3}, 1 + \sqrt{3}$

$3x + \dfrac{2}{x}, \left(3x + \dfrac{2}{x}\right)^2$

---

(_____)(_____) = 0.
$u = $ _____ or $u = $ _____.
Since $u = x^{-2}$ and $u = 1$ or $u = 16$, we have
$x^{-2} = $ _____ or $x^{-2} = $ _____

$\dfrac{1}{x^2} = $ _____ or $\dfrac{1}{x^2} = $ _____

$x^2 = $ _____ or $x^2 = $ _____

$x = $ _____ or $x = $ _____

$x = $ _____ or $x = $ _____.

The solutions are _____.

**4** $(x^2 + 2x)^2 - (x^2 + 2x) = 6$
Let $x^2 + 2x = u$; then $(x^2 + 2x)^2 = $ _____. Substitute $u$ in the equation
to obtain _____ $= 0$. Then
(_____)(_____) = 0.
$u = $ _____ or $u = $ _____.
Since $u = x^2 + 2x$ and $u = 3$ or $u = -2$, we have
$x^2 + 2x = $ _____ or $x^2 + 2x = $ _____
_____ $= 0$ or _____ $= 0$
$(x + 3)($ _____ $) = 0$ or $x = \dfrac{-2 \pm \sqrt{4 - 4(1)(\underline{\phantom{x}})}}{2}$
$x = $ _____ or $x = $ _____
$x = $ _____ or $x = $ _____.
The solutions are _____.

**5** $\dfrac{x^2}{x + 1} + \dfrac{2(x + 1)}{x^2} = 3$

Letting $u = \dfrac{x^2}{x + 1}$ and $\dfrac{x + 1}{x^2} = $ _____, the resulting equation

is _____ $= 0$ or $u^2 - $ _____ $+ 2 = 0$. Then
(_____)(_____) = 0.
$u = $ _____ or $u = $ _____.
Since $u = \dfrac{x^2}{x + 1}$ and $u = 1$ or $u = 2$, we have

$\dfrac{x^2}{x + 1} = $ _____ or $\dfrac{x^2}{x + 1} = $ _____

_____ $= 0$ or _____ $= 0$

$x = \dfrac{1 \pm \sqrt{5}}{2}$ or $x = $ _____.

The solutions are _____.

**6** $\left(3x + \dfrac{2}{x}\right)^2 + 6\left(3x + \dfrac{2}{x}\right) + 5 = 0$

Let $u = $ _____ ; then $u^2 = $ _____. Substituting $u$ in the

$u^2 + 6u + 5$

$u + 1$

$-1, -5$

$-1, -5$

$3x^2 + x + 2, 3x^2 + 5x + 2$

$\dfrac{-1 \pm \sqrt{23}i}{6}, \dfrac{-5 \pm 1}{6}$

$\dfrac{-1 - \sqrt{23}i}{6}, \dfrac{-1 + \sqrt{23}i}{6}, -1, -\dfrac{2}{3}$

equation, we have _____ = 0, so

$(u + 5)(\underline{\phantom{xx}}) = 0$. Then

$u = \underline{\phantom{xx}}$ or $u = \underline{\phantom{xx}}$.

Since $u = 3x + \dfrac{2}{x}$ and $u = -1$ or $u = -5$, we have

$3x + \dfrac{2}{x} = \underline{\phantom{xx}}$ or $3x + \dfrac{2}{x} = \underline{\phantom{xx}}$.

The quadratic equations formed are

_____ = 0 or _____ = 0

$x = \underline{\phantom{xxxx}}$ or $x = \underline{\phantom{xxxx}}$.

The solutions are _____.

## SOLUTIONS TO SELECTED ODD PROBLEMS  Section 6.5, text pages 307–308

**1**   $x^4 - 13x^2 + 36 = 0$
$(x^2)^2 - 13x^2 + 36 = 0$
Let $u = x^2$, then $u^2 - 13u + 36 = 0$
$(u - 9)(u - 4) = 0$
$u = 9$    or    $u = 4$
$x^2 = 9$    or    $x^2 = 4$
$x = \pm\sqrt{9}$     $x = \pm\sqrt{4}$
$x = \pm 3$      $x = \pm 2$
Therefore, the solutions are $-3, -2, 2,$ and $3$.

**5**   $y^4 - 29y^2 + 100 = 0$
$(y^2)^2 - 29y^2 + 100 = 0$
Let $u = y^2$, then $u^2 - 29u + 100 = 0$
$(u - 25)(u - 4) = 0$
$u = 25$    or    $u = 4$
$y^2 = 25$    or    $y^2 = 4$
$y = \pm\sqrt{25}$     $y = \pm\sqrt{4}$
$y = \pm 5$      $y = \pm 2$
Therefore, the solutions are $-5, -2, 2,$ and $5$.

**9**   $6y^{-2} + 13y^{-1} - 5 = 0$
$6(y^{-1})^2 + 13y^{-1} - 5 = 0$
Let $u = y^{-1}$, then $6u^2 + 13u - 5 = 0$
$(3u - 1)(2u + 5) = 0$
$u = \dfrac{1}{3}$ or $u = -\dfrac{5}{2}$
or
$y^{-1} = \dfrac{1}{3}$    $y^{-1} = -\dfrac{5}{2}$
$\dfrac{1}{y} = \dfrac{1}{3}$    $\dfrac{1}{y} = -\dfrac{5}{2}$
$y = 3$     $y = -\dfrac{2}{5}$
Therefore, the solutions are $-\dfrac{2}{5}$ and $3$.

**13**   $t^{-4} - 20t^{-2} + 64 = 0$
$(t^{-2})^2 - 20t^{-2} + 64 = 0$
Let $u = t^{-2}$, then $u^2 - 20u + 64 = 0$
$(u - 16)(u - 4) = 0$
$u = 16$    or    $u = 4$
$t^{-2} = 16$    or    $t^{-2} = 4$
$t^2 = \dfrac{1}{16}$     $t^2 = \dfrac{1}{4}$
$t = \pm\sqrt{\dfrac{1}{16}}$     $t = \pm\sqrt{\dfrac{1}{4}}$
$t = \pm\dfrac{1}{4}$     $t = \pm\dfrac{1}{2}$
Therefore, the solutions are $-\dfrac{1}{2}, -\dfrac{1}{4}, \dfrac{1}{4},$ and $\dfrac{1}{2}$.

**17**   $(y^2 + 2y)^2 - 2(y^2 + 2y) = 3$
Let $u = y^2 + 2y$, then $u^2 - 2u = 3$
$u^2 - 2u - 3 = 0$
$(u - 3)(u + 1) = 0$
$u = 3$ or    $u = -1$
$y^2 + 2y = 3$ or    $y^2 + 2y = -1$
$y^2 + 2y - 3 = 0$    $y^2 + 2y + 1 = 0$
$(y + 3)(y - 1) = 0$    $(y + 1)^2 = 0$
$y = -3$ or $y = 1$    $y = -1$
Therefore, the solutions are $-3, -1,$ and $1$.

**21**   $(m^2 + 2m)^2 + m^2 + 2m - 12 = 0$
Let $u = m^2 + 2m$, then $u^2 + u - 12 = 0$
$(u + 4)(u - 3) = 0$
$u = -4$    or    $u = 3$
$m^2 + 2m = -4$   or   $m^2 + 2m = 3$
$m^2 + 2m + 1 = -4 + 1$    $m^2 + 2m - 3 = 0$
$(m + 1)^2 = -3$    $(m + 3)(m - 1) = 0$
$m + 1 = \pm\sqrt{-3}$    $m = -3$
$m = -1 \pm \sqrt{3}i$    $m = 1$
Therefore, the solutions are $-3, 1, -1 - \sqrt{3}i,$ and $-1 + \sqrt{3}i$.

**25** $\left(3x - \dfrac{2}{x}\right)^2 + 6\left(3x - \dfrac{2}{x}\right) + 5 = 0$

Let $u = 3x - \dfrac{2}{x}$, then

$u^2 + 6u + 5 = 0$
$(u + 1)(u + 5) = 0$
$u = -1$ or $u = -5$

$3x - \dfrac{2}{x} = -1$ or $\quad 3x - \dfrac{2}{x} = -5$

$3x^2 - 2 = -x \qquad\qquad 3x^2 - 2 = -5x$
$3x^2 + x - 2 = 0 \qquad 3x^2 + 5x - 2 = 0$
$(3x - 2)(x + 1) = 0 \quad (3x - 1)(x + 2) = 0$

$x = \dfrac{2}{3}$ or $x = -1 \qquad x = \dfrac{1}{3}$ or $x = -2$

Therefore, the solutions are $-2, -1, \tfrac{1}{3}$, and $\tfrac{2}{3}$.

**29** $\dfrac{m+1}{m} + 2 = \dfrac{3m}{m+1}$

$\dfrac{m+1}{m} + 2 = 3\left(\dfrac{m}{m+1}\right)$

Let $u = \dfrac{m+1}{m}$, so that $\dfrac{1}{u} = \dfrac{m}{m+1}$

$u + 2 = 3\left(\dfrac{1}{u}\right)$

$u^2 + 2u = 3$
$u^2 + 2u - 3 = 0$
$(u + 3)(u - 1) = 0$
$u = -3$ or $u = 1$

$\dfrac{m+1}{m} = -3$ or $\dfrac{m+1}{m} = 1$

$m + 1 = -3m \qquad m + 1 = m$
$4m = -1 \qquad\quad$ No solution here.

$m = -\dfrac{1}{4}$

Therefore, the solution is $-\tfrac{1}{4}$.

## 6.6 Equations Involving Radicals

**SEMIPROGRAMMED PROBLEMS**

In problems 1–10, solve each equation (find only the real solutions) and check the solution when necessary.

|  |  |
|---|---|
| | **1** $\sqrt{x} + 2 = 5$ |
| | $\sqrt{x} = 3$ |
| 9, 9 | $(\sqrt{x})^2 = \underline{\quad}$, so $x = \underline{\quad}$. |
| 3, 9 | Check: $\sqrt{9} \stackrel{?}{=} \underline{\quad}$. Yes. Therefore, the solution is $\underline{\quad}$. |
| | **2** $\sqrt{2x + 1} - 4 = 2$ |
| | $\sqrt{2x + 1} = \underline{\quad}$ |
| 6 | $(\sqrt{2x + 1})^2 = (\underline{\quad})^2$ |
| 6 | |
| 36 | $2x + 1 = \underline{\quad}$ |
| 35 | $2x = \underline{\quad}$ |
| | $x = \tfrac{35}{2}$ |
| 2, $\tfrac{35}{2}$ | Check: $\sqrt{2(\tfrac{35}{2}) + 1} - 4 \stackrel{?}{=} \underline{\quad}$. Yes. Thus, the solution is $\underline{\quad}$. |
| | **3** $\sqrt[3]{2y + 1} = 3$ |
| 27, 27, 26 | $(\sqrt[3]{2y + 1})^3 = \underline{\quad}$, so $2y + 1 = \underline{\quad}$ or $2y = \underline{\quad}$ |
| 13 | or $y = \underline{\quad}$. |
| 13 | The solution is $\underline{\quad}$. |
| | **4** $\sqrt[4]{x^3 + 7} = 2$ |
| 16, 16, 9 | $(\sqrt[4]{x^3 + 7})^4 = \underline{\quad}$ or $x^3 + 7 = \underline{\quad}$ or $x^3 = \underline{\quad}$, |
| $\sqrt[3]{9}$ | so $x = \underline{\quad}$. |

# CHAPTER 6 NONLINEAR EQUATIONS AND INEQUALITIES

|  |  |
|---|---|
| 2 | Check: $\sqrt[4]{(\sqrt[3]{9})^3 + 7} \stackrel{?}{=}$ _____. |
| $\sqrt[3]{9}$ | Yes. Therefore, the solution is _____. |

5    $\sqrt{x^2 - 5} = x + 1$

$x + 1$        $(\sqrt{x^2 - 5})^2 = (\underline{\hspace{2cm}})^2$

$x^2 + 2x + 1$        $x^2 - 5 = \underline{\hspace{2cm}}$

6        $-2x = \underline{\hspace{1cm}}$

$-3$        $x = \underline{\hspace{1cm}}$

Check: $\sqrt{(-3)^2 - 5} \stackrel{?}{=} -3 + 1$

$\sqrt{\underline{\hspace{1cm}}} \stackrel{?}{=} -2$

4        $2 \stackrel{?}{=} -2.$

solution    No. Therefore, the equation has no _____.

6    $\sqrt{x + 2} + \sqrt{3 - x} = 3$

$\sqrt{x + 2} = 3 - \sqrt{3 - x}$

$3 - \sqrt{3 - x}$        $(\sqrt{x + 2})^2 = (\underline{\hspace{2cm}})^2$

$3 - x$        $x + 2 = 9 - 6\sqrt{3 - x} + \underline{\hspace{1cm}}$

$2x - 10$        $\underline{\hspace{2cm}} = -6\sqrt{3 - x}$

$x - 5$        $\underline{\hspace{2cm}} = -3\sqrt{3 - x}$

$-3\sqrt{3 - x}$        $(x - 5)^2 = (\underline{\hspace{2cm}})^2$

$9(3 - x)$        $x^2 - 10x + 25 = \underline{\hspace{1cm}}$

$x^2 - x - 2$        $\underline{\hspace{2cm}} = 0$

$x - 2, x + 1$    $(\underline{\hspace{1cm}})(\underline{\hspace{1cm}}) = 0$

$2, -1$        $x = \underline{\hspace{1cm}}$ or $x = \underline{\hspace{1cm}}$

Check:

$x = 2$:

$\sqrt{2 + 2} + \sqrt{3 - 2} \stackrel{?}{=} 3$. Yes.

$x = -1$:

$\sqrt{-1 + 2} + \sqrt{3 - (-1)} \stackrel{?}{=} 3$. Yes.

$-1, 2$    The solutions are _____.

7    $\sqrt{2x - 5} = \sqrt[4]{x^2 - 7x + 13}$

$\sqrt[4]{x^2 - 7x + 13}$        $(\sqrt{2x - 5})^4 = (\underline{\hspace{2cm}})^4$

$x^2 - 7x + 13$        $(2x - 5)^2 = \underline{\hspace{2cm}}$

$4x^2 - 20x + 25$        $\underline{\hspace{2cm}} = x^2 - 7x + 13$

$3x^2 - 13x + 12$        $\underline{\hspace{2cm}} = 0$

$x - 3, 3x - 4$    $(\underline{\hspace{1cm}})(\underline{\hspace{1cm}}) = 0$

$3, \frac{4}{3}$        $x = \underline{\hspace{1cm}}$ or $x = \underline{\hspace{1cm}}$

Check:

$x = \frac{4}{3}$:

$\sqrt{2(\frac{4}{3}) - 5} \stackrel{?}{=} \sqrt[4]{(\frac{4}{3})^2 - 7(\frac{4}{3}) + 13}$. No.

$x = 3$:

$\sqrt{2(3) - 5} \stackrel{?}{=} \sqrt[4]{(3)^2 - 7(3) + 13}$. Yes.

3    Therefore, the solution is _____.

## 6.6 SEMIPROGRAMMED PROBLEMS

$1 - \sqrt{x+1}$
$1 - \sqrt{x+1}$
$1 - 2\sqrt{x+1} + x + 1$
$x + 1$
$-2\sqrt{x+1}$
$4(x+1)$
$x^2 - 2x - 3$
$x - 3, x + 1$
$3, -1$












$-1$



$3 + \sqrt{2x-6}$
$4x + 5$
$\sqrt{2x-6}$
$\sqrt{2x-6}$
$9(2x-6)$
$18x - 54$
$0$
$x - 5$
$x - 5$
$11, 5$



Yes


Yes
$5, 11$


$-x - 1$
$(-x-1)^2$
$x^2 + 2x + 1$
$0$
$x - 2$
$-3, 2$




Yes

8  $\sqrt{2x+3} + \sqrt{x+1} - 1 = 0$
$\sqrt{2x+3} = $ _____
$(\sqrt{2x+3})^2 = ($_____$)^2$
$2x + 3 = $ _____
_____ $= -2\sqrt{x+1}$
$(x+1)^2 = ($_____$)^2$
$x^2 + 2x + 1 = $ _____
_____ $= 0$
(_____)(_____) $= 0$
$x = $ _____ or $x = $ _____
Check:
$x = -1$:
$\sqrt{2(-1)+3} + \sqrt{-1+1} - 1 \stackrel{?}{=} 0$. Yes.
$x = 3$:
$\sqrt{2(3)+3} + \sqrt{3+1} - 1 \stackrel{?}{=} 0$. No.
The solution is _____.

9  $\sqrt{4x+5} = 3 + \sqrt{2x-6}$
$(\sqrt{4x+5})^2 = ($_____$)^2$
_____ $= 9 + 6\sqrt{2x-6} + 2x - 6$
$2x + 2 = 6($_____$)$
$x + 1 = 3($_____$)$
$(x+1)^2 = $ _____.
$x^2 + 2x + 1 = $ _____.
$x^2 - 16x + 55 = $ _____
$(x - 11)($_____$) = 0$
$x - 11 = 0$  or  _____ $= 0$
$x = $ _____  or  $x = $ _____
Check:
$x = 5$:
$\sqrt{4(5)+5} \stackrel{?}{=} 3 + \sqrt{2(5)-6}$. _____.
$x = 11$:
$\sqrt{4(11)+5} \stackrel{?}{=} 3 + \sqrt{2(11)-6}$. _____.
The solutions are _____.

10  $x + 1 + \sqrt{x+7} = 0$
$\sqrt{x+7} = $ _____
$(\sqrt{x+7})^2 = $ _____
$x + 7 = $ _____
$x^2 + x - 6 = $ _____
$(x + 3)($_____$) = 0$
$x = $ _____  or  $x = $ _____
Check:
$x = -3$:
$-3 + 1 + \sqrt{-3+7} \stackrel{?}{=} 0$. _____.

No
−3

$x = 2$:
$2 + 1 + \sqrt{2 + 7} \stackrel{?}{=} 0.$ _____
The solution is _____.

## SOLUTIONS TO SELECTED ODD PROBLEMS   Section 6.6, text pages 312–313

**1**   $\sqrt{x} - 2 = 3$
$\sqrt{x} = 5$
$(\sqrt{x})^2 = (5)^2$
$x = 25$
Check: $\sqrt{25} - 2 = 3$
$5 - 2 = 3$

**5**   $\sqrt{2w + 5} - 4 = 0$
$\sqrt{2w + 5} = 4$
$(\sqrt{2w + 5})^2 = 4^2$
$2w + 5 = 16$
$2w = 11$
$w = \frac{11}{2}$
Check: $\sqrt{2(\frac{11}{2}) + 5} - 4 = 0$
$\sqrt{16} - 4 = 0$

**9**   $\sqrt{11 - x} = \sqrt{x + 6}$
$(\sqrt{11 - x})^2 = (\sqrt{x + 6})^2$
$11 - x = x + 6$
$-2x = -5$
$x = \frac{5}{2}$
Check: $\sqrt{11 - \frac{5}{2}} = \sqrt{\frac{5}{2} + 6}$
$\sqrt{\frac{17}{2}} = \sqrt{\frac{17}{2}}$

**13**   $\sqrt{x^2 + 3x} = x + 1$
$(\sqrt{x^2 + 3x})^2 = (x + 1)^2$
$x^2 + 3x = x^2 + 2x + 1$
$3x = 2x + 1$
$x = 1$
Check: $\sqrt{1^2 + 3(1)} = 1 + 1$
$\sqrt{4} = 2$

**17**   $\sqrt[3]{3b - 4} = 2$
$(\sqrt[3]{3b - 4})^3 = 2^3$
$3b - 4 = 8$
$3b = 12$
$b = 4$
Check: $\sqrt[3]{3(4) - 4} = 2$
$\sqrt[3]{8}$

**21**   $5 + \sqrt[4]{x - 5} = 0$
$\sqrt[4]{x - 5} = -5$
$(\sqrt[4]{x - 5})^4 = (-5)^4$
$x - 5 = 625$
$x = 630$
Check: $5 + \sqrt[4]{630 - 5} = 0$
$5 + \sqrt[4]{625} = 0$
$5 + 5 \neq 0$
Therefore, no solution.

**25**   $2 + \sqrt[4]{7x - 5} = 6$
$\sqrt[4]{7x - 5} = 4$
$(\sqrt[4]{7x - 5})^4 = 4^4$
$7x - 5 = 256$
$7x = 261$
$x = \frac{261}{7}$
Check: $2 + \sqrt[4]{7(\frac{261}{7}) - 5} = 6$
$2 + \sqrt[4]{256} = 6$
$2 + 4 = 6$

**29**   $\sqrt{p + 12} = 2 + \sqrt{p}$
$(\sqrt{p + 12})^2 = (2 + \sqrt{p})^2$
$p + 12 = 4 + 4\sqrt{p} + p$
$8 = 4\sqrt{p}$
$2 = \sqrt{p}$
$2^2 = (\sqrt{p})^2$
$4 = p$
Check: $\sqrt{4 + 12} = 2 + \sqrt{4}$
$\sqrt{16} = 2 + 2$
$4 = 4$

**33**   $\sqrt{5t + 1} = 1 + \sqrt{3t}$
$(\sqrt{5t + 1})^2 = (1 + \sqrt{3t})^2$
$5t + 1 = 1 + 2\sqrt{3t} + 3t$
$2t = 2\sqrt{3t}$
$t = \sqrt{3t}$
$t^2 = (\sqrt{3t})^2$
$t^2 = 3t$
$t^2 - 3t = 0$
$t(t - 3) = 0$
$t = 0$ or $t = 3$
Check: $t = 0$   $\sqrt{5(0) + 1} = 1 + \sqrt{3(0)}$
$\sqrt{1} = 1 + \sqrt{0}$
$1 = 1$
$t = 3$   $\sqrt{5(3) + 1} = 1 + \sqrt{3(3)}$
$\sqrt{16} = 1 + \sqrt{9}$
$4 = 1 + 3$

**37**   $\sqrt{t + 2} + \sqrt{t - 3} = 5$
$\sqrt{t + 2} = 5 - \sqrt{t - 3}$
$(\sqrt{t + 2})^2 = (5 - \sqrt{t - 3})^2$
$t + 2 = 25 - 10\sqrt{t - 3} + t - 3$
$10\sqrt{t - 3} = 20$
$\sqrt{t - 3} = 2$
$(\sqrt{t - 3})^2 = 2^2$
$t - 3 = 4$
$t = 7$
Check: $\sqrt{7 + 2} + \sqrt{7 - 3} = 5$
$\sqrt{9} + \sqrt{4} = 5$
$3 + 2 = 5$

41
$$2\sqrt{1-3y} = 2 - \sqrt{2-4y}$$
$$(2\sqrt{1-3y})^2 = (2-\sqrt{2-4y})^2$$
$$4(1-3y) = 4 - 4\sqrt{2-4y} + 2 - 4y$$
$$4 - 12y = 6 - 4\sqrt{2-4y} - 4y$$
$$4\sqrt{2-4y} = 2 + 8y$$
$$2\sqrt{2-4y} = 1 + 4y$$
$$(2\sqrt{2-4y})^2 = (1+4y)^2$$
$$4(2-4y) = 1 + 8y + 16y^2$$
$$8 - 16y = 1 + 8y + 16y^2$$
$$16y^2 + 24y - 7 = 0$$
$$(4y-1)(4y+7) = 0$$
$$y = \tfrac{1}{4} \text{ or } y = -\tfrac{7}{4}$$

Check: $y = \tfrac{1}{4}$   $2\sqrt{1-3(\tfrac{1}{4})} = 2 - \sqrt{2-4(\tfrac{1}{4})}$
$$2\sqrt{\tfrac{1}{4}} = 2 - \sqrt{1}$$
$$2(\tfrac{1}{2}) = 2 - 1$$
$$1 = 1$$

$y = -\tfrac{7}{4}$   $2\sqrt{1-3(-\tfrac{7}{4})} = 2 - \sqrt{2-4(-\tfrac{7}{4})}$
$$2\sqrt{\tfrac{25}{4}} = 2 - \sqrt{9}$$
$$2(\tfrac{5}{2}) = 2 - 3$$
$$5 \neq -1$$

Therefore, $\tfrac{1}{4}$ is the only solution.

45
$$r = \sqrt{\frac{V}{\pi h}}, \text{ for } V$$
$$r^2 = \left(\sqrt{\frac{V}{\pi h}}\right)^2$$
$$r^2 = \frac{V}{\pi h}$$
$$(r^2)\pi h = \left(\frac{V}{\pi h}\right)\pi h$$
$$\pi r^2 h = V$$

## 6.7 Equations Involving Rational Exponents

### SEMIPROGRAMMED PROBLEMS

In problems 1–8, solve each equation.

| | |
|---|---|
| $\sqrt[3]{x}$ | **1**   $x^{1/3} = 2$ |
| $\sqrt[3]{x}$ | The equation can be rewritten as: \_\_\_\_ = 2 so that |
| 8 | (\_\_\_\_)$^3$ = $2^3$ or |
| | $x = $ \_\_\_\_. |
| $\sqrt[5]{y}$ | **2**   $y^{1/5} = -2$ |
| $\sqrt[5]{y}$ | The equation can be rewritten as: \_\_\_\_ = $-2$ so that |
| $-32$ | (\_\_\_\_)$^5$ = $(-2)^5$ or |
| | $y = $ \_\_\_\_. |
| $\sqrt[4]{z}$ | **3**   $z^{1/4} = 3$ |
| $\sqrt[4]{z}$ | The equation can be rewritten as: \_\_\_\_ = 3 so that |
| 81 | (\_\_\_\_)$^4$ = $3^4$ or |
| | $z = $ \_\_\_\_. |
| | Check: |
| 81, 3, 81 | (\_\_\_\_)$^{1/4}$ = \_\_\_\_. Therefore, the solution is \_\_\_\_. |
| | **4**   $(x-1)^{2/3} = 4$ |
| | The equation can be rewritten as: |
| $(x-1)^2$ | $\sqrt[3]{\underline{\phantom{xxxx}}} = 4$ so that |
| $(x-1)^2$ | $(\sqrt[3]{\underline{\phantom{xxxx}}})^3 = (4)^3$ or |
| $(x-1)^2$ | \_\_\_\_ = 64 |
| $(x-1)$ | \_\_\_\_ = $\pm\sqrt{64}$. |

8, 8
8
9, −7
−7, 9

$(5z + 2)^{2/5}$

$(5z + 2)^{2/5}$

$5z + 2$
$(5z + 2)^2$
$5z + 2$
32, 32
30, −34
6, $-\frac{34}{5}$

Yes

Yes
6, $-\frac{34}{5}$

$u^2$
$2u^2 + 5u − 3$
$2u − 1, u + 3$
$\frac{1}{2}, −3$

$\frac{1}{2}, −3$
$\frac{1}{2}, −3$
$\frac{1}{8}, −27$

$\frac{1}{4}, \frac{1}{2}$
3

9, −3
−15
−27, $\frac{1}{8}$

$u$
$u^2$
$u^2 + u − 6, u + 3, u − 2$
−3, 2

$x − 1 = +(\_\_\_)$ or $x − 1 = −(\_\_\_)$
$x = 1 + \_\_\_.$  $x = 1 − 8$
$x = \_\_\_$  $x = \_\_\_$
The solutions are _____.

5 $\quad (5z + 2)^{-2/5} = \dfrac{1}{4}$

$\dfrac{1}{_____} = \dfrac{1}{4}$

$_____ = 4$

This equation can be rewritten as
$\sqrt[5]{(\_\_\_\_\_)^2} = 4$ so that
$_____ = (4)^5 = 1024$ or
$_____ = \pm\sqrt{1024} = \pm 32.$
Thus, $5z + 2 = +(\_\_\_)$ or $5z + 2 = −(\_\_\_)$
$5z = \_\_\_$ or $5z = \_\_\_$
$z = \_\_\_$ or $z = \_\_\_$
Check:
$z = 6$:
$[5(6) + 2]^{-2/5} \stackrel{?}{=} \frac{1}{4}. \_\_\_.$
$z = -\frac{34}{5}$:
$[5(-\frac{34}{5}) + 2]^{-2/5} \stackrel{?}{=} \frac{1}{4}. \_\_\_.$
Therefore, the solutions are _____.

6 $\quad 2x^{2/3} + 5x^{1/3} − 3 = 0$
Let $u = x^{1/3}$; then $x^{2/3} = \_\_\_$. Substitute $u$ in the equation to obtain $_____ = 0$.
$(\_\_\_\_\_)(\_\_\_\_\_) = 0$
$u = \_\_\_$ or $u = \_\_\_$.
Since $u = x^{1/3}$ and $u = \frac{1}{2}$ or $u = −3$, we have
$x^{1/3} = \_\_\_$ or $x^{1/3} = \_\_\_$
$\sqrt[3]{x} = \_\_\_$ or $\sqrt[3]{x} = \_\_\_$
$x = \_\_\_$ or $x = \_\_\_$.
Check:
$2(\frac{1}{8})^{2/3} + 5(\frac{1}{8})^{1/3} − 3 \stackrel{?}{=} 0$
$2(\_\_\_) + 5(\_\_\_) − 3 \stackrel{?}{=} 0$
$\frac{1}{2} + \frac{5}{2} − \_\_\_ \stackrel{?}{=} 0.$ Yes
$2(−27)^{2/3} + 5(−27)^{1/3} − 3 \stackrel{?}{=} 0$
$2(\_\_\_) + 5(\_\_\_) − 3 \stackrel{?}{=} 0$
$18 + (\_\_\_) − 3 \stackrel{?}{=} 0.$ Yes
Therefore, the solutions are _____.

7 $\quad x^2 + 3x + (x^2 + 3x)^{1/2} − 6 = 0$
Let $(x^2 + 3x)^{1/2} = u$ or $\sqrt{x^2 + 3x} = \_\_\_$.
Then $x^2 + 3x = \_\_\_$. Substitute $u$ in the equation to obtain
$_____ = 0,$ so $(\_\_\_\_\_)(\_\_\_\_\_) = 0.$ Then
$u = \_\_\_$ or $u = \_\_\_$.

## 6.7 SEMIPROGRAMMED PROBLEMS

9, 4

$x^2 + 3x - 9, x^2 + 3x - 4$

$\dfrac{-3 - 3\sqrt{5}}{2}, \dfrac{-3 + 3\sqrt{5}}{2}$

$-4, 1$

Since $u = \sqrt{x^2 + 3x}$ and $u = -3$ or $u = 2$, we have the equations
$$x^2 + 3x = \underline{\phantom{xx}} \quad \text{or} \quad x^2 + 3x = \underline{\phantom{xx}}$$
$$\underline{\phantom{xxxxxxxx}} = 0 \quad \text{or} \quad \underline{\phantom{xxxxxxxx}} = 0$$
$$x = \underline{\phantom{xxxxx}} \quad \text{or} \quad x = \underline{\phantom{xxxxx}}$$
$$x = \underline{\phantom{xx}} \quad \text{or} \quad x = \underline{\phantom{xx}}.$$

*Check:*

Let $(x^2 + 3x)^{1/2} = \sqrt{x^2 + 3x}$. So for $x = 1$:
$1^2 + 3(1) + \sqrt{1^2 + 3(1)} - 6 \stackrel{?}{=} 0$. Yes.

$x = -4$:
$$(-4)^2 + 3(-4) + \sqrt{(-4)^2 + 3(-4)} - 6 \stackrel{?}{=} 0$$
$$4 + \sqrt{(\underline{\phantom{xx}})} - 6 \stackrel{?}{=} 0. \text{ Yes.}$$

$x = \dfrac{-3 + 3\sqrt{5}}{2}$:

$$\left(\dfrac{-3 + 3\sqrt{5}}{2}\right)^2 + 3\left(\dfrac{-3 + 3\sqrt{5}}{2}\right)$$
$$+ \sqrt{\left(\dfrac{-3 + 3\sqrt{5}}{2}\right)^2 + 3\left(\dfrac{-3 + 3\sqrt{5}}{2}\right)} - 6 \stackrel{?}{=} 0$$
$$9 + \sqrt{9} - 6 \stackrel{?}{=} 0. \text{ No.}$$

$x = \dfrac{-3 - 3\sqrt{5}}{2}$:

$$\left(\dfrac{-3 - 3\sqrt{5}}{2}\right)^2 + 3\left(\dfrac{-3 - 3\sqrt{5}}{2}\right)$$
$$+ \sqrt{\left(\dfrac{-3 - 3\sqrt{5}}{2}\right)^2 + 3\left(\dfrac{-3 - 3\sqrt{5}}{2}\right)} - 6 \stackrel{?}{=} 0$$
$$9 + \sqrt{9} - 6 \stackrel{?}{=} 0. \text{ No.}$$

$-4, 1$

The solutions are $\underline{\phantom{xxxxx}}$.

**8** $2x^2 + x - 4(2x^2 + x + 4)^{1/2} = 1$

The equation can be written in the form
$(2x^2 + x + 4) - 4(2x^2 + x + 4)^{1/2} = 5$.

$\sqrt{2x^2 + x + 4}$

$u^2$

Let $(2x^2 + x + 4)^{1/2} = u$, or $\underline{\phantom{xxxxxx}} = u$. Then
$2x^2 + x + 4 = \underline{\phantom{xx}}$.

Substitute $u$ in the equation to obtain

$u^2 - 4u - 5$

$u - 5, u + 1$

$5, -1$

$u^2 - 4u = 5$ or $\underline{\phantom{xxxxxx}} = 0$.
$(\underline{\phantom{xx}})(\underline{\phantom{xx}}) = 0$
$u = \underline{\phantom{xx}}$ or $u = \underline{\phantom{xx}}$.

Since $u = \sqrt{2x^2 + x + 4}$ and $u = 5$ or $u = -1$, we have

25, 1

$2x^2 + x - 21, 2x^2 + x + 3$

$2x^2 + x + 4 = \underline{\phantom{xx}}$ or $2x^2 + x + 4 = \underline{\phantom{xx}}$
$\underline{\phantom{xxxxxxxx}} = 0$ or $\underline{\phantom{xxxxxxxx}} = 0$
$(2x + 7)(x - 3) = 0$ or $x = \dfrac{-1 \pm \sqrt{1 - 24}}{4}$

$-\frac{7}{2}, 3$

$\dfrac{-1-\sqrt{23}i}{4}, \dfrac{-1+\sqrt{23}i}{4}$

21, 25

25

$-\frac{7}{2}, 3$

$x = \underline{\hspace{1cm}}$ or $x = \underline{\hspace{1cm}}$

$x = \underline{\hspace{2cm}}$ or $x = \underline{\hspace{2cm}}$.

Check:

Let $(2x^2 + x + 4)^{1/2} = \sqrt{2x^2 + x + 4}$. So for $x = -\frac{7}{2}$:

$2(-\frac{7}{2})^2 + (-\frac{7}{2}) - 4\sqrt{2(-\frac{7}{2})^2 + (-\frac{7}{2}) + 4} \stackrel{?}{=} 1$

$\underline{\hspace{1cm}} - 4\sqrt{\underline{\hspace{1cm}}} \stackrel{?}{=} 1$. Yes.

$x = 3$:

$2(9) + 3 - 4\sqrt{2(9) + 3 + 4} \stackrel{?}{=} 1$

$21 - 4\sqrt{\underline{\hspace{1cm}}} \stackrel{?}{=} 1$. Yes.

$x = \dfrac{-1 - \sqrt{23}i}{4}$:

$2\left(\dfrac{-1-\sqrt{23}i}{4}\right)^2 + \left(\dfrac{-1-\sqrt{23}i}{4}\right)$

$-4\sqrt{2\left(\dfrac{-1-\sqrt{23}i}{4}\right)^2 + \left(\dfrac{-1+\sqrt{23}i}{4}\right)} + 4 \stackrel{?}{=} 1$. No.

$x = \dfrac{-1 + \sqrt{23}i}{4}$:

$2\left(\dfrac{-1+\sqrt{23}i}{4}\right)^2 + \left(\dfrac{-1+\sqrt{23}i}{4}\right)$

$-4\sqrt{2\left(\dfrac{-1+\sqrt{23}i}{4}\right)^2 + \left(\dfrac{-1-\sqrt{23}i}{4}\right)} + 4 \stackrel{?}{=} 1$. No.

Therefore, the solutions are $\underline{\hspace{2cm}}$.

## SOLUTIONS TO SELECTED ODD PROBLEMS   Section 6.7, text page 317

1  $y^{1/5} = -3$
   $(y^{1/5})^5 = (-3)^5$
   $y = -243$

5  $x^{2/3} = 16$
   $(x^{2/3})^3 = 16^3$
   $x^2 = 4096$
   $x = \pm 64$

9  $(x + 1)^{3/5} = 1$
   $[(x + 1)^{3/5}]^5 = 1^5$
   $(x + 1)^3 = 1$
   $x + 1 = 1$
   $x = 0$

13  $(3y - 7)^{4/3} = 1$
    $[(3y - 7)^{4/3}]^3 = 1^3$
    $(3y - 7)^4 = 1$
    $3y - 7 = \pm 1$
    $3y - 7 = 1$ or $3y - 7 = -1$
    $y = \frac{8}{3}$ or $y = 2$

17  $(2x - 1)^{-1/3} = 8$
    $[(2x - 1)^{-1/3}]^{-3} = 8^{-3}$
    $2x - 1 = \dfrac{1}{8^3}$
    $2x - 1 = \dfrac{1}{512}$
    $2x = \dfrac{513}{512}$
    $x = \dfrac{513}{1024}$

21  $y^{2/3} + 2y^{1/3} = 8$
    Let $u = y^{1/3}$
    $u^2 + 2u - 8 = 0$
    $(u + 4)(u - 2) = 0$
    $u = -4$    $u = 2$
    $y^{1/3} = -4$    $y^{1/3} = 2$
    $y = -64$    $y = 8$

**25**  $x^3 - 9x^{3/2} + 8 = 0$
Let $u = x^{3/2}$
$u^2 - 9u + 8 = 0$
$(u - 8)(u - 1) = 0$
$u = 8 \qquad u = 1$
$x^{3/2} = 8 \qquad x^{3/2} = 1$
$x^3 = 64 \qquad x^3 = 1$
$x = 4 \qquad x = 1$
Check:
$4^3 - 9(4)^{3/2} + 8 = 0$
$64 - 9(8) + 8 = 0$
$64 - 72 + 8 = 0$
$1^3 - 9(1)^{3/2} + 8 = 0$
$1 - 9(1) + 8 = 0$
$1 - 9 + 8 = 0$

**29**  $(m + 20)^{1/2} - 4(m + 20)^{1/4} + 3 = 0$
Let $u = (m + 20)^{1/4}$
$u^2 - 4u + 3 = 0$
$(u - 3)(u - 1) = 0$
$u = 3 \qquad u = 1$
$(m + 20)^{1/4} = 3 \qquad (m + 20)^{1/4} = 1$
$m + 20 = 81 \qquad m + 20 = 1$
$m = 61 \qquad m = -19$
Check:
$(61 + 20)^{1/2} - 4(61 + 20)^{1/4} + 3 = 0$
$9 - 4(3) + 3 = 0$
$9 - 12 + 3 = 0$
$(-19 + 20)^{1/2} - 4(-19 + 20)^{1/4} + 3 = 0$
$1^{1/2} - 4(1)^{1/4} + 3 = 0$
$1 - 4 + 3 = 0$

## 6.8 Nonlinear Inequalities

### SEMIPROGRAMMED PROBLEMS

In problems 1–6, solve each quadratic inequality.

**1**  $(x - 1)(x - 3) < 0$

*Step 1*  Set $(x - 1)(x - 3) = 0$
Solve the equation so that

0, 0    $x - 1 = $ _____ or $x - 3 = $ _____
1, 3    $x = $ _____ or $x = $ _____.

*Step 2*  The numbers 1 and 3 from step 1 separate the number line into

three    _____ parts, denoted by $A$, $B$, and $C$.

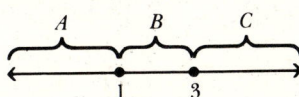

*Step 3*  Select numbers from each part and check them as follows:

0 − 1, 3    $A$: If $x = 0$, then (_____)$(0 - 3) = $ _____.
2 − 3, −1    $B$: If $x = 2$, then $(2 - 1)($_____$) = $ _____.
4 − 1, 4 − 3, 3    $C$: If $x = 4$, then (_____)(_____) = _____.

*Step 4*  From step 3, we see that the algebraic sign of each part is:

+ + + 0 − − − 0 + + +
  1       3

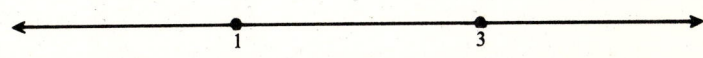

Thus, the solution set consists of all real numbers $x$ such that

$1 < x < 3$    _____.

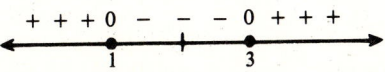

The graph is _____.

**2**  $x(x - 3) \leq 0$

*Step 1*  Set _____ = 0.
$x(x - 3)$    Solve the equation so that

0, 0    $x = $ _____ or $x - 3 = $ _____
0, 3    $x = $ _____ or $x = $ _____.

**134** CHAPTER 6 NONLINEAR EQUATIONS AND INEQUALITIES

three

$-1(-1-3), 4$
$1(1-3), -2$
$4(4-3), 4$

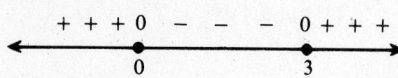

$0 \leq x \leq 3$

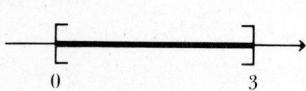

$(x-1)(x-4)$

$0, 0$
$1, 4$
three

$(0-1)(0-4), 4$
$(2-1)(2-4), -2$
$(5-1)(5-4), 4$

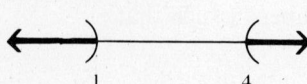

$x < 1$ or $x > 4$

$(x+4)(x-3)$

$0, 0$
$-4, 3$

---

*Step 2* The numbers 0 and 3 separate the number line into _____ parts, denoted by A, B, and C.

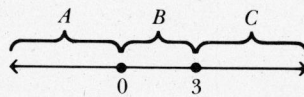

*Step 3* Select numbers from each part and check them as follows:
A: If $x = -1$, then _____ = _____.
B: If $x = 1$, then _____ = _____.
C: If $x = 4$, then _____ = _____.

*Step 4* From step 3, we see that the algebraic sign of each part is:

Thus, the solution set consists of all real numbers $x$ such that _____.

The graph is _____.

**3** $x^2 - 5x + 4 > 0$

*Step 1* Set $x^2 - 5x + 4 =$ _____ $= 0$.
Solve the equation so that
$x - 1 =$ _____ or $x - 4 =$ _____
$x =$ _____ or $x =$ _____.

*Step 2* The numbers 1 and 4 separate the number line into _____ parts, denoted by A, B, and C.

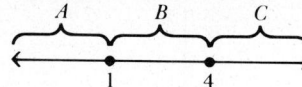

*Step 3* Select numbers from each part and check them as follows:
A: If $x = 0$, then _____ = _____.
B: If $x = 2$, then _____ = _____.
C: If $x = 5$, then _____ = _____.

*Step 4* The algebraic sign of each part is:

Thus, the solution set consists of all real numbers $x$ such that _____.

The graph is _____.

**4** $x^2 + x - 12 \geq 0$

*Step 1* Set $x^2 + x - 12 =$ _____ $= 0$.
Solve the equation so that
$x + 4 =$ _____ or $x - 3 =$ _____
$x =$ _____ or $x =$ _____.

three

$(-5+4)(-5-3), 8$
$(0+4)(0-3), -12$
$(4+4)(4-3), 8$

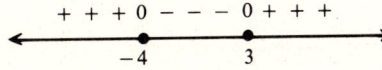

$x \leq -4$ or $x \geq 3$

$(5x-2)(2x+3)$

$0, 0$

$\frac{2}{5}, -\frac{3}{2}$

three

$(-10-2)(-4+3), 12$
$(0-2)(0+3), -6$
$(5-2)(2+3), 15$

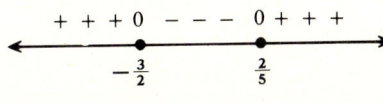

$-\frac{3}{2} < x < \frac{2}{5}$

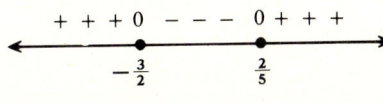

$(2x+1)(x-1)$

$0, 0$

$-\frac{1}{2}, 1$

---

6.8 SEMIPROGRAMMED PROBLEMS **135**

*Step 2* The numbers $-4$ and $3$ separate the number line into _____ parts, denoted by $A$, $B$, and $C$.

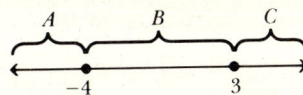

*Step 3* Select numbers from each part and check them as follows:
A: If $x = -5$, then _____ = ____.
B: If $x = 0$, then _____ = ____.
C: If $x = 4$, then _____ = ____.

*Step 4* The algebraic sign of each part is:

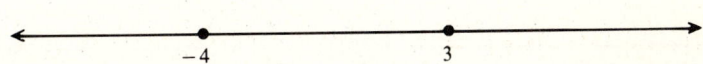

Thus, the solution set consists of all real numbers $x$ such that _____.

The graph is _____.

5 $10x^2 + 11x < 6$

*Step 1* Set $10x^2 + 11x - 6 = $ _____ $= 0$.
Solve the equation so that
$5x - 2 = $ ____ or $2x + 3 = $ ____
$x = $ ____ or $x = $ ____.

*Step 2* The numbers $\frac{2}{5}$ and $-\frac{3}{2}$ separate the number line into _____ parts, denoted by $A$, $B$, and $C$.

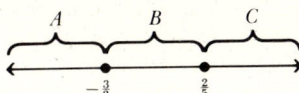

*Step 3* Select numbers from each part and check them as follows:
A: If $x = -2$, then _____ = ____.
B: If $x = 0$, then _____ = ____.
C: If $x = 1$, then _____ = ____.

*Step 4* The algebraic sign of each part is:

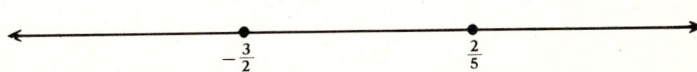

The solution set consists of all real numbers $x$ such that _____.

The graph is _____.

6 $2x^2 - x - 1 \geq 0$

*Step 1* Set $2x^2 - x - 1 = $ _____ $= 0$.
Solve the equation so that
$2x + 1 = $ ____ or $x - 1 = $ ____
$x = $ ____ or $x = $ ____.

three

$(-2+1)(-1-1), 2$
$(0+1)(0-1), -1$
$(4+1)(2-1), 5$

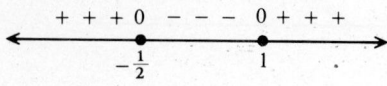

$x \leq -\tfrac{1}{2}$ or $x \geq 1$

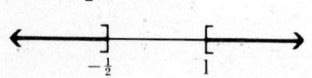

$x + 3$

$-3$

three

$-12 - 2, 14$

$0 - 2, -\tfrac{2}{3}$

$3 - 2, \tfrac{1}{4}$

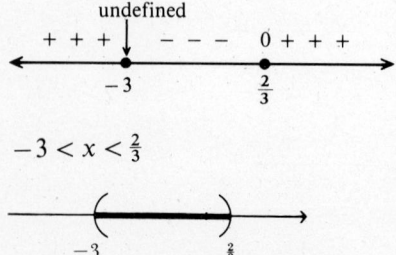

$-3 < x < \tfrac{2}{3}$

*Step 2* The numbers $-\tfrac{1}{2}$ and 1 separate the number line into _____ parts, denoted by A, B, and C.

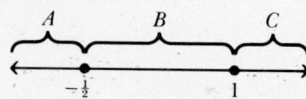

*Step 3* Select numbers from each part and check them as follows:
A: If $x = -1$, then _____ = ____.
B: If $x = 0$, then _____ = ____.
C: If $x = 2$, then _____ = ____.

*Step 4* The algebraic sign of each part is:

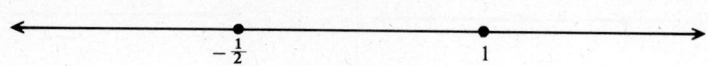

The solution set is all real $x$ such that _____.
The graph is _____.

In problems 7–9, solve each rational inequality.

7  $\dfrac{3x - 2}{x + 3} < 0$

*Step 1* Set $3x - 2 = 0$ and _____ = 0.
Solve these equations for $x$ so that
$x = \tfrac{2}{3}$ or $x =$ _____.

*Step 2* The numbers $-3$ and $\tfrac{2}{3}$ divide the number line into _____ parts, which we denote by A, B, and C.

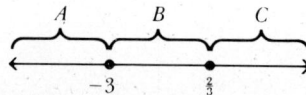

*Step 3* Select numbers in each part and check them as follows:
A: If $x = -4$, then $\dfrac{\phantom{xxxx}}{-4 + 3} =$ ____.
B: If $x = 0$, then $\dfrac{\phantom{xxxx}}{0 + 3} =$ ____.
C: If $x = 1$, then $\dfrac{\phantom{xxxx}}{1 + 3} =$ ____.

*Step 4* The algebraic sign of each part is:

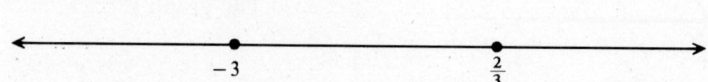

The solution set is all real $x$ such that _____.
The graph is _____.

$x + 2$

$-2$
$-2$

$-3 - 4, 7$

$0 - 4, -2$

$5 - 4, \frac{1}{7}$

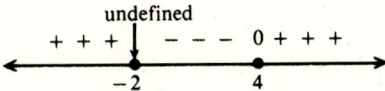

$x < -2$ or $x > 4$

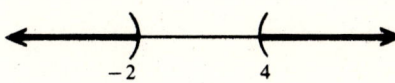

$x - 2$

$2$

three

$-5 + 1, \frac{4}{3}$

$0 + 1, -\frac{1}{2}$

$15 + 1, 16$

---

**6.8 SEMIPROGRAMMED PROBLEMS**    137

**8**   $\dfrac{x-4}{x+2} > 0$

*Step 1*   Set $x - 4 = 0$ and _____ = 0.
Solve these equations so that
$x = 4$ or $x = $ _____.

*Step 2*   The numbers 4 and _____ divide the number line into three parts, which we denote by $A$, $B$, and $C$.

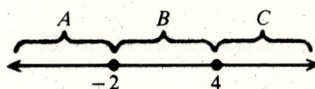

*Step 3*   Select numbers in each part and check them as follows:

$A$:   If $x = -3$, then $\dfrac{\phantom{xxx}}{-3+2} = $ _____.

$B$:   If $x = 0$, then $\dfrac{\phantom{xxx}}{0+2} = $ _____.

$C$:   If $x = 5$, then $\dfrac{\phantom{xxx}}{5+2} = $ _____.

*Step 4*   The algebraic sign of each part is:

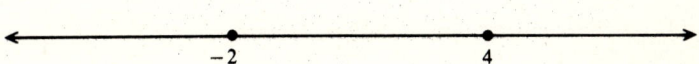

The solution set is all real $x$ such that _____.

The graph is _____.

**9**   $\dfrac{5x+1}{x-2} \geq 0$

*Step 1*   Set $5x + 1 = 0$ and _____ = 0.
Solve these equations so that
$x = -\frac{1}{5}$ or $x = $ _____.

*Step 2*   The numbers $-\frac{1}{5}$ and 2 divide the number line into _____ parts, which we denote by $A$, $B$, and $C$.

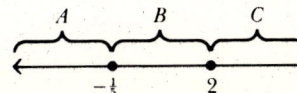

*Step 3*   Select numbers in each part and check them as follows:

$A$:   If $x = -1$, then $\dfrac{\phantom{xxx}}{-1-2} = $ _____.

$B$:   If $x = 0$, then $\dfrac{\phantom{xxx}}{0-2} = $ _____.

$C$:   If $x = 3$, then $\dfrac{\phantom{xxx}}{3-2} = $ _____.

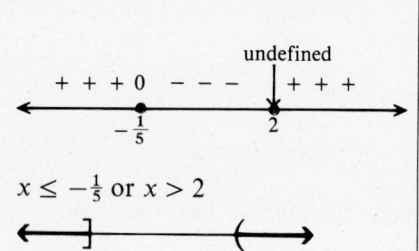

$x \leq -\frac{1}{5}$ or $x > 2$

*Step 4* The algebraic sign of each part is:

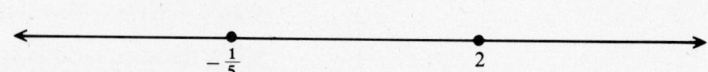

The solution set is all real $x$ such that _____.

The graph is _____.

## SOLUTIONS TO SELECTED ODD PROBLEMS  Section 6.8, text page 327

**1**  $x^2 + x - 2 < 0$
$x^2 + x - 2 = (x + 2)(x - 1) = 0$
When $x = -2$ or $x = 1$,

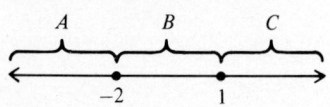

A:  for $x = -3$, $(-3)^2 + (-3) - 2 = 4 > 0$
B:  for $x = 0$, $(0)^2 + (0) - 2 = -2 < 0$
C:  for $x = 2$, $(2)^2 + (2) - 2 = 4 > 0$
Thus, $x^2 + x - 2 < 0$ in B,
or $-2 < x < 1$.
The graph is

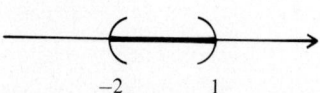

**9**  $x^2 - x \leq 20$
$x^2 - x - 20 \leq 0$
$x^2 - x - 20 = (x + 4)(x - 5) = 0$
When $x = -4$ or $x = 5$,

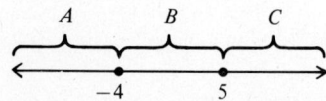

A:  for $x = -5$, $(-5)^2 - (-5) - 20 = 10 > 0$
B:  for $x = 0$, $(0)^2 - (0) - 20 = -20 < 0$
C:  for $x = 6$, $(6)^2 - (6) - 20 = 10 > 0$
Thus, $x^2 - x - 20 < 0$ in B,
so $-4 \leq x \leq 5$.
The graph is

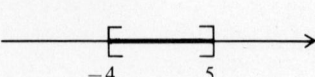

**17**  $2x^2 - 11x + 13 \geq -2x^2 + 9x - 12$
$4x^2 - 20x + 25 \geq 0$
$(2x - 5)^2 \geq 0$
This inequality is true for all real values of $x$.
The graph is

**5**  $x^2 + 2x - 3 > 0$
$x^2 + 2x - 3 = (x + 3)(x - 1) = 0$
When $x = -3$ or $x = 1$,

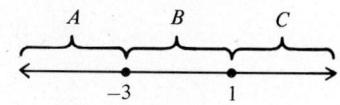

A:  for $x = -4$, $(-4)^2 + 2(-4) - 3 = 5 > 0$
B:  for $x = 0$, $(0)^2 + 2(0) - 3 = -3 < 0$
C:  for $x = 2$, $(2)^2 + 2(2) - 3 = 5 > 0$
Thus, $x^2 + 2x - 3 > 0$ in A or C,
or $x < -3$ or $x > 1$.
The graph is

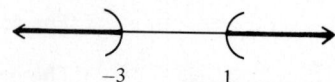

**13**  $3x^2 - 2x - 5 > 0$
$3x^2 - 2x - 5 = (3x - 5)(x + 1) = 0$
When $x = -1$ or $x = \frac{5}{3}$,

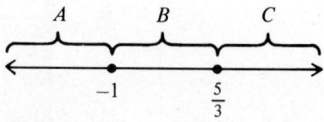

A:  for $x = -2$, $3(-2)^2 - 2(-2) - 5 = 11 > 0$
B:  for $x = 0$, $3(0)^2 - 2(0) - 5 = -5 < 0$
C:  for $x = 2$, $3(2)^2 - 2(2) - 5 = 3 > 0$
Thus, $3x^2 - 2x - 5 > 0$ in A or C,
so $x < -1$ or $x > \frac{5}{3}$.
The graph is

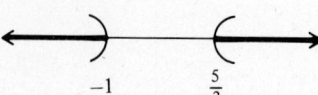

**21**  $16x^2 - 24x < -9$
$16x^2 - 24x + 9 < 0$
$(4x - 3)^2 < 0$
Since the square of any real number is never negative, this inequality has no solution.

**25** $\dfrac{x+1}{x-2} > 0$

$x + 1 = 0 \qquad x - 2 = 0$
$\quad x = -1 \qquad\quad x = 2$

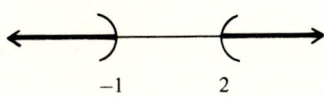

A: for $x = -2$, $\dfrac{-2+1}{-2-2} = \dfrac{1}{4} > 0$

B: for $x = 0$, $\dfrac{0+1}{0-2} = -\dfrac{1}{2} < 0$

C: for $x = 3$, $\dfrac{3+1}{3-2} = 4 > 0$

Thus, $\dfrac{x+1}{x-2} > 0$ in $A$ or $C$,

so $x < -1$ or $x > 2$.
The graph is

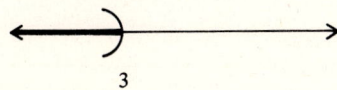

**29** $\dfrac{5x-1}{3x-2} \geq 0$

$5x - 1 = 0 \qquad 3x - 2 = 0$
$\quad x = \dfrac{1}{5} \qquad\quad x = \dfrac{2}{3}$

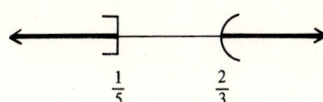

A: for $x = 0$, $\dfrac{5(0)-1}{3(0)-2} = \dfrac{1}{2} > 0$

B: for $x = \dfrac{1}{3}$, $\dfrac{5(\frac{1}{3})-1}{3(\frac{1}{3})-2} = -\dfrac{2}{3} < 0$

C: for $x = 1$, $\dfrac{5(1)-1}{3(1)-2} = 4 > 0$

Thus, $\dfrac{5x-1}{3x-2} > 0$ in $A$ or $C$

Also, $\dfrac{5x-1}{3x-2} = 0$ when $5x - 1 = 0$

and $3x - 2 \neq 0$, or $x = \dfrac{1}{5}$.

The solution is

$x \leq \dfrac{1}{5}$ or $x > \dfrac{2}{3}$.

The graph is

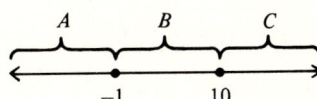

**33** $\qquad\dfrac{x+1}{x-3} \leq 1$

$\dfrac{x+1}{x-3} - 1 \leq 0$

$\dfrac{x+1-1(x-3)}{x-3} \leq 0$

$\dfrac{4}{x-3} \leq 0$

This fraction cannot equal 0 since its numerator is 4. Therefore, we solve

$\dfrac{4}{x-3} < 0.$

This fraction is negative when its denominator is negative, that is
$x - 3 < 0$ or $x < 3$.
The graph is

**37** $\sqrt{x^2 - 9x - 10}$ represents a real number for
$x^2 - 9x - 10 \geq 0$
$(x+1)(x-10) \geq 0$
$x + 1 = 0 \qquad x - 10 = 0$
$\quad x = -1 \qquad\quad x = 10$

A: for $x = -2$, $(-2)^2 - 9(-2) - 10 = 12 > 0$
B: for $x = 0$, $(0)^2 - 9(0) - 10 = -10 < 0$
C: for $x = 11$, $(11)^2 - 9(11) - 10 = 12 > 0$
Thus, $x^2 - 9x - 10 \geq 0$ for
$x \leq -1$ or $x \geq 10$.

## CUMULATIVE REVIEW PROBLEM SET  Chapters 1–6

1. Translate the following statement into symbols: The sum of $x$ and 3 is greater than or equal to 7 less than the product of 2 and $x$.

2. Evaluate $-|-3|-(-3)$.

3. Simplify $\dfrac{(2x^2y^3)^4}{(2x^3y)^3}$.

4. Perform each operation.
   (a) $(a+3)(a^2-3a+9)$
   (b) $(u^4-3uv^3-2v^4) \div (u^2-uv-v^2)$

5. Determine all values of the variable for which the fraction $\dfrac{x^2-1}{x^2-5x-36}$ is not defined.

6. Perform the indicated operations and simplify the results.
   (a) $\dfrac{x^3-1}{x^2-9y^2} \div \dfrac{x^2+x+1}{x+3y}$
   (b) $\dfrac{5}{m^2-n^2} + \dfrac{2}{m^2+2mn+n^2}$

7. Simplify $\dfrac{\dfrac{1}{u}+1}{u+2+\dfrac{1}{u}}$.

8. Solve each equation.
   (a) $\dfrac{x}{a}-\dfrac{2}{a-y}=2$, for $y$
   (b) $|5-x|=11$

9. Solve each inequality.
   (a) $\tfrac{1}{2}(2x+5) \le 2-x$
   (b) $|3x-2| \le 4$

10. A nurse wishes to administer a shot containing a 10% solution. How much of a 15% solution must she mix with 10 cubic centimeters of an 8% solution in order to obtain the 10% solution?

11. Rewrite each expression so that it contains only positive exponents and simplify.
    (a) $\left(\dfrac{x^3}{3y^{-4}}\right)^{-1}\left(\dfrac{x^{-2}y^3}{9^{-1}}\right)^{-2}$
    (b) $\left(\dfrac{4x^{-6}y^{-10}}{9z^{-4}}\right)^{-1/2}$

12. Use properties of radicals to simplify each expression.
    (a) $\sqrt[3]{-54x^4y^{11}}$
    (b) $\dfrac{\sqrt{250x^7}}{\sqrt{2x}}$

13. Rationalize each denominator and simplify the result.
    (a) $\dfrac{15}{\sqrt{5}}$
    (b) $\dfrac{x}{\sqrt{x}+y}$

14. Perform the indicated operations.
    (a) $(2-3i)+(7+4i)$
    (b) $(-7+i)-(3-2i)$
    (c) $(3+4i)(2-3i)$
    (d) $\dfrac{1+i}{2-i}$

15. Solve each equation by the indicated method.
    (a) $3x^2+14x-5=0$   (Factoring)
    (b) $2x^2+3x-4=0$   (Completing the Square)
    (c) $3x^2+x+5=0$   (Quadratic Formula)

16. Solve by using an appropriate substitution.
    $3y^{-4}-2y^{-2}-1=0$

17. Solve and check the solutions.
    $\sqrt{3x-2}-\sqrt{x}=2$

**18** Solve and check the solutions.
$(2y - 1)^{3/4} = 27$

**19** Solve each inequality and illustrate the solution set on a number line.

(a) $x^2 + x - 56 \geq 0$  (b) $\dfrac{2x - 1}{x + 2} \leq 0$

**20** Find a number such that the sum of the number and its reciprocal is 4.

---

## Answers

1  $x + 3 \geq 2x - 7$   2  0   3  $\dfrac{2y^9}{x}$   4(a) $a^3 + 27$   (b) $u^2 + uv + 2v^2$   5  $-4, 9$

6(a) $\dfrac{x-1}{x-3y}$   (b) $\dfrac{7m + 3n}{(m-n)(m+n)^2}$   7  $\dfrac{1}{u+1}$   8(a) $y = \dfrac{2a^2 - ax + 2a}{2a - x}$   (b) $-6, 16$

9(a) $x \leq -\tfrac{1}{4}$   (b) $-\tfrac{2}{3} \leq x \leq 2$   10  4 cubic centimeters   11(a) $\dfrac{x}{27y^{10}}$   (b) $\dfrac{3x^3y^5}{2z^2}$

12(a) $-3xy^3 \sqrt[3]{2xy^2}$   (b) $5x^3\sqrt{5}$   13(a) $3\sqrt{5}$   (b) $\dfrac{x\sqrt{x} - xy}{x - y^2}$   14(a) $9 + i$

(b) $-10 + 3i$   (c) $18 - i$   (d) $\tfrac{1}{5} + \tfrac{3}{5}i$   15(a) $-5, \tfrac{1}{3}$   (b) $\dfrac{-3 \pm \sqrt{41}}{4}$   (c) $\dfrac{-1 \pm \sqrt{59}i}{6}$

16  $\pm 1, \pm \sqrt{3}i$   17  9   18  41   19(a) $x \leq -8$ or $x \geq 7$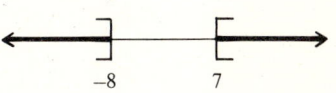

(b) $-2 < x \leq \tfrac{1}{2}$   20  $2 \pm \sqrt{3}$

---

## CHAPTER 6 TESTS

### Chapter Test A

**1** Solve each quadratic equation by the factoring method.
(a) $x^2 - 5x + 4 = 0$   (b) $2x^2 + 5x - 3 = 0$   (c) $6x^2 + 11x = 10$

**2** Solve each quadratic equation by completing the square.
(a) $x^2 + 2x - 5 = 0$   (b) $x^2 + 4x + 7 = 0$   (c) $3x^2 - 6x - 7 = 0$

**3** Solve each quadratic equation by using the quadratic formula.
(a) $2x^2 - 3x - 2 = 0$   (b) $x^2 + 6x = 5$   (c) $3x^2 - x + 1 = 0$

**4** Use the quadratic discriminant $b^2 - 4ac$ to determine the types of solutions of each quadratic equation.
(a) $6x^2 + x - 1 = 0$   (b) $4x^2 - 12x + 9 = 0$   (c) $2x^2 + 3x + 5 = 0$

**5** Solve each equation for $x$.
(a) $x^4 - 13x^2 + 36 = 0$   (b) $x^{-1} - 3x^{-1/2} + 2 = 0$   (c) $\sqrt{x + 2} = 3$
(d) $x^2 + 2x + (x^2 + 2x - 3)^{1/2} = 3$   (e) $\sqrt{2x + 1} = \sqrt{7x - 3} - 2$

**6** Find the solution set of each inequality.
(a) $x^2 - 3x > 4$   (b) $x^2 - x - 2 \leq 0$   (c) $\dfrac{x-1}{x+1} > 0$   (d) $\dfrac{2x+1}{x+3} \leq 0$

**7** A rectangular lawn whose length is 5 feet greater than its width is surrounded by a sidewalk 3 feet wide. If the area of the sidewalk is 54 square feet less than the area of the lawn, what are the dimensions of the lawn?

## Solutions

**1** **(a)** $x^2 - 5x + 4 = 0$
$(x - 1)(x - 4) = 0$
$x - 1 = 0 \quad x - 4 = 0$
$\quad x = 1 \quad\quad x = 4$

**(b)** $2x^2 + 5x - 3 = 0$
$(2x - 1)(x + 3) = 0$
$2x - 1 = 0 \quad x + 3 = 0$
$\quad x = \tfrac{1}{2} \quad\quad x = -3$

**(c)** $6x^2 + 11x = 10$
$6x^2 + 11x - 10 = 0$
$(3x - 2)(2x + 5) = 0$
$3x - 2 = 0 \quad 2x + 5 = 0$
$\quad x = \tfrac{2}{3} \quad\quad x = -\tfrac{5}{2}$

**2** **(a)** $x^2 + 2x - 5 = 0$
$x^2 + 2x = 5$
$x^2 + 2x + 1 = 5 + 1$
$(x + 1)^2 = 6$
$x + 1 = \pm\sqrt{6}$
$x = -1 \pm \sqrt{6}$
$x = -1 - \sqrt{6} \quad x = -1 + \sqrt{6}$

**(b)** $x^2 + 4x + 7 = 0$
$x^2 + 4x = -7$
$x^2 + 4x + 4 = -7 + 4$
$(x + 2)^2 = -3$
$x + 2 = \pm\sqrt{-3}$
$x = -2 \pm \sqrt{3}i$
$x = -2 - \sqrt{3}i \quad x = -2 + \sqrt{3}i$

**(c)** $3x^2 - 6x - 7 = 0$
$x^2 - 2x - \dfrac{7}{3} = 0$
$x^2 - 2x = \dfrac{7}{3}$
$x^2 - 2x + 1 = \dfrac{7}{3} + 1$
$(x - 1)^2 = \dfrac{10}{3}$
$x - 1 = \pm\sqrt{\dfrac{10}{3}} = \pm\sqrt{\dfrac{30}{9}}$
$x = 1 \pm \dfrac{\sqrt{30}}{3}$
$x = \dfrac{3 - \sqrt{30}}{3} \quad x = \dfrac{3 + \sqrt{30}}{3}$

**3** **(a)** $2x^2 - 3x - 2 = 0$
$a = 2, b = -3, c = -2$
$x = \dfrac{-(-3) \pm \sqrt{(-3)^2 - 4(2)(-2)}}{2(2)}$
$x = \dfrac{3 \pm \sqrt{25}}{4}$
$x = \dfrac{3 \pm 5}{4}$
$x = \dfrac{3 - 5}{4} = -\dfrac{1}{2}$
$x = \dfrac{3 + 5}{4} = 2$

**(b)** $x^2 + 6x = 5$
$x^2 + 6x - 5 = 0$
$a = 1, b = 6, c = -5$
$x = \dfrac{-6 \pm \sqrt{6^2 - 4(1)(-5)}}{2(1)}$
$x = \dfrac{-6 \pm \sqrt{56}}{2}$
$x = \dfrac{-6 \pm 2\sqrt{14}}{2}$
$x = -3 - \sqrt{14} \quad x = -3 + \sqrt{14}$

**(c)** $3x^2 - x + 1 = 0$
$a = 3, b = -1, c = 1$
$x = \dfrac{-(-1) \pm \sqrt{(-1)^2 - 4(3)(1)}}{2(3)}$
$x = \dfrac{1 \pm \sqrt{-11}}{6}$
$x = \dfrac{1 \pm \sqrt{11}i}{6}$
$x = \dfrac{1 - \sqrt{11}i}{6} \quad x = \dfrac{1 + \sqrt{11}i}{6}$

**4** **(a)** $6x^2 + x - 1 = 0$
$a = 6, b = 1, c = -1$
$b^2 - 4ac = 1^2 - 4(6)(-1)$
$\quad = 25 > 0$
Two real and unequal solutions.

**(b)** $4x^2 - 12x + 9 = 0$
$a = 4, b = -12, c = 9$
$b^2 - 4ac = (-12)^2 - 4(4)(9)$
$\quad = 0$
One real solution.

**(c)** $2x^2 + 3x + 5 = 0$
$a = 2, b = 3, c = 5$
$b^2 - 4ac = 3^2 - 4(2)(5)$
$\quad = -31 < 0$
Two complex solutions.

5  (a) $x^4 - 13x^2 + 36 = 0$
$(x^2)^2 - 13x^2 + 36 = 0$
Let $u = x^2$
$u^2 - 13u + 36 = 0$
$(u - 4)(u - 9) = 0$
$u = 4 \quad\quad u = 9$
$x^2 = 4 \quad\quad x^2 = 9$
$x = \pm\sqrt{4} \quad x = \pm\sqrt{9}$
$x = \pm 2 \quad\quad x = \pm 3$

(b) $x^{-1} - 3x^{-1/2} + 2 = 0$
$(x^{-1/2})^2 - 3x^{-1/2} + 2 = 0$
Let $u = x^{-1/2}$
$u^2 - 3u + 2 = 0$
$(u - 1)(u - 2) = 0$
$u = 1 \quad\quad u = 2$
$x^{-1/2} = 1 \quad x^{-1/2} = 2$
$x^{1/2} = 1 \quad x^{1/2} = \frac{1}{2}$
$x = 1^2 \quad\quad x = (\frac{1}{2})^2$
$x = 1 \quad\quad x = \frac{1}{4}$

Check:
$x = 1 \quad (1)^{-1} - 3(1)^{-1/2} + 2 = 1 - 3 + 2 = 0$
$x = \frac{1}{4} \quad (\frac{1}{4})^{-1} - 3(\frac{1}{4})^{-1/2} + 2 = 4 - 6 + 2 = 0$

(c) $\sqrt{x + 2} = 3$
$(\sqrt{x + 2})^2 = 3^2$
$x + 2 = 9$
$x = 7$
Check: $\sqrt{7 + 2} = \sqrt{9} = 3$

(d) $x^2 + 2x + (x^2 + 2x - 3)^{1/2} = 3$
Let $u = (x^2 + 2x - 3)^{1/2}$
so that
$u^2 = x^2 + 2x - 3$
$u^2 + 3 = x^2 + 2x$
$u^2 + 3 + u = 3$
$u^2 + u = 0$
$u(u + 1) = 0$
$u = 0 \quad u + 1 = 0$
$\quad\quad\quad u = -1$
$(x^2 + 2x - 3)^{1/2} = 0 \quad (x^2 + 2x - 3)^{1/2} = -1$
$x^2 + 2x - 3 = 0^2 \quad$ No solution here, since a
$(x + 3)(x - 1) = 0 \quad$ square root cannot be
$x = -3 \quad x = 1 \quad$ negative.

Check:
$x = -3$
$(-3)^2 + 2(-3) + [(-3)^2 + 2(-3) - 3]^{1/2} = 3$
$9 - 6 + [9 - 6 - 3]^{1/2} = 3$
$3 + 0 = 3$
$x = 1$
$1^2 + 2(1) + [1^2 + 2(1) - 3]^{1/2} = 3$
$1 + 2 + [1 + 2 - 3]^{1/2} = 3$
$3 + 0 = 3$

(e) $\sqrt{2x + 1} = \sqrt{7x - 3} - 2$
$(\sqrt{2x + 1})^2 = (\sqrt{7x - 3} - 2)^2$
$2x + 1 = 7x - 3 - 4\sqrt{7x - 3} + 4$
$4\sqrt{7x - 3} = 5x$
$(4\sqrt{7x - 3})^2 = (5x)^2$
$16(7x - 3) = 25x^2$
$25x^2 - 112x + 48 = 0$
$(25x - 12)(x - 4) = 0$
$x = \frac{12}{25} \quad x = 4$

Check: $x = 4 \quad \sqrt{2(4) + 1} = \sqrt{7(4) - 3} - 2$
$\sqrt{9} = \sqrt{25} - 2$
$3 = 5 - 2$
$x = \frac{12}{25} \quad \sqrt{2(\frac{12}{25}) + 1} = \sqrt{7(\frac{12}{25}) - 3} - 2$
$\sqrt{\frac{49}{25}} = \sqrt{\frac{9}{25}} - 2$
$\frac{7}{5} = \frac{3}{5} - 2$
$\frac{7}{5} \neq -\frac{7}{5}$

Therefore, the only solution is 4.

6  (a) $x^2 - 3x > 4$
$x^2 - 3x - 4 > 0$
$(x + 1)(x - 4) > 0$
$x + 1 = 0 \quad x - 4 = 0$
$x = -1 \quad x = 4$

A: $x = -2, (-2)^2 - 3(-2) - 4 = 6 > 0$
B: $x = 0, 0^2 - 3(0) - 4 = -4 < 0$
C: $x = 5, 5^2 - 3(5) - 4 = 6 > 0$
$x^2 - 3x - 4 > 0$ in A or C,
so $x < -1$ or $x > 4$

(b) $x^2 - x - 2 \le 0$
$(x + 1)(x - 2) \le 0$
$x + 1 = 0 \quad x - 2 = 0$
$x = -1 \quad x = 2$

A: $x = -2, (-2)^2 - (-2) - 2 = 4 > 0$
B: $x = 0, 0^2 - 0 - 2 = -2 < 0$
C: $x = 3, 3^2 - 3 - 2 = 4 > 0$
$x^2 - x - 2 < 0$ in B, so that
$x^2 - x - 2 \le 0$ for $-1 \le x \le 2$.

(c) $\dfrac{x - 1}{x + 1} > 0$

$x - 1 = 0$ if $x = 1$ and $x + 1 = 0$ if $x = -1$.
The numbers $-1$ and $1$ divide the number line into three parts, A, B, and C.

A: $x = -2, \dfrac{-2 - 1}{-2 + 1} = 3 > 0$

B: $x = 0, \dfrac{0 - 1}{0 + 1} = -1 < 0$

C: $x = 2, \dfrac{2 - 1}{2 + 1} = \dfrac{1}{3} > 0$

$\dfrac{x - 1}{x + 1} > 0$ on A or C, so

$x < -1$ or $x > 1$.

(d) $\dfrac{2x + 1}{x + 3} \le 0$

$2x + 1 = 0$ if $x = -\tfrac{1}{2}$ and $x + 3 = 0$ if $x = -3$.
Notice that $-3$ is not a solution of

$\dfrac{2x + 1}{x + 3} \le 0$.

The numbers $-3$ and $-\tfrac{1}{2}$ divide the number line into three parts, A, B, and C.

A: $x = -4, \dfrac{2(-4) + 1}{-4 + 3} = 7 > 0$.

B: $x = -1, \dfrac{2(-1) + 1}{-1 + 3} = -\dfrac{1}{2} < 0$.

C: $x = 0, \dfrac{0 + 1}{0 + 3} = \dfrac{1}{3} > 0$.

$\dfrac{2x + 1}{x + 3} \le 0$ on B, so $-3 < x \le -\tfrac{1}{2}$.

7  Let $x = $ width of lawn and
$x + 5 = $ length of lawn.
Therefore, Area of total rectangle
$= (x + 11)(x + 6)$
$= x^2 + 17x + 66$
Area of lawn
$= x(x + 5)$
$= x^2 + 5x$
Area of sidewalk
$= (x^2 + 17x + 66) - (x^2 + 5x)$
$= 12x + 66$
Therefore, $x^2 + 5x - 54 = 12x + 66$
$x^2 - 7x - 120 = 0$
$(x + 8)(x - 15) = 0$
$x \ne -8 \quad x = 15$ and $x + 5 = 20$
The dimensions are 15 feet and 20 feet.

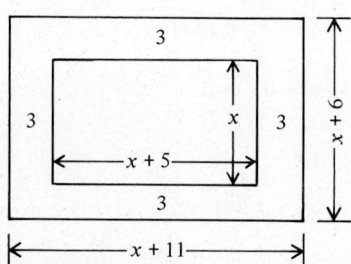

# Chapter Test B

*Multiple Choice:* Select the *one* correct answer for each of the following questions.

1. The solutions of $x^2 - 4 = 0$ are _____.
   (a) $2i, -2i$     (b) $3$     (c) $9, 4$     (d) $-2, 2$

2. The solutions of $(x - \frac{1}{3})^2 = \frac{16}{9}$ are _____.
   (a) $-1, \frac{5}{3}$     (b) $1, \frac{5}{3}$     (c) $1, -\frac{5}{3}$     (d) $-1, -\frac{5}{3}$

3. Solving $x^2 - 4x - 21 = 0$ by the factoring method, we obtain the solutions _____.
   (a) $-3, -7$     (b) $3, -7$     (c) $-2, 6$     (d) $-3, 7$

4. Solving $\dfrac{x^2 - 4}{4} = x - 1$ by the factoring method, we obtain the solutions _____.
   (a) $0, 4$     (b) $-4, 0$     (c) $1, -4$     (d) none of these

5. Solving $2x^2 - 5x + 1 = 0$ by completing the square, we obtain the solutions _____.
   (a) $5 + \sqrt{17}i, 5 - \sqrt{17}i$     (b) $\dfrac{2 + \sqrt{17}}{4}, \dfrac{2 - \sqrt{17}}{4}$
   (c) $\dfrac{5 - \sqrt{17}}{4}, \dfrac{5 + \sqrt{17}}{4}$     (d) none of these

6. Solving $x^2 - 9x + 3 = 0$ by completing the square, we obtain the solutions _____.
   (a) $\dfrac{9 - 69i}{2}, \dfrac{9 + 69i}{2}$     (b) $\dfrac{3 - \sqrt{69}}{2}, \dfrac{3 + \sqrt{69}}{2}$
   (c) $\dfrac{9 - \sqrt{69}}{2}, \dfrac{9 + \sqrt{69}}{2}$     (d) none of these

7. Solving $x^2 - 2x = 15$ by the quadratic formula, we obtain the solutions _____.
   (a) $3, -5$     (b) $-3, -5$     (c) $-3, 5$     (d) none of these

8. Solving $\dfrac{x^2 - 3}{3} + \dfrac{x}{4} = 2$ by the quadratic formula, we obtain the solutions _____.
   (a) $\dfrac{-3 - 3\sqrt{65}}{8}, \dfrac{-3 + 3\sqrt{65}}{8}$     (b) $\dfrac{3 - 3\sqrt{65}}{8}, \dfrac{3 + 3\sqrt{65}}{8}$
   (c) $\dfrac{3 - 3\sqrt{65}i}{8}, \dfrac{3 + 3\sqrt{65}i}{8}$     (d) none of these

9. The value of the discriminant of the quadratic equation $3x^2 - 4x + 1 = 0$ is _____.
   (a) $4$     (b) $-4$     (c) $3$     (d) $1$

10. After evaluating the discriminant of the quadratic equation $x^2 + 14x + 49 = 0$, we conclude that the roots are _____.
    (a) complex     (b) real and equal     (c) unequal     (d) none of these

11. After evaluating the discriminant of the quadratic equation $5x^2 + 2x + 2 = 0$, we conclude that the roots are _____.
    (a) real and equal     (b) complex     (c) real and unequal     (d) none of these

12. The equation $x^{-2} + x^{-1} - 6 = 0$ can be expressed in quadratic form by the substitution $u = $ _____.
    (a) $x^{-2}$     (b) $x^{-1} - 6$     (c) $x^{-1}$     (d) $x^{-2} - 6$

13. Solving the equation $x^{-2} + x^{-1} - 6 = 0$, we find that $x = $ _____.
    (a) $-\frac{1}{3}, \frac{1}{2}$     (b) $-3, 2$     (c) $-2, 3$     (d) $-\frac{1}{2}, \frac{1}{3}$

14. The solution of $\sqrt{x - 2} = 3$ is _____.
    (a) $7$     (b) $5$     (c) $1$     (d) $11$

**15** The solution of $\sqrt[4]{x+8} = \sqrt[4]{2x}$ is _____.
   (a) 8   (b) 4   (c) 12   (d) 16

**16** The solution of $\sqrt{2y+1} = 1 - \sqrt{y}$ is _____.
   (a) 0, 4   (b) 0   (c) 4   (d) undefined

**17** The solution of $z^{2/5} = 4$ is _____.
   (a) $4^{2/5}$   (b) 2   (c) $\pm 32$   (d) no solution

**18** The solution set of the inequality $10x^2 + 11x - 6 < 0$ is _____.
   (a) $\frac{3}{2} < x < \frac{2}{5}$   (b) $-\frac{3}{2} < x < \frac{2}{5}$   (c) $-\frac{3}{2} < x < \frac{5}{2}$   (d) $\frac{1}{2} < x < \frac{3}{2}$

**19** The solution set of the inequality $\frac{x-3}{x+3} > 0$ is _____.
   (a) $x > 3$ and $x < -3$
   (b) $x < 3$ and $x > -3$
   (c) $x > 3$ or $x < -3$
   (d) $x < 3$ or $x > -3$

**20** If the product of two positive consecutive integers is 56, then an equation that can be used to find these numbers is _____, in which $x$ represents the smaller integer.
   (a) $x^2 - 15x + 56 = 0$
   (b) $x^2 - x - 56 = 0$
   (c) $x^2 + x + 56 = 0$
   (d) $x^2 + x - 56 = 0$

---

## Answers

1 d   2 a   3 d   4 a   5 c   6 c   7 c   8 a   9 a   10 b
11 b   12 c   13 a   14 d   15 a   16 b   17 c   18 b   19 c
20 d

# 7 Graphing Linear Systems of Equations and Inequalities

In this chapter we extend the notion of representing points on a line by a number to representing points in the plane by an ordered pair of numbers. After completing the appropriate sections, the student should be able to:

1. Graph ordered pairs of numbers on a Cartesian coordinate system.
2. Find the distance between two points in the plane.
3. Graph linear equations in two unknowns.
4. Find the slope of a line determined by two given points.
5. Use slopes to determine if two given lines are parallel or perpendicular.
6. Apply the different forms of the equation of a line.
7. Graph linear inequalities in two unknowns.

## 7.1 The Cartesian Coordinate System

### SEMIPROGRAMMED PROBLEMS

In problems 1–4, graph the given ordered pairs.

1. $(0, 2), (-1, -3), (-2, 1), (3, -2)$

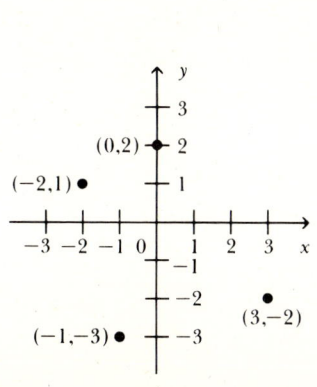
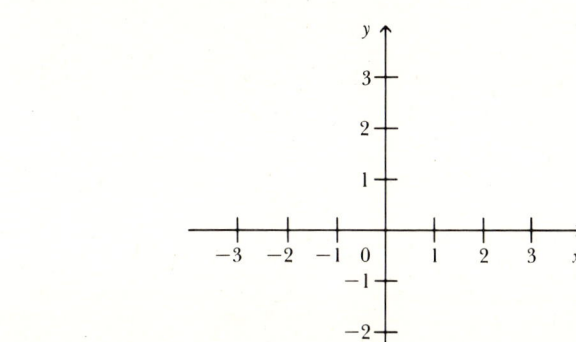

**148** CHAPTER 7 GRAPHING LINEAR SYSTEMS OF EQUATIONS AND INEQUALITIES

**2** $(-2, 1), (-1, 2), (3, 3), (0, 4)$

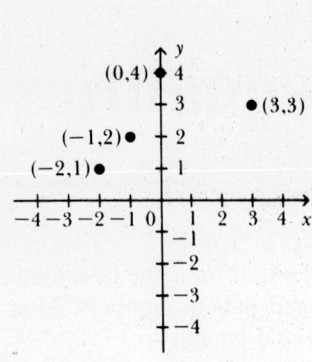
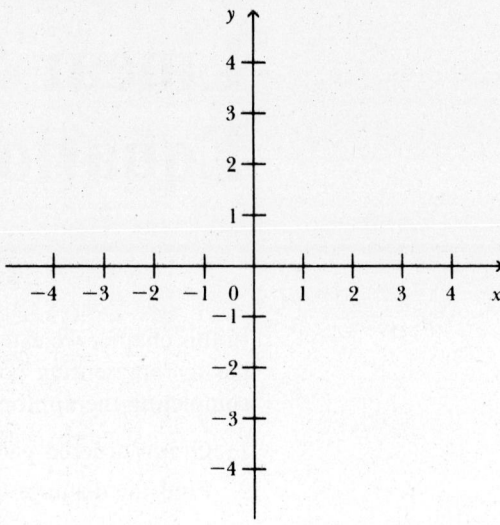

**3** $(1, -2), (1, 0), (1, 1), (2, -2), (2, 0), (2, 1), (3, -2), (3, 0), (3, 1)$

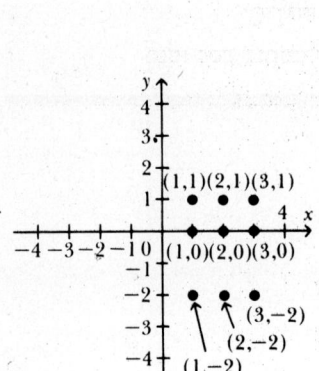
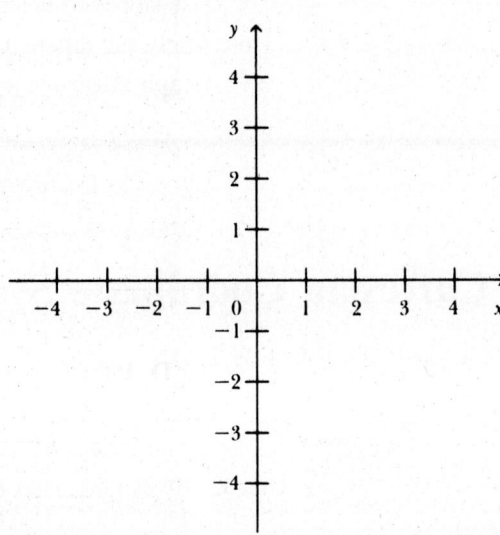

**4** $(2, -3), (2, 1), (4, -3), (4, 1)$

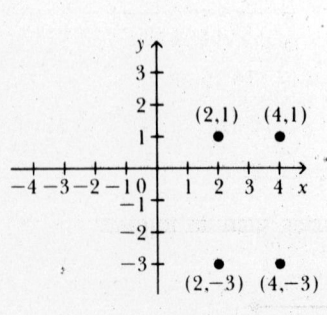
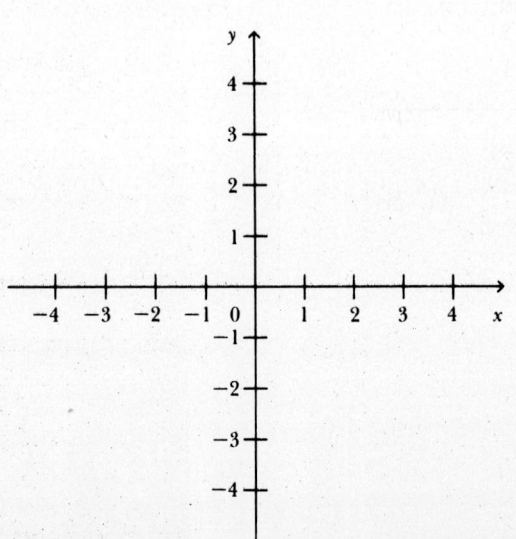

In problems 5–9, indicate what quadrant or coordinate axis contains the point.

| | |
|---|---|
| quadrant I | 5 (2, 5) lies in/on _____. |
| quadrant II | 6 (−3, 6) lies in/on _____. |
| the x axis | 7 (7, 0) lies in/on _____. |
| quadrant IV | 8 (3, −11) lies in/on _____. |
| quadrant III | 9 (−7, −8) lies in/on _____. |

In problems 10 and 11, find the ordered pair that satisfies $3x + 2y = 12$ for the given value of $x$.

| | |
|---|---|
| | 10   $x = 2$ |
| 2 | $3(\underline{\phantom{xx}}) + 2y = 12$ |
| 6 | $\underline{\phantom{xx}} + 2y = 12$ |
| 6 | $2y = \underline{\phantom{xx}}$ |
| 3 | $y = \underline{\phantom{xx}}$ |
| (2, 3) | Thus, $(x, y) = \underline{\phantom{xx}}$ |
| | 11   $x = −4$ |
| −4 | $3(\underline{\phantom{xx}}) + 2y = 12$ |
| −12 | $\underline{\phantom{xx}} + 2y = 12$ |
| 24 | $2y = \underline{\phantom{xx}}$ |
| 12 | $y = \underline{\phantom{xx}}$ |
| (−4, 12) | Thus, $(x, y) = \underline{\phantom{xx}}$ |

## SOLUTIONS TO SELECTED ODD PROBLEMS   Section 7.1, text pages 336–338

1  (a)  (1, 2) lies in quadrant I
   (b)  (−1, 1) lies in quadrant II
   (c)  (−1, −2) lies in quadrant III
   (d)  (3, −2) lies in quadrant IV
   (e)  (0, 1) lies on $y$ axis
   (f)  (−6, 0) lies on $x$ axis
   (g)  (3, 0) lies on $x$ axis
   (h)  (0, −4) lies on $y$ axis

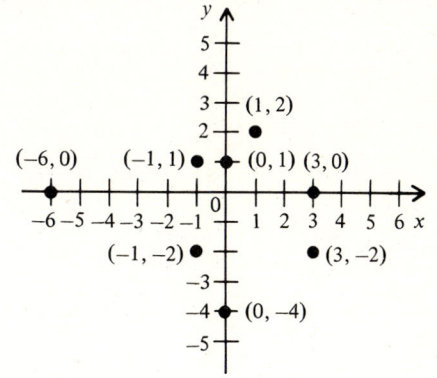

5  Let $(x, y)$ represent the fourth vertex. Since the opposite sides of a parallelogram are equal and parallel, then $x = 10$ and $y = 6$. Thus, the fourth vertex is the point (10, 6).

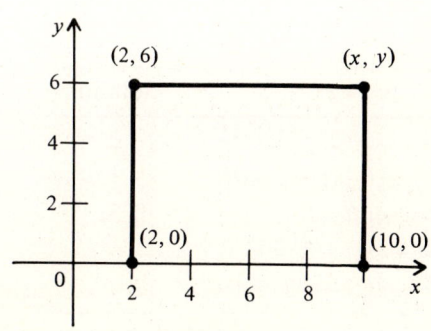

**150** CHAPTER 7 GRAPHING LINEAR SYSTEMS OF EQUATIONS AND INEQUALITIES

**9**    $3x - 2y = 6$
         $-2y = -3x + 6$
           $y = \frac{3}{2}x - 3$

| $x$ | $y = \frac{3}{2}x - 3$ | Solution |
|---|---|---|
| $-2$ | $y = \frac{3}{2}(-2) - 3 = -6$ | $(-2, -6)$ |
| $-1$ | $y = \frac{3}{2}(-1) - 3 = -\frac{9}{2}$ | $(-1, -\frac{9}{2})$ |
| $0$ | $y = \frac{3}{2}(0) - 3 = -3$ | $(0, -3)$ |
| $1$ | $y = \frac{3}{2}(1) - 3 = -\frac{3}{2}$ | $(1, -\frac{3}{2})$ |
| $2$ | $y = \frac{3}{2}(2) - 3 = 0$ | $(2, 0)$ |

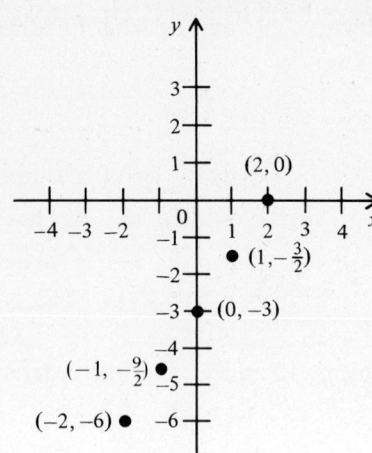

**13**

| $x$ | $y = \frac{1}{2}x + 2$ | Solution |
|---|---|---|
| $-2$ | $y = \frac{1}{2}(-2) + 2 = 1$ | $(-2, 1)$ |
| $-1$ | $y = \frac{1}{2}(-1) + 2 = \frac{3}{2}$ | $(-1, \frac{3}{2})$ |
| $0$ | $y = \frac{1}{2}(0) + 2 = 2$ | $(0, 2)$ |
| $1$ | $y = \frac{1}{2}(1) + 2 = \frac{5}{2}$ | $(1, \frac{5}{2})$ |
| $2$ | $y = \frac{1}{2}(2) + 2 = 3$ | $(2, 3)$ |

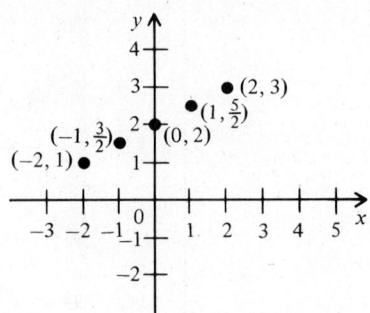

**17**    $2y = 4x + 7$
       $y = 2x + \frac{7}{2}$

| $x$ | $y = 2x + \frac{7}{2}$ | Solution |
|---|---|---|
| $-2$ | $y = 2(-2) + \frac{7}{2} = -\frac{1}{2}$ | $(-2, -\frac{1}{2})$ |
| $-1$ | $y = 2(-1) + \frac{7}{2} = \frac{3}{2}$ | $(-1, \frac{3}{2})$ |
| $0$ | $y = 2(0) + \frac{7}{2} = \frac{7}{2}$ | $(0, \frac{7}{2})$ |
| $1$ | $y = 2(1) + \frac{7}{2} = \frac{11}{2}$ | $(1, \frac{11}{2})$ |
| $2$ | $y = 2(2) + \frac{7}{2} = \frac{15}{2}$ | $(2, \frac{15}{2})$ |

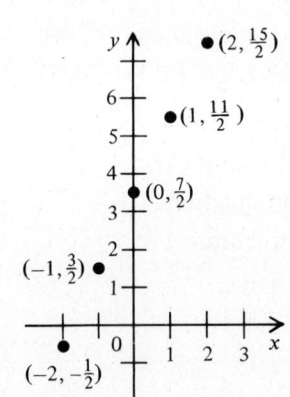

**21**    $3x + 5y = 15$
         $5y = -3x + 15$
           $y = -\frac{3}{5}x + 3$

| $x$ | $y = -\frac{3}{5}x + 3$ | Solution |
|---|---|---|
| $-2$ | $y = -\frac{3}{5}(-2) + 3 = \frac{21}{5}$ | $(-2, \frac{21}{5})$ |
| $-1$ | $y = -\frac{3}{5}(-1) + 3 = \frac{18}{5}$ | $(-1, \frac{18}{5})$ |
| $0$ | $y = -\frac{3}{5}(0) + 3 = 3$ | $(0, 3)$ |
| $1$ | $y = -\frac{3}{5}(1) + 3 = \frac{12}{5}$ | $(1, \frac{12}{5})$ |
| $2$ | $y = -\frac{3}{5}(2) + 3 = \frac{9}{5}$ | $(2, \frac{9}{5})$ |

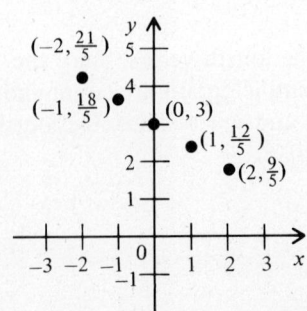

## 7.2 SEMIPROGRAMMED PROBLEMS

25  Since $y = x$, the missing number equals the given number in each case. Thus, the completed ordered pairs are $(0, 0), (0, 0), (1, 1), (-1, -1)$

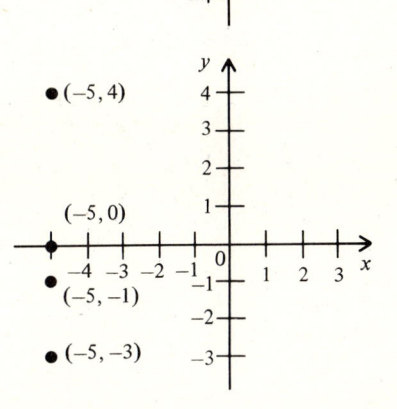

29  Since $x = -5$, the missing number is $-5$ in each case. Thus, the completed ordered pairs are $(-5, 0), (-5, -1), (-5, -3), (-5, 4)$

33  $y + 3x = 4$

37  $x = -3$

## 7.2 Distance Between Two Points

### SEMIPROGRAMMED PROBLEMS

1  The distance between two points whose coordinates are $(x_1, y_1)$ and $(x_2, y_2)$ is expressed by the formula

$d = $ _____.

$\sqrt{(x_2 - x_1)^2 + (y_2 - y_1)^2}$

In problems 2–6, find the distance between each set of points.

2  $(1, 5)$ and $(3, 4)$

Consider

1, 5
3, 4

$x_1 = $ ___, $y_1 = $ ___ and

$x_2 = $ ___, $y_2 = $ ___

$4 - 5, 1, \sqrt{5}$

$d = \sqrt{(3-1)^2 + (\underline{\quad})^2} = \sqrt{4 + \underline{\quad}} = \underline{\quad}$.

3  $(-4, -1)$ and $(2, -2)$

Consider

$-4, -1$
$2, -2$

$x_1 = $ ___, $y_1 = $ ___ and

$x_2 = $ ___, $y_2 = $ ___

$36, 1, \sqrt{37}$

$d = \sqrt{[2-(-4)]^2 + [-2-(-1)]^2} = \sqrt{\underline{\quad} + \underline{\quad}} = \underline{\quad}$.

4  $(5, -3)$ and $(7, 7)$

Consider

$5, -3$
$7, 7$

$x_1 = $ ___, $y_1 = $ ___ and

$x_2 = $ ___, $y_2 = $ ___

$2, 10, 104, 2\sqrt{26}$

$d = \sqrt{(\underline{\quad})^2 + (\underline{\quad})^2} = \sqrt{\underline{\quad}} = \underline{\quad}$.

**152** CHAPTER 7 GRAPHING LINEAR SYSTEMS OF EQUATIONS AND INEQUALITIES

5    $(1, -7)$ and $(5, 3)$

Consider

1, −7

5, 3

4, 10, 116, $2\sqrt{29}$

$x_1 = \underline{\quad}, y_1 = \underline{\quad}$ and

$x_2 = \underline{\quad}, y_2 = \underline{\quad}$

$d = \sqrt{(\underline{\quad})^2 + (\underline{\quad})^2} = \sqrt{\underline{\quad}} = \underline{\quad}$.

6    $(-5, 12)$ and $(0, 0)$

Consider

−5, 12

0, 0

5, −12, 169, 13

$x_1 = \underline{\quad}, y_1 = \underline{\quad}$ and

$x_2 = \underline{\quad}, y_2 = \underline{\quad}$

$d = \sqrt{(\underline{\quad})^2 + (\underline{\quad})^2} = \sqrt{\underline{\quad}} = \underline{\quad}$.

In problems 7 and 8, find the coordinates of the midpoint $M$ of the line segment $\overline{PQ}$.

7    $P = (2, 3)$ and $Q = (4, 5)$

$\dfrac{y_1 + y_2}{2}$

5, 3, 4

$M = \left(\dfrac{x_1 + x_2}{2}, \underline{\quad}\right)$

$= \left(\dfrac{2+4}{2}, \dfrac{3+\underline{\quad}}{2}\right) = (\underline{\quad}, \underline{\quad})$

8    $P = (-3, 1)$ and $Q = (7, -11)$

$\dfrac{x_1 + x_2}{2}$

−3, 2, −5

$M = \left(\underline{\quad}, \dfrac{y_1 + y_2}{2}\right)$

$= \left(\dfrac{\underline{\quad} + 7}{2}, \dfrac{1 + (-11)}{2}\right) = (\underline{\quad}, \underline{\quad})$

In problems 9 and 10, find the center $(h, k)$ and the radius $r$ of the circle with the given equation, and sketch the graph.

9    $(x - 2)^2 + (y + 3)^2 = 16$.

Writing this equation in standard form, we have

−3, 4

(2, −3), 4

$(x - 2)^2 + [y - (\underline{\quad})]^2 = (\underline{\quad})^2$,

so that $(h, k) = \underline{\quad}$ and $r = \underline{\quad}$.

The graph is

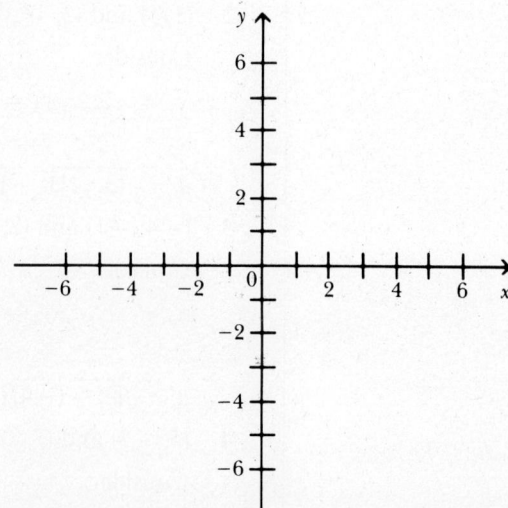

10    $x^2 + y^2 - 2x + 4y = 4$.

First, rewrite the equation in standard form by completing the square.

$y^2 + 4y$

1, $y^2 + 4y + 4$, 5

$y + 2$

$-2, 3$

$(1, -2), 3$

Thus,
$$(x^2 - 2x) + (\underline{\phantom{xx}}) = 4$$
$$(x^2 - 2x + \underline{\phantom{xx}}) + (\underline{\phantom{xxxx}}) = 4 + \underline{\phantom{xx}}$$
$$(x - 1)^2 + (\underline{\phantom{xxxx}})^2 = 9$$
$$(x - 1)^2 + [y - (\underline{\phantom{xx}})]^2 = (\underline{\phantom{xx}})^2,$$
so that $(h, k) = \underline{\phantom{xxxx}}$ and $r = \underline{\phantom{xx}}$.

The graph is

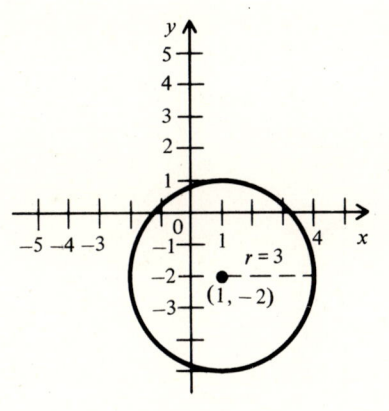
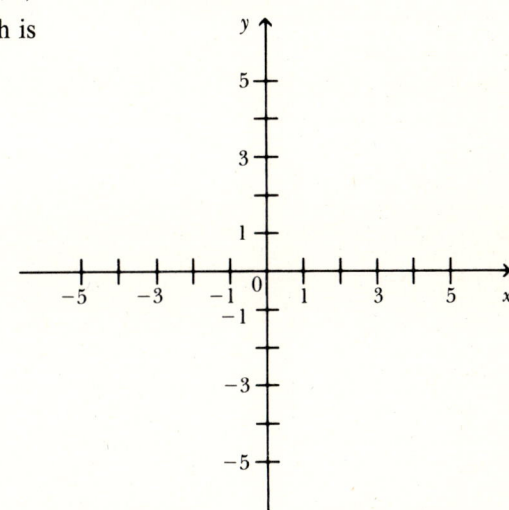

## SOLUTIONS TO SELECTED ODD PROBLEMS  Section 7.2, text pages 346–347

1  $d = \sqrt{(7 - 1)^2 + (10 - 2)^2}$
  $= \sqrt{36 + 64}$
  $= \sqrt{100}$
  $= 10$

5  $d = \sqrt{(6 - 4)^2 + [2 - (-3)]^2}$
  $= \sqrt{4 + 25}$
  $= \sqrt{29}$

9  Since $y_1 = y_2$,
  $d = |x_2 - x_1|$
  $= |7 - 3|$
  $= 4.$

13  Since $x_1 = x_2$,
  $d = |y_2 - y_1|$
  $= |3 - (-1)|$
  $= 4.$

17  $|\overline{AB}| = \sqrt{(-5 + 6)^2 + (1 - 5)^2} = \sqrt{1 + 16} = \sqrt{17}$
  $|\overline{AC}| = \sqrt{(-5 + 2)^2 + (1 - 4)^2} = \sqrt{9 + 9} = \sqrt{18}$
  $|\overline{BC}| = \sqrt{(-6 + 2)^2 + (5 - 4)^2} = \sqrt{16 + 1} = \sqrt{17}$
  Since $|\overline{AB}| = |\overline{BC}|$, $\triangle ABC$ is isosceles.

21  $|\overline{AB}| = \sqrt{(5 - 2)^2 + (-3 - 4)^2} = \sqrt{9 + 49} = \sqrt{58}$
  $|\overline{AC}| = \sqrt{(5 - 2)^2 + (-3 - 5)^2} = \sqrt{9 + 64} = \sqrt{73}$
  $|\overline{BC}| = |4 - 5| = 1$
  Since no two sides are equal, $\triangle ABC$ is not isosceles.

25  $|\overline{AB}| = \sqrt{(-3 - 3)^2 + (1 - 10)^2} = \sqrt{36 + 81} = \sqrt{117}$
  $|\overline{AC}| = |-3 - 3| = 6$
  $|\overline{BC}| = |10 - 1| = 9$
  Since $(\sqrt{117})^2 = 6^2 + 9^2$
  or $117 = 36 + 81$
  $\triangle ABC$ is a right triangle.

29  $|\overline{AB}| = \sqrt{(-2 - 1)^2 + (2 - 4)^2} = \sqrt{9 + 4} = \sqrt{13}$
  $|\overline{AD}| = \sqrt{(-2 - 0)^2 + (2 + 1)^2} = \sqrt{4 + 9} = \sqrt{13}$
  $|\overline{BC}| = \sqrt{(1 - 3)^2 + (4 - 1)^2} = \sqrt{4 + 9} = \sqrt{13}$
  $|\overline{CD}| = \sqrt{(3 - 0)^2 + (1 + 1)^2} = \sqrt{9 + 4} = \sqrt{13}$

Therefore, all four sides are equal. Also
$|\overline{AC}| = \sqrt{(-2-3)^2 + (2-1)^2} = \sqrt{25+1} = \sqrt{26}$.
Since $(\sqrt{26})^2 = (\sqrt{13})^2 + (\sqrt{13})^2$ then $|\overline{AC}|^2 = |\overline{AB}|^2 + |\overline{BC}|^2$, so that $\triangle ABC$ is a right triangle and $\angle ABC$ is a right angle. Hence, $ABCD$ is a square.

33  $M = \left(\dfrac{-2+4}{2}, \dfrac{3+(-2)}{2}\right) = (1, \tfrac{1}{2})$

37  $d = \sqrt{(93.72 - 81.31)^2 + (61.37 - 74.01)^2}$
$= \sqrt{(12.41)^2 + (-12.64)^2}$
$= \sqrt{154.01 + 159.77}$
$= \sqrt{313.78}$
$= 17.71$

41  $(x+1)^2 + (y-2)^2 = 4$
$[x-(-1)]^2 + (y-2)^2 = 2^2$
Center $= (h, k) = (-1, 2)$
Radius $= r = 2$

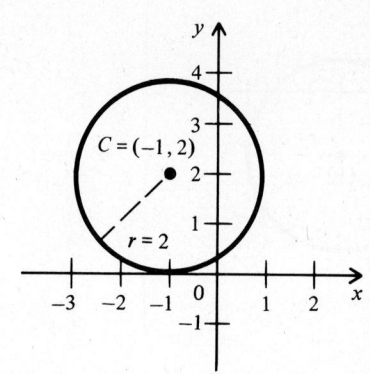

45  $2x^2 + 2y^2 + 12x - 8y + 13 = 0$
$2(x^2 + 6x) + 2(y^2 - 4y) = -13$
$(x^2 + 6x + 9) + (y^2 - 4y + 4) = -\dfrac{13}{2} + 9 + 4$
$(x+3)^2 + (y-2)^2 = \dfrac{13}{2}$
$[x-(-3)]^2 + (y-2)^2 = \left(\dfrac{\sqrt{26}}{2}\right)^2$
Center $= (h, k) = (-3, 2)$
Radius $= r = \dfrac{\sqrt{26}}{2}$

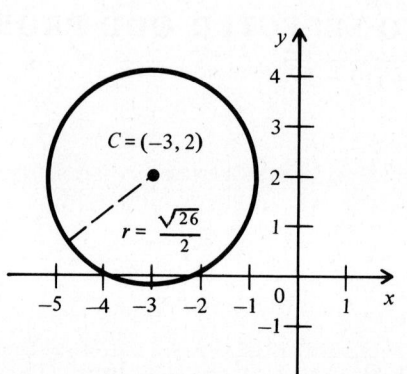

## 7.3  Graphs of Linear Equations

### SEMIPROGRAMMED PROBLEMS

In problems 1–4, graph the given equation by the following procedure:

*Step 1*  Find three points on the graph of the given equation. (Choose three different values of $x$ and calculate the corresponding values of $y$.)
*Step 2*  Plot the three points on a Cartesian coordinate system.
*Step 3*  Draw a straight line through the three points.

1  $y = -3x$
If $x = 0$, then $y = $ _____. Thus, the point $(0, 0)$ is on the _____. If $x = 1$, then $y = $ _____, so the point $(1, -3)$ is on the graph of the _____. If $x = -1$, then $y = $ _____, so that the point _____ is on the graph.

0, graph
−3, line
3, (−1, 3)

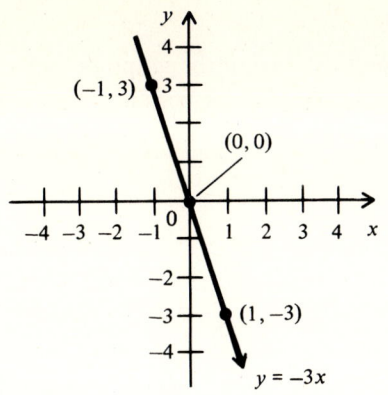

−2
line, 0
graph, 1
(1, 1)

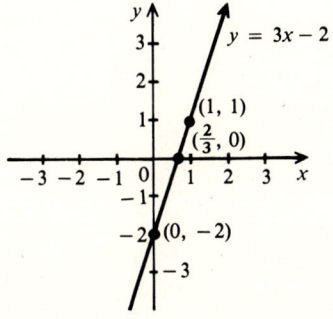

1
−2, (1, −2)
−5, (2, −5)

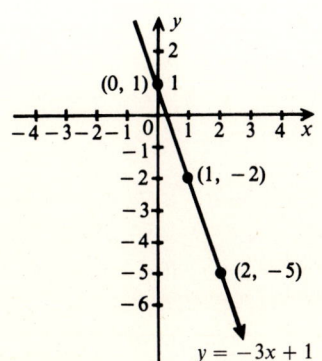

The graph is

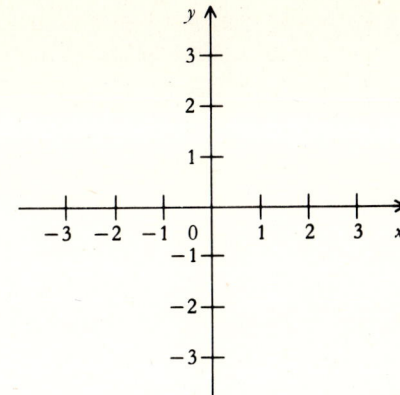

**2**  $y = 3x - 2$

If $x = 0$, then $y =$ _____, so that the point $(0, -2)$ is on the graph of the _____. If $x = \frac{2}{3}$, then $y =$ _____, so that the point $(\frac{2}{3}, 0)$ is on the _____ of the line. If $x = 1$, then $y =$ _____, so that the point _____ is on the graph. The graph is

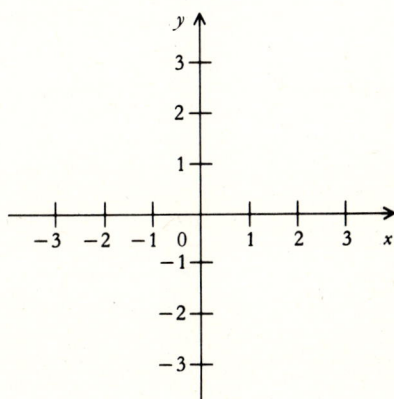

**3**  $y = -3x + 1$

If $x = 0$, then $y =$ _____, so that the point $(0, 1)$ is on the graph. If $x = 1$, then $y =$ _____, so that the point _____ is on the graph. If $x = 2$, then $y =$ _____, so that the point _____ is on the graph. The graph is

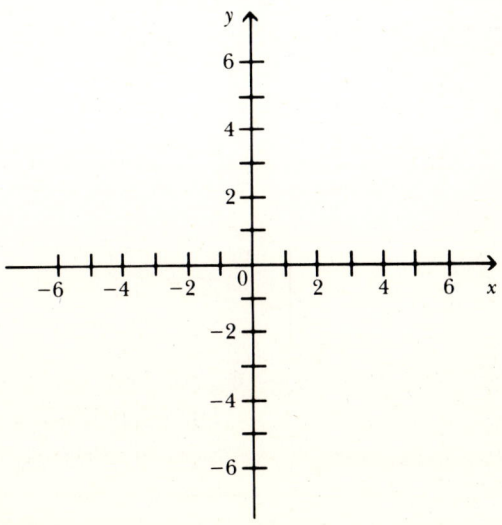

**156**  CHAPTER 7  GRAPHING LINEAR SYSTEMS OF EQUATIONS AND INEQUALITIES

2
line, 0
line, 4
(−3, 4)

**4**  $y = -\frac{2}{3}x + 2$

If $x = 0$, then $y =$ _____, so that the point (0, 2) is on the graph of the _____. If $x = 3$, then $y =$ _____, so that the point (3, 0) is on the graph of the _____. If $x = -3$, then $y =$ _____, so that the point _____ is on the graph. The graph is

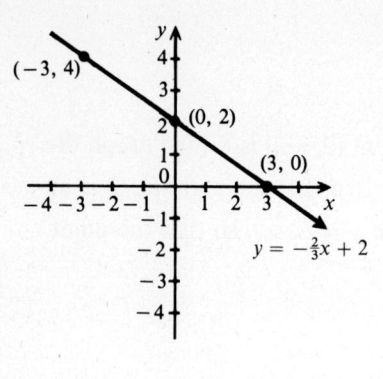

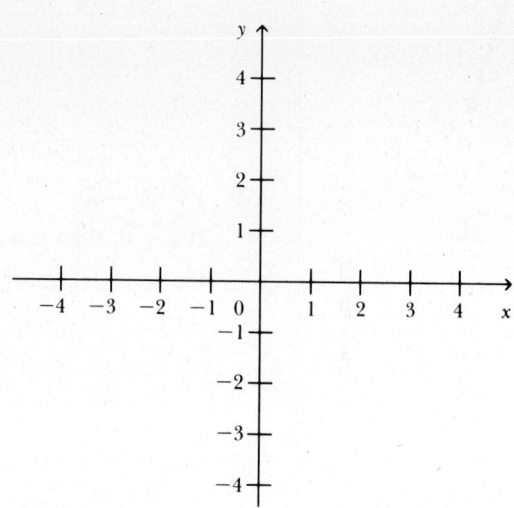

In problems 5–9, find the $x$ intercept and the $y$ intercept of each equation and sketch the graph.

6, 6
−2, −2

**5**  $y = 3x + 6$

If $x = 0$, then $y =$ _____, so the $y$ intercept is _____. If $y = 0$, then $x =$ _____, so the $x$ intercept is _____. The graph is

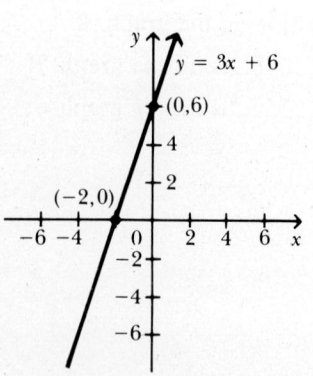

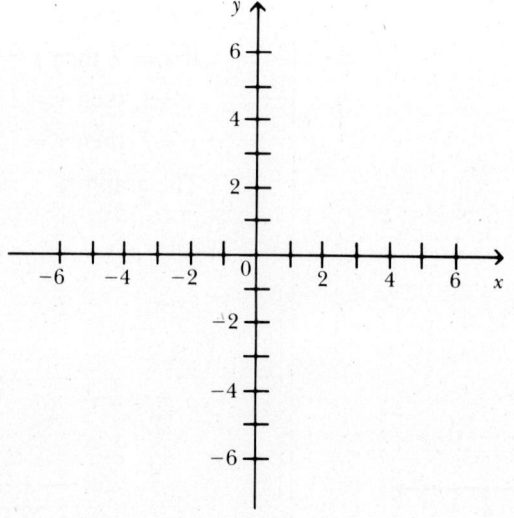

$\frac{2}{3}, \frac{2}{3}$
−2, −2

**6**  $x - 3y + 2 = 0$

If $x = 0$, then $y =$ _____, so that the $y$ intercept is _____. If $y = 0$, then $x =$ _____, so that the $x$ intercept is _____.

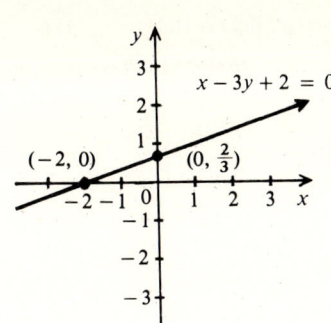

−3, −3
4, 4

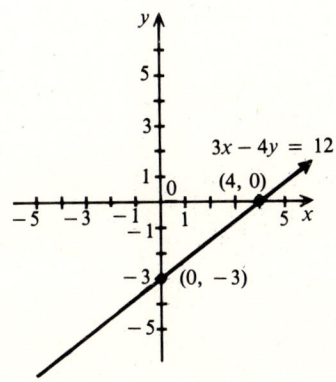

2
2
x, 2
x

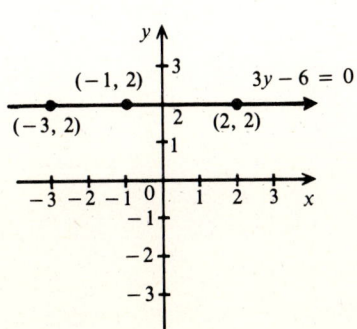

The graph is

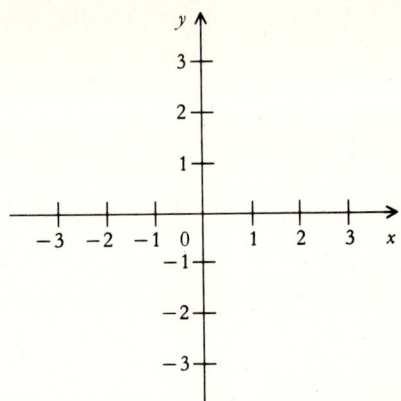

**7**  $3x - 4y = 12$

If $x = 0$, then $y = $ _____, so that the $y$ intercept is _____. If $y = 0$, then $x = $ _____, so that the $x$ intercept is _____. The graph is

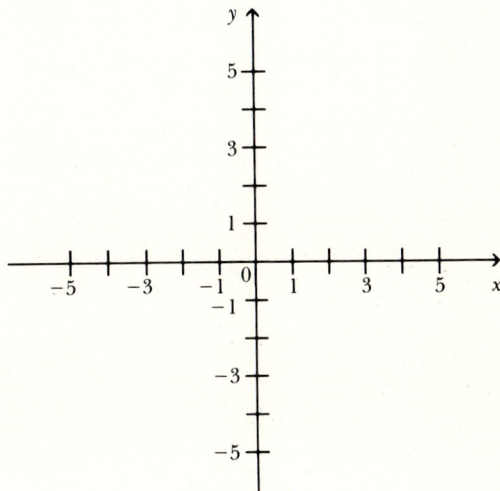

**8**  $3y - 6 = 0$

This equation can be written as $y = $ _____. This line is the set of all points whose $y$ coordinate is _____. The graph is a horizontal line parallel to the _____ axis and whose $y$ intercept is _____. There is no _____ intercept. The graph is

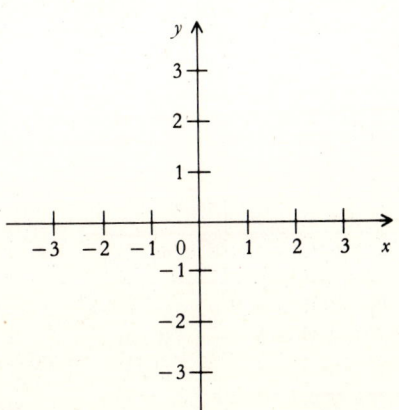

**158** CHAPTER 7 GRAPHING LINEAR SYSTEMS OF EQUATIONS AND INEQUALITIES

−3, x  
y  
−3, y

**9** $x + 3 = 0$

The line $x + 3 = 0$ or $x =$ _____ is the set of all points whose _____ coordinate is −3. Its graph is a vertical line parallel to the _____ axis and whose x intercept is _____. There is no _____ intercept. The graph is

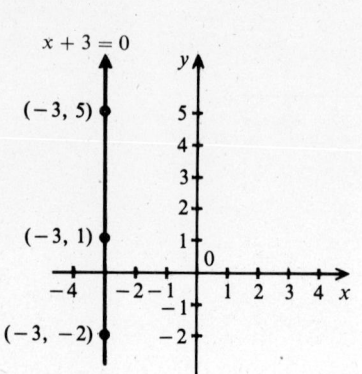
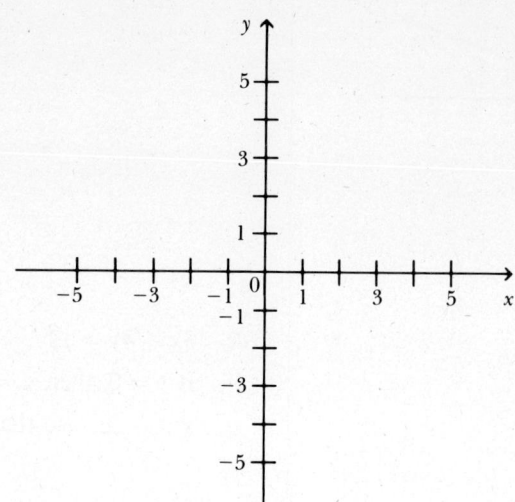

## SOLUTIONS TO SELECTED ODD PROBLEMS  Section 7.3, text pages 354–355

**1**

| x | y = 2x | (x, y) |
|---|---|---|
| −1 | y = 2(−1) = −2 | (−1, −2) |
| 0 | y = 2(0) = 0 | (0, 0) |
| 1 | y = 2(1) = 2 | (1, 2) |

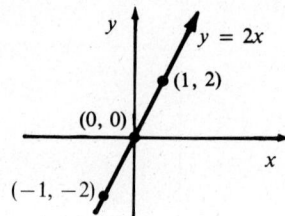

**5**

| x | y = −4x | (x, y) |
|---|---|---|
| −1 | y = −4(−1) = 4 | (−1, 4) |
| 0 | y = −4(0) = 0 | (0, 0) |
| 1 | y = −4(1) = −4 | (1, −4) |

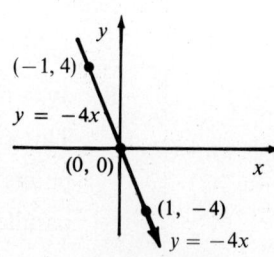

**9**

| x | y = 3x + 1 | (x, y) |
|---|---|---|
| −2 | y = 3(−2) + 1 = −5 | (−2, −5) |
| 0 | y = 3(0) + 1 = 1 | (0, 1) |
| 1 | y = 3(1) + 1 = 4 | (1, 4) |

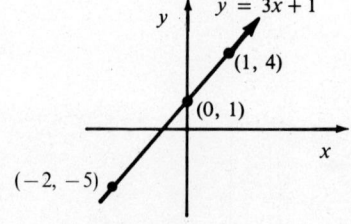

**13**

| x | y = −2x + 5 | (x, y) |
|---|---|---|
| −2 | y = −2(−2) + 5 = 9 | (−2, 9) |
| 0 | y = −2(0) + 5 = 5 | (0, 5) |
| 1 | y = −2(1) + 5 = 3 | (1, 3) |

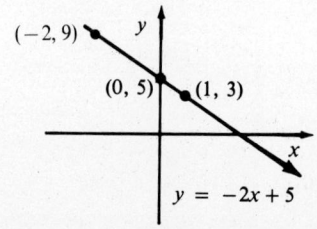

17    $2x + 3y = -6$
       For $y = 0$, $2x = -6$
                  $x = -3$, the $x$ intercept.
       For $x = 0$, $3y = -6$
                  $y = -2$, the $y$ intercept.

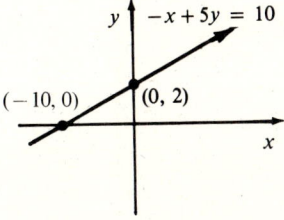

21    $-x + 5y = 10$
       For $y = 0$, $-x = 10$
                  $x = -10$, the $x$ intercept.
       For $x = 0$,
                $5y = 10$
                 $y = 2$, the $y$ intercept.

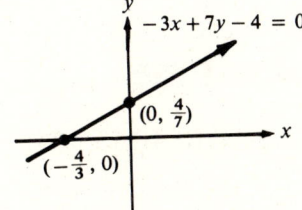

25    $-3x + 7y - 4 = 0$
       For $y = 0$, $-3x - 4 = 0$
                  $-3x = 4$
                  $x = -\frac{4}{3}$, the $x$ intercept.
       For $x = 0$,   $7y - 4 = 0$
                       $7y = 4$
                       $y = \frac{4}{7}$, the $y$ intercept.

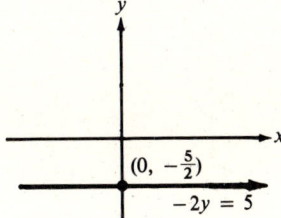

29    $-2y = 5$
         $y = -\frac{5}{2}$, the $y$ intercept
      This line is parallel to the $x$ axis and has no $x$ intercept.

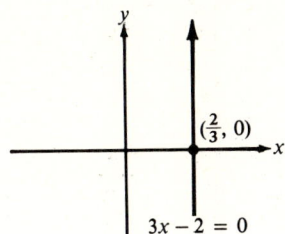

33    $3x - 2 = 0$
         $3x = 2$
         $x = \frac{2}{3}$, the $x$ intercept
      This line is parallel to the $y$ axis and has no $y$ intercept.

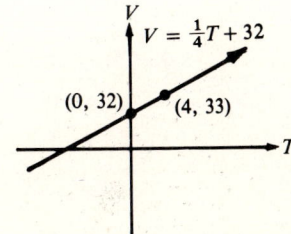

37    $V = \frac{1}{4}T + 32$
       For $T = 0$,
           $V = \frac{1}{4}(0) + 32 = 32$
       For $T = 4$,
           $V = \frac{1}{4}(4) + 32 = 33$
       For $T = 5$
           $V = \frac{1}{4}(5) + 32 = 33\frac{1}{4}$
      Therefore, $V = 33\frac{1}{4}$ cubic centimeters when $T = 5°C$

**41** (a) $V = 10{,}000 - 2{,}000t$
For $t = 0$,
$V = 10{,}000 - 2{,}000(0) = 10{,}000$
For $t = 5$,
$V = 10{,}000 - 2{,}000(5) = 0$
(b) For $t = 3$,
$V = 10{,}000 - 2{,}000(3) = 4{,}000$
Therefore, the value is $4,000.

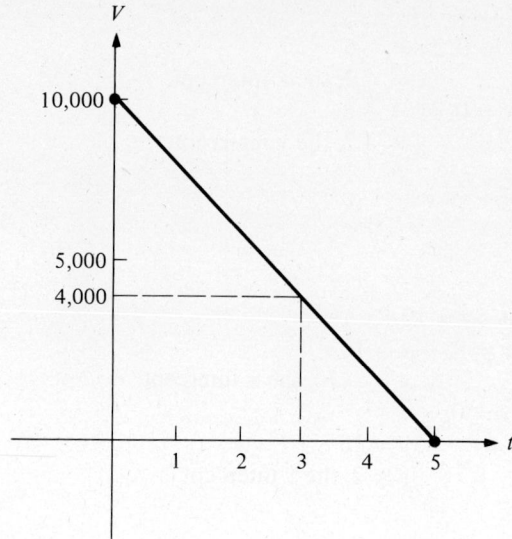

## 7.4 The Slope of a Line

### SEMIPROGRAMMED PROBLEMS

In problems 1–5, find the slope of the line containing each pair of points.

**1** (3, 2) and (1, −8)
Consider
$x_1 = $ ___, $y_1 = $ ___ and
$x_2 = $ ___, $y_2 = $ ___

3, 2
1, −8

$m = \dfrac{\phantom{xxx}}{1 - 3} = \dfrac{\phantom{xxx}}{-2} = $ ___.

−8 − 2, −10, 5

**2** (2, 5) and (−4, 3)
Consider
$x_1 = $ ___, $y_1 = $ ___ and
$x_2 = $ ___, $y_2 = $ ___

2, 5
−4, 3

$m = \dfrac{\phantom{xxx}}{-4 - 2} = $ ___ = ___.

$3 - 5, \tfrac{-2}{-6}, \tfrac{1}{3}$

**3** (−5, 2) and (3, −7)
Consider
$x_1 = $ ___, $y_1 = $ ___ and
$x_2 = $ ___, $y_2 = $ ___

−5, 2
3, −7

$m = \dfrac{-7 - 2}{\phantom{xxx}} = $ ___.

$3 + 5, -\tfrac{9}{8}$

**4** (5, 7) and (9, 7)
Consider
$x_1 = $ ___, $y_1 = $ ___ and
$x_2 = $ ___, $y_2 = $ ___

5, 7
9, 7

$m = \dfrac{\phantom{xxx}}{9 - 5} = \dfrac{\phantom{xxx}}{4} = $ ___.

7 − 7, 0, 0

3, 6
3, 8
3 − 3, undefined

**5** (3, 6) and (3, 8)

Consider

$x_1 = \underline{\phantom{xx}}$, $y_1 = \underline{\phantom{xx}}$ and

$x_2 = \underline{\phantom{xx}}$, $y_2 = \underline{\phantom{xx}}$

$m = \dfrac{8 - 6}{\underline{\phantom{xxx}}} = \underline{\phantom{xxx}}$.

In problems 6 and 7, sketch the line that contains the given point $P$ and has slope $m$.

**6** $P = (0, -3)$ and $m = \frac{3}{4}$

We start at the point $P = (0, -3)$ and move 4 units to the right and _____ up to obtain another point $Q$ on the line, where

3
3, (4, 0)

$Q = (0 + 4, -3 + \underline{\phantom{xx}}) = \underline{\phantom{xxx}}$. The graph is

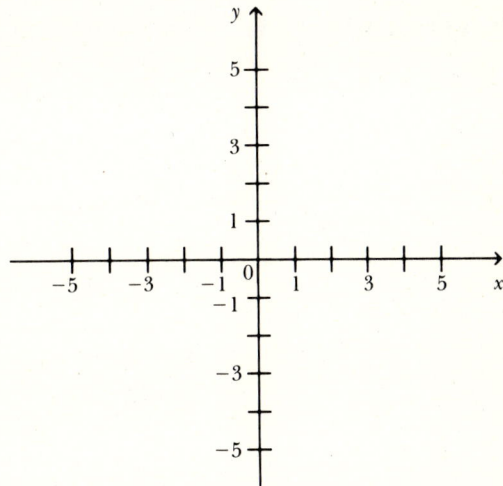

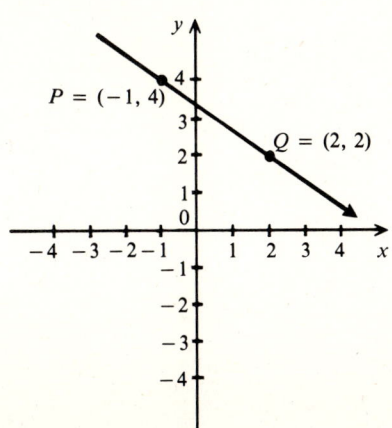

down
4 − 2, (2, 2)

**7** $P = (-1, 4)$ and $m = -\frac{2}{3}$

We start at the point $P = (-1, 4)$ and move 3 units to the right and 2 units _____ to obtain another point $Q$ on the line where

$Q = (-1 + 3, \underline{\phantom{xxx}}) = \underline{\phantom{xxx}}$. The graph is

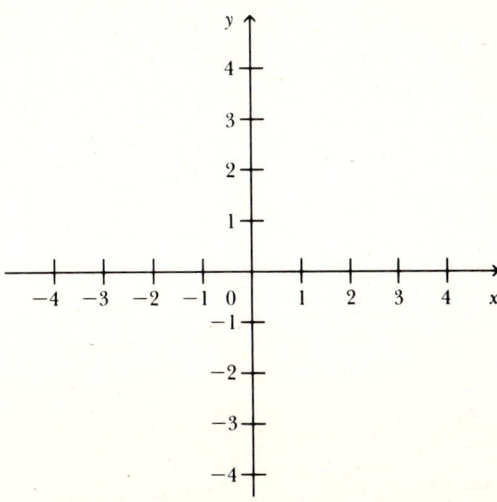

In problems 8–10, determine if line $L_1$ is parallel or perpendicular (or neither) to line $L_2$.

$-\frac{5}{4}$

**8** The slope of the line $L_1$, containing points $P_1 = (-5, 4)$ and $P_2 = (7, -11)$, is _____, and the slope of the line $L_2$, containing the

**162** CHAPTER 7 GRAPHING LINEAR SYSTEMS OF EQUATIONS AND INEQUALITIES

$-\frac{5}{4}$
parallel

$-\frac{3}{2}$
$\frac{2}{3}, -1$
perpendicular

$2$
$-\frac{5}{2}$
$-1$, neither

points $P_3 = (12, 25)$ and $P_4 = (0, 40)$, is _____. Therefore, the line $L_1$ is _____ to the line $L_2$.

9  The slope of the line $L_1$, containing the points $P_1 = (0, 7)$ and $P_2 = (8, -5)$ is _____, and the slope of the line $L_2$, containing the point $P_3 = (5, 5)$ and $P_4 = (2, 3)$, is _____. Since $(-\frac{3}{2})(\frac{2}{3}) = $ _____, the line $L_1$ is _____ to the line $L_2$.

10  The slope of the line $L_1$, containing the points $P_1 = (3, 2)$ and $P_2 = (5, 6)$, is _____, and the slope of the line $L_2$, containing the points $P_3 = (-1, 9)$ and $P_4 = (1, 4)$, is _____. Since $2 \neq -\frac{5}{2}$ and $(2)(-\frac{5}{2}) \neq $ _____, then $L_1$ is _____ parallel or perpendicular to $L_2$.

## SOLUTIONS TO SELECTED ODD PROBLEMS  Section 7.4, text pages 362–363

1  $m = \dfrac{4 - (-1)}{-3 - 2}$
$= \dfrac{5}{-5}$
$= -1$

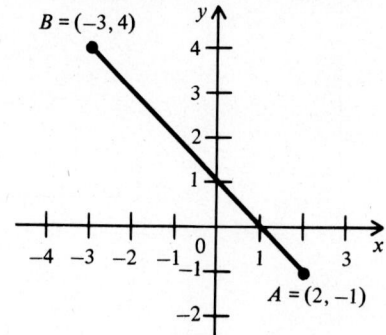

5  $m = \dfrac{-1 - 2}{6 - 0}$
$= \dfrac{-3}{6}$
$= -\dfrac{1}{2}$

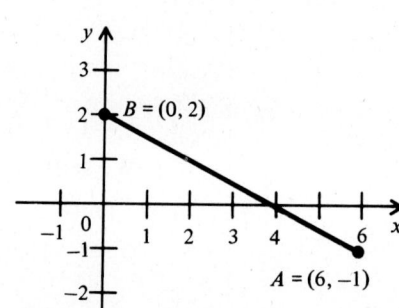

9  $m = \dfrac{-1 - 3}{4 - 4}$
$= \dfrac{-4}{0}$
undefined

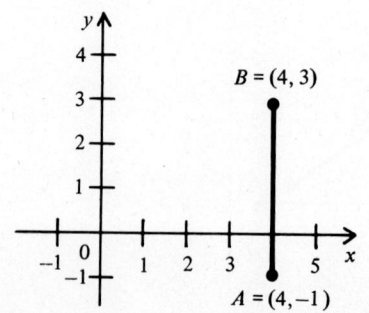

13  $P = (2, -1)$ and
$m = 2 = \frac{2}{1}$
Start at $P$ and move 1 unit to the right and 2 units up to obtain $Q = (2 + 1, -1 + 2) = (3, 1)$.

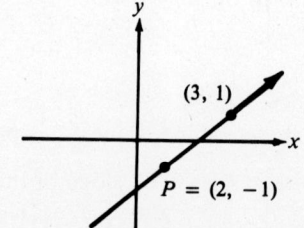

## 7.4 SOLUTIONS TO SELECTED ODD PROBLEMS

**17** $P = (2, 3)$ and
$$m = -\frac{3}{7} = \frac{-3}{7}$$
Start at $P$ and move 7 units to the right and 3 units down to obtain $Q = (2 + 7, 3 - 3) = (9, 0)$.

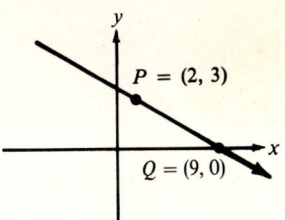

**21** Slope of $\overline{AB}$ is $m_1 = \frac{8 - 4}{3 - 2} = 4$.

Slope of $\overline{CD}$ is $m_2 = \frac{1 - (-3)}{5 - 4} = 4$.

Since $m_1 = m_2$, the lines are parallel.

**25** Slope of $\overline{AB}$ is $m_1 = \frac{8 - 4}{3 - 2} = 4$.

Slope of $\overline{CD}$ is $m_2 = \frac{1 - (-2)}{-4 - 8} = -\frac{1}{4}$.

Since $m_1 m_2 = (4)\left(-\frac{1}{4}\right) = -1$,

the lines are perpendicular.

**29** Slope of $\overline{AB}$ is $m_1 = \frac{3 - 1}{6 - 2} = \frac{1}{2}$.

Slope of $\overline{CD}$ is $m_2 = \frac{k - 1}{4 - 3} = k - 1$.

(a) If parallel, then $m_1 = m_2$

or $k - 1 = \frac{1}{2}$

$k = \frac{3}{2}$.

(b) If perpendicular, then $m_1 m_2 = -1$

or $(k - 1)\frac{1}{2} = -1$

$k - 1 = -2$

$k = -1$.

**33** Slope of $\overline{AB}$ is $m_1 = \frac{4 - 1}{0 - 2} = -\frac{3}{2}$.

Slope of $\overline{CD}$ is $m_2 = \frac{k - 1}{2 - (-3)} = \frac{k - 1}{5}$.

(a) If parallel, $m_1 = m_2$

or $\frac{k - 1}{5} = -\frac{3}{2}$

$k - 1 = -\frac{15}{2}$

$k = -\frac{13}{2}$

(b) If perpendicular, $m_2 = -\frac{1}{m_1}$

or $\frac{k - 1}{5} = -\frac{1}{\left(-\frac{3}{2}\right)} = \frac{2}{3}$

$k - 1 = \frac{10}{3}$

$k = \frac{13}{3}$

**37** Slope of $\overline{AB}$ is $m_1 = \frac{4 - 2}{1 - (-4)} = \frac{2}{5}$

Slope of $\overline{BC}$ is $m_2 = \frac{4 - (-1)}{1 - 3} = -\frac{5}{2}$.

Since $m_1 m_2 = \left(\frac{2}{5}\right)\left(-\frac{5}{2}\right) = -1$,

$\overline{AB}$ is perpendicular to $\overline{BC}$ so that $\triangle ABC$ is a right triangle.

**41** Let $m_1 = \frac{4 - 1}{2 - 1} = 3$

and $m_2 = \frac{4 - 2}{2 - 3} = -2$

Since $m_1 \neq m_2$, points are not collinear.

**45** Slope of $\overline{AB}$ is $\frac{\frac{8}{3} - (-1)}{3 - (-4)} = \frac{\frac{11}{3}}{7} = \frac{11}{21}$

Slope of $\overline{AD}$ is $\frac{-1 - (-9)}{-4 - 2} = \frac{8}{-6} = -\frac{4}{3}$

**164** CHAPTER 7 GRAPHING LINEAR SYSTEMS OF EQUATIONS AND INEQUALITIES

Slope of $\overline{BC}$ is $\dfrac{\frac{8}{3}-(-4)}{3-8} = \dfrac{\frac{20}{3}}{-5} = -\dfrac{4}{3}$

Slope of $\overline{CD}$ is $\dfrac{-4-(-9)}{8-2} = \dfrac{5}{6}$

Therefore, $\overline{AD}$ is parallel to $\overline{BC}$, so that $ABCD$ is a trapezoid.

## 7.5 Equations of Lines

### SEMIPROGRAMMED PROBLEMS

In problems 1–4, find the equation of the line with the given point and given slope. Sketch the graph of the line.

**1**   $m = -\frac{1}{2}$ and $(x_1, y_1) = (1, 3)$

3, 1

In point–slope form, the equation is $y - \underline{\quad} = -\frac{1}{2}(x - \underline{\quad})$.

The graph is

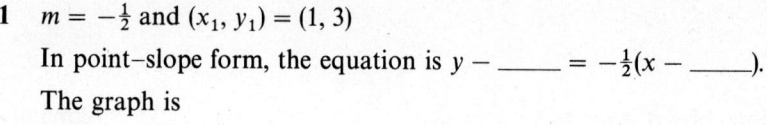

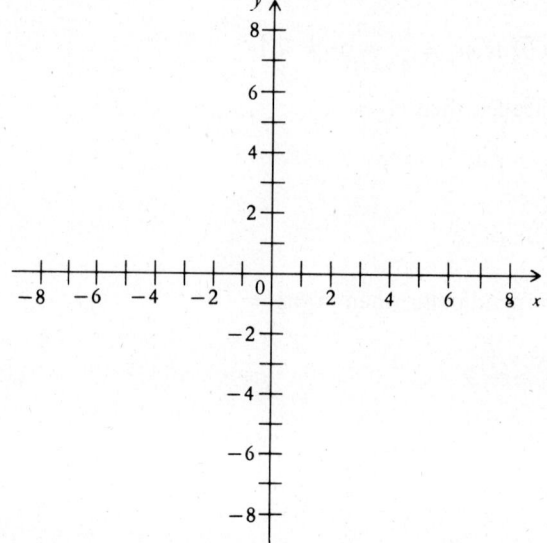

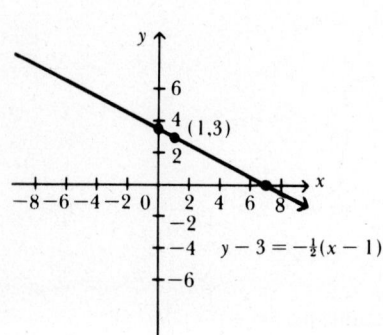

**2**   $m = \frac{2}{3}$ and $(x_1, y_1) = (-5, 2)$

$x + 5$

In point–slope form, the equation is $y - 2 = \frac{2}{3}(\underline{\quad\quad})$.

The graph is

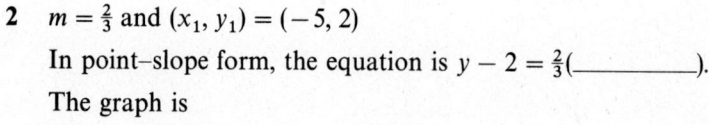

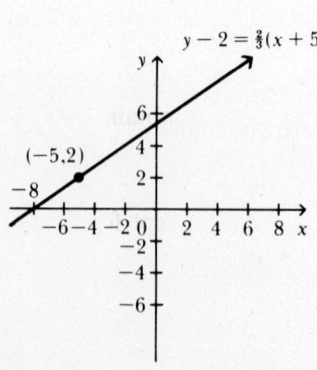

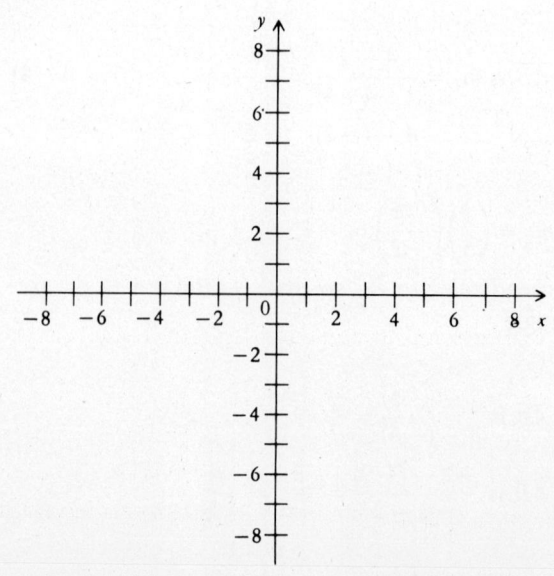

**3** $m = 0$ and $(x_1, y_1) = (3, -2)$

The point–slope form of the equation is

$y - (-2) = 0(\underline{\phantom{x-3}})$ or $y = \underline{\phantom{-2}}$.

$x - 3$, $-2$

The graph is

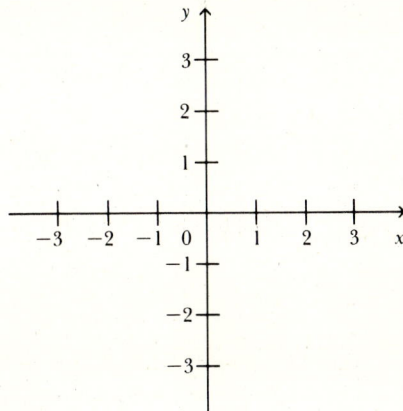

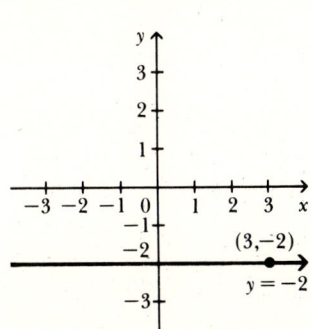

**4** $m =$ undefined and $(x_1, y_1) = (2, 1)$

The point–slope form of the equation does not apply in this case. However, $m$ undefined means that the line is parallel to the $y$ axis; that is, its equation is $x = \underline{\phantom{2}}$. The graph is

2

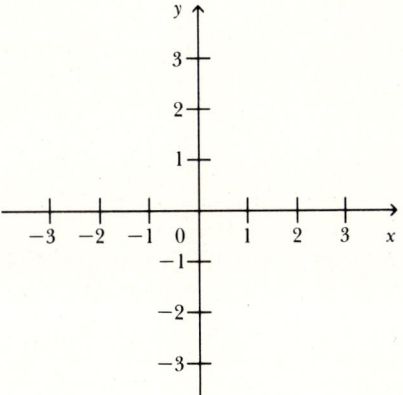

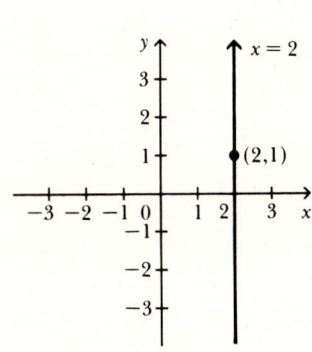

In problems 5 and 6, use the slope–intercept form to find an equation of the line with the given slope $m$ and given $y$ intercept $b$.

**5** $m = 3$, $b = 7$

Substituting these values in the equation $y = mx + b$, we obtain

$3x + 7$

$y = \underline{\phantom{3x+7}}$.

**6** $m = -\frac{1}{2}$, $b = 5$

Substituting these values in the equation $y = mx + b$, we obtain

$-\frac{1}{2}x + 5$

$y = \underline{\phantom{-x/2+5}}$.

In problems 7 and 8, write each equation as an equivalent equation in slope-intercept form, and determine the slope, the $y$ intercept, and the $x$ intercept. Sketch the graph of the line.

**7** $5x + 3y = 6$

$-\frac{5}{3}x + 2$, $-\frac{5}{3}$

Solving for $y$ explicitly, we obtain $y = \underline{\phantom{xxxx}}$. $\underline{\phantom{x}}$ is the

2

slope and $\underline{\phantom{x}}$ is the $y$ intercept. To find the $x$ intercept, set $y = 0$ in

$\frac{6}{5}$

the original equation to obtain $5x = 6$ or $x = \underline{\phantom{x}}$.

**166** CHAPTER 7 GRAPHING LINEAR SYSTEMS OF EQUATIONS AND INEQUALITIES

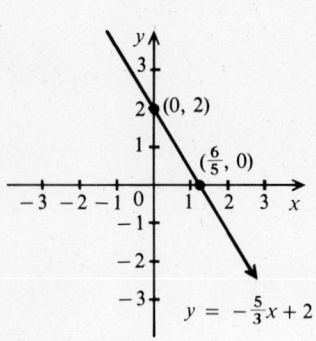

$\frac{2}{3}x - \frac{5}{3}, \frac{2}{3}$

$-\frac{5}{3}$

$\frac{5}{2}$

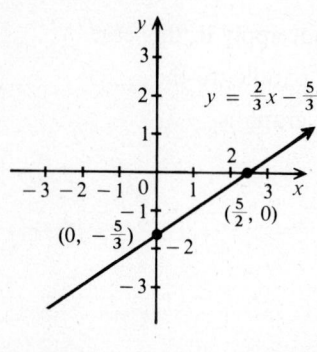

The graph is

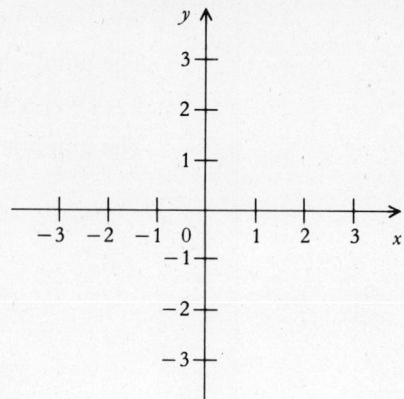

**8** $2x - 3y - 5 = 0$

Solving for $y$ explicitly, we obtain $y =$ _____. _____ is the slope and _____ is the $y$ intercept. To find the $x$ intercept, set $y = 0$ in the original equation to obtain $2x - 5 = 0$ or $x =$ _____. The graph is

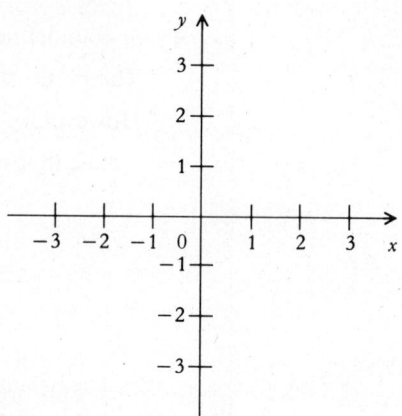

In problems 9 and 10, find an equation of the line $L$ in the following forms: (a) point–slope form, (b) slope–intercept form, and (c) general form.

$\frac{3}{2}x + \frac{7}{2}, \frac{3}{2}$

$\frac{3}{2}$

$\frac{3}{2}$

$2, \frac{3}{2}, 1$

$y, \frac{3}{2}(x-1), \frac{3}{2}x + \frac{1}{2}$

$2$

$3x + 1, 3x - 2y + 1$

$-\frac{1}{4}x + \frac{3}{2}, -\frac{1}{4}$

$4$

$4$

$3, 4, -2$

$3, 4(x+2)$

**9** $L$ contains the point $(1, 2)$ and is parallel to the line $L_1$ whose equation is $2y = 3x + 7$. We obtain the slope of $L_1$ by solving the equation $2y = 3x + 7$ for $y$. That is, $y =$ _____, so that $m_1 =$ _____. By the parallelism condition, the slope $m$ of $L$ is $m = m_1 =$ _____.

(a) Since $L$ has slope $m =$ _____ and contains the point $(1, 2)$, its equation in point–slope form is $y -$ _____ $=$ _____ $(x -$ _____$)$.

(b) To obtain the slope–intercept form of $L$, solve the equation in (a) for _____. Thus, $y = 2 +$ _____ or $y =$ _____.

(c) Multiplying both sides of the equation in (b) by _____, we obtain $2y =$ _____ or _____ $= 0$.

**10** $L$ contains the point $(-2, 3)$ and is perpendicular to the line $L_1$ whose equation is $x + 4y = 6$. We obtain the slope of $L_1$ by solving the equation $x + 4y = 6$ for $y$. That is, $y =$ _____, so $m_1 =$ _____.

By the perpendicularity condition, the slope $m$ of $L$ is $m = -\dfrac{1}{m_1} =$ _____.

(a) Since $L$ has slope $m =$ _____ and contains the point $(-2, 3)$, its equation in point–slope form is $y -$ _____ $=$ _____ $(x -$ _____$)$ or $y -$ _____ $=$ _____.

# 7.5 SOLUTIONS TO SELECTED ODD PROBLEMS    167

$y$, $4(x + 2)$, $4x + 11$

$4x - y + 11$

(b) To obtain the slope–intercept form of $L$, solve the equation in (a) for ____. Thus, $y = 3 +$ _____ or $y =$ _____.

(c) The general form of $L$ is obtained from (b) and is

_____ $= 0$.

## SOLUTIONS TO SELECTED ODD PROBLEMS   Section 7.5, text pages 371–372

1  $(x_1, y_1) = (-1, 2)$ and $m = 5$
Substituting in
$y - y_1 = m(x - x_1)$
$y - 2 = 5[x - (-1)]$
$y - 2 = 5(x + 1)$

9  $(x_1, y_1) = (-1, -5)$ and $m = 0$
Since $m = 0$, the line is a horizontal line and its equation is $y = -5$

17  $(x_1, y_1) = (-3, 4)$ with undefined slope. Since the slope is undefined, the line is a vertical line through $(-3, 4)$ and its equation is $x = -3$.

21  $y - 1 = -2(x - 2)$
$y - 1 = -2x + 4$
$y = -2x + 5$
so $m = -2$ and $b = 5$.
Let $y = 0$, so that
$0 = -2x + 5$
$2x = 5$
$x = \frac{5}{2}$, the $x$ intercept.

25  $-2x + y = 0$
$y = 2x$
or  $y = 2x + 0$
so $m = 2$ and $b = 0$.
Let $y = 0$, so that
$0 = 2x$
$x = 0$, the $x$ intercept.

29  Slope of $\overline{P_1P_2}$ is $m_1 = \frac{7 - (-3)}{5 - 4} = \frac{10}{1} = 10$.
Therefore, the slope of $L$ is $m = 10$.

(a) Since $P = \left(-\frac{1}{2}, 5\right)$, by substituting in
$y - y_1 = m(x - x_1)$, we obtain
$y - 5 = 10\left[x - \left(-\frac{1}{2}\right)\right]$
$y - 5 = 10\left(x + \frac{1}{2}\right)$

(b) Solve the equation in part (a) for $y$ to obtain
$y = 5 + 10\left(x + \frac{1}{2}\right)$
$y = 10x + 10$

(c) $10x - y + 10 = 0$

5  $(x_1, y_1) = (5, -1)$ and $m = -\frac{3}{7}$
Substituting in
$y - y_1 = m(x - x_1)$
$y - (-1) = -\frac{3}{7}(x - 5)$
$y + 1 = -\frac{3}{7}(x - 5)$

13  $m = -3$ and $b = 5$
Substituting in
$y = mx + b$
$y = -3x + 5$

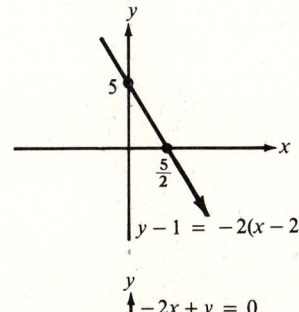

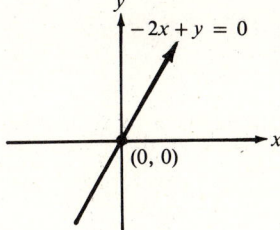

33  Slope of $\overline{P_1P_2}$ is $= \frac{3 - 0}{-2 - 3} = -\frac{3}{5}$
Therefore, the slope of $L$ is $m = -\frac{1}{\left(-\frac{3}{5}\right)} = \frac{5}{3}$

(a) Since $P = (0, 0)$ and $m = \frac{5}{3}$, we have
$y - 0 = \frac{5}{3}(x - 0)$

(b) $y = \frac{5}{3}x$

(c) $3y = 5x$, so $5x - 3y = 0$

**37**  $3x - 2y = 7$
$-2y = -3x + 7$
$y = \frac{3}{2}x - \frac{7}{2}$

The slope of this line is $\frac{3}{2}$.

(a) The slope of line $L$ is $m = -\frac{1}{\frac{3}{2}} = -\frac{2}{3}$ and contains the point $(-1, 0)$

Therefore, $y - 0 = -\frac{2}{3}[x - (-1)]$

$y - 0 = -\frac{2}{3}(x + 1)$

(b) $y = -\frac{2}{3}x - \frac{2}{3}$

(c) $3y = -2x - 2$, so $2x + 3y + 2 = 0$

**41** (a)  $3x + ky + 2 = 0$ and $6x - 5y + 3 = 0$
$ky = -3x - 2$ $\qquad\qquad -5y = -6x - 3$
$y = -\frac{3}{k}x - \frac{2}{k}$ $\qquad\qquad y = \frac{6}{5}x + \frac{3}{5}$
$m_1 = -\frac{3}{k}$ $\qquad\qquad\qquad m_2 = \frac{6}{5}$

These lines are parallel if $m_1 = m_2$, that is

$-\frac{3}{k} = \frac{6}{5}$

$6k = -15$

$k = -\frac{15}{6} = -\frac{5}{2}$

(b) $y = (2 - k)x + 2$ and $y = 3x - 1$
$m_1 = 2 - k$ $\qquad\qquad m_2 = 3$

These lines are perpendicular if $m_1 m_2 = -1$, that is
$(2 - k)3 = -1$
$6 - 3k = -1$
$3k = 7$
$k = \frac{7}{3}$

**45** $x$ intercept $= a = -3$
$y$ intercept $= b = -1$

Substituting in $\frac{x}{a} + \frac{y}{b} = 1$,

$\frac{x}{-3} + \frac{y}{-1} = 1$

**49** Let $(V_1, N_1) = (10, 75)$ and $(V_2, N_2) = (12, 80)$

Therefore, $m = \frac{N_2 - N_1}{V_2 - V_1} = \frac{80 - 75}{12 - 10} = \frac{5}{2}$

(a) Using the point $(10, 75)$, we obtain the equation

$N - 75 = \frac{5}{2}(V - 10)$

$N - 75 = \frac{5}{2}V - 25$

$N = \frac{5}{2}V + 50$

(b) If $N = 90$,

$90 = \frac{5}{2}V + 50$

$\frac{5}{2}V = 40$

$V = 16$

Therefore, her speed is 16 feet per second.

# 7.6 Graphs of Linear Inequalities

## SEMIPROGRAMMED PROBLEMS

In problems 1–4, sketch the region containing the points that satisfy the given inequality.

**1** $x > 3$

First, we sketch the graph of the associated linear equation $x = 3$. Next, we test one point, say, (4, 1), to determine the solution region. Since the point (4, 1) _____ the inequality $x > 3$, the region _____ the point (4, 1) is the solution region. The graph is

satisfies

containing

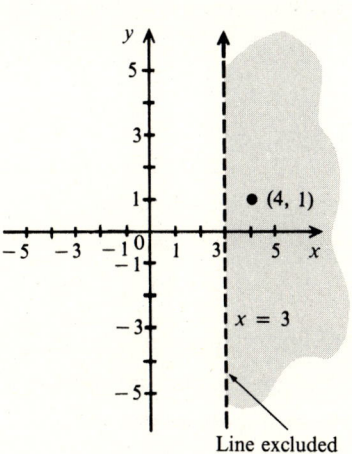

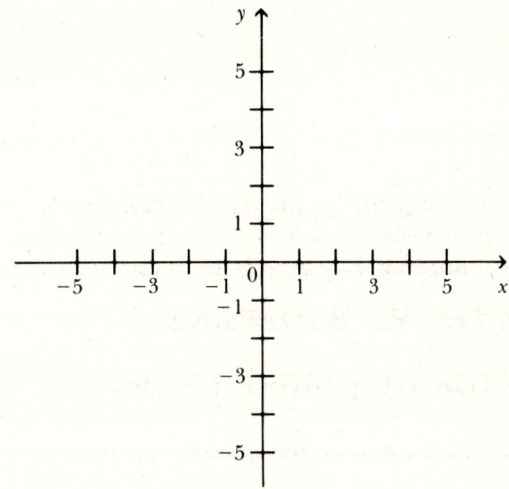

**2** $y \geq x + 1$

First, we sketch the graph of the associated linear equation _____. Next, we test one point not on the graph of the linear equation, say, (4, 2), to determine the solution region. Since for $x = 4$ and $y = 2$, we have $2 \not\geq 4 + 1$, so that the solution region is the region _____ the point (4, 2). The graph is

$y = x + 1$

not containing

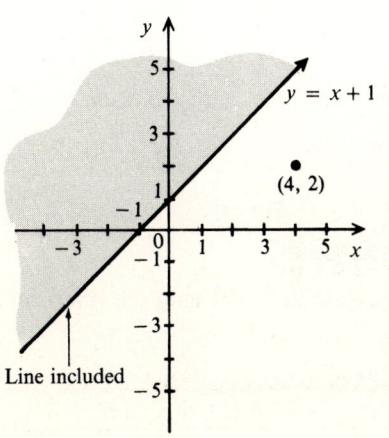

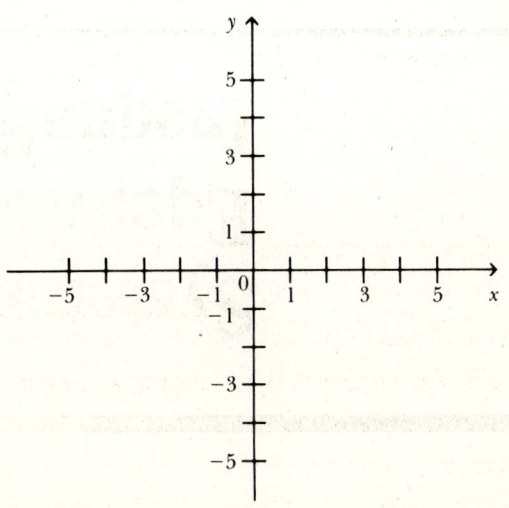

**3** $y > -x + 3$

First, we sketch the graph of the associated linear equation

$y = -x + 3$ _____. Next, we test one point not on the graph of the linear equation, say, (4, 2), to determine the solution region. Since for $x = 4$ and $y = 2$, we have $2 > -4 + 1$, so that the solution region is

containing the region _____ the point (4, 2). The graph is

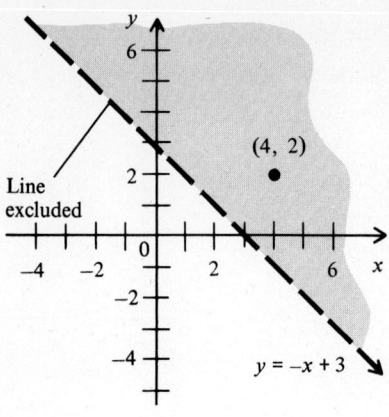

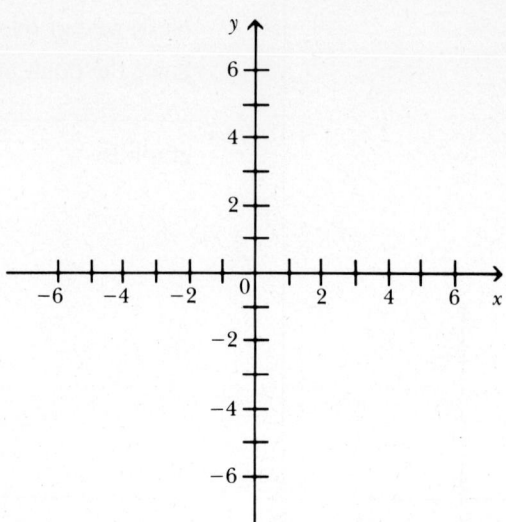

**4** $y \leq -2$

First, we sketch the graph of the associated linear equation _____.

$y = -2$ Next, we test one point, say, (4, −5) to determine the solution region.

satisfies Since the point (4, −5) _____ the inequality $y \leq -2$, the

containing region _____ the point (4, −5) is the solution region.

The graph is

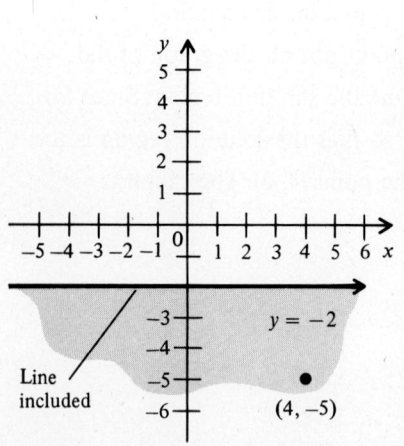

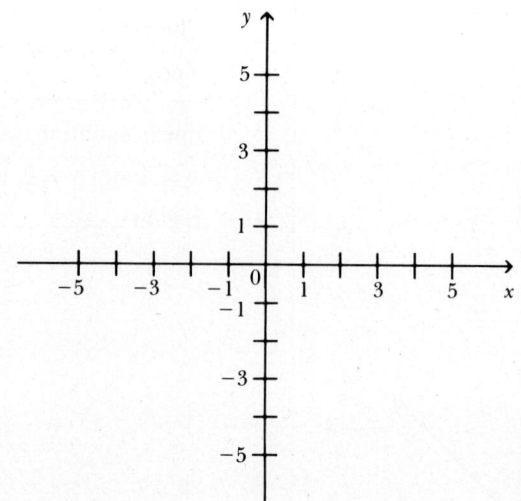

In problems 5 and 6, write an inequality whose solution is the given graph.

**5** First, we determine the equation of the dashed line. Using the intercept form, we obtain

$\frac{2}{3}x + 2$  $\qquad \frac{x}{-3} + \frac{y}{2} = 1$, or $y =$ _____.

## 7.6 SOLUTIONS TO SELECTED ODD PROBLEMS

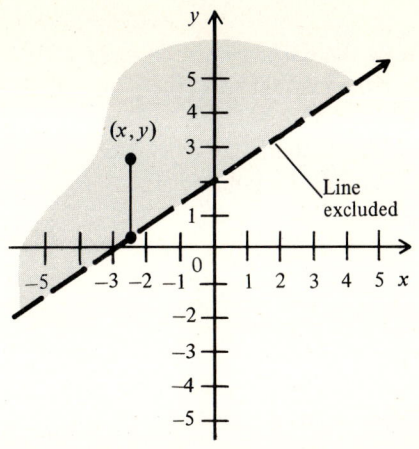

| | |
|---|---|
| greater | The graph consists of all points $(x, y)$ whose $y$ coordinate is _____ than the $y$ coordinate of the point on the line with the same $x$ coordinate. Therefore, the inequality is $y$ _____ $\frac{2}{3}x + 2$. |
| $>$ | |
| 6, $-\frac{2}{3}x + 4$ | 6  The equation of the line is $\dfrac{x}{\_\_\_} + \dfrac{y}{4} = 1$ or $y = $ _____. |

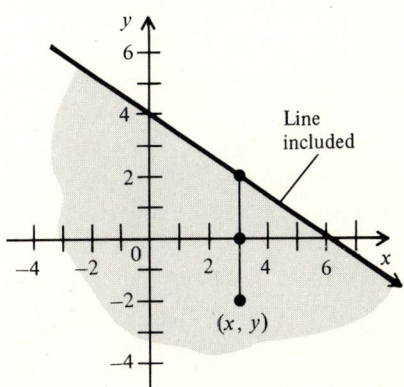

| | |
|---|---|
| less | The graph consists of all points $(x, y)$ which lie on the line or whose $y$ coordinate is _____ than the $y$ coordinate of the point on the line with the same $x$ coordinate. Therefore, the inequality is |
| $\leq$ | $y$ _____ $-\frac{2}{3}x + 4$. |

## SOLUTIONS TO SELECTED ODD PROBLEMS  Section 7.6, text pages 376–378

1  $y \leq 2x + 5$
   Step 1.  Draw the solid line $y = 2x + 5$.
   Step 2.  Test the inequality at the point $(3, 2)$ to find that $2 \leq 2(3) + 5$ is true. Therefore, shade the half-plane containing the point $(3, 2)$.

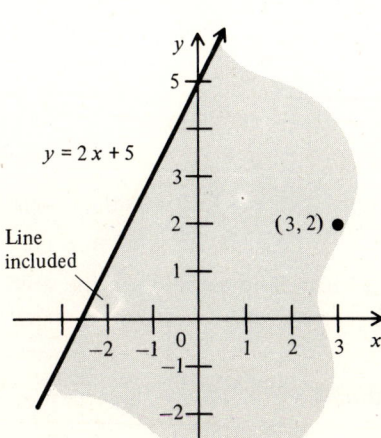

**172** CHAPTER 7 GRAPHING LINEAR SYSTEMS OF EQUATIONS AND INEQUALITIES

**5**    $y > -2x + 3$
     Step 1.      Draw the dashed line $y = -2x + 3$.
     Step 2.      Test the inequality at the point $(3, 1)$ to find that $1 > -2(3) + 3$ is true. Therefore, shade the half-plane containing the point $(3, 1)$.

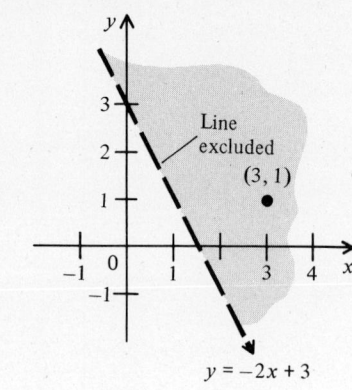

**9**    $3y \leq -4$
     Step 1.      Draw the solid line $3y = -4$.
     Step 2.      Test the inequality at the point $(0, 1)$ to find that $1 \leq -\frac{4}{3}$ is false. Therefore, shade the half-plane not containing the point $(0, 1)$.

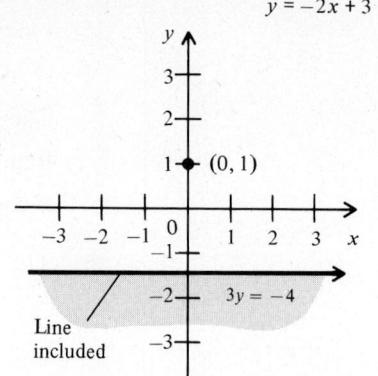

**13**    $3x - 4y > 12$
     Step 1.      Draw the dashed line $3x - 4y = 12$.
     Step 2.      Test the inequality at the point $(2, -5)$ to find that $3(2) - 4(-5) > 12$ is true. Therefore, shade the half-plane containing the point $(2, -5)$.

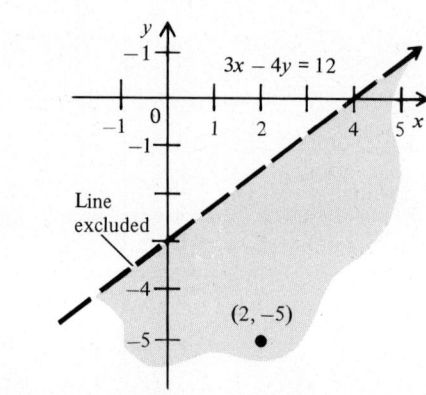

**17**    $3x + 9 < 0$
     Step 1.      Draw the dashed line $3x + 9 = 0$.
     Step 2.      Test the inequality at the point $(1, 0)$ to find that $3(1) + 9 < 0$ is false. Therefore, shade the half-plane not containing the point $(1, 0)$.

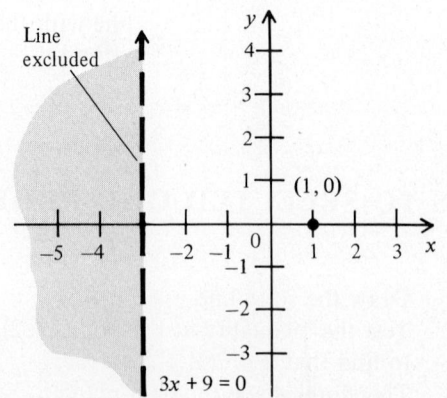

**21**    Since the solid line shown is parallel to the $y$ axis and has $x$ intercept $= 2$, its equation is $x = 2$. Thus, the half-plane shown is the graph of (a) $x \geq 2$.

**25**    Using the intercept form, the equation of the solid line is
$$\frac{x}{3} + \frac{y}{3} = 1 \text{ or } y = -x + 3.$$
The half-plane shown lies above this line, so the inequality is $y \geq -x + 3$, or $x + y \geq 3$.

**29** The equation of the solid line is

$\dfrac{x}{2} + \dfrac{y}{4} = 1$ or $y = -2x + 4$.

The half–plane shown lies above this line, so the inequality is $y \geq -2x + 4$ or $2x + y \geq 4$.

## CUMULATIVE REVIEW PROBLEM SET  Chapters 1–7

1. Use the rule for the order of operations to evaluate $\dfrac{8^2 + 12 \div 6 - 2^2}{10 - 2^3}$.

2. Rewrite 238,400,000 in scientific notation.

3. Perform the indicated operation.
   (a) $(3x^2y - 2xy + 3y^2) + (5xy^2 - x^2y) - (2x^2y + 2xy - y^2)$
   (b) $(x + 3y)(x^2 - xy + 3y^2)$
   (c) $(x^5 - 32y^5) \div (x - 2y)$

4. Factor each expression.
   (a) $100x^3y - 25xy^3$
   (b) $2u^3v + 10u^2v^2 - 28uv^3$

5. Determine if $\dfrac{9x}{11y^2}$ and $\dfrac{27x^3y^2}{33x^2y^4}$ are equivalent fractions.

6. Perform the indicated operations.
   (a) $\dfrac{4t^2 - 9}{9t^2 - 4} \div \dfrac{2t - 3}{3t + 2}$
   (b) $\dfrac{7}{x^2 + 8x + 15} - \dfrac{4}{x^2 + 2x - 3}$

7. Simplify $\dfrac{\dfrac{1}{x} + \dfrac{1}{x + 3}}{\dfrac{1}{x} - \dfrac{1}{x + 3}}$.

8. Solve each equation.
   (a) $\dfrac{x + 1}{x - 1} + \dfrac{x - 5}{x^2 - 1} = \dfrac{x}{x + 1}$
   (b) $|3 - 2x| = 7$

9. Solve each inequality and graph its solution set.
   (a) $|2x - 4| \leq 6$
   (b) $|5 - x| > 3$

10. Two cars leave from the same point at the same time traveling in opposite directions. If one car is traveling 10 miles per hour slower than the other car and after 2 hours they are 220 miles apart, find the rate of speed of each car.

11. Rewrite each expression so that it contains only positive exponents and simplify.
    (a) $\left[\dfrac{x^3y^{-2}z^{-4}}{y^3z^{-6}}\right]^{-1}$
    (b) $\dfrac{(w^{-5/3})^{2/5}}{w^{-2}}$

12. Simplify each radical expression.
    (a) $\sqrt[4]{\dfrac{16u^{12}}{81v^8}}$
    (b) $\sqrt[3]{\sqrt[4]{x^{24}}}$

13. Perform the indicated operations.
    (a) $\sqrt[3]{16x^2} + \sqrt[3]{54x^2}$
    (b) $(2\sqrt{x} + \sqrt{y})(\sqrt{x} - 3\sqrt{y})$

14. Rationalize the denominator of $\dfrac{\sqrt{5} + 2\sqrt{3}}{\sqrt{5} - \sqrt{3}}$.

15  Perform the indicated operations.
  (a)  $(-2 + 7i) - (-3 + 2i)$
  (b)  $\dfrac{7 - 4i}{2 + 3i}$

16  Solve each quadratic equation.
  (a)  $2x^2 - 3x - 3 = 0$
  (b)  $2x^2 - x - 15 = 0$

17  Solve by using an appropriate substitution.
  $t^{-4} - 17t^{-2} + 16 = 0$

18  Solve each inequality and graph its solution set.
  (a)  $x^2 - 2x \le 24$
  (b)  $\dfrac{2x - 1}{x + 3} \ge 0$

19  Find the distance between the points $(-3, 5)$ and $(5, -1)$.

20  Find the center $C$ and radius $r$ of the circle $x^2 + y^2 - 4x + 2y - 11 = 0$.

21  Find the $x$ and $y$ intercepts and use them to graph $3y - 6x = 12$.

22  Find the slope of the line segment $\overline{AB}$.
  (a)  $A = (-3, 4)$ and $B = (2, -3)$
  (b)  $A = (-5, 7)$ and $B = (4, 7)$

23  Determine whether the line containing the points $A$ and $B$ is parallel or perpendicular to the line $y = 2x - 3$.
  (a)  $A = (0, 5)$ and $B = (3, 11)$
  (b)  $A = (-2, 2)$ and $B = (4, -1)$

24  Find the equation of the line $L$ in (i) point–slope form, (ii) slope–intercept form and (iii) general form.
  (a)  $L$ contains $P = (-1, 2)$ and is parallel to the line $3x - 4y = 8$
  (b)  $L$ contains $P = (-3, 0)$ and is perpendicular to the line containing the points $(0, 4)$ and $(6, 1)$.

25  Sketch the graph of each inequality.
  (a)  $y \le -3x + 2$
  (b)  $2x - y < 5$

---

## Answers

1  31    2  $2.384 \times 10^8$    3(a)  $4y^2 - 4xy + 5xy^2$    (b)  $x^3 + 2x^2y + 9y^3$
(c)  $x^4 + 2x^3y + 4x^2y^2 + 8xy^3 + 16y^4$    4(a)  $25xy(2x - y)(2x + y)$    (b)  $2uv(u - 2v)(u + 7v)$
5  yes    6(a)  $\dfrac{2t + 3}{3t - 2}$    (b)  $\dfrac{3x - 27}{(x - 1)(x + 3)(x + 5)}$    7  $\dfrac{2x + 3}{3}$
8(a)  No solution, 1 is an extraneous solution    (b)  $-2, 5$
9(a)  $-1 \le x \le 5$  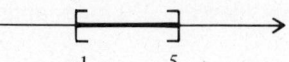    (b)  $x < 2$  or  $x > 8$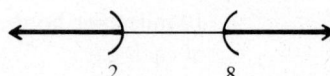

10  50 miles per hour, 60 miles per hour    11(a)  $\dfrac{y^5}{x^3z^2}$    (b)  $w^{4/3}$    12(a)  $\dfrac{2|u^3|}{3v^2}$    (b)  $x^2$

13(a)  $5\sqrt[3]{2x^2}$    (b)  $2x - 5\sqrt{xy} - 3y$    14  $\dfrac{11 + 3\sqrt{15}}{2}$    15(a)  $1 + 5i$    (b)  $\dfrac{2}{13} - \dfrac{29}{13}i$

16(a)  $\dfrac{3 - \sqrt{33}}{4}, \dfrac{3 + \sqrt{33}}{4}$    (b)  $-\dfrac{5}{2}, 3$    17  $-1, -\dfrac{1}{4}, \dfrac{1}{4}, 1$

18(a)  $-4 \le x \le 6$    (b)  $x < -3$  or  $x \ge \dfrac{1}{2}$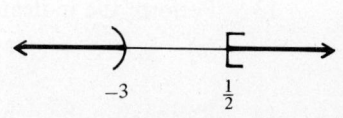

19  10    20  $C = (2, -1), r = 4$

21  x intercept = −2
    y intercept = 4

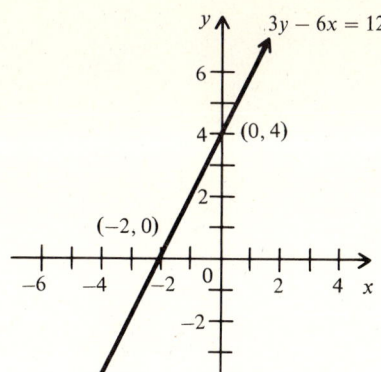

22(a)  −$\frac{7}{5}$   (b)  0      23(a)  parallel   (b)  perpendicular
24(a)  (i)  $y - 2 = \frac{3}{4}(x + 1)$     (b)  (i)  $y - 0 = 2(x + 3)$
       (ii) $y = \frac{3}{4}x + \frac{11}{4}$      (ii) $y = 2x + 6$
       (iii) $3x - 4y + 11 = 0$              (iii) $2x - y + 6 = 0$

25(a) 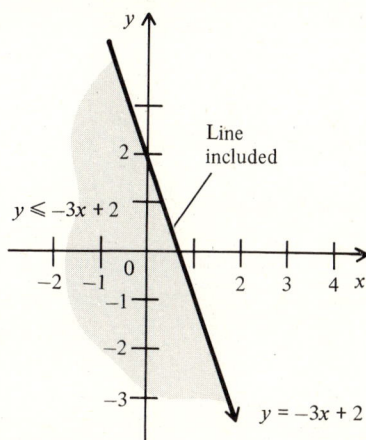   (b)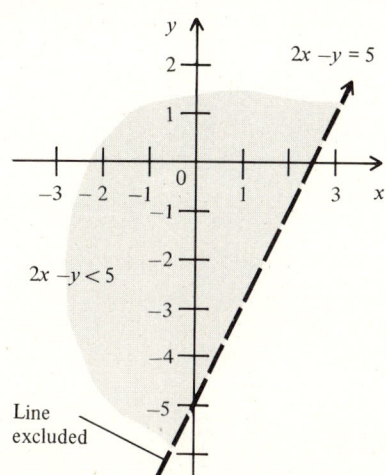

# CHAPTER 7 TESTS

## Chapter Test A

1  Find the missing number so that the ordered pair $(x, y)$ will satisfy the equation $2x + y = 5$.
   (a)  $(1, y)$                                   (b)  $(x, -3)$

2  Find the distance between each pair of points.
   (a)  $(1, 1)$ and $(4, 5)$                      (b)  $(-3, 6)$ and $(2, -3)$

3  Find the equation of the line which satisfies each condition.
   (a)  Its slope is 3 and it contains the point $(2, -5)$.
   (b)  It is parallel to the line $y + 3x = 2$ and it contains the point $(1, 1)$.
   (c)  It is perpendicular to the line $2y - 3x + 1 = 0$ and it contains the point $(-2, 2)$.
   (d)  It is parallel to the $x$ axis and it contains the point $(2, 5)$.

4  Given the line whose equation is $3x - 5y - 15 = 0$:
   (a)  Find its slope.
   (b)  Find its $y$ intercept and its $x$ intercept.
   (c)  Find the equation of the line in slope-intercept form.
   (d)  Find the equation of the line in intercept form.

**5** Sketch the graphs of the linear equations.
   (a) $y = 2x + 1$   (b) $x = 3$   (c) $y = -2$

**6** Find the equation of the line that contains each pair of points.
   (a) $(1, 2)$ and $(-1, 5)$   (b) $(-3, 4)$ and $(4, 4)$   (c) $(2, 0)$ and $(0, 3)$

**7** Graph the following linear inequalities.
   (a) $y > 2x$   (b) $y \leq x + 2$

---

## Solutions

**1** (a) $x = 1$, so that
$$2(1) + y = 5$$
$$2 + y = 5$$
$$y = 3$$

(b) $y = -3$, so that
$$2x + (-3) = 5$$
$$2x = 8$$
$$x = 4$$

**2** (a) $d = \sqrt{(4-1)^2 + (5-1)^2}$
$= \sqrt{9 + 16}$
$= \sqrt{25}$
$= 5$

(b) $d = \sqrt{(2+3)^2 + (-3-6)^2}$
$= \sqrt{25 + 81}$
$= \sqrt{106}$

**3** (a) $m = 3$ and $(x_1, y_1) = (2, -5)$
Therefore,
$$y - (-5) = 3(x - 2)$$
$$y + 5 = 3(x - 2)$$

(b) $y + 3x = 2$
$y = -3x + 2$
so $m = -3$ and $(x_1, y_1) = (1, 1)$
and $y - 1 = -3(x - 1)$

(c) $2y - 3x + 1 = 0$
$2y = 3x - 1$
$y = \dfrac{3}{2}x - \dfrac{1}{2}$

so $m = -\dfrac{1}{\frac{3}{2}} = -\dfrac{2}{3}$ and $(x_1, y_1) = (-2, 2)$

so $y - 2 = -\dfrac{2}{3}(x + 2)$

(d) Since the line is parallel to the $x$ axis its slope is $m = 0$. $(x_1, y_1) = (2, 5)$ so $y - 5 = 0(x - 2)$ or $y = 5$.

**4** (a) $3x - 5y - 15 = 0$
$5y = 3x - 15$
$y = \dfrac{3}{5}x - 3$

Therefore, $m = \dfrac{3}{5}$

(b) $3x - 5y - 15 = 0$
Let $y = 0$, $3x - 15 = 0$
$x = 5$, the $x$ intercept.
Let $x = 0$, $-5y - 15 = 0$
$y = -3$, the $y$ intercept.

(c) From part (a) we have $y = \dfrac{3}{5}x - 3$.

(d) From part (b), we have $a = 5$ and $b = -3$, so that
$$\dfrac{x}{5} + \dfrac{y}{-3} = 1.$$

**5** (a)

| $x$ | $y = 2x + 1$ | $(x, y)$ |
|---|---|---|
| $-1$ | $y = 2(-1) + 1 = -1$ | $(-1, -1)$ |
| $0$ | $y = 2(0) + 1 = 1$ | $(0, 1)$ |
| $2$ | $y = 2(2) + 1 = 5$ | $(2, 5)$ |

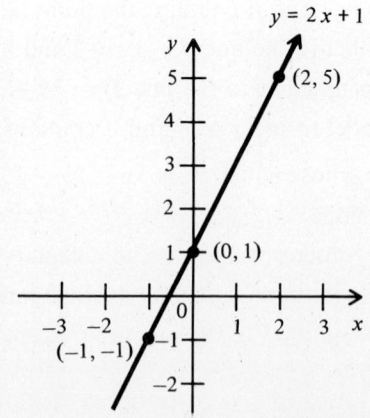

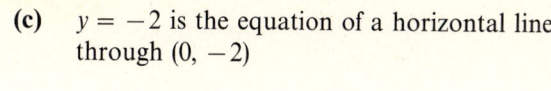

(b) $x = 3$ is the equation of a vertical line through $(3, 0)$

(c) $y = -2$ is the equation of a horizontal line through $(0, -2)$

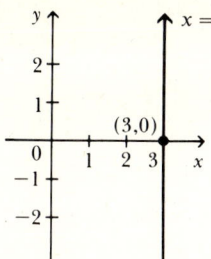
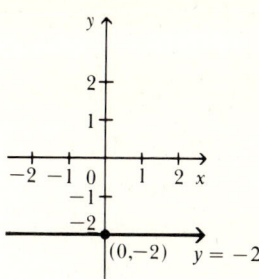

6  (a) $(1, 2)$ and $(-1, 5)$

$$m = \frac{5 - 2}{-1 - 1} = -\frac{3}{2}$$

Let $(x_1, y_1) = (1, 2)$

$$y - 2 = -\frac{3}{2}(x - 1)$$

or $y = -\frac{3}{2}x + \frac{7}{2}$.

(b) $(-3, 4)$ and $(4, 4)$

$$m = \frac{4 - 4}{-3 - 4} = 0$$

So equation is $y = 4$.

(c) $(2, 0)$ and $(0, 3)$
$x$ intercept $= 2$
$y$ intercept $= 3$

So $\dfrac{x}{2} + \dfrac{y}{3} = 1$

or $y = -\dfrac{3}{2}x + 3$

7  (a) $y > 2x$
   Step 1. Draw the dashed line $y = 2x$.
   Step 2. Test the inequality at the point $(2, 1)$ to find that $1 > 2(2)$ is false. Therefore, shade the half–plane not containing $(2, 1)$.

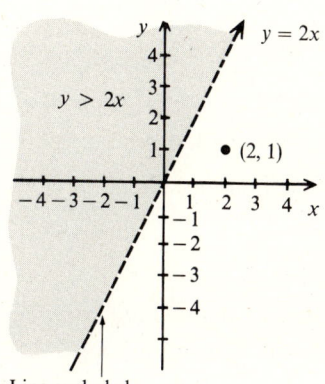

7  (b) $y \leq x + 2$
   Step 1. Draw the solid line $y = x + 2$.
   Step 2. Test the inequality at the point $(3, 2)$ to find that $2 \leq 3 + 2$ is true. Therefore, shade the half–plane containing $(3, 2)$.

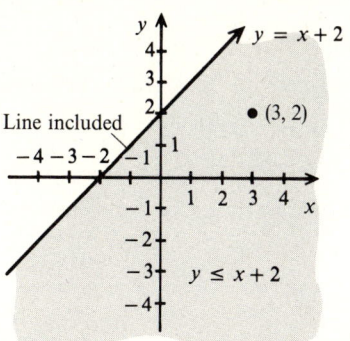

## Chapter Test B

*Multiple Choice:* Select the *one* correct answer for each of the following questions.

1  The coordinate axes divide the plane into four disjoint regions called quadrants. The second quadrant includes all points $(x, y)$ such that _____.

   (a)  $x$ is positive and $y$ is positive
   (b)  $x$ is negative and $y$ is negative
   (c)  $x$ is positive and $y$ is negative
   (d)  $x$ is negative and $y$ is positive

**2** The point $(-3, -4)$ lies in quadrant _____.
  (a) I
  (b) IV
  (c) II
  (d) III

**3** The distance between the points $(x_1, y_1)$ and $(x_2, y_2)$ is given by the formula $d = $ _____.
  (a) $\sqrt{(x_1 - y_1)^2 + (x_2 - y_2)^2}$
  (b) $\sqrt{(x_1 + y_1)^2 + (x_2 + y_2)^2}$
  (c) $\sqrt{(x_2 - x_1)^2 + (y_2 - y_1)^2}$
  (d) $\sqrt{(x_2 + x_1)^2 + (y_2 + y_1)^2}$

**4** The distance between the points $(1, 5)$ and $(3, 4)$ is _____.
  (a) $\sqrt{5}$
  (b) $\sqrt{3}$
  (c) 4
  (d) none of these

**5** The distance between the point $(3, 0)$ and $(7, 0)$ is _____.
  (a) $-4$
  (b) 10
  (c) $\sqrt{10}$
  (d) 4

**6** If two points $(x, y)$ lie on the same vertical line, that is, if $x_1 = x_2$, then the distance is _____.
  (a) $x_1 - x_2$
  (b) $|y_2 - y_1|$
  (c) $|x_1 - x_2|$
  (d) 0

**7** The slope of the line containing the points $(-5, 2)$ and $(3, 7)$ is _____.
  (a) $\frac{3}{4}$
  (b) $\frac{5}{8}$
  (c) $-5$
  (d) $-\frac{5}{8}$

**8** The slope of the line containing the points $(-3, 2)$ and $(-1, 2)$ is _____.
  (a) 0
  (b) $-4$
  (c) undefined
  (d) none of these

**9** The slope of the line $x - 3y + 2 = 0$ is _____.
  (a) $-\frac{1}{3}$
  (b) 3
  (c) $\frac{1}{3}$
  (d) undefined

**10** The $x$ and $y$ intercepts of the line $y = 2x + 3$ are _____.
  (a) $-\frac{3}{2}, 3$
  (b) $-4, 3$
  (c) $2, -\frac{2}{3}$
  (d) none of these

**11** The equation of the line containing the point $(1, 3)$ and with slope $-\frac{1}{2}$ is _____.
  (a) $y - 3 = -\frac{1}{2}(x - 1)$
  (b) $y + 3 = -\frac{1}{2}(x + 1)$
  (c) $y - 1 = -2(x + 1)$
  (d) none of these

**12** The equation of the line with slope 0 and containing the point $(2, 5)$ is _____.
  (a) $y = 2$
  (b) $y - 5 = x - 2$
  (c) $y = 5$
  (d) $y + 5 = x + 2$

**13** Which of the following pair of lines are parallel? _____
  (a) $y = -x + 1$ and $2x - 2y = 1$
  (b) $3x - y = 5$ and $2x - y = 3$
  (c) $y = -2x + 3$ and $4x + 2y = 7$
  (d) $x + 5y = 6$ and $-x + 4y = 7$

**14** Which of the following pair of lines are perpendicular? _____
  (a) $3x - y + 4 = 0$ and $5x + 3y = 6$
  (b) $x - 2y = 5$ and $3x - y = 3$
  (c) $2x - y = 7$ and $3x - y = 4$
  (d) $2y = 3x + 7$ and $y - 2 = -\frac{2}{3}(x - 1)$

**15** The shaded region shown in the given figure is the graph of _____.
  (a) $y \leq x + 1$
  (b) $y < x + 1$
  (c) $y \geq x + 1$
  (d) $y > x + 1$

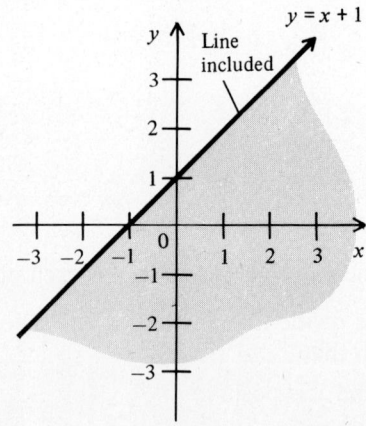

**16** The shaded region shown in the given figure is the graph of _____.

(a) $y < -\frac{2}{3}x - 2$
(b) $y \leq -\frac{2}{3}x - 2$
(c) $y > -\frac{2}{3}x - 2$
(d) $y \geq -\frac{2}{3}x - 2$

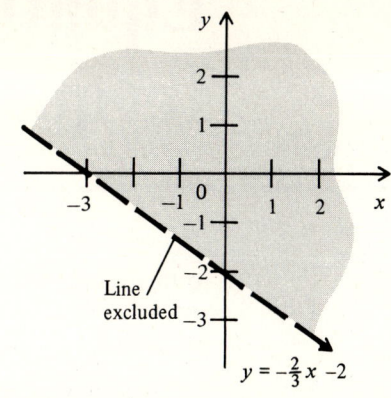

---

## Answers

| 1 d | 2 d | 3 c | 4 a | 5 d | 6 b | 7 b | 8 a | 9 c | 10 a |
| 11 a | 12 c | 13 c | 14 d | 15 a | 16 c | | | | |

# 8 Systems of Linear Equations and Inequalities

In this chapter we investigate methods of solving systems of linear equations and inequalities. After completing the appropriate sections, the student should be able to:

1. Solve systems of linear equations in two or three unknowns by the substitution method and the elimination method.
2. Evaluate determinants of order 2 and 3.
3. Solve systems of linear equations by using determinants (Cramer's rule).
4. Use systems of linear equations to solve word problems.
5. Graph systems of linear inequalities in two unknowns.

## 8.1 Systems of Linear Equations in Two Variables

### SEMIPROGRAMMED PROBLEMS

In problems 1–3, sketch the graph of each system of linear equations. Use the graph to determine whether each system is dependent, inconsistent, or independent. If the system is independent, determine the coordinates of the solution graphically.

1. $\begin{cases} x + y = 2 & (1) \\ -2x - 2y = 3 & (2) \end{cases}$

   Rewrite the system by putting each of the two equations in slope-intercept form:

   $-x - \frac{3}{2}$     $y = -x + 2$ and $y = $ _____.

   Since the slopes of equations (1) and (2) are equal and the $y$ intercepts

   parallel     are different, the lines are _____, so the system does not

   inconsistent     have a solution and the system is _____. The graph is

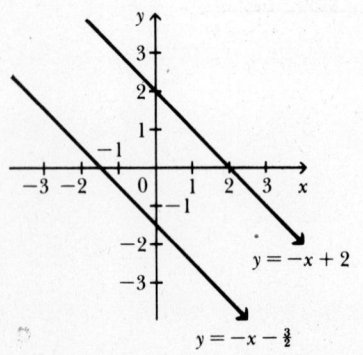

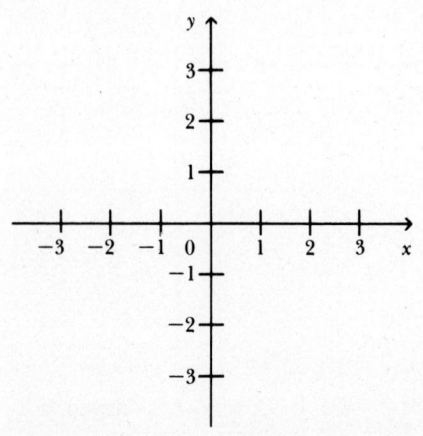

2. $\begin{cases} y - 3x = 2 & (1) \\ 2y = 6x + 4 & (2) \end{cases}$

Rewrite the system by putting each of the two equations in slope-intercept form:

$y - 3x = 2$ and $2y = 6x + 4$ or

$y = $ _____ and $y = 3x + 2$.

$3x + 2$

Since the two equations (1) and (2) have identical slope-intercept forms, the two lines _____ and the system is _____. The graph is

coincide, dependent

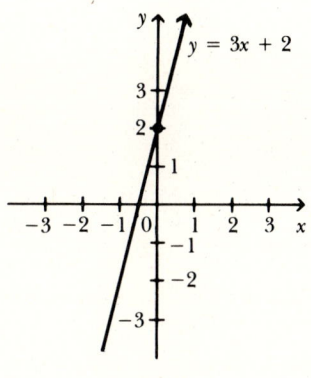

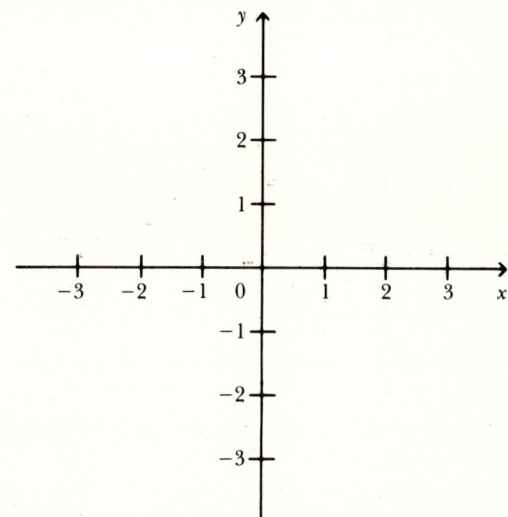

3  $\begin{cases} 2x + y = 7 & (1) \\ x + y = 5 & (2) \end{cases}$

The system can be expressed in slope-intercept form as

$y = -2x + 7$ and $y = $ _____.

$-x + 5$

Since the slopes of equations (1) and (2) are different, the lines _____ and the system is _____. The graph is

intersect, independent

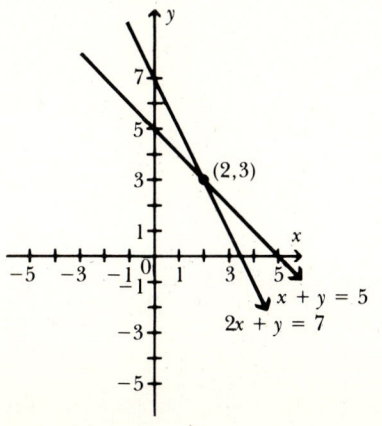

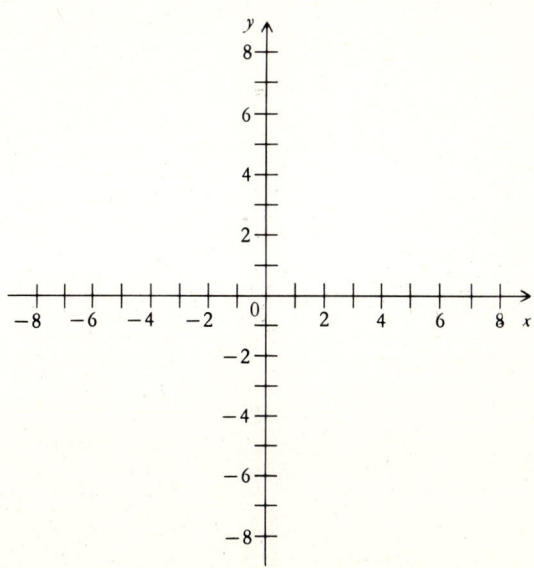

From the graph we see that the solution of the system is the ordered pair _____.

(2, 3)

# 182 CHAPTER 8 SYSTEMS OF LINEAR EQUATIONS AND INEQUALITIES

In problems 4–6, solve each system by the substitution method and check the solution.

**4** $\begin{cases} 3x - y = 12 & (1) \\ 5x + 3y = 34 & (2) \end{cases}$

Solving equation (1) for $y$, we have $y = $ _____. Substituting $y$ from equation (1) into equation (2), we obtain $5x + 3(\underline{\phantom{xx}}) = 34$ or $5x + 9x - \underline{\phantom{xx}} = 34$.

Then, $14x = \underline{\phantom{xx}}$ or $x = \underline{\phantom{xx}}$.

Substituting $x = 5$ in $y = 3x - 12$, we obtain $y = 3(5) - 12$.

Then $y = \underline{\phantom{xx}}$. The solution of the given statement is _____.

*Check:* In (1): $(3)(5) - 3 = \underline{\phantom{xx}} - 3 = \underline{\phantom{xx}}$

In (2): $(5)(5) + (3)(3) = 25 + 9 = \underline{\phantom{xx}}$

**5** $\begin{cases} x - 9y = 11 & (1) \\ 7x + 2y = 12 & (2) \end{cases}$

Solving equation (1) for $x$, we have $x = \underline{\phantom{xx}}$. Substituting $x$ from equation (1) into equation (2), we obtain $7(\underline{\phantom{xx}}) + 2y = 12$ or $63y + \underline{\phantom{xx}} + 2y = 12$. So $65y = \underline{\phantom{xx}}$ or $y = \underline{\phantom{xx}}$.

Substituting $-1$ for $y$ in $x = 9y + 11$, we obtain $x = 9(-1) + 11$ or $x = \underline{\phantom{xx}}$. The solution of the given system is _____.

*Check:* In (1): $2 - 9(-1) = 2 + 9 = \underline{\phantom{xx}}$

In (2): $7(2) + 2(-1) = 14 - 2 = \underline{\phantom{xx}}$

**6** $\begin{cases} 2x - 3y = -6 & (1) \\ 2x - y = -4 & (2) \end{cases}$

Solving equation (2) for $y$, we obtain $y = \underline{\phantom{xx}}$. Substituting $y$ from equation (2) into equation (1), we obtain $2x - 3(\underline{\phantom{xx}}) = -6$, so $-4x - 12 = \underline{\phantom{xx}}$. Therefore, $-4x = \underline{\phantom{xx}}$ or $x = \underline{\phantom{xx}}$.

Substituting $-\frac{3}{2}$ for $x$ in $y = 2x + 4$, we obtain $y = 2(-\frac{3}{2}) + 4$ or $y = \underline{\phantom{xx}}$. The solution of the given system is _____.

*Check:* In (1): $2(-\frac{3}{2}) - 3(1) = -3 - 3 = \underline{\phantom{xx}}$

In (2): $2(-\frac{3}{2}) - (1) = -3 - 1 = \underline{\phantom{xx}}$

In problems 7–10, solve each system by the elimination method.

**7** $\begin{cases} x + y = 3 & (1) \\ 2x - y = 3 & (2) \end{cases}$

Adding equations (1) and (2), we obtain $3x = \underline{\phantom{xx}}$ or $x = \underline{\phantom{xx}}$.

Substituting 2 for $x$ in equation (1), we have $2 + y = 3$ or $y = \underline{\phantom{xx}}$.

The solution is _____.

**8** $\begin{cases} 2x + 3y = 2 & (1) \\ x + 2y = 0 & (2) \end{cases}$

Multiplying equation (2) by 2, we obtain the equivalent system

$\begin{cases} 2x + 3y = 2 & (3) \\ 2x + \underline{\phantom{xx}} = 0. & (4) \end{cases}$

Subtracting equation (3) from equation (4), we have $y = \underline{\phantom{xx}}$.

Substituting $-2$ for $y$ in equation (2), we obtain $x + 2(\underline{\phantom{xx}}) = 0$ or $x = \underline{\phantom{xx}}$. The solution is _____.

---

Margin answers:

$3x - 12$
$3x - 12,$
$36$
$70, 5$

$3, (5, 3)$
$15, 12$
$34$

$9y + 11$
$9y + 11$
$77, -65, -1$

$2, (2, -1)$
$11$
$12$

$2x + 4$
$2x + 4$
$-6, 6, -\frac{3}{2}$

$1, (-\frac{3}{2}, 1)$
$-6$
$-4$

$6, 2$
$1$
$(2, 1)$

$4y$
$-2$
$-2$
$4, (4, -2)$

# 8.1 SOLUTIONS TO SELECTED ODD PROBLEMS

3, 2

32, 2
2
2
−6
−3, (2, −3)

$\dfrac{1}{x}, \dfrac{1}{y}$
−1
−7

−8
−7
−15, $-\dfrac{15}{2}$

−1, $-\dfrac{17}{2}$

$-\dfrac{2}{15}, -\dfrac{2}{17}$

$\left(-\dfrac{2}{15}, -\dfrac{2}{17}\right)$

9 $\begin{cases} 4x + 2y = 2 & (1) \\ 2x - 3y = 13 & (2) \end{cases}$

Multiplying equation (1) by ____ and equation (2) by ____, we obtain the equivalent system

$\begin{cases} 12x + 6y = 6 & (3) \\ 4x - 6y = 26 & (4) \end{cases}$

Adding equations (3) and (4), we have $16x =$ ____ or $x =$ ____.

Substituting ____ for $x$ in equation (1), we have

$4(\underline{\phantom{xx}}) + 2y = 2$

$\phantom{4(\_\_\_) +} 2y =$ ____

or $y =$ ____. The solution is ____.

10 $\begin{cases} -\dfrac{1}{x} + \dfrac{1}{y} = -1 & (1) \\ \dfrac{10}{x} - \dfrac{8}{y} = -7 & (2) \end{cases}$

This system can be expressed as a system that is linear in form if we substitute $u =$ ____ and $v =$ ____. The new system is written

$\begin{cases} -u + v = \underline{\phantom{xx}} & (3) \\ 10u - 8v = \underline{\phantom{xx}}. & (4) \end{cases}$

Multiplying equation (3) by 8, we obtain the equivalent system

$\begin{cases} -8u + 8v = \underline{\phantom{xx}} & (5) \\ 10u - 8v = \underline{\phantom{xx}} & (6) \end{cases}$

Adding equations (5) and (6), we have $2u =$ ____ or $u =$ ____.

Substituting $-\dfrac{15}{2}$ for $u$ in equation (3), we obtain

$-\left(-\dfrac{15}{2}\right) + v =$ ____ or $v =$ ____. Therefore,

$x = \dfrac{1}{u} =$ ____ and $y = \dfrac{1}{v} =$ ____.

The solution is ____.

## SOLUTIONS TO SELECTED ODD PROBLEMS  Section 8.1, text pages 390–391

1 The graphs intersect at one point, therefore the system is independent. The solution is $(\tfrac{1}{3}, 0)$.

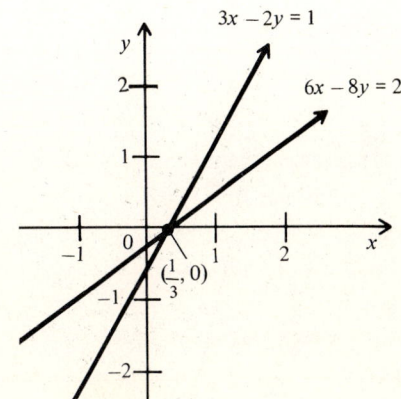

**184** CHAPTER 8 SYSTEMS OF LINEAR EQUATIONS AND INEQUALITIES

**5** The graphs intersect at one point, therefore the system is independent. The solution is (2, 1).

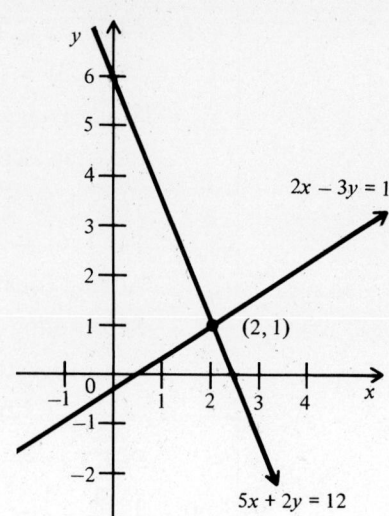

**9** $\begin{cases} 3x - y = -2 \\ x + y = 6 \end{cases}$

From the second equation: $y = 6 - x$
Therefore, $3x - (6 - x) = -2$
$$4x - 6 = -2$$
$$4x = 4$$
$$x = 1$$
and $y = 6 - 1$
$y = 5.$
The solution is (1, 5).

**13** $\begin{cases} 13a + 11b = 21 \\ 7a + 6b = -3 \end{cases}$

From the second equation: $7a = -6b - 3$
$$a = \frac{-6b - 3}{7}$$
Therefore, $13\left(\dfrac{-6b - 3}{7}\right) + 11b = 21$
$$13(-6b - 3) + 77b = 147$$
$$-b - 39 = 147$$
$$b = -186$$
and $a = \dfrac{-6(-186) - 3}{7}$
$$a = 159.$$
The solution is (159, −186).

**17** $\begin{cases} 0.5x - 1.2y = 0.3 \\ 0.7x + 1.5y = 3.6 \end{cases}$

First multiply each equation by 10:
$\begin{cases} 5x - 12y = 3 \\ 7x + 15y = 36 \end{cases}$

From the first equation: $5x = 12y + 3$

or $x = \dfrac{12y + 3}{5}$

so that $7\left(\dfrac{12y + 3}{5}\right) + 15y = 36$
$$84y + 21 + 75y = 180$$
$$159y = 159$$
$$y = 1$$
and $x = \dfrac{12(1) + 3}{5}$
$$x = 3.$$
The solution is (3, 1).

**21** $\begin{cases} u + 3v = 9 \\ u - v = 1 \end{cases}$

Subtracting, we obtain $4v = 8$
$$v = 2,$$
so that $u - 2 = 1$
$$u = 3.$$
The solution is (3, 2).

**25** $\begin{cases} 3x + 2y = 4 \\ 5x + 3y = 7 \end{cases}$

Multiplying the first equation by 3 and the second by 2:

$\begin{cases} 9x + 6y = 12 \\ 10x + 6y = 14 \end{cases}$ (subtract)

$\qquad\qquad x = 2.$

Therefore, $3(2) + 2y = 4$
$\qquad\qquad 2y = -2$
$\qquad\qquad y = -1.$

The solution is $(2, -1)$.

**33** $\begin{cases} \frac{1}{2}x + \frac{1}{3}y = 13 \\ \frac{1}{5}x + \frac{1}{8}y = 5 \end{cases}$

Clearing fractions:

$\begin{cases} 3x + 2y = 78 \\ 8x + 5y = 200, \end{cases}$

so that

$\begin{cases} 15x + 10y = 390 \\ 16x + 10y = 400 \end{cases}$ (subtract)

$\qquad\qquad x = 10$

and
$3(10) + 2y = 78$
$\qquad 2y = 48$
$\qquad y = 24.$

The solution is $(10, 24)$.

**41** $\begin{cases} \dfrac{2}{x} - \dfrac{1}{y} = 9 \\ \dfrac{5}{x} - \dfrac{3}{y} = 14 \end{cases}$

Let $u = \dfrac{1}{x}$ and $v = \dfrac{1}{y}$:

$\begin{cases} 2u - v = 9 \\ 5u - 3v = 14 \end{cases}$

Multiplying the first equation by $-3$:

$\begin{cases} -6u + 3v = -27 \\ \phantom{-}5u - 3v = \phantom{-}14 \end{cases}$ (add)

$\qquad -u = -13$
$\qquad u = 13$
$2(13) - v = 9$
$\qquad v = 17$

$u = \dfrac{1}{x}$ and $v = \dfrac{1}{y}$

$13 = \dfrac{1}{x} \qquad 17 = \dfrac{1}{y}$

$x = \dfrac{1}{13} \qquad y = \dfrac{1}{17}$

The solution is $(\frac{1}{13}, \frac{1}{17})$.

**29** $\begin{cases} -3x + y = 3 \\ \phantom{-}4x + 2y = 10 \end{cases}$

Multiplying the first equation by 2:

$\begin{cases} -6x + 2y = \phantom{0}6 \\ \phantom{-}4x + 2y = 10 \end{cases}$ (subtract)

$\qquad 10x = 4$

$\qquad x = \frac{2}{5}.$

Therefore, $-3(\frac{2}{5}) + y = 3$
$\qquad -\frac{6}{5} + y = 3$
$\qquad y = \frac{21}{5}.$

The solution is $(\frac{2}{5}, \frac{21}{5})$.

**37** $\begin{cases} 0.5x + 0.2y = 0.9 \\ 0.3x - 0.4y = -0.5 \end{cases}$

so that

$\begin{cases} 5x + 2y = 9 \\ 3x - 4y = -5 \end{cases}$

or

$\begin{cases} 10x + 4y = 18 \\ \phantom{1}3x - 4y = -5 \end{cases}$ (add)

$\qquad 13x = 13$

$\qquad x = 1.$

Therefore, $5(1) + 2y = 9$
$\qquad 2y = 4$
$\qquad y = 2$

The solution is $(1, 2)$.

**45** $\begin{cases} \dfrac{5}{x} + \dfrac{2}{y} = 1 \\ \dfrac{13}{x} + \dfrac{8}{y} = 11 \end{cases}$

Let $u = \dfrac{1}{x}, v = \dfrac{1}{y}$

$\begin{cases} 5u + 2v = 1 \\ 13u + 8v = 11 \end{cases}$

Multiplying the first equation by 4:

$\begin{cases} 20u + 8v = \phantom{1}4 \\ 13u + 8v = 11 \end{cases}$ (subtract)

$\qquad 7u = -7$
$\qquad u = -1$

and $5(-1) + 2v = 1$
$\qquad 2v = 6$
$\qquad v = 3$

$u = \dfrac{1}{x}$ and $v = \dfrac{1}{y}$

$-1 = \dfrac{1}{x} \qquad 3 = \dfrac{1}{y}$

$x = -1 \qquad y = \dfrac{1}{3}$

Therefore, the solution is $(-1, \frac{1}{3})$.

# 8.2 Systems of Linear Equations in Three Variables

## SEMIPROGRAMMED PROBLEMS

In problems 1 and 2, use the substitution method to solve each system of three unknowns. Check your solution.

**1** $\begin{cases} x + y + z = 6 & (1) \\ 2x + y - z = 2 & (2) \\ x + y - 2z = -3 & (3) \end{cases}$

We are seeking an ordered triple of numbers $(x, y, z)$ that

satisfies  _____ all three equations simultaneously. Solving for $z$ in

$6 - x - y$  equation (1), we get $z =$ _____. Then, substituting for $z$ from equation (1) into equations (2) and (3), we obtain

2   $\begin{cases} 2x + y - (6 - x - y) = \underline{\quad} & (4) \\ x + y - 2(6 - x - y) = \underline{\quad}. & (5) \end{cases}$
$-3$

Equations (4) and (5) can be solved as a linear system containing two unknowns, that is,

$\begin{cases} 3x + 2y = 8 & (6) \\ 3x + 3y = 9 & (7) \end{cases}$

1, 2   so that $y =$ ____ and $x =$ ____. Now substitute 2 for $x$ and 1 for $y$ in
3    the equation $z = 6 - x - y$ to obtain $z =$ ____. The solution of the
(2, 1, 3)  given system is _____.
6     *Check:* In (1): $2 + 1 + 3 =$ ____
2          In (2): $4 + 1 - 3 =$ ____
$-3$        In (3): $2 + 1 - 6 =$ ____

**2** $\begin{cases} 2x + 3y + z = 6 & (1) \\ x - 2y + 3z = -3 & (2) \\ 3x + y - z = 8 & (3) \end{cases}$

$-3 + 2y - 3z$   Solving for $x$ in equation (2), we get $x =$ _____.
Substituting for $x$ in equation (1), we obtain

$-3 + 2y - 3z$   $2(\underline{\quad\quad\quad}) + 3y + z = 6$ or $7y - 5z = 12$. (4)
Substituting for $x$ in equation (3), we obtain

8   $3(-3 + 2y - 3z) + y - z =$ ____ or $7y - 10z = 17$. (5)
Solving the two linear equations (4) and (5) in the new system, we

$-1, 1$  obtain $z =$ ____ and $y =$ ____. Substituting 1 for $y$ and $-1$ for $z$ in
2     the equation $x = -3 + 2y - 3z$, we obtain $x =$ ____. The solution of
$(2, 1, -1)$  the given system is _____.
6     *Check:* In (1): $4 + 3 - 1 =$ ____
$-3$         In (2): $2 - 2 - 3 =$ ____
8           In (3): $6 + 1 + 1 =$ ____

In problems 3 and 4, use the elimination method to find the solution of the system.

**3** $\begin{cases} x + y + z = 6 & (1) \\ 2x - y + z = 3 & (2) \\ x - y + z = 2 & (3) \end{cases}$

## SOLUTIONS TO SELECTED ODD PROBLEMS  Section 8.3, text pages 400–401

**1** $\begin{vmatrix} 7 & 1 \\ -5 & 3 \end{vmatrix} = 7(3) - (-5)(1) = 26$

**5** $\begin{vmatrix} -3 & -1 \\ -5 & \frac{1}{2} \end{vmatrix} = -3(\frac{1}{2}) - (-5)(-1) = -\frac{13}{2}$

**9** $\begin{vmatrix} 1 & 0 \\ 0 & 1 \end{vmatrix} = (1)(1) - (0)(0) = 1$

**13** $\begin{vmatrix} 1 & 0 & 0 \\ 0 & 1 & 0 \\ 0 & 0 & 1 \end{vmatrix} = 1\begin{vmatrix} 1 & 0 \\ 0 & 1 \end{vmatrix} - 0\begin{vmatrix} 0 & 0 \\ 0 & 1 \end{vmatrix} - 0\begin{vmatrix} 0 & 0 \\ 1 & 0 \end{vmatrix}$

$= 1[1(1) - 0(0)] - 0[0(1) - 0(0)] - 0[0(0) - 1(0)]$
$= 1(1) - 0(0) - 0(0)$
$= 1$

**17** $\begin{vmatrix} 2 & 3 & 5 \\ 9 & 4 & 2 \\ 11 & -6 & 2 \end{vmatrix} = 2\begin{vmatrix} 4 & 2 \\ -6 & 2 \end{vmatrix} - 9\begin{vmatrix} 3 & 5 \\ -6 & 2 \end{vmatrix} + 11\begin{vmatrix} 3 & 5 \\ 4 & 2 \end{vmatrix}$

$= 2[4(2) - (-6)2] - 9[3(2) - (-6)5] + 11[3(2) - 4(5)]$
$= 2(20) - 9(36) + 11(-14)$
$= -438$

**21** $\begin{vmatrix} x & x \\ 5 & 3 \end{vmatrix} = 2$

$3x - 5x = 2$
$-2x = 2$
$x = -1$

**25** $\begin{vmatrix} x & 5 \\ 4 & 2-x \end{vmatrix} = -x^2 + 3$

$x(2-x) - 4(5) = -x^2 + 3$
$2x - x^2 - 20 = -x^2 + 3$
$2x = 23$
$x = \frac{23}{2}$

## 8.4  Cramer's Rule

### SEMIPROGRAMMED PROBLEMS

$\begin{vmatrix} a_1 & b_1 \\ a_2 & b_2 \end{vmatrix}, \begin{vmatrix} c_1 & b_1 \\ c_2 & b_2 \end{vmatrix}, \begin{vmatrix} a_1 & c_1 \\ a_2 & c_2 \end{vmatrix}$

**1** The solution of the system of equations

$\begin{cases} a_1 x + b_1 y = c_1 \\ a_2 x + b_2 y = c_2 \end{cases}$

by Cramer's rule is $x = \dfrac{D_x}{D}$ and $y = \dfrac{D_y}{D}$ where

$D = \underline{\hspace{2cm}}, D_x = \underline{\hspace{2cm}}, D_y = \underline{\hspace{2cm}}.$

In problems 2–5, use Cramer's rule to solve each given system.

**2** $\begin{cases} 2x - 3y = 3 \\ x + 4y = 7 \end{cases}$

11      $D = \begin{vmatrix} 2 & -3 \\ 1 & 4 \end{vmatrix} = \underline{\hspace{1cm}}$

33      $D_x = \begin{vmatrix} 3 & -3 \\ 7 & 4 \end{vmatrix} = \underline{\hspace{1cm}}$

11      $D_y = \begin{vmatrix} 2 & 3 \\ 1 & 7 \end{vmatrix} = \underline{\hspace{1cm}}$

3, 1    $x = \dfrac{D_x}{D} = \underline{\hspace{1cm}}$ and $y = \dfrac{D_y}{D} = \underline{\hspace{1cm}}.$

(3, 1)  The solution is $\underline{\hspace{2cm}}.$

## CHAPTER 8 SYSTEMS OF LINEAR EQUATIONS AND INEQUALITIES

$-29$

$-107$

$-23$

$\frac{107}{29}, \frac{23}{29}$

$(\frac{107}{29}, \frac{23}{29})$

$7$

$-19$

$8$

$-\frac{19}{7}, \frac{8}{7}$

$(-\frac{19}{7}, \frac{8}{7})$

$x - 2y = 6$

$-5$

$-22$

$4$

$\frac{22}{5}, -\frac{4}{5}$

$(\frac{22}{5}, -\frac{4}{5})$

---

**3** $\begin{cases} 5x + 7y = 24 \\ 2x - 3y = 5 \end{cases}$

$D = \begin{vmatrix} 5 & 7 \\ 2 & -3 \end{vmatrix} = \underline{\quad}$

$D_x = \begin{vmatrix} 24 & 7 \\ 5 & -3 \end{vmatrix} = \underline{\quad}$

$D_y = \begin{vmatrix} 5 & 24 \\ 2 & 5 \end{vmatrix} = \underline{\quad}$

$x = \dfrac{D_x}{D} = \underline{\quad}$ and $y = \dfrac{D_y}{D} = \underline{\quad}$.

The solution is $\underline{\quad}$.

**4** $\begin{cases} 2x + 3y = -2 \\ x + 5y = 3 \end{cases}$

$D = \begin{vmatrix} 2 & 3 \\ 1 & 5 \end{vmatrix} = \underline{\quad}$

$D_x = \begin{vmatrix} -2 & 3 \\ 3 & 5 \end{vmatrix} = \underline{\quad}$

$D_y = \begin{vmatrix} 2 & -2 \\ 1 & 3 \end{vmatrix} = \underline{\quad}$

$x = \dfrac{D_x}{D} = \underline{\quad}$ and $y = \dfrac{D_y}{D} = \underline{\quad}$.

The solution is $\underline{\quad}$.

**5** $\begin{cases} 3x - y = -14 \\ \dfrac{x - 2y}{3} = 2 \end{cases}$

Rewrite the second equation without fractions to obtain the equivalent system:

$\begin{cases} 3x - y = 14 \\ \underline{\qquad\qquad} \end{cases}$.

$D = \begin{vmatrix} 3 & -1 \\ 1 & -2 \end{vmatrix} = \underline{\quad}$

$D_x = \begin{vmatrix} 14 & -1 \\ 6 & -2 \end{vmatrix} = \underline{\quad}$

$D_y = \begin{vmatrix} 3 & 14 \\ 1 & 6 \end{vmatrix} = \underline{\quad}$

$x = \dfrac{D_x}{D} = \underline{\quad}$ and $y = \dfrac{D_y}{D} = \underline{\quad}$.

The solution is $\underline{\quad}$.

**6** The linear system in three unknowns

$\begin{cases} a_1 x + b_1 y + c_1 z = d_1 \\ a_2 x + b_2 y + c_2 z = d_2 \\ a_3 x + b_3 y + c_3 z = d_3 \end{cases}$

## 8.4 SEMIPROGRAMMED PROBLEMS

has the solution $x = \dfrac{D_x}{D}$, $y = \dfrac{D_y}{D}$, $z = \dfrac{D_z}{D}$, where

$\begin{vmatrix} a_1 & b_1 & c_1 \\ a_2 & b_2 & c_2 \\ a_3 & b_3 & c_3 \end{vmatrix}$, $\begin{vmatrix} d_1 & b_1 & c_1 \\ d_2 & b_2 & c_2 \\ d_3 & b_3 & c_3 \end{vmatrix}$

$D = $ _____ , $D_x = $ _____

$\begin{vmatrix} a_1 & d_1 & c_1 \\ a_2 & d_2 & c_2 \\ a_3 & d_3 & c_3 \end{vmatrix}$, $\begin{vmatrix} a_1 & b_1 & d_1 \\ a_2 & b_2 & d_2 \\ a_3 & b_3 & d_3 \end{vmatrix}$

$D_y = $ _____ , $D_x = $ _____

In problems 7–10, use Cramer's rule to solve each system.

7 $\begin{cases} 2x + 3y + z = 4 \\ x + 5y - 2z = -1 \\ 3x - 4y + 4z = -1 \end{cases}$

−25

$D = \begin{vmatrix} 2 & 3 & 1 \\ 1 & 5 & -2 \\ 3 & -4 & 4 \end{vmatrix} = $ _____

75

$D_x = \begin{vmatrix} 4 & 3 & 1 \\ -1 & 5 & -2 \\ -1 & -4 & 4 \end{vmatrix} = $ _____

−50

$D_y = \begin{vmatrix} 2 & 4 & 1 \\ 1 & -1 & -2 \\ 3 & -1 & 4 \end{vmatrix} = $ _____

−100

$D_z = \begin{vmatrix} 2 & 3 & 4 \\ 1 & 5 & -1 \\ 3 & -4 & -1 \end{vmatrix} = $ _____

−3, 2, 4

$x = \dfrac{D_x}{D} = $ _____ , $y = \dfrac{D_y}{D} = $ _____ , $z = \dfrac{D_z}{D} = $ _____

(−3, 2, 4)

The solution is _____ .

8 $\begin{cases} 3x - y - 4z = 7 \\ 2x + 3y + 5z = 8 \\ 5x - 2y - 6z = 10 \end{cases}$

15

$D = \begin{vmatrix} 3 & -1 & -4 \\ 2 & 3 & 5 \\ 5 & -2 & -6 \end{vmatrix} = $ _____

30

$D_x = \begin{vmatrix} 7 & -1 & -4 \\ 8 & 3 & 5 \\ 10 & -2 & -6 \end{vmatrix} = $ _____

45

$D_y = \begin{vmatrix} 3 & 7 & -4 \\ 2 & 8 & 5 \\ 5 & 10 & -6 \end{vmatrix} = $ _____

−15

$D_z = \begin{vmatrix} 3 & -1 & 7 \\ 2 & 3 & 8 \\ 5 & -2 & 10 \end{vmatrix} = $ _____

2, 3, −1

$x = \dfrac{D_x}{D} = $ _____ , $y = \dfrac{D_y}{D} = $ _____ , $z = \dfrac{D_z}{D} = $ _____

(2, 3, −1)

The solution is _____ .

## CHAPTER 8 SYSTEMS OF LINEAR EQUATIONS AND INEQUALITIES

7

6

10

$-2$

$\frac{6}{7}, \frac{10}{7}, -\frac{2}{7}$

$(\frac{6}{7}, \frac{10}{7}, -\frac{2}{7})$

9 $\begin{cases} x + y + z = 2 \\ 2x - y + z = 0 \\ x + 2y - z = 4 \end{cases}$

$D = \begin{vmatrix} 1 & 1 & 1 \\ 2 & -1 & 1 \\ 1 & 2 & -1 \end{vmatrix} = \underline{\phantom{aa}}$

$D_x = \begin{vmatrix} 2 & 1 & 1 \\ 0 & -1 & 1 \\ 4 & 2 & -1 \end{vmatrix} = \underline{\phantom{aa}}$

$D_y = \begin{vmatrix} 1 & 2 & 1 \\ 2 & 0 & 1 \\ 1 & 4 & -1 \end{vmatrix} = \underline{\phantom{aa}}$

$D_z = \begin{vmatrix} 1 & 1 & 2 \\ 2 & -1 & 0 \\ 1 & 2 & 4 \end{vmatrix} = \underline{\phantom{aa}}$

$x = \dfrac{D_x}{D} = \underline{\phantom{aa}}, y = \dfrac{D_y}{D} = \underline{\phantom{aa}}, z = \dfrac{D_z}{D} = \underline{\phantom{aa}}$

The solution is _____.

23

23

23

23

1, 1, 1

(1, 1, 1)

10 $\begin{cases} 2x + y - 3z = 0 \\ 3x - 2y + 4z = 5 \\ 4x - y - 2z = 1 \end{cases}$

$D = \begin{vmatrix} 2 & 1 & -3 \\ 3 & -2 & 4 \\ 4 & -1 & -2 \end{vmatrix} = \underline{\phantom{aa}}$

$D_x = \begin{vmatrix} 0 & 1 & -3 \\ 5 & -2 & 4 \\ 1 & -1 & -2 \end{vmatrix} = \underline{\phantom{aa}}$

$D_y = \begin{vmatrix} 2 & 0 & -3 \\ 3 & 5 & 4 \\ 4 & 1 & -2 \end{vmatrix} = \underline{\phantom{aa}}$

$D_z = \begin{vmatrix} 2 & 1 & 0 \\ 3 & -2 & 5 \\ 4 & -1 & 1 \end{vmatrix} = \underline{\phantom{aa}}$

$x = \dfrac{D_x}{D} = \underline{\phantom{aa}}, y = \dfrac{D_y}{D} = \underline{\phantom{aa}}, z = \dfrac{D_z}{D} = \underline{\phantom{aa}}$

The solution is _____.

## SOLUTIONS TO SELECTED ODD PROBLEMS Section 8.4, text pages 405–406

1 $D = \begin{vmatrix} 2 & -1 \\ 1 & 1 \end{vmatrix} = 2 + 1 = 3$

$D_x = \begin{vmatrix} 0 & -1 \\ 1 & 1 \end{vmatrix} = 0 + 1 = 1$

$D_y = \begin{vmatrix} 2 & 0 \\ 1 & 1 \end{vmatrix} = 2 - 0 = 2$

$x = \dfrac{D_x}{D} = \dfrac{1}{3}, y = \dfrac{D_y}{D} = \dfrac{2}{3}$

The solution is $(\frac{1}{3}, \frac{2}{3})$.

5 $D = \begin{vmatrix} 1 & 1 \\ 2 & -2 \end{vmatrix} = -2 - 2 = -4$

$D_x = \begin{vmatrix} 30 & 1 \\ 25 & -2 \end{vmatrix} = -60 - 25 = -85$

$D_y = \begin{vmatrix} 1 & 30 \\ 2 & 25 \end{vmatrix} = 25 - 60 = -35$

$x = \dfrac{D_x}{D} = \dfrac{-85}{-4} = \dfrac{85}{4}, y = \dfrac{D_y}{D} = \dfrac{-35}{-4} = \dfrac{35}{4}$

The solution is $(\frac{85}{4}, \frac{35}{4})$.

9. $D = \begin{vmatrix} 8 & -2 \\ 3 & -5 \end{vmatrix} = -40 + 6 = -34$

$D_z = \begin{vmatrix} 52 & -2 \\ 45 & -5 \end{vmatrix} = -260 + 90 = -170$

$D_w = \begin{vmatrix} 8 & 52 \\ 3 & 45 \end{vmatrix} = 360 - 156 = 204$

$z = \dfrac{D_x}{D} = \dfrac{-170}{-34} = 5,\ w = \dfrac{D_y}{D} = \dfrac{204}{-34} = -6$. The solution is $(5, -6)$.

13. $D = \begin{vmatrix} 2 & -1 & 1 \\ -1 & 2 & -1 \\ 3 & 1 & 2 \end{vmatrix} = 2(4 + 1) - (-1)(-2 - 1) + 3(1 - 2) = 4$

$D_r = \begin{vmatrix} 3 & -1 & 1 \\ 1 & 2 & -1 \\ -1 & 1 & 2 \end{vmatrix} = 3(4 + 1) - 1(-2 - 1) - 1(1 - 2) = 19$

$D_s = \begin{vmatrix} 2 & 3 & 1 \\ -1 & 1 & -1 \\ 3 & -1 & 2 \end{vmatrix} = 2(2 - 1) - (-1)(6 + 1) + 3(-3 - 1) = -3$

$D_t = \begin{vmatrix} 2 & -1 & 3 \\ -1 & 2 & 1 \\ 3 & 1 & -1 \end{vmatrix} = 2(-2 - 1) - (-1)(1 - 3) + 3(-1 - 6) = -29$

$r = \dfrac{D_r}{D} = \dfrac{19}{4},\ s = \dfrac{D_s}{D} = \dfrac{-3}{4} = -\dfrac{3}{4},\ t = \dfrac{D_t}{D} = \dfrac{-29}{4} = -\dfrac{29}{4}$. The solution is $\left(\dfrac{19}{4}, -\dfrac{3}{4}, -\dfrac{29}{4}\right)$.

17. $D = \begin{vmatrix} 2 & 3 & 1 \\ 1 & -2 & 3 \\ 3 & 1 & -1 \end{vmatrix} = 2(2 - 3) - 1(-3 - 1) + 3(9 + 2) = 35$

$D_u = \begin{vmatrix} 6 & 3 & 1 \\ -3 & -2 & 3 \\ 8 & 1 & -1 \end{vmatrix} = 6(2 - 3) - (-3)(-3 - 1) + 8(9 + 2) = 70$

$D_v = \begin{vmatrix} 2 & 6 & 1 \\ 1 & -3 & 3 \\ 3 & 8 & -1 \end{vmatrix} = 2(3 - 24) - 1(-6 - 8) + 3(18 + 3) = 35$

$D_w = \begin{vmatrix} 2 & 3 & 6 \\ 1 & -2 & -3 \\ 3 & 1 & 8 \end{vmatrix} = 2(-16 + 3) - 1(24 - 6) + 3(-9 + 12) = -35$

$u = \dfrac{D_u}{D} = \dfrac{70}{35} = 2,\ v = \dfrac{D_v}{D} = \dfrac{35}{35} = 1,\ w = \dfrac{D_w}{D} = \dfrac{-35}{35} = -1$. The solution is $(2, 1, -1)$.

## 8.5 Applications Involving Linear Systems

### SEMIPROGRAMMED PROBLEMS

In problems 1–5, use a system of linear equations to solve the word problems.

1. A man said to his son, "I am now 4 times as old as you, but in 20 years I shall be only twice as old as you." Find the age of each. Let $x$ years = the man's age and $y$ years = the son's age. The values of $x$ and $y$ are found by solving the system

$x = 4y$
$x + 20 = 2(y + 20)$

$\begin{cases} \underline{\phantom{xxxxxxxx}} \\ \underline{\phantom{xxxxxxxxxxxx}} \end{cases}$

## CHAPTER 8 SYSTEMS OF LINEAR EQUATIONS AND INEQUALITIES

The system is written
$$\begin{cases} x = 4y \\ x - 2y = 20. \end{cases}$$
Substituting $4y$ for $x$ in the second equation, we obtain $4y - 2y = 20$ or ____ 2y, 10 = 20 or $y =$ ____ 10.
Substituting ____ 10 for $y$ in the first equation, we have
$x = 4(\_\_\_\_)$ 10 or $x =$ ____ 40.
Therefore, the man is ____ 40 years old and the son is ____ 10 years old.

2. The sum of two numbers is 12. If one of the numbers is multiplied by 6 and the other by 4, the sum of the products is 62. Find the numbers. Let $x =$ the first number and $y =$ the second number. Then the values of $x$ and $y$ are found by solving the system
$$\begin{cases} x + y = 12 \\ 6x + 4y = 62 \end{cases}$$
Multiplying the first equation by 6, we obtain the equivalent system
$$\begin{cases} 6x + 6y = 72 \\ 4x + 6y = 62. \end{cases}$$
Subtracting the second equation from the first in the new system, we have $2x =$ ____ 10 or $x =$ ____ 5.
Substituting ____ 7 for $x$ in the equation $x + y = 12$, we obtain
____ 7 $+ y = 12$ or $y =$ ____ 5.
Therefore, the numbers are ____ 7 and ____ 5.

3. If part of $125,000 is invested at 5% and part at 7%, find the amount invested at each rate if the total income is $7,050. Let $x =$ amount invested at 5% and $y =$ amount invested at ____ 7%.
The system is
$$\begin{cases} 0.05x + 0.07y = 7,050 \\ x + y = \_\_\_\_ \end{cases}$$ 125,000
or
$$\begin{cases} 5x + 7y = 705,000 \\ 5x + 5y = \_\_\_\_ \end{cases}$$ 625,000
$2y =$ ____ 80,000
$y =$ ____ 40,000.
Hence, $x +$ ____ = 125,000 or $x =$ ____. 40,000, 85,000 Thus, the amount invested at 5% is ____ $85,000 and the amount invested at 7% is ____ $40,000.

4. A merchant bought 800 pounds of coffee for $875. If he paid $1.00 a pound for part of the coffee and $1.25 for the rest, how much did he buy at each price? Let $x =$ the amount purchased at $1.00 and $y =$ the amount purchased at ____ $1.25. The system is
$$\begin{cases} x + 1.25y = 875 \\ x + y = \_\_\_\_. \end{cases}$$ 800

75, 300
300, 500
500
300




$y + 51$



$y + 51$


$y + 51$
510
560
16
16, 67
67, 16

Subtracting, we have

$0.25y = $ _____ or $y = $ _____

$x + $ _____ $= 800$ or $x = $ _____.

Hence, the amount purchased at $1.00 is _____ pounds and the amount purchased at $1.25 is _____.

5 In a collection of dimes and quarters there are 51 more dimes than quarters, and the value of the collection is $10.70. Find the number of dimes and quarters in the collection. Let $x = $ the number of dimes and $y = $ the number of quarters. The system is

$$\begin{cases} 0.10x + 0.25y = 10.70 \\ x = \underline{\phantom{xxxx}} \end{cases}$$

or

$$\begin{cases} 10x + 25y = 1,070 \\ x = \underline{\phantom{xxxx}} \end{cases}.$$

Substituting for $x$ in the first equation, we have

$10(\underline{\phantom{xxxx}}) + 25y = 1,070$

$35y + \underline{\phantom{xx}} = 1,070$

$35y = \underline{\phantom{xx}}$

$y = \underline{\phantom{xx}}$

so that $x = $ _____ $+ 51$ or $x = $ _____.

Hence, there are _____ dimes and _____ quarters in the collection.

---

**SOLUTIONS TO SELECTED ODD PROBLEMS**  Section 8.5, text pages 410–413

1  Let $x = $ number multiplied by 5
and $y = $ number multiplied by 8.
Then $\begin{cases} x + y = 12 \\ 5x + 8y = 75 \end{cases}$
so that $x = 12 - y$
and $5(12 - y) + 8y = 75$
$60 + 3y = 75$
$y = 5$
$x = 12 - 5 = 7$
The numbers are 5 and 7.

5  Let $x = $ number of prey
and $y = $ number of predators.
Then $\begin{cases} x - y = 4,200 \\ x + y = 5,650 \end{cases}$
so that  $2x = 9,850$
$x = 4,925$
and $4,925 + y = 5,650$
$y = 725$
There are 4,925 prey and 725 predators.

9  Let $q = $ number of quarters
and $d = $ number of dimes.
Then $\begin{cases} d + q = 18 \\ 10d + 25q = 360 \end{cases}$
or $\begin{cases} 10d + 10q = 180 \\ 10d + 25q = 360 \end{cases}$
$\overline{\phantom{xxxxx} 15q = 180}$
$q = 12$
and $d + 12 = 18$
$d = 6$
There are 12 quarters and 6 dimes.

13  Let $x = $ number of $5 bills
and $y = $ number of $10 bills.
Then $\begin{cases} x + y = 78 \\ 5x + 10y = 465 \end{cases}$
or $\begin{cases} 10x + 10y = 780 \\ 5x + 10y = 465 \end{cases}$
$\overline{\phantom{xx} 5x \phantom{xxxxx} = 315}$
$x = 63$
and  $63 + y = 78$
$y = 15$
There are 63 $5 bills and 15 $10 bills.

17  Let $x$ = time for John to do the job alone
and $y$ = time for Tom to do the job alone.
Then $\begin{cases} y = 2x \\ \dfrac{4}{x} + \dfrac{4}{y} = 1 \end{cases}$

or $\dfrac{4}{x} + \dfrac{4}{2x} = 1$

$\dfrac{6}{x} = 1$

$x = 6$

and $y = 2(6) = 12.$

It takes John 6 hours and Tom 12 hours to do the job alone.

21  Let $x$ = amount invested at 10.5%
and $y$ = amount invested at 13.5%.
Then $\begin{cases} x + y = 40{,}000 \\ 0.105x + 0.135y = 4{,}650 \end{cases}$

or $\begin{cases} 135x + 135y = 5{,}400{,}000 \\ 105x + 135y = 4{,}650{,}000 \end{cases}$

$\phantom{or\ }30x \phantom{+ 135y} = 750{,}000$

$\phantom{or\ \ \ }x = 25{,}000$

and $25{,}000 + y = 40{,}000$

$\phantom{and\ 25{,}000 + }y = 15{,}000.$

She invested \$25,000 at 10.5% and \$15,000 at 13.5%.

25  Let $x$ = amount loaned at 18%
and $y$ = amount loaned at 19.5%.
Then $\begin{cases} x + y = 30{,}000 \\ 0.18x - 0.195y = 1{,}650 \end{cases}$

or $\begin{cases} 195x + 195y = 5{,}850{,}000 \\ 180x - 195y = 1{,}650{,}000 \end{cases}$

$\overline{\phantom{or\ }375x \phantom{+ 195y} = 7{,}500{,}000}$

$x = 20{,}000$

and $20{,}000 + y = 30{,}000$

$y = 10{,}000$

There was \$20,000 loaned at 18% and \$10,000 loaned at 19.5%.

29  Let $x$ = number of type I suits sold,
$y$ = number of type II suits sold,
and $z$ = number of type III suits sold.
Then $\begin{cases} x + y + z = 80 \\ 80x + 90y + 95z = 6{,}825 \\ 75x + 80y + 85z = 6{,}250. \end{cases}$

First, eliminate $z$'s:

$95x + 95y + 95z = 7{,}600 \qquad 85x + 85y + 85z = 6{,}800$
$\underline{80x + 90y + 95z = 6{,}825} \qquad \underline{75x + 80y + 85z = 6{,}250}$
$15x + \phantom{0}5y \phantom{+ 95z} = \phantom{0}775 \qquad 10x + \phantom{0}5y \phantom{+ 85z} = \phantom{0}550$

or $\begin{cases} 3x + y = 155 \\ 2x + y = 110 \end{cases}$

$\phantom{or\ \ }\overline{\phantom{3}x \phantom{+ y} = \phantom{1}45}$

and $2(45) + y = 110$

$y = 20.$

so that $45 + 20 + z = 80$

$z = 15$

There were 45 type I suits sold, 20 type II suits sold, and 15 type III suits sold.

## 8.6 Systems of Linear Inequalities

### SEMIPROGRAMMED PROBLEMS

In problems 1–4, sketch the region determined by the given system.

1  $\begin{cases} y > x - 2 \\ y > -x + 3 \end{cases}$

8.6 SEMIPROGRAMMED PROBLEMS 199

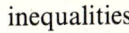

$y = x - 2$
$y = -x + 3$

inequalities

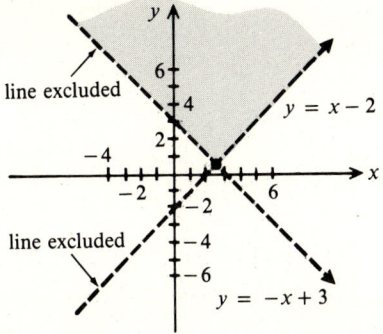

The solution set of $y > x - 2$ consists of all points that lie above the line _____, and the solution set of $y > -x + 3$ consists of all points that lie above the line _____. Hence, the solution set of the given system is the set of points that satisfy both _____ simultaneously. The graph is

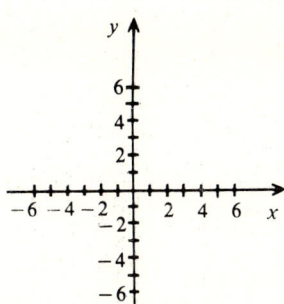

2 $\begin{cases} y \leq 2x \\ y > -3 \end{cases}$

The solution set of $y \leq 2x$ consists of all points that lie below or coincide with the line _____, and the solution set of $y > -3$ consists of all points that lie _____ the line $y = -3$. Hence, the solution set of the given system consists of all points that satisfy both inequalities simultaneously. The graph is

$y = 2x$
above

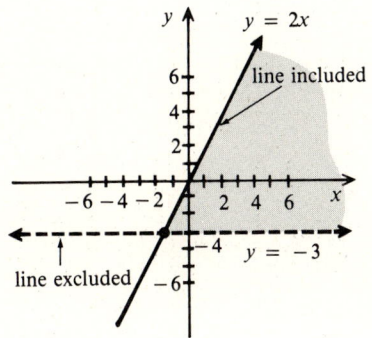

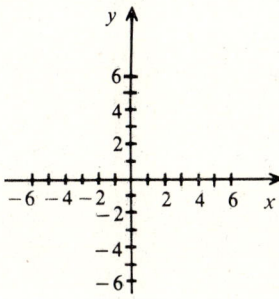

3 $\begin{cases} 2x + y \geq 2 \\ -x + 3y \leq 4 \end{cases}$

The solution set of $2x + y \geq 2$ consists of all points that lie above or coincide with the line _____, and the solution set of $-x + 3y \leq 4$ consists of all points that lie below or coincide with the line _____. Therefore, the solution set of the given system consists of all points that satisfy ____ inequalities simultaneously. The graph is

$2x + y = 2$

$-x + 3y = 4$
both

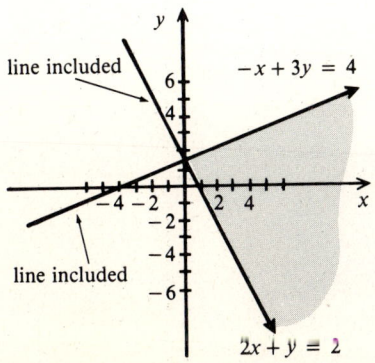

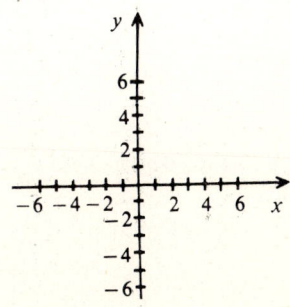

**200** CHAPTER 8 SYSTEMS OF LINEAR EQUATIONS AND INEQUALITIES

$y = -6, x = -3$

4  $\begin{cases} y < -6 \\ x < -3. \end{cases}$

The solution set of this system consists of all points that lie below the line _____ and to the left of the line _____. The graph is

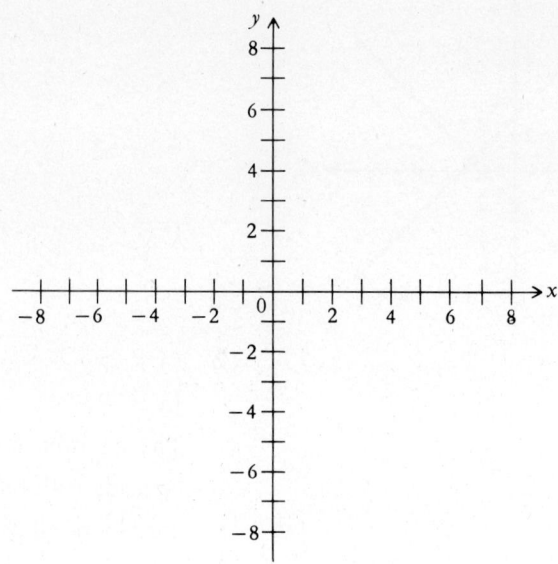

## SOLUTIONS TO SELECTED ODD PROBLEMS  Section 8.6, text pages 415–416

1  $\begin{cases} x + y < 2 \\ 2x - y < -1 \end{cases}$

First, sketch the graphs of $x + y = 2$ and $2x - y = -1$. The graph of $x + y < 2$ is the half-plane below the line $x + y = 2$ and the graph of $2x - y < -1$ is the half-plane above the line $2x - y = -1$. The graph of the system is the region where these two half-planes overlap.

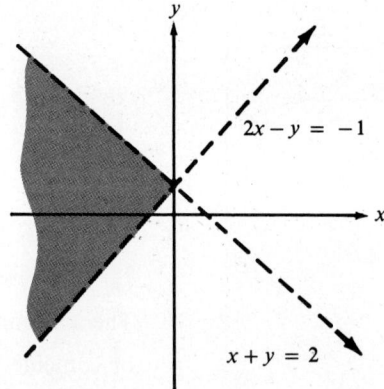

5  $\begin{cases} x + 2y \leq 12 \\ x - y > 6 \end{cases}$

First, sketch the graphs of $x + 2y = 12$ and $x - y = 6$. The graph of $x + 2y \leq 12$ is the half-plane below and including the line $x + 2y = 12$ and the graph of $x - y > 6$ is the half-plane below the line $x - y = 6$. The graph of the system is the region where these two half-planes overlap.

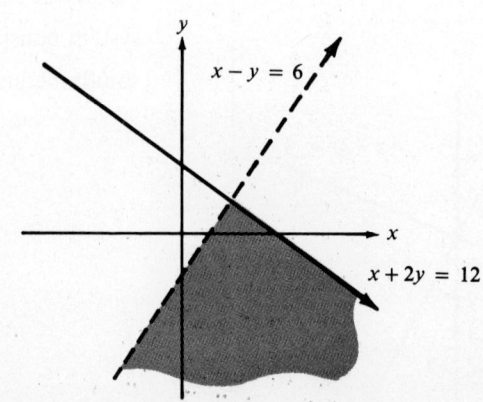

9  $\begin{cases} 2x - 3y \leq -3 \\ 5x - 2y > 9 \end{cases}$

First, sketch the graphs of $2x - 3y = -3$ and $5x - 2y = 9$. The graph of $2x - 3y \leq -3$ is the half-plane above and including the line $2x - 3y = -3$. The graph of $5x - 2y > 9$ is the half-plane below the line $5x - 2y = 9$. The graph of the system is the region where these two half-planes overlap.

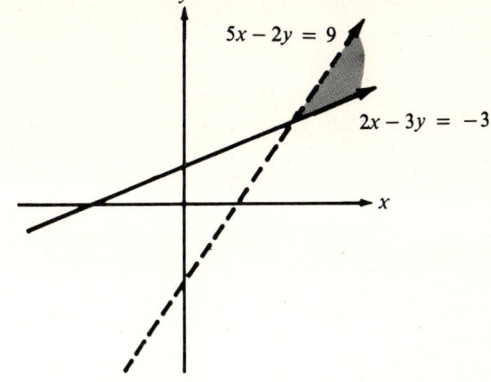

13  $\begin{cases} x \geq 0, y \geq 0 \\ 2x + y \leq 2 \\ x + 2y \leq 2 \end{cases}$

The restrictions $x \geq 0$ and $y \geq 0$ limit the graph to the first quadrant, including the $x$ and $y$ axis. Now graph the lines $2x + y = 2$ and $x + 2y = 2$. The graph of $2x + y \leq 2$ is the half-plane below and including the line $2x + y = 2$ and the graph of $x + 2y \leq 2$ is the half-plane below and including the line $x + 2y = 2$. The graph of the system is the region in the first quadrant where these two half-planes overlap.

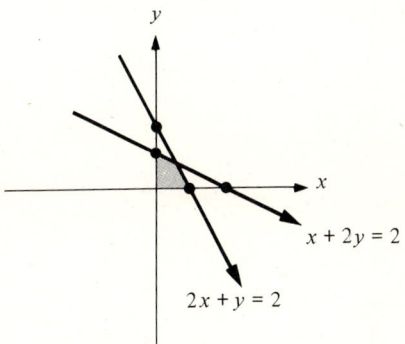

17  Since $x$ and $y$ represent the number of shares of the stocks, then $x \geq 0$ and $y \geq 0$. Therefore, the system is

$\begin{cases} x \geq 0, y \geq 0 \\ x + y \leq 3{,}000 \\ 56x + 86y \leq 200{,}000 \end{cases}$

The restrictions $x \geq 0$ and $y \geq 0$ limit the graph to the first quadrant including the $x$ and $y$ axis. First, graph the lines $x + y = 3{,}000$ and $56x + 86y = 200{,}000$. The graph of $x + y \leq 3{,}000$ is the half-plane below and including the line $x + y = 3{,}000$ and the graph of $56x + 86y \leq 200{,}000$ is the half-plane below and including the line $56x + 86y = 200{,}000$. The graph of the system is the region in the first quadrant where the two half-planes overlap.

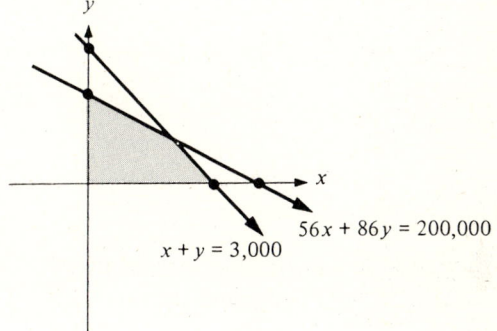

# CUMULATIVE REVIEW PROBLEM SET  Chapters 1–8

1  Evaluate each expression.

   (a)  $4^2 \div 2^3 + 3 \times 5 - 1$  (b)  $|-13| + |13|$  (c)  $\dfrac{x - y}{2}$, for $x = 5$ and $y = -3$

2  Perform the indicated operations.

   (a)  $(-w^3 + 2w^2 - 3w + 7) - (-3w^3 + w^2 - 3)$  (b)  $(2t - u)(4t^2 + 2tu + u^2)$

3  Factor each expression.

   (a)  $x^3 + x^2y + x^2 + xy + 2x + 2y$  (b)  $6x^4y + 3x^3y^2 - 9x^2y^3$

**4** Reduce each fraction to lowest terms.

(a) $\dfrac{24x^3y^5}{36x^6y^2}$

(b) $\dfrac{9x^2 - 81y^2}{3x^2 - 18xy + 27y^2}$

**5** Perform the indicated operations.

(a) $\dfrac{2x^2 + x - 3}{3x - 6x^2} \cdot \dfrac{2x^2 + x - 1}{x^2 - 1}$

(b) $y - \dfrac{1}{1-y} - \dfrac{y^3}{y^2-1}$

**6** Simplify $\dfrac{\dfrac{1}{2x-1} - \dfrac{1}{2x+1}}{\dfrac{x}{4x^2-1}}$.

**7** Solve each equation.

(a) $5(x+2) - 3(x-1) = 4(x+1) + 3$

(b) $\dfrac{3z-b}{2} + \dfrac{z}{3} - b = 1$, for $z$

(c) $|3x - 4| = |-2|$

**8** Graph each solution on a number line.

(a) $\{x \mid x < -2\} \cup \{x \mid x > 4\}$

(b) $\{x \mid x \geq 0\} \cap \{x \mid x < 7\}$

**9** Solve each inequality.

(a) $-3x \leq 15$

(b) $|2x + 3| > 4$

**10** Rewrite each expression so that it contains only positive exponents and simplify.

(a) $(2x^{-2}y^0)^{-3}$

(b) $\dfrac{1 + 2^{-3}}{1 - 4^{-1}}$

(c) $\dfrac{(x^{-2}y^{-1/3})^{-4}}{(x^{-2}y^{-1/3})^2}$

**11** Use the properties of radicals to simplify each expression.

(a) $\sqrt{128x^5y^3}$

(b) $\dfrac{\sqrt[3]{54u^7v^{13}}}{\sqrt[3]{-2uv^2}}$

**12** Perform the indicated operations and simplify the result.

(a) $\sqrt{63x^3} + 2\sqrt{112x^3} - \sqrt{252x}$

(b) $(\sqrt{5x} - \sqrt{3})(\sqrt{5x} + \sqrt{3})$

**13** Solve each equation by the indicated method.

(a) $3x^2 - x - 2 = 0$ (Factoring)

(b) $5x^2 + 3x + 1 = 0$ (Quadratic Formula)

**14** Solve each equation and check the solutions.

(a) $\sqrt{3x + 10} - 2 = \sqrt{2x - 1}$

(b) $(2t - 6)^{2/3} = 4$

**15** Solve each inequality.

(a) $6x^2 + 13x \geq 5$

(b) $\dfrac{3x + 2}{1 - x} > 1$

**16** Find the distance between the two points.

(a) $(-1, 6)$ and $(4, 6)$

(b) $(2, -3)$ and $(-3, -4)$

**17** Find the equation of line $L$.

(a) $L$ has slope $m = \tfrac{2}{3}$ and contains $P = (3, -4)$

(b) $L$ is perpendicular to the line $y = \tfrac{3}{5}x - 2$ and contains $P = (-1, 7)$.

**18** Solve each system by the indicated method.

(a) $\begin{cases} 2x - 3y = 1 \\ x + 4y = 6 \end{cases}$ (by substitution)

(b) $\begin{cases} 4x + 5y = 2 \\ 3x - 2y = 13 \end{cases}$ (by Cramer's rule)

(c) $\begin{cases} 3x - 2y + z = 11 \\ x + 4y + 3z = 7 \\ 2x - 3y - z = 4 \end{cases}$ (by elimination)

19  Sketch the graph of the system of inequalities.
$$\begin{cases} 2x - 3y \le 6 \\ 4x + 5y > 20 \end{cases}$$

20  Use a system of equations to solve.
Dan has 38 coins worth $4 in nickels, dimes, and quarters. If there are twice as many nickels as dimes, how many of each kind of coin does Dan have?

---

## Answers

1(a) 16  (b) 26  (c) 4   2(a) $2w^3 + w^2 - 3w + 10$   (b) $8t^3 - u^3$

3(a) $(x + y)(x^2 + x + 2)$   (b) $3x^2y(2x + 3y)(x - y)$   4(a) $\dfrac{2y^3}{3x^3}$   (b) $\dfrac{3(x + 3y)}{x - 3y}$

5(a) $-\dfrac{2x + 3}{3x}$   (b) $\dfrac{1}{y^2 - 1}$   6 $\dfrac{2}{x}$   7(a) 3   (b) $z = \dfrac{9b + 6}{11}$   (c) $\tfrac{2}{3}$, 2

8(a) ←———)——(———→  at $-2$, $4$   (b) ←———[———)——→  at $0$, $7$

9(a) $x \ge -5$   (b) $x < -\tfrac{7}{2}$ or $x > \tfrac{1}{2}$   10(a) $\dfrac{x^6}{8}$   (b) $\tfrac{3}{2}$   (c) $x^{12}y^2$   11(a) $8x^2y\sqrt{2xy}$

(b) $-3u^2v^3\sqrt[3]{v^2}$   12(a) $(11x - 6)\sqrt{7x}$   (b) $5x - 3$   13(a) $-\tfrac{2}{3}$, 1

(b) $\dfrac{-3 - \sqrt{11}i}{10}, \dfrac{-3 + \sqrt{11}i}{10}$   14(a) 5, 13   (b) $-1, 7$   15(a) $x \le -\tfrac{5}{2}$ or $x \ge \tfrac{1}{3}$

(b) $-\tfrac{1}{4} < x < 1$   16(a) 5   (b) $\sqrt{26}$   17(a) $y + 4 = \tfrac{2}{3}(x - 3)$   (b) $y - 7 = -\tfrac{5}{3}(x + 1)$

18(a) $(2, 1)$   (b) $(3, -2)$   (c) $(2, -1, 3)$   19

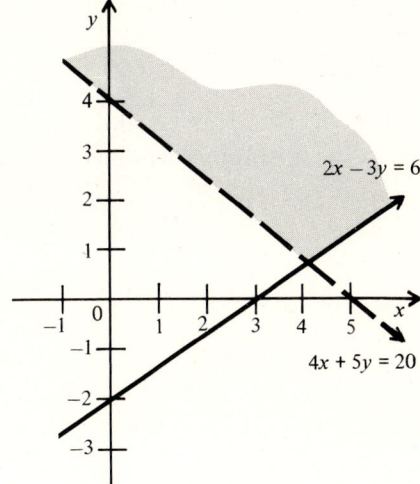

20  20 nickels, 10 dimes, 8 quarters

---

# CHAPTER 8  TESTS

## Chapter Test A

1  Use graphs to determine whether the system is dependent, inconsistent, or independent.

(a) $\begin{cases} 2x + 5y = 10 \\ -3x + 4y = 12 \end{cases}$   (b) $\begin{cases} y = 2x - 3 \\ 4x - 2y = 6 \end{cases}$

## CHAPTER 8 SYSTEMS OF LINEAR EQUATIONS AND INEQUALITIES

**2** Solve each system by the substitution method.

(a) $\begin{cases} 6x - y = 4 \\ 4x + 3y = -1 \end{cases}$

(b) $\begin{cases} x + y + z = 3 \\ 2x - y + z = 2 \\ x + 3y - 2z = 2 \end{cases}$

**3** Solve each system by the elimination method.

(a) $\begin{cases} 5x - 2y = 12 \\ 2x + 3y = 1 \end{cases}$

(b) $\begin{cases} x + 3y - 2z = 4 \\ 5x - y + 6z = -4 \\ 3x + 2y - 4z = -1 \end{cases}$

**4** Evaluate each determinant.

(a) $\begin{vmatrix} 2 & -3 \\ 4 & 7 \end{vmatrix}$

(b) $\begin{vmatrix} 1 & 3 & -2 \\ 2 & 4 & 1 \\ -1 & 2 & 5 \end{vmatrix}$

**5** Solve each system in problem 2 by Cramer's rule.

**6** Sketch the graph of each system.

(a) $\begin{cases} x \geq 3 \\ 2x - 3y < 6 \end{cases}$

(b) $\begin{cases} x \geq 0, y \geq 0 \\ x + 2y \leq 4 \\ 5x + 2y \leq 10 \end{cases}$

**7** Part of $30,000 is invested at 6% and the remaining part is invested at 5%. The annual income from both investments is $1,700. Use a system of equations to find the amount invested at each rate.

---

## Solutions

**1** (a) The graphs intersect at one point. Therefore, the system is independent.

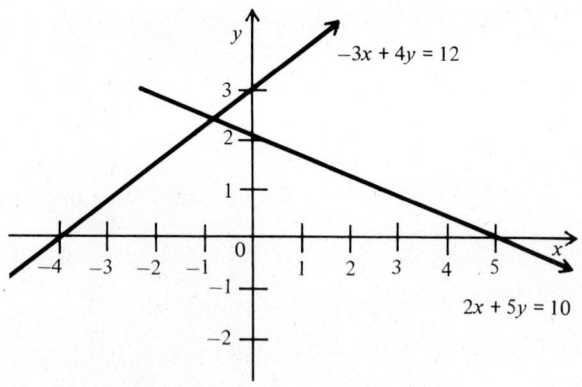

(b) The equations have the same graph. Therefore, the system is dependent.

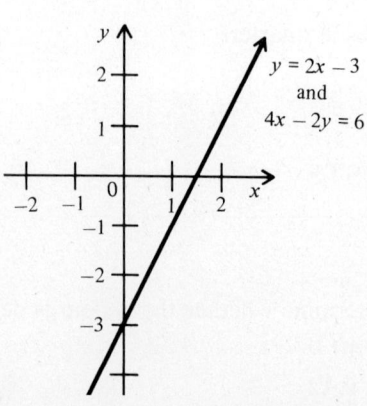

**2 (a)** $\begin{cases} 6x - y = 4 \\ 4x + 3y = -1 \end{cases}$

$$6x - y = 4$$
$$y = 6x - 4$$
$$4x + 3(6x - 4) = -1$$
$$22x - 12 = -1$$
$$22x = 11$$
$$x = \tfrac{1}{2}$$
$$y = 6(\tfrac{1}{2}) - 4 = -1$$

The solution is $(\tfrac{1}{2}, -1)$.

**(b)** $\begin{cases} x + y + z = 3 \\ 2x - y + z = 2 \\ x + 3y - 2z = 2 \end{cases}$

$$x + y + z = 3$$
$$z = 3 - x - y$$
$$2x - y + (3 - x - y) = 2$$
$$x + 3y - 2(3 - x - y) = 2$$

So we now have:
$\begin{cases} x - 2y = -1 \\ 3x + 5y = 8 \end{cases}$

$$x - 2y = -1$$
$$x = 2y - 1$$
$$3(2y - 1) + 5y = 8$$
$$11y - 3 = 8$$
$$11y = 11$$
$$y = 1$$
$$x = 2(1) - 1 = 1$$
$$z = 3 - 1 - 1 = 1$$

Therefore, the solution is $(1, 1, 1)$.

**3 (a)** $\begin{cases} 5x - 2y = 12 \\ 2x + 3y = 1 \end{cases}$ $\xrightarrow[\text{multiply by 2}]{\text{multiply by 3}}$ $\begin{cases} 15x - 6y = 36 \\ 4x + 6y = 2 \end{cases}$ (add)

$$19x = 38$$
$$x = 2$$
$$2(2) + 3y = 1$$
$$3y = -3$$
$$y = -1$$

The solution is $(2, -1)$.

**(b)** $\begin{cases} x + 3y - 2z = 4 \\ 5x - y + 6z = -4 \\ 3x + 2y - 4z = -1 \end{cases}$ $\xrightarrow{\text{multiply by 3}}$ $\begin{cases} 3x + 9y - 6z = 12 \\ 5x - y + 6z = -4 \end{cases}$ (add)

$$8x + 8y = 8$$
or $x + y = 1$

$\begin{cases} x + 3y - 2z = 4 \\ 5x - y + 6z = -4 \\ 3x + 2y - 4z = -1 \end{cases}$ $\xrightarrow{\text{multiply by } -2}$ $\begin{cases} -2x - 6y + 4z = -8 \\ 3x + 2y - 4z = -1 \end{cases}$ (add)

$$x - 4y = -9$$

We now have:
$\begin{cases} x + y = 1 \\ x - 4y = -9 \end{cases}$ (subtract)

$$5y = 10$$
$$y = 2$$
$$x + 2 = 1$$
$$x = -1.$$

Using $x + 3y - 2z = 4$ and substituting $x = -1, y = 2$:
$$-1 + 3(2) - 2z = 4$$
$$5 - 2z = 4$$
$$z = \tfrac{1}{2}$$

The solution is $(-1, 2, \tfrac{1}{2})$.

4  (a) $\begin{vmatrix} 2 & -3 \\ 4 & 7 \end{vmatrix} = 2(7) - (4)(-3)$

$= 14 + 12$

$= 26$

(b) $\begin{vmatrix} 1 & 3 & -2 \\ 2 & 4 & 1 \\ -1 & 2 & 5 \end{vmatrix} = 1 \begin{vmatrix} 4 & 1 \\ 2 & 5 \end{vmatrix} - 2 \begin{vmatrix} 3 & -2 \\ 2 & 5 \end{vmatrix} + (-1) \begin{vmatrix} 3 & -2 \\ 4 & 1 \end{vmatrix}$

$= 1[4(5) - 2(1)] - 2[3(5) - 2(-2)] - 1[3(1) - 4(-2)]$
$= 1(18) - 2(19) - 1(11)$
$= -31$

5  (a) $D = \begin{vmatrix} 6 & -1 \\ 4 & 3 \end{vmatrix} = 18 + 4 = 22$

$D_x = \begin{vmatrix} 4 & -1 \\ -1 & 3 \end{vmatrix} = 12 - 1 = 11$

$D_y = \begin{vmatrix} 6 & 4 \\ 4 & -1 \end{vmatrix} = -6 - 16 = -22$

$x = \dfrac{D_x}{D} = \dfrac{11}{22} = \dfrac{1}{2}, \; y = \dfrac{D_y}{D} = \dfrac{-22}{22} = -1$

(b) $D = \begin{vmatrix} 1 & 1 & 1 \\ 2 & -1 & 1 \\ 1 & 3 & -2 \end{vmatrix} = 1 \begin{vmatrix} -1 & 1 \\ 3 & -2 \end{vmatrix} - 2 \begin{vmatrix} 1 & 1 \\ 3 & -2 \end{vmatrix} + 1 \begin{vmatrix} 1 & 1 \\ -1 & 1 \end{vmatrix} = 1(-1) - 2(-5) + 1(2) = 11$

$D_x = \begin{vmatrix} 3 & 1 & 1 \\ 2 & -1 & 1 \\ 2 & 3 & -2 \end{vmatrix} = 3 \begin{vmatrix} -1 & 1 \\ 3 & -2 \end{vmatrix} - 2 \begin{vmatrix} 1 & 1 \\ 3 & -2 \end{vmatrix} + 2 \begin{vmatrix} 1 & 1 \\ -1 & 1 \end{vmatrix} = 3(-1) - 2(-5) + 2(2) = 11$

$D_y = \begin{vmatrix} 1 & 3 & 1 \\ 2 & 2 & 1 \\ 1 & 2 & -2 \end{vmatrix} = 1 \begin{vmatrix} 2 & 1 \\ 2 & -2 \end{vmatrix} - 2 \begin{vmatrix} 3 & 1 \\ 2 & -2 \end{vmatrix} + 1 \begin{vmatrix} 3 & 1 \\ 2 & 1 \end{vmatrix} = 1(-6) - 2(-8) + 1(1) = 11$

$D_z = \begin{vmatrix} 1 & 1 & 3 \\ 2 & -1 & 2 \\ 1 & 3 & 2 \end{vmatrix} = 1 \begin{vmatrix} -1 & 2 \\ 3 & 2 \end{vmatrix} - 2 \begin{vmatrix} 1 & 3 \\ 3 & 2 \end{vmatrix} + 1 \begin{vmatrix} 1 & 3 \\ -1 & 2 \end{vmatrix} = 1(-8) - 2(-7) + 1(5) = 11$

$x = \dfrac{D_x}{D} = \dfrac{11}{11} = 1, \; y = \dfrac{D_y}{D} = \dfrac{11}{11} = 1, \; z = \dfrac{D_z}{D} = \dfrac{11}{11} = 1$

6  (a) First, sketch the graphs of $x = 3$ and $2x - 3y = 6$. The graph of $x > 3$ is the half-plane to the right of $x = 3$ and including $x = 3$. The graph of $2x - 3y < 6$ is the half-plane above the line $2x - 3y = 6$. The graph of the system is the region where these two half-planes overlap.

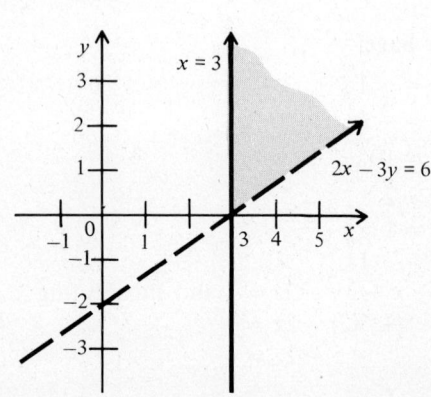

(b) Since $x \geq 0$ and $y \geq 0$, the graph is limited to the first quadrant, including the $x$ and $y$ axis. The graph of $x + 2y \leq 4$ is the half–plane below and including the line $x + 2y = 4$, and the graph of $5x + 2y \leq 10$ is the half–plane below and including the line $5x + 2y = 10$. The graph of the system is the region in the first quadrant where these two half–planes intersect.

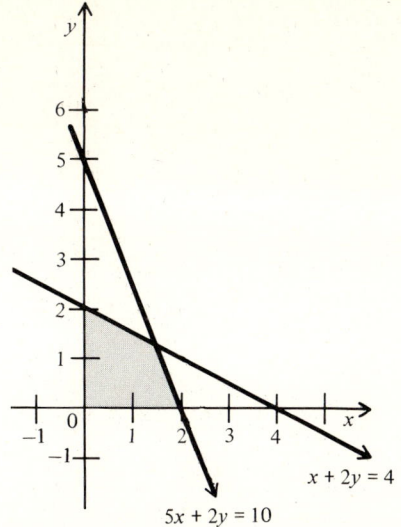

7   Let $x$ = amount invested at 6% and $y$ = amount invested at 5%.
Therefore, $\begin{cases} x + y = 30{,}000 \\ 0.06x + 0.05y = 1{,}700 \end{cases}$ or $\begin{cases} 6x + 6y = 180{,}000 \\ 6x + 5y = 170{,}000 \end{cases}$
$$y = 10{,}000$$
$$x + 10{,}000 = 30{,}000$$
$$x = 20{,}000$$
Therefore, $20,000 was invested at 6% and $10,000 was invested at 5%.

## Chapter Test B

*Multiple Choice:* Select the *one* correct answer for each of the following questions.

1   The system of equations illustrated by the graph is _____.
   (a) independent
   (b) inconsistent
   (c) dependent
   (d) none of these

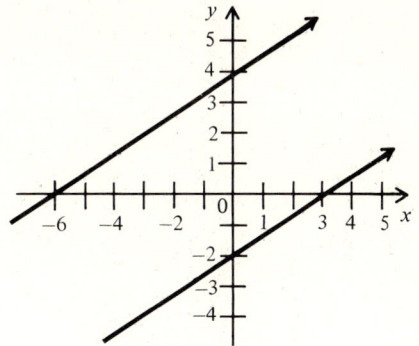

2   The system of equations illustrated by the graph is _____.
   (a) independent
   (b) inconsistent
   (c) dependent
   (d) none of these

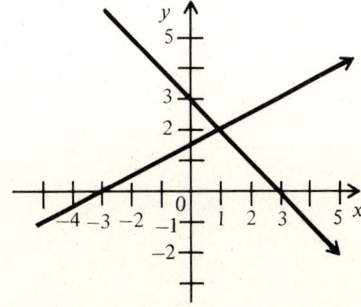

**208** CHAPTER 8 SYSTEMS OF LINEAR EQUATIONS AND INEQUALITIES

3. The solution of $\begin{cases} 5x + 3y = 34 \\ 3x - y = 12 \end{cases}$ is _____.
   (a) (5, 3)   (b) (3, 5)   (c) (−5, 3)   (d) no solution

4. The solution of $\begin{cases} x + 2y = 3 \\ 2x + 4y = 5 \end{cases}$ is _____.
   (a) (1, 1)   (b) $(4, -\tfrac{1}{2})$   (c) (−3, 3)   (d) no solution

5. The solution of $\begin{cases} \dfrac{3}{x} - \dfrac{2}{y} = 19 \\ \dfrac{1}{x} + \dfrac{1}{y} = 23 \end{cases}$ is _____.
   (a) $(-\tfrac{1}{10}, -\tfrac{1}{13})$   (b) $(-\tfrac{1}{10}, \tfrac{1}{13})$   (c) $(\tfrac{1}{10}, \tfrac{1}{13})$   (d) $(\tfrac{1}{13}, \tfrac{1}{10})$

6. The solution of $\begin{cases} x + y + z = 6 \\ 2x - y - z = 0 \\ x - y + 2z = 7 \end{cases}$ is _____.
   (a) (2, 1, 3)   (b) (3, 1, 2)   (c) (1, 2, 3)   (d) no solution

7. The value of $\begin{vmatrix} 3 & 1 \\ -2 & 4 \end{vmatrix}$ is _____.
   (a) 11   (b) 14   (c) 7   (d) none of these

8. The value of $\begin{vmatrix} 2 & 3 & 1 \\ 1 & 5 & -2 \\ 3 & -4 & 4 \end{vmatrix}$ is _____.
   (a) −30   (b) 24   (c) −19   (d) −25

9. Using Cramer's rule to solve the system $\begin{cases} 2x - 3y = 3 \\ x + 4y = 7 \end{cases}$ we find $D_x$ and $D_y$ equal _____, respectively.
   (a) 33 and 11   (b) −33 and 11   (c) 11 and 33   (d) −11 and 33

10. Using Cramer's rule to solve the system $\begin{cases} 3x - 2y + z = z \\ x + y - 2z = 0 \\ -x + 3y + z = 3 \end{cases}$ we find the value of $D$ is _____.
    (a) 23   (b) 15   (c) 11   (d) none of these

11. The system of inequalities whose solution set is illustrated by the graph is _____.
    (a) $\begin{cases} x + y \leq 1 \\ y - x \geq 1 \end{cases}$
    (b) $\begin{cases} x + y \geq 1 \\ y - x \geq 1 \end{cases}$
    (c) $\begin{cases} x + y \geq 1 \\ x - y \leq 1 \end{cases}$
    (d) $\begin{cases} x + y \leq 1 \\ x - y \leq 1 \end{cases}$

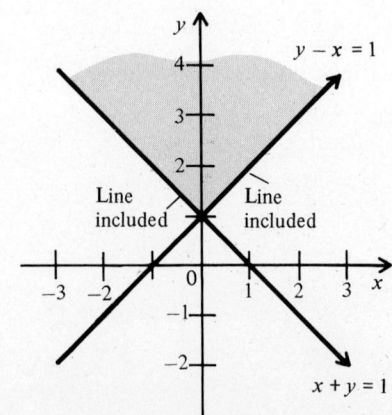

**12** The system of inequalities whose solution set is illustrated by the graph is _____.

(a) $\begin{cases} y < 2 \\ y \le x - 2 \end{cases}$

(b) $\begin{cases} y < 2 \\ y \ge x - 2 \end{cases}$

(c) $\begin{cases} y > 2 \\ y \le x - 2 \end{cases}$

(d) $\begin{cases} y > 2 \\ y \ge x - 2 \end{cases}$

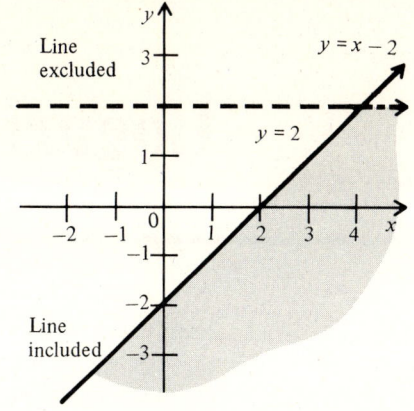

**13** The system of inequalities whose solution set is illustrated by the graph is _____.

(a) $\begin{cases} y \ge -x - 2 \\ y \le x \end{cases}$

(b) $\begin{cases} y \le -x - 2 \\ y \le x \end{cases}$

(c) $\begin{cases} y \ge -x - 2 \\ y \ge x \end{cases}$

(d) $\begin{cases} y \le -x - 2 \\ y \ge x \end{cases}$

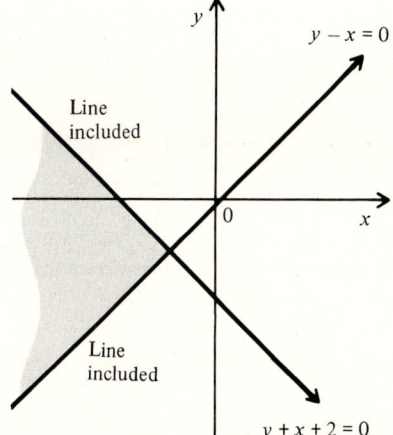

**14** The sum of two numbers is 17 and their difference is 1. A system of equations that can be used to find these numbers is _____.

(a) $\begin{cases} y = x + 17 \\ y = x - 1 \end{cases}$
(b) $\begin{cases} x + y = 17 \\ x - y = 1 \end{cases}$
(c) $\begin{cases} x - y = 17 \\ x + y = 1 \end{cases}$
(d) $\begin{cases} x = 17 + 1 \\ y = 17 - 1 \end{cases}$

---

## Answers

1 b  2 a  3 a  4 d  5 d  6 a  7 b  8 d  9 a  10 a
11 b  12 a  13 d  14 b

# 9 Logarithms

In this chapter we use the algebra of exponents studied in Chapter 5 to develop the algebra of logarithms. After completing the appropriate sections in this chapter, the student should be able to:

1. Solve exponential equations.
2. Change exponential equations to logarithmic form and vice versa.
3. Apply the properties of logarithms.
4. Use the table of common logarithms.
5. Use a calculator to evaluate expressions involving logarithms and exponential expressions.
6. Work applications involving exponential expressions.

## 9.1 Exponential Equations and Logarithms

### SEMIPROGRAMMED PROBLEMS

In problems 1–3, solve each exponential equation.

$3^3, 3, 3$

1  $3^x = 27$
Since $3^x = 27 = $ _____, then $x = $ _____. The solution is _____.

$4, 4, -1$
$-1$

2  $2^{3-x} = 16$
Since $2^{3-x} = 16 = 2$____, then $3 - x = $ _____ or $x = $ _____.
The solution is _____.

$3x + 9$
$25, \frac{25}{3}, \frac{25}{3}$

3  $4^{3x-8} = 2^{3x+9}$
Since $4^{3x-8} = 2^{6x-16} = 2^{3x+9}$, then $6x - 16 = $ _____, or
$3x = $ _____ or $x = $ _____. The solution is _____.

In problems 4–6, write each exponential statement as an equivalent logarithmic statement.

4

4  $2^4 = 16$ is equivalent to $\log_2 16 = $ _____.

5

5  $5^{-2} = \frac{1}{25}$ is equivalent to $\log$____ $\frac{1}{25} = -2$.

$\log_{4/9} \frac{27}{8} = -\frac{3}{2}$

6  $(\frac{4}{9})^{-3/2} = \frac{27}{8}$ is equivalent to _____.

In problems 7–9, write each logarithmic statement as an equivalent exponential statement.

2

7  $\log_6 36 = 2$ is equivalent to $6$____ $= 36$.

$\frac{1}{243}$

8  $\log_3 \frac{1}{243} = -5$ is equivalent to $3^{-5} = $ _____.

$100^{-3/2} = 0.001$

9  $\log_{100} 0.001 = -\frac{3}{2}$ is equivalent to _____

In problems 10–12, find the values of the logarithms.

16, 4, 4
4

**10** $\log_2 16$

Let $\log_2 16 = x$, so $2^x = $ _____ $= 2$——. Then $x = $ _____.

Therefore, $\log_2 16 = $ _____.

100, 2
2

**11** $\log_{10} 100$

Let $\log_{10} 100 = x$, so $10^x = $ _____ $= 10^2$. Then $x = $ _____.

Therefore, $\log_{10} 100 = $ _____.

$\frac{1}{64}, 8^{-2}, -2$
$-2$

**12** $\log_8 \frac{1}{64}$

Let $\log_8 \frac{1}{64} = x$, so $8^x = $ _____ $= $ _____. Then $x = $ _____.

Therefore, $\log_8 \frac{1}{64} = $ _____.

In problems 13–15, solve each equation.

$2^5$

35, 35

**13** $\log_2(x - 3) = 5$

$\log_2(x - 3) = 5$ is equivalent to _____ $= x - 3$ or $32 = x - 3$, so $x = $ _____. The solution is _____.

$3^2$

10, 5, 5

**14** $\log_3(2x - 1) = 2$

$\log_3(2x - 1) = 2$ is equivalent to _____ $= 2x - 1$ or $2x - 1 = 9$, so so $2x = $ _____ or $x = $ _____. The solution is _____.

$x^{-2}, x^2,$
$\frac{1}{7}$

**15** $\log_x 49 = -2$

$\log_x 49 = -2$ is equivalent to _____ $= 49$ or _____ $= \frac{1}{49}$ so that $x = $ _____, since by definition the base must be positive.

## SOLUTIONS TO SELECTED ODD PROBLEMS  Section 9.1, text pages 424–425

**1**  $5^x = 25$
$5^x = 5^2$
$x = 2$

**5**  $3^{u-5} = 27$
$3^{u-5} = 3^3$
$u - 5 = 3$
$u = 8$

**9**  $3^{4c-5} = 81$
$3^{4c-5} = 3^4$
$4c - 5 = 4$
$c = \frac{9}{4}$

**13**  $2^{x^2 - 6x} = (\frac{1}{2})^3$
$2^{x^2 - 6x} = 2^{-3}$
$x^2 - 6x = -3$
$x^2 - 6x + 9 = 9 - 3$
$(x - 3)^2 = 6$
$x - 3 = \pm\sqrt{6}$
$x = 3 \pm \sqrt{6}$

| | Exponential Expression $(b^x = C)$ | Equivalent Logarithmic Expression $(x = \log_b C)$ |
|---|---|---|
| 17 | $10^5 = 100{,}000$ | $5 = \log_{10} 100{,}000$ |
| 21 | $6^{-2} = \frac{1}{36}$ | $-2 = \log_6 \frac{1}{36}$ |
| 25 | $(100)^{-3/2} = 0.001$ | $-\frac{3}{2} = \log_{100} 0.001$ |
| 29 | $x^3 = a$ | $3 = \log_x a$ |

| | Logarithmic Expression $(\log_b C = x)$ | Equivalent Exponential Expression $(b^x = C)$ |
|---|---|---|
| 33 | $\log_{27} 9 = \frac{2}{3}$ | $27^{2/3} = 9$ |
| 37 | $\log_{1/3} 9 = -2$ | $(\frac{1}{3})^{-2} = 9$ |
| 41 | $\log_{\sqrt{16}} 2 = \frac{1}{2}$ | $(\sqrt{16})^{1/2} = 2$ |

## 212  CHAPTER 9  LOGARITHMS

**45**    $x = \log_2 64$
$2^x = 64$
$2^x = 2^6$
$x = 6$

**49**    $x = \log_9 \frac{1}{3}$
$9^x = \frac{1}{3}$
$3^{2x} = 3^{-1}$
$2x = -1$
$x = -\frac{1}{2}$

**53**    $x = \log_5 \frac{1}{125}$
$5^x = \frac{1}{125} = \frac{1}{5^3}$
$5^x = 5^{-3}$
$x = -3$

**57**    $x = \log_{10} \frac{1}{10{,}000}$
$10^x = \frac{1}{10{,}000} = \frac{1}{10^4}$
$10^x = 10^{-4}$
$x = -4$

**61**    $x = \log_7 1$
$7^x = 1$
$7^x = 7^0$
$x = 0$

**65**    $\log_{10}(x+1) = 1$
$x + 1 = 10^1$
$x + 1 = 10$
$x = 9$

**69**    $\log_5 N = 2$
$N = 5^2$
$N = 25$

**73**    $\log_b 3 = \frac{1}{5}$
$3 = b^{1/5}$
$3^5 = (b^{1/5})^5$
$243 = b$

**77**    $\text{pH} = -\log_{10}[\text{H}^+]$
$\text{pH} = -\log_{10} 10^{-5.6}$
$-\text{pH} = \log_{10} 10^{-5.6}$
$10^{-\text{pH}} = 10^{-5.6}$
$-\text{pH} = -5.6$
$\text{pH} = 5.6$

**81**    $M = \log_{10}\left(\dfrac{I}{I_0}\right)$
$M = \log_{10}\left(\dfrac{100 I_0}{I_0}\right)$
$M = \log_{10} 100$
$10^M = 100 = 10^2$
$M = 2$

## 9.2 Basic Properties of Logarithms

### SEMIPROGRAMMED PROBLEMS

In problems 1–5, use the properties of logarithms to rewrite each expression as a sum or difference of multiples of logarithms.

| | |
|---|---|
| $\log_5 11$ | **1**   $\log_5(7)(11)$ <br> $\log_5(7)(11) = \log_5 7 + \underline{\hspace{1cm}}$ |
| $\log_2 7$ | **2**   $\log_2 \tfrac{13}{7}$ <br> $\log_2 \tfrac{13}{7} = \log_2 13 - \underline{\hspace{1cm}}$ |
| $\log_3 7$ | **3**   $\log_3 7^{21}$ <br> $\log_3 7^{21} = 21(\underline{\hspace{1cm}})$ |
| $\log_4 y^4$ <br> $4$ | **4**   $\log_4 x^3 y^4$ <br> $\log_4 x^3 y^4 = \log_4 x^3 + \underline{\hspace{1cm}}$ <br> $= 3\log_4 x + \underline{\hspace{1cm}} \log_4 y$ |
| $\dfrac{1}{3}$ <br> $\log_3 y^2$ <br> $\tfrac{2}{3}$ | **5**   $\log_3 \sqrt[3]{\dfrac{x}{y^2}}$ <br> $\log_3 \sqrt[3]{\dfrac{x}{y^2}} = \log_3 \left(\dfrac{x}{y^2}\right)^{1/3} = (\underline{\hspace{1cm}}) \log_3 \dfrac{x}{y^2}$ <br> $= \tfrac{1}{3} \log_3 x - \tfrac{1}{3}(\underline{\hspace{1cm}})$ <br> $= \tfrac{1}{3} \log_3 x - \underline{\hspace{1cm}} \log_3 y$ |

In problems 6–10, use the properties of logarithms to write each expression as a single logarithm.

| | |
|---|---|
| $7^2$ <br> $\dfrac{5^3}{7^2}, \log_3 \dfrac{125}{49}$ | **6**   $3 \log_3 5 - 2 \log_3 7$ <br> $3 \log_3 5 - 2 \log_3 7 = \log_3 5^3 - \log_3 \underline{\hspace{1cm}}$ <br> $= \log_3 \underline{\hspace{1cm}} = \underline{\hspace{1cm}}$ |
| $\log_a y^3,\ x^5 y^3$ | **7**   $5 \log_a x + 3 \log_a y$ <br> $5 \log_a x + 3 \log_a y = \log_a x^5 + \underline{\hspace{1cm}} = \log_a \underline{\hspace{1cm}}$ |

## 9.2 SEMIPROGRAMMED PROBLEMS

$\sqrt{x}$, $x^{5/2}$, $\frac{5}{2}\log_a x$

8  $\log_a x^3 - \log_a \sqrt{x}$
$\log_a x^3 - \log_a \sqrt{x} = \log_a \dfrac{x^3}{\underline{\phantom{xx}}} = \log_a \underline{\phantom{xx}} = \underline{\phantom{xx}}$

9  $\log_c \dfrac{a}{\sqrt{x}} - \log_c \dfrac{\sqrt{x}}{a}$

$\dfrac{\sqrt{x}}{a}, \dfrac{a^2}{x}$

$\log_c \dfrac{a}{\sqrt{x}} - \log_c \dfrac{\sqrt{x}}{a} = \log_c \dfrac{\dfrac{a}{\sqrt{x}}}{\underline{\phantom{xx}}} = \log_c \underline{\phantom{xx}}$

10  $\log_a(x^2 - y^2) - \log_a(x - y)$

$x - y$

$\log_a(x^2 - y^2) - \log_a(x - y) = \log_a \dfrac{x^2 - y^2}{\underline{\phantom{xx}}}$

$x - y$

$= \log_a \dfrac{(x-y)(x+y)}{\underline{\phantom{xx}}}$

$x + y$

$= \log_a (\underline{\phantom{xxxx}})$

In problems 11–13, let $\log_{10} 2 = 0.3010$, $\log_{10} 3 = 0.4771$, and $\log_{10} 7 = 0.8451$. Use the properties of logarithms to find each value.

11  $\log_{10} \frac{7}{2}$

$\log_{10} 2$
0.3010
0.5441

$\log_{10} \frac{7}{2} = \log_{10} 7 - \underline{\phantom{xx}}$
$= 0.8451 - \underline{\phantom{xx}}$
$= \underline{\phantom{xx}}$

12  $\log_{10} 3^5$

$\log_{10} 3^5 = 5 \log_{10} 3$

0.4771
2.3855

$= 5(\underline{\phantom{xx}})$
$= \underline{\phantom{xx}}$

13  $\log_{10} 8$

$\log_{10} 2$
0.3010
0.9030

$\log_{10} 8 = \log_{10} 2^3 = 3 \underline{\phantom{xx}}$
$= 3(\underline{\phantom{xx}})$
$= \underline{\phantom{xx}}$

In problems 14 and 15, solve each equation.

14  $\log_2(x - 1) + \log_2 \frac{1}{3} = 3$

3
$\frac{1}{3}(x-1) = 8$, 25

24
8
25

Write the left side of the equation as a single logarithmic expression, and obtain $\log_2 \frac{1}{3}(x - 1) = \underline{\phantom{xx}}$. In exponential form, this equation is $\underline{\phantom{xx}}$. Solving for $x$, we have $x = \underline{\phantom{xx}}$.
Check: $\log_2(25 - 1) + \log_2 \frac{1}{3} \stackrel{?}{=} 3$
$\log_2 \underline{\phantom{xx}} + \log_2 \frac{1}{3} \stackrel{?}{=} 3$
$\log_2 \underline{\phantom{xx}} \stackrel{?}{=} 3$  Yes.
Hence, the solution is $\underline{\phantom{xx}}$.

15  $\log_{10}(x^2 - 121) - \log_{10}(x + 11) = 1$
Write the left side of the equation as a single logarithmic expression,

$\log_{10}\left(\dfrac{x^2 - 121}{x + 11}\right)$

and obtain the equation $\underline{\phantom{xxxxxxxx}} = 1$. This

**214** CHAPTER 9 LOGARITHMS

| | equation can be written in a simpler form as |
|---|---|
| $\log_{10}(x-11) = 1$ | _____. In exponential form this equation is |
| $x - 11 = 10, 21$ | _____. Solving for $x$, we have $x = $ ___. |
| | Check: $x = 21$: |
| | $\log_{10}(441 - 121) - \log_{10}(21 + 11) \stackrel{?}{=} 1$ |
| 320 | $\log_{10}\underline{\phantom{xx}} - \log_{10} 32 \stackrel{?}{=} 1$ |
| 32 | $\log_{10} \dfrac{320}{\underline{\phantom{xx}}} \stackrel{?}{=} 1$ |
| 10 | $\log_{10} \underline{\phantom{xx}} = 1$ |
| 21 | The solution is ___. |

## SOLUTIONS TO SELECTED ODD PROBLEMS   Section 9.2, text pages 430–432

**1**  $\log_4 5y = \log_4 5 + \log_4 y$    **5**  $\log_5 \dfrac{x}{3} = \log_5 x - \log_5 3$    **9**  $\log_7 3^5 = 5 \log_7 3$

**13**  $\log_4 \sqrt{w} = \log_4 w^{1/2} = \tfrac{1}{2} \log_4 w$

**17**  $\log_{11} x^4 y^2 = \log_{11} x^4 + \log_{11} y^2 = 4 \log_{11} x + 2 \log_{11} y$

**21**  $\log_5 \dfrac{a^2}{b^4} = \log_5 a^2 - \log_5 b^4 = 2 \log_5 a - 4 \log_5 b$

**25**  $\log_4 \dfrac{u^4 v^5}{\sqrt[4]{z^3}} = \log_4 u^4 v^5 - \log_4 \sqrt[4]{z^3}$    **29**  $\log_5 4 - \log_5 3 = \log_5 \tfrac{4}{3}$

$\phantom{25}\quad = \log_4 u^4 + \log_4 v^5 - \log_4 z^{3/4}$

$\phantom{25}\quad = 4 \log_4 u + 5 \log_4 v - \dfrac{3}{4} \log_4 z$

**33**  $\log_3 5 + \log_3 z - \log_3 y = \log_3 5z - \log_3 y$    **37**  $3 \log_2 x + 7 \log_2 y = \log_2 x^3 + \log_2 y^7$

$\phantom{33}\quad = \log_3 \dfrac{5z}{y}$    $\phantom{37}\quad = \log_2 x^3 y^7$

**41**  $\tfrac{1}{2} \log_4 a - 3 \log_4 b - 4 \log_4 z = \log_4 a^{1/2} - \log_4 b^3 - \log_4 z^4$

$\phantom{41}\quad = \log_4 \sqrt{a} - (\log_4 b^3 + \log_4 z^4)$

$\phantom{41}\quad = \log_4 \sqrt{a} - \log_4 b^3 z^4$

$\phantom{41}\quad = \log_4 \dfrac{\sqrt{a}}{b^3 z^4}$

**45**  $\log_e(y^2 - 25) - \log_e(y - 5) = \log_e \dfrac{y^2 - 5}{y - 5}$    **49**  $\log_p \sqrt[5]{p} = \log_p p^{1/5} = \dfrac{1}{5}$

$\phantom{45}\quad = \log_e(y + 5)$

**53**  $\log_{10} 18 = \log_{10}(2)(3)^2$    **57**  $\log_{10} 5 = \log_{10} \tfrac{10}{2}$    **61**  $\log_{10} 0.5 = \log_{10} \tfrac{1}{2}$

$\phantom{53}\quad = \log_{10} 2 + 2 \log_{10} 3$    $\phantom{57}\quad = \log_{10} 10 - \log_{10} 2$    $\phantom{61}\quad = \log_{10} 1 - \log_{10} 2$

$\phantom{53}\quad = 0.3010 + 2(0.4771)$    $\phantom{57}\quad = 1 - 0.3010$    $\phantom{61}\quad = 0 - 0.3010$

$\phantom{53}\quad = 1.2552$    $\phantom{57}\quad = 0.6990$    $\phantom{61}\quad = -0.3010$

**65**  $\log_3 w - \log_3 2 = 2$    **69**  $\log_4(y + 2) - \log_4(y - 1) = 2$

$\phantom{65}\quad \log_3 \dfrac{w}{2} = 2$    $\phantom{69}\quad \log_4 \dfrac{y + 2}{y - 1} = 2$

$\phantom{65}\quad \dfrac{w}{2} = 3^2$    $\phantom{69}\quad \dfrac{y + 2}{y - 1} = 4^2$

$\phantom{65}\quad \dfrac{w}{2} = 9$    $\phantom{69}\quad y + 2 = 16(y - 1)$

$\phantom{65}\quad w = 18$    $\phantom{69}\quad y + 2 = 16y - 16$

$\phantom{69}\quad y = \dfrac{6}{5}$

73 $\log_b b^r = r \log_b b$
$= r(1)$
$= r$

77 $\log_3 \left[ \dfrac{x^3 + 1}{x^3 - x^2 + x} \right]$
$= \log_3 \left[ \dfrac{(x+1)(x^2 - x + 1)}{x(x^2 - x + 1)} \right]$
$= \log_3 \left[ \dfrac{x+1}{x} \right] = \log_3(x+1) - \log_3 x$

## 9.3 Common Logarithms

### SEMIPROGRAMMED PROBLEMS

In problems 1–5, express each number in scientific notation and give the characteristic value $n$ of the logarithm of each number.

| | |
|---|---|
| $10^2$, 2 | **1** $234 = 2.34 \times$ \_\_\_\_, so that $n =$ \_\_\_\_. |
| $10^3$, 3 | **2** $4{,}300 = 4.3 \times$ \_\_\_\_, so that $n =$ \_\_\_\_. |
| $10^0$, 0 | **3** $7.31 = 7.31 \times$ \_\_\_\_, so that $n =$ \_\_\_\_. |
| $10^{-4}$, $-4$ | **4** $0.000512 = 5.12 \times$ _____, so that $n =$ \_\_\_\_. |
| $10^{-1}$, $-1$ | **5** $0.264 = 2.64 \times$ _____, so that $n =$ \_\_\_\_. |

In problems 6–9, use Table I in Appendix A to find the value of each logarithm. Indicate the characteristic and the mantissa.

| | |
|---|---|
| | **6** $\log 700$ |
| 2 | $\log 700 = \log 7 +$ \_\_\_\_. |
| 2, 0.8451 | The characteristic is \_\_\_\_ and the mantissa is _____, so |
| 2, 2.8451 | $\log 700 = 0.8451 +$ \_\_\_\_ $=$ _____. |
| | **7** $\log 3960$ |
| 3.96 | $\log 3960 = \log$ _____ $+ 3$ |
| 3, 0.5977 | The characteristic is \_\_\_\_ and the mantissa is _____, so |
| 3, 3.5977 | $\log 3960 = 0.5977 +$ \_\_\_\_ $=$ _____. |
| | **8** $\log 0.538$ |
| $-1$ | $\log 0.538 = \log 5.38 +$ \_\_\_\_ |
| $-1$, 0.7308 | The characteristic is \_\_\_\_ and the mantissa is _____, so |
| $(-1)$, $-0.2692$ | $\log 0.538 = 0.7308 +$ \_\_\_\_ $=$ _____. |
| | **9** $\log 0.00351$ |
| $-3$ | $\log 0.00351 = \log 3.51 +$ \_\_\_\_ |
| $-3$, 0.5453 | The characteristic is \_\_\_\_ and the mantissa is _____, so |
| $-3$, $-2.4547$ | $\log 0.00351 = 0.5453 + ($ \_\_\_\_ $) =$ _____. |

In problems 10 and 11, use Table I in Appendix A and interpolation to approximate each logarithm.

**10** $\log 1.234$

0

The characteristic of $\log 1.234$ is \_\_\_\_. To find $\log 1.234$, set up the following table:

| | |
|---|---|
| 0.0934 | 1.240   _____ |
| | 1.234      ? |
| 0.0899 | 1.230   _____ |

Compute the differences indicated by the pairings below and obtain

$$0.010\left\{0.004\left\{\begin{matrix}1.240 & 0.0934\\ 1.234 & ?\\ 1.230 & 0.0899\end{matrix}\right\}d\right\}0.0035$$

Set up the proportion $\dfrac{0.004}{0.010} = \dfrac{d}{0.0035}$, so that $d =$ _____.

| | |
|---|---|
| 0.0014 | |

Then log 1.234 = 0.0899 + $d$ = 0.0899 + _____

| | |
|---|---|
| 0.0014 | |
| 0.0913 | |

Therefore, log 1.234 = _____.

**11** log 0.003522

The characteristic of log 0.003522 is _____. To find log 0.003522, set up the following table:

| | |
|---|---|
| −3 | |

| 0.003520 | _____ |
| 0.003522 | ? |
| 0.003530 | _____ |

| | |
|---|---|
| −2.4535 | |
| −2.4522 | |

Compute the differences indicated by the pairings below and obtain

$$0.000010\left\{0.000008\left\{\begin{matrix}0.003520 & -2.4535\\ 0.003522 & ?\\ 0.003530 & -2.4522\end{matrix}\right\}d\right\}-0.0013$$

Set up the proportion and obtain $\dfrac{0.000008}{0.000010} = \dfrac{d}{-0.0013}$, so that

$d =$ _____. Then

log 0.003522 = −2.4522 + $d$ = _____.

| | |
|---|---|
| −0.0010 | |
| −2.4532 | |
| −2.4532 | |

Therefore, log 0.003522 = _____.

In problems 12–14, find the values of the given antilogarithms by using Table I in Appendix A.

**12** antilog 4.5977

| | |
|---|---|
| 4.5977 | |
| 4 | |
| 0.5977 | |
| log $10^4$ | |
| 39,600 | |
| 39,600 | |

Let $x =$ antilog 4.5977, so that log $x =$ _____,

log $x$ = 0.5977 + _____.

Since log 3.96 = _____,

log $x$ = log 3.96 + _____

       = log(3.96 × $10^4$) = log(_____).

Therefore, $x =$ antilog 4.5977 = _____.

**13** antilog[0.5079 + (−4)]

| | |
|---|---|
| 0.5079 | |
| 0.5079 | |
| log $10^{-4}$ | |
| log(0.000322) | |
| 0.000322 | |

Let $x =$ antilog[0.5079 + (−4)], so that log $x =$ _____ + (−4).

Since log 3.22 = _____,

log $x$ = log 3.22 + _____

       = log(3.22 × $10^{-4}$) = _____.

Therefore, $x =$ antilog[0.5079 + (−4)] = _____.

**14** antilog(−2.1002)

| | |
|---|---|
| −2.1002 | |

Let $x =$ antilog(−2.1002), so that log $x =$ _____.

Since the mantissa is always positive, we have (−2.1002 + 3) − 3 =

| | |
|---|---|
| 0.8998 | |
| 0.8998 | |

_____ + (−3), so that log $x$ = [0.8998 + (−3)].

Since log 7.94 = _____,

# 9.3 SOLUTIONS TO SELECTED ODD PROBLEMS 217

| | |
|---|---|
| log $10^{-3}$ | log $x$ = log 7.94 + _____ = log(7.94 × $10^{-3}$) |
| 0.00794 | = log(_____). |
| 0.00794 | Therefore, $x$ = antilog($-2.1002$) = _____. |

In problems 15 and 16, use interpolation to find each antilogarithm.

**15** antilog 3.6129

| | |
|---|---|
| 3.6129 | Let $x$ = antilog 3.6129 so that log $x$ = _____. |
| | From the tables, we find successive mantissa entries such that the given |
| 0.6128, 0.6138 | mantissa lies between _____ and _____. Now, set up the following table: |
| 4,100 | 3.6128 _____ |
| | 3.6129  $x$ |
| 4,110 | 3.6138 _____ |

Compute the differences indicated by the pairings below and obtain

$$0.0010 \left\{ 0.0001 \left\{ \begin{matrix} 3.6128 & 4,100 \\ 3.6129 & x \\ 3.6138 & 4,110 \end{matrix} \right\} d \right\} 10$$

| | |
|---|---|
| 1 | Set up the proportion $\dfrac{d}{10} = \dfrac{0.0001}{0.0010}$, so that $d$ = _____. Hence, |
| | $x = 4{,}100 + d$ |
| 1 | $= 4{,}100 +$ _____ |
| 4,101 | $=$ _____. |

**16** antilog($-2.4542$)

| | |
|---|---|
| $-2.4542$ | Let $x$ = antilog($-2.4542$), so that log $x$ = _____. |
| | Since the mantissa must be positive, we write |
| | $-2.4542 = (-2.4542 + 3) - 3$ |
| 0.5458 | $=$ _____ $+ (-3)$. |
| | Thus, log $x = 0.5458 + (-3)$. |
| | To find antilog 0.5458, we use interpolation, so that if |
| 0.5458 | $s$ = antilog 0.5458, then log $s$ = _____. |
| 3.51 | 0.5453 _____ |
| | 0.5458  $s$ |
| 3.52 | 0.5465 _____ |

Compute the differences indicated by the pairings below and obtain

$$0.0012 \left\{ 0.0005 \left\{ \begin{matrix} 0.5453 & 3.51 \\ 0.5458 & s \\ 0.5465 & 3.52 \end{matrix} \right\} d \right\} 0.01$$

| | |
|---|---|
| 0.004 | Set up the proportion $\dfrac{d}{0.01} = \dfrac{0.0005}{0.0012}$, so that $d$ = _____. |
| 3.514 | Hence, $s = 3.51 + 0.004 =$ _____. Therefore, $x =$ |
| 0.003514 | antilog[$0.5458 + (-3)$] = _____. |

**SOLUTIONS TO SELECTED ODD PROBLEMS**  Section 9.3, text pages 439–440

**1** $3.782 \times 10^3$, characteristic is 3    **5** $3.75 \times 10^5$, characteristic is 5

**9** $2.71312 \times 10^{-4}$, characteristic is $-4$    **13** $3.14 \times 10^{-5}$, characteristic is $-5$

**17**   $5.41 \times 10^{-6}$, characteristic is $-6$

**21**   $\log 317 = \log 3.17 + 2 = 0.5011 = 2.5011$

**25**   $\log 17.1 = \log 1.71 + 1 = 0.2330 + 1 = 1.2330$

**29**   $\log 1.18 = \log 1.18 + 0 = 0.0719 + 0 = 0.0719$

**33**   $\log 0.0713 = \log 7.13 + (-2) = 0.8531 + (-2) = -1.1469$

**37**   $\log 0.000007 = \log 7 + (-6) = 0.8451 + (-6) = -5.1549$

**41**
$$0.1 \left[ 0.06 \left[ \begin{array}{l} \log 79.5 = 1.9004 \\ \log 79.56 = \ ? \\ \log 79.6 = 1.9009 \end{array} \right] d \right] 0.0005$$

$\dfrac{d}{0.0005} = \dfrac{0.06}{0.1}$, so that $d = 0.0003$

Therefore, $\log 79.56 = \log 79.5 + d = 1.9004 + 0.0003 = 1.9007$.

**45**
$$0.001 \left[ 0.0005 \left[ \begin{array}{l} \log 0.572 = -0.2426 \\ \log 0.5725 = \ ? \\ \log 0.573 = -0.2418 \end{array} \right] d \right] 0.0008$$

$\dfrac{d}{0.0008} = \dfrac{0.0005}{0.001}$, so that $d = 0.0004$

Therefore, $\log 0.5725 = \log 0.572 + d = -0.2426 + 0.0004 = -0.2422$.

**49**   Let $x = $ antilog $0.4133$, so that
$\log x = 0.4133$
$\log x = \log 2.59$.
Therefore, antilog $0.4133 = 2.59$.

**53**   Let $x = $ antilog $2.7427$, so that
$\log x = 2.7427$
$\phantom{\log x} = 0.7427 + 2$
$\phantom{\log x} = \log 5.53 + \log 10^2$
$\phantom{\log x} = \log(5.53 \times 10^2)$
$\phantom{\log x} = \log 553$.
Therefore, antilog $2.7427 = 553$

**57**   Let $x = $ antilog$[0.7348 + (-1)]$, so that
$\log x = 0.7348 + (-1)$
$\phantom{\log x} = \log 5.43 + \log 10^{-1}$
$\phantom{\log x} = \log(5.43 \times 10^{-1})$
$\phantom{\log x} = \log(0.543)$.
Therefore, antilog$[0.7348 + (-1)] = 0.543$.

**61**   Let $x = $ antilog$(-1.6289) = $ antilog$[0.3711 + (-2)]$, so that
$\log x = 0.3711 + (-2)$
$\phantom{\log x} = \log 2.35 + \log 10^{-2}$
$\phantom{\log x} = \log(2.35 \times 10^{-2})$
$\phantom{\log x} = \log(0.0235)$.
Therefore, antilog$(-1.6289) = 0.0235$.

**65**   Let $x = $ antilog$(-0.1574) = $ antilog$[0.8426 + (-1)]$, so that
$\log x = 0.8426 + (-1)$
$\phantom{\log x} = \log 6.96 + \log 10^{-1}$
$\phantom{\log x} = \log(6.96 \times 10^{-1})$
$\phantom{\log x} = \log 0.696$.
Therefore, antilog$(-0.1574) = 0.696$.

**69**   Let $x = $ antilog $0.1452$, so that $\log x = 0.1452$.
$$0.01 \left[ d \left[ \begin{array}{l} \log 1.39 = 0.1430 \\ \log x = 0.1452 \\ \log 1.4 = 0.1461 \end{array} \right] 0.0022 \right] 0.0031$$

$\dfrac{d}{0.01} = \dfrac{0.0022}{0.0031}$, so that $d = 0.007$

Therefore, antilog $0.1452 = 1.39 + d = 1.39 + 0.007 = 1.397$.

73  Let $x = \text{antilog}[0.2259 + (-2)]$, so that $\log x = 0.2259 + (-2)$.

$$0.0001 \begin{bmatrix} d \begin{bmatrix} \log 0.0168 = 0.2253 + (-2) \\ \log x = 0.2259 + (-2) \\ \log 0.0169 = 0.2279 + (-2) \end{bmatrix} 0.0006 \end{bmatrix} 0.0026$$

$\dfrac{d}{0.0001} = \dfrac{0.0006}{0.0026}$, so that $d = 0.00002$

Therefore, $\text{antilog}[0.2259 + (-2)] = 0.0168 + 0.00002 = 0.01682$.

## 9.4 Using a Calculator to Evaluate Logarithmic and Exponential Expressions

### SEMIPROGRAMMED PROBLEMS

In problems 1–13, round off answers to four significant digits.

| | |
|---|---|
| log | 1 Using a calculator with a $\boxed{\log}$ key, we find the value $\log x$ by entering the number $x$ and then pressing the _____ key. |
| 2.262 | 2 To find $\log 183$, enter 183 and then press the $\boxed{\log}$ key to obtain $\log 183 = $ _____. |
| $-1.377$ | 3 To find $\log 0.042$, enter 0.042 and then press the $\boxed{\log}$ key to obtain $\log 0.042 = $ _____. |
| $y^x$ | 4 To find antilogarithms, we can use the $\boxed{y^x}$ key. Thus, to find antilog $r$, enter 10 and press the _____ key. Then enter the number $r$ and press the $\boxed{=}$ key to obtain the value of antilog $r$. |
| 2.3142, 206.2 | 5 To find antilog 2.3142, enter 10 and press the $\boxed{y^x}$ key. Next, enter _____ and press the $\boxed{=}$ key to obtain antilog $2.3142 = $ _____. |
| $-1.117$ | 6 To find antilog$(-1.117)$, enter 10 and press the $\boxed{y^x}$ key. Next, enter |
| 0.07638 | _____ and press the $\boxed{=}$ key to obtain antilog $(-1.117) = $ _____. |
| ln | 7 Using a calculator with an $\boxed{\ln}$ key, we find the value $\ln x$ by entering the number $x$ and then pressing the _____ key. |
| ln | 8 To find $\ln 17.3$, enter 17.3 and press the _____ key to obtain |
| 2.851 | $\ln 17.3 = $ _____. |
| 0.0928 | 9 To find $\ln 0.0928$, enter _____ and press the $\boxed{\ln}$ key to obtain |
| $-2.377$ | $\ln 0.0928 = $ _____. |
|  | 10 To find $e^{2.3}$, enter $e = 2.7182818$ and press the $\boxed{y^x}$ key. Next, enter 2.3 and press the $\boxed{=}$ key to obtain $e^{2.3} = $ _____. |
| 9.974 |  |
|  | 11 To find $e^{-0.317}$, enter $e = 2.7182818$ and press the $\boxed{y^x}$ key. Next, |
| $-0.317$, 0.7283 | enter _____ and press the $\boxed{=}$ key to obtain $e^{-0.317} = $ _____. |
| $y^x$ | 12 To find $(3.21)^{0.78}$, enter 3.21 and press the _____ key. Next, enter 0.78 |
| $=$, 2.484 | and press the _____ key to obtain $(3.21)^{0.78} = $ _____. |
|  | 13 To find $\sqrt[4]{26.8}$, we have |
| 2.275 | $\sqrt[4]{26.8} = (26.8)^{1/4} = (26.8)^{0.25} = $ _____. |

## CHAPTER 9  LOGARITHMS

In problems 14–17, use a calculator and the properties of logarithms to solve each equation.

**14**  $3^x = 11$

log 11        Write $\log 3^x =$ _____, so that

log 11        $x \log 3 =$ _____ or $x = \dfrac{\log 11}{\log 3}$.

1.0414, 2.1828      Using a calculator, we find that $x = \dfrac{(\underline{\hspace{1cm}})}{0.4771} =$ _____.

**15**  $5^{-x} = 9$

log 9        Write $\log 5^{-x} =$ _____, so that

log 9        $-x \log 5 =$ _____ or $x = -\dfrac{\log 9}{\log 5}$.

0.9542, −1.365      Using a calculator, we find that $x = -\dfrac{\underline{\hspace{1cm}}}{0.6990} =$ _____.

**16**  $e^{2x+1} = 4$

$\ln e^{2x+1}, 2x+1$      Write _____ $= \ln 4$, so that (_____) $\ln e = \ln 4$ or

$2x + 1, 1$      _____ $= \ln 4$ since $\ln e =$ ____.

Using a calculator, we find that

1.386      $2x + 1 =$ _____. Solving for $x$, we obtain

0.386, 0.193      $2x =$ _____ or $x =$ _____.

**17**  $2^{2x-3} = 3^{x+1}$

$\log 3^{x+1}$      Write $\log 2^{2x-3} =$ _____ or

$(x+1) \log 3$      $(2x-3) \log 2 =$ _____, so that

$x \log 3$      $2x \log 2 - 3 \log 2 =$ _____ $+ \log 3$.

Solving for $x$, we have

$x \log 3$      $2x \log 2 -$ _____ $= 3 \log 2 + \log 3$

$2 \log 2 - \log 3$      (_____)$x = 3 \log 2 + \log 3$

$2 \log 2 - \log 3$      $x = \dfrac{3 \log 2 + \log 3}{\underline{\hspace{1cm}}}$.

Using a calculator, we find that

0.1249, 11.05      $x = \dfrac{1.3802}{\underline{\hspace{1cm}}} =$ _____.

## SOLUTIONS TO SELECTED ODD PROBLEMS  Section 9.4, text pages 444–445

**1**  Enter 32.94 and press the $\boxed{\log}$ key to obtain $\log 32.94 = 1.518$.

**5**  Enter 603.75 and press the $\boxed{\log}$ key to obtain $\log 603.75 = 2.781$.

**9**  Enter 0.003561 and press the $\boxed{\log}$ key to obtain $\log 0.003561 = -2.448$.

**13**  Enter 10, press the $\boxed{y^x}$ key, enter 1.9281, and press the $\boxed{=}$ key to obtain antilog $1.9281 = 84.74$.

**17**  Enter 10, press the $\boxed{y^x}$ key, enter 1.47372, and press the $\boxed{=}$ key to obtain antilog $1.47372 = 29.77$.

**21**  Enter 10, press the $\boxed{y^x}$ key, enter −1.4837, and press the $\boxed{=}$ key to obtain antilog$(-1.4837) = 0.0328$.

**25**  Enter 7,324 and press the $\boxed{\ln}$ key to obtain $\ln 7{,}324 = 8.899$.

**29**  Enter 0.5342 and press the $\boxed{\ln}$ key to obtain $\ln 0.5342 = -0.6270$.

**33**  Enter 2, press the $\boxed{\sqrt{\phantom{x}}}$ key, then press the $\boxed{e^x}$ key to obtain $e^{\sqrt{2}} = 4.113$.

37  Enter $-3.1$ and press the $\boxed{e^x}$ key to obtain $e^{-3.1} = 0.04505$.

41  Enter 0.4014, press the $\boxed{y^x}$ key, enter 3.2, and press the $\boxed{=}$ key to obtain $(0.4014)^{3.2} = 0.05388$.

45  Enter 3.912 and press the $\boxed{x^2}$ key to obtain $(3.912)^2 = 15.30$.

49  $\sqrt[4]{7.18} = (7.18)^{1/4} = (7.18)^{0.25}$
Enter 7.18, press the $\boxed{y^x}$ key, enter 0.25, and press the $\boxed{=}$ key to obtain $\sqrt[4]{7.18} = 1.637$.

53  $\sqrt[4]{0.293} = (0.293)^{1/4} = (0.293)^{0.25}$
Enter 0.293, press the $\boxed{y^x}$ key, enter 0.25, and press the $\boxed{=}$ key to obtain $\sqrt[4]{0.293} = 0.7357$.

57  $\dfrac{\ln 5}{2 \ln 1.15} = \dfrac{1.6094379}{2(0.13976194)} = 5.758$

61  $2x = 7$
$\log 2^x = \log 7$
$x \log 2 = \log 7$
$x = \dfrac{\log 7}{\log 2}$
$x = 2.807$

65  $e^{2x} = 3$
$\ln e^{2x} = \ln 3$
$2x \ln e = \ln 3$
$2x = \ln 3 \quad (\ln e = 1)$
$x = \dfrac{\ln 3}{2}$
$x = 0.5493$

69  $3^{5-2t} = 8^{t-4}$
$\log 3^{5-2t} = \log 8^{t-4}$
$(5 - 2t) \log 3 = (t - 4) \log 8$
$5 \log 3 - 2t \log 3 = t \log 8 - 4 \log 8$
$t \log 8 + 2t \log 3 = 5 \log 3 + 4 \log 8$
$(\log 8 + 2 \log 3)t = 5 \log 3 + 4 \log 8$
$t = \dfrac{5 \log 3 + 4 \log 8}{2 \log 3 + \log 8}$
$t = 3.229$

73  $\text{pH} = -\log[1.6 \times 10^{-8}]$
$= -[\log 1.6 + (-8)]$
$= -[0.20 + (-8)]$
$= -(-7.8)$
$= 7.8$

77  $\log_5 7 = \dfrac{\log 7}{\log 5} = 1.2091$

## 9.5 Applications

### SEMIPROGRAMMED PROBLEMS

In problems 1 and 2, consider the application of compound interest, which states that if $P$ dollars represents the amount invested at an annual interest rate $r$, the amount $S$ accumulated in $n$ years, when interest is compounded $t$ times per year, is given by $S = P\left(1 + \dfrac{r}{t}\right)^{nt}$.

1  Suppose that \$1,000 is invested at a nominal annual interest rate of 12% compounded every 4 months. How much money is accumulated after 4 years? Here, $P =$ _____, $r =$ _____, $t = 3$, and $n = 4$, so

1,000, 0.12

$S = 1,000\left(1 + \dfrac{\phantom{xxx}}{3}\right)^{3 \times 4}$

0.12

$= 1,000(\underline{\phantom{xxx}})^{12}$.

1.04

Using a calculator, we find that $S =$ _____.

1,601.03

Therefore, the amount accumulated is _____.

\$1,601.03

**222** CHAPTER 9 LOGARITHMS

**2** If $5,000 is invested at a nominal annual interest rate of 10% compounded every 3 months, how much money is accumulated after 5 years? Here, $P = $ _____, $r = $ _____

5,000, 0.10

$t = $ _____, and $n = $ _____, so

4, 5

$$S = 5{,}000\left(1 + \frac{\rule{1cm}{0.15mm}}{4}\right)^{5 \times 4}$$

0.10

$$= 5{,}000(\rule{1cm}{0.15mm})^{20}.$$

1.025

Using a calculator, we find that $S = $ _____.

8,193.08

Therefore, the amount accumulated is _____.

$8,193.08

In problems 3 and 4, use the formula $R = \left(1 + \frac{r}{t}\right)^t - 1$ to find the effective simple interest $R$ that corresponds to a compound rate $r$ when compounded $t$ times per year.

**3** Find the effective simple annual interest rate corresponding to a nominal annual interest rate of 12% compounded quarterly.

0.12

Here, $r = $ _____ and $t = 4$, so

0.12

$$R = \left(1 + \frac{\rule{1cm}{0.15mm}}{4}\right)^4 - 1$$

1.03

$$= (\rule{1cm}{0.15mm})^4 - 1.$$

0.1255

Using a calculator, we find that $R = $ _____. Thus, the effective

12.55%

simple annual interest rate is _____.

**4** Find the effective simple annual interest rate corresponding to a nominal annual interest rate of 9% compounded every 4 months.

0.09, 3

Here, $r = $ _____ and $t = $ _____, so that

0.03, 3

$R = (1 + \rule{1cm}{0.15mm})^{\rule{0.5cm}{0.15mm}} - 1$

1.03

$= (\rule{1cm}{0.15mm})^3 - 1$

0.092727

$= \rule{1cm}{0.15mm}$.

9.27%

Thus, the effective simple annual interest rate is _____.

In problems 5 and 6, use the formula $P = S\left(1 + \frac{r}{t}\right)^{-nt}$ to find the present value $P$ of an amount $S$ to be received $n$ years in the future if investments during this period are earning a nominal annual interest rate $r$ when compounded $t$ times a year.

**5** Find the present value of $10,000 to be paid to you in 10 years if investments are earning a nominal annual interest rate of 8%

0.08

compounded quarterly. Here, $S = 10{,}000$, $r = $ _____, $n = 10$, and $t = 4$, so

0.08

$$P = 10{,}000\left(1 + \frac{\rule{1cm}{0.15mm}}{4}\right)^{-(10 \times 4)}$$

1.02

$$= 10{,}000(\rule{1cm}{0.15mm})^{-40}.$$

4,528.90

Using a calculator, we find that $P = $ _____.

$4,528.90

Thus, the present value is _____.

**6** Suppose that someone owes you $50,000 to be paid to you in 4 years. What is the present value of this money if it could be invested at a

## 9.5 SEMIPROGRAMMED PROBLEMS

| | |
|---|---|
| 50,000, 0.105 | nominal annual interest rate of 10.5% compounded semiannually? Here, $S = \underline{\qquad}$, $r = \underline{\qquad}$, |
| 4, 2 | $n = \underline{\qquad}$, and $t = \underline{\qquad}$, so |
| 0.0525, $-8$ | $P = 50,000(1 + \underline{\qquad})^{\underline{\qquad}}$ |
| 1.0525 | $= 50,000(\underline{\qquad})^{-8}$. |
| 33,204.21 | Using a calculator, we find that $P = \underline{\qquad}$. |
| $33,204.21 | Thus, the present value is $\underline{\qquad}$. |

In problems 7 and 8, use the formula $P = P_0 e^{kt}$ to find the population $P$ after time $t$ for an initial population $P_0$ and where $k$ is the rate of growth per unit of time.

**7** The population of a country was 12,000,000 in 1982 and is growing at the rate of 3% per year. Find its population in 1990. Here,

| | |
|---|---|
| 0.03, 8 | $P_0 = 12,000,000$, $k = \underline{\qquad}$, and $t = \underline{\qquad}$, so |
| 0.03 | $P = 12,000,000 e^{(\underline{\qquad})8}$ |
| 0.24 | $= 12,000,000 e^{\underline{\qquad}}$. |
| 15,254,990 | Using a calculator, we find that $P = \underline{\qquad}$. |
| 15,254,990 | Therefore, the population will be $\underline{\qquad}$ in 1990. |

**8** A biologist finds the number of bacteria in a culture is 5,000 and is growing at the rate of 6% per hour. How many bacteria will be in the culture after 9 hours? Here, $P_0 = \underline{\qquad}$,

| | |
|---|---|
| 5,000 | $k = \underline{\qquad}$, and $t = 9$, so |
| 0.06 | $P = P_0 e^{kt} = \underline{\qquad} e^{(\underline{\qquad})9}$ |
| 5,000, 0.06 | $= \underline{\qquad} e^{\underline{\qquad}}$. |
| 5,000, 0.54 | Using a calculator, we find that $P = \underline{\qquad}$. |
| 8,580 | Therefore, there are $\underline{\qquad}$ bacteria in the culture after 9 hours. |
| 8,580 | |

**9** Newton's law of cooling states that under certain conditions, the temperature $T$ of an object is given by the equation $T = 75e^{-2t}$, where $t$ is the time in hours. Find the temperature of an object after 1.5 hours.

| | |
|---|---|
| 1.5 | Here, $t = \underline{\qquad}$, so |
| 1.5 | $T = 75e^{-2(\underline{\qquad})}$ |
| $-3$ | $= 75e^{\underline{\qquad}}$. |
| 3.734 | Using a calculator, we find that $T = \underline{\qquad}$. Therefore, the |
| 3.734 | temperature after 1.5 hours was $\underline{\qquad}$ degrees. |

**10** A fully charged electrical condenser is allowed to discharge for $t$ seconds. The remaining charge $Q$ is given by the formula $Q = 750(2.7)^{-0.4t}$. What is the charge after 10 seconds? Here,

| | |
|---|---|
| 10 | $t = \underline{\qquad}$, so that |
| 10 | $Q = 750(2.7)^{-0.4(\underline{\qquad})}$ |
| $-4$ | $= 750(2.7)^{\underline{\qquad}}$. |
| 14.11 | Using a calculator, we find that $Q = \underline{\qquad}$. Therefore, the charge |
| 14.11 | after 10 seconds is $\underline{\qquad}$. |

## SOLUTIONS TO SELECTED ODD PROBLEMS Section 9.5, text pages 449–451

**1** $S = P\left(1 + \dfrac{r}{t}\right)^{nt}$

$= 7{,}000\left(1 + \dfrac{0.13}{1}\right)^{3(1)}$

$= 7{,}000(1.13)^3$

$= 10{,}100.28$

The amount accumulated is $10,100.28.

**9** $P = S\left(1 + \dfrac{r}{t}\right)^{-nt}$

$= 10{,}000\left(1 + \dfrac{0.145}{1}\right)^{-3(1)}$

$= 10{,}000(1.145)^{-3}$

$= 6{,}661.68$

The present value is $6,661.68.

**17** $T = 75e^{-2t}$
$T = 75e^{-2(2.5)}$
$T = 75(0.0067379)$
$T = 0.505$

The temperature after 2.5 hours is 0.505°C.

**5** $R = \left(1 + \dfrac{r}{t}\right)^t - 1$

$= \left(1 + \dfrac{0.11}{1}\right)^1 - 1$

$= (1.11) - 1$

$= 0.11$

The effective simple annual interest rate is 11%.

**13** $P = P_0 e^{kt}$
$= 2{,}000{,}000 e^{0.04(10)}$
$= 2{,}000{,}000 e^{0.4}$
$= 2{,}983{,}649$

The population in 1993 will be 2,983,649.

**21** $\log(1 - r) = \dfrac{1}{t} \log \dfrac{W}{P}$

$\log(1 - r) = \dfrac{1}{5} \log \dfrac{5{,}600}{11{,}400}$

$= \dfrac{1}{5}(-0.3087)$

$= -0.0617$

$1 - r = \text{antilog}(-0.0617)$
$= 0.867$
$r = 0.13253$

The annual rate of depreciation is 13.253%.

**25** (a) $N = 50 - 25(0.8)^t$
$= 50 - 25(0.8)^4$
$= 39.76$

The worker can produce approximately 40 items per day.

(b) $N = 50 - 25(0.8)^t$
$45 = 50 - 25(0.8)^t$
$-5 = -25(0.8)^t$
$(0.8)^t = 0.2$
$\log(0.8)^t = \log 0.2$
$t \log 0.8 = \log 0.2$

$t = \dfrac{\log 0.2}{\log 0.8} = 7.213$

The worker needs approximately 7.2 weeks experience.

## CUMULATIVE REVIEW PROBLEM SET  Chapters 1–9

**1** Evaluate $3x^3 - 2x^2y - 4y^5$ for $x = 2$, $y = -1$.

**2** Perform the indicated operations.

(a) $(x - 2y)(x^3 + 2x^2y - xy^2 - y^3)$

(b) $(x^3 - 2x^2y + y^3) \div (x - y)$

**3** Factor each expression completely.

(a) $32x^3y - 4y$

(b) $6au - bv + 3av - 2bu$

**4** Perform the indicated operations and simplify.

(a) $\dfrac{2x^3 - 8x}{3x^3 + 9x^2y} \div \dfrac{x^2 + 4x + 4}{x^3 - 9xy^2}$

(b) $\dfrac{3}{2x^2 - x - 1} - \dfrac{2}{4x^2 - 1}$

**5** Solve each equation.

(a) $\dfrac{1}{y(y - 1)} - \dfrac{1}{y - 1} = \dfrac{1}{y}$

(b) $|2 - 3x| = 5$

**6** Solve each inequality and graph the solution set.
(a) $3(y-4) < 2(5-y)$
(b) $|4x-3| \geq 5$

**7** Rewrite each expression so that it contains only positive exponents and simplify.
(a) $3^{-2} \cdot 3^{-1}$
(b) $\left(\dfrac{x^{-1/2}y^2}{x^{3/2}y^{-3}}\right)^{-4}$

**8** Simplify each radical expression.
(a) $\sqrt[3]{16x^7y^5}$
(b) $\dfrac{\sqrt{50x^5y^7}}{\sqrt{2xy^4}}$

**9** Perform the indicated operations and simplify.
(a) $\sqrt{75}+\sqrt{48}$
(b) $(\sqrt{5}+\sqrt{3})(2\sqrt{5}-\sqrt{3})$

**10** Rationalize the denominator of each expression.
(a) $\dfrac{5}{\sqrt{7}}$
(b) $\dfrac{a-\sqrt{b}}{a+\sqrt{b}}$

**11** Perform the indicated operations and write the results in the form $a+bi$.
(a) $(2+3i)+(7-5i)$
(b) $(-3+4i)-(-2-6i)$
(c) $(3+5i)(2-7i)$
(d) $\dfrac{4-3i}{2+5i}$

**12** Solve each equation.
(a) $2x^2-3x+2=0$
(b) $y^4-10y^2+9=0$
(c) $\sqrt{x-1}=2\sqrt{x}$
(d) $(3z-1)^{3/5}=8$

**13** Use an equation to solve the following problem: Find a number such that the sum of the number and its reciprocal is 6.

**14** Find the distance between the points $(\tfrac{1}{2}, 1)$ and $(-1, -2)$.

**15** Find the center and the radius of the circle $x^2+y^2-6x+8y-24=0$.

**16** Find the equation of each line described.
(a) slope is $\tfrac{2}{3}$ and $y$ intercept is 3
(b) $x$ intercept is 2 and $y$ intercept is 5
(c) the line is parallel to the line $3x-4y=7$ and contains the point $(3, -2)$

**17** Complete each sentence.
(a) The graph of the inequality $4x-3y>2$ is the half-plane _____ the line $4x-3y=2$.
(b) If a system of linear equations is inconsistent, then the graph of the system is two _____ lines.

**18** Solve each system.
(a) $\begin{cases} 3x-5y=4 \\ 2x+3y=9 \end{cases}$
(b) $2x+y-z=6$
$x+3y+2z=3$
$-x+y+4z=-5$

**19** Evaluate each determinant.
(a) $\begin{vmatrix} 3 & 4 \\ -1 & 2 \end{vmatrix}$
(b) $\begin{vmatrix} 1 & 3 & 4 \\ -1 & 2 & 5 \\ 2 & -1 & 6 \end{vmatrix}$

**20** Solve each equation.
(a) $4^{2x-1}=8^{2-x}$
(b) $\log_5 x + \log_5(x+4)=1$

**21** Write the following in equivalent exponential form.
(a) $\log_2 16 = 4$
(b) $\log_c x = b$

# CHAPTER 9  LOGARITHMS

**22** Use the properties of logarithms to rewrite as single logarithms.
   (a) $2\log_b x + 3\log_b y$
   (b) $\log_c t - \frac{1}{2}\log_c u$

**23** If $\log_5 2 = 0.4307$ and $\log_5 3 = 0.6826$, find the following.
   (a) $\log_5 6$
   (b) $\log_5 \frac{3}{2}$
   (c) $\log_5 8$

**24** If antilog $0.3010 = 2$, find the following.
   (a) antilog $2.3010$
   (b) antilog $[0.3010 + (-3)]$

**25** The population of a city was 200,000 in 1985 and has been growing according to the equation $P = P_0 e^{kt}$ at 2.5% per year. What is the population of the city expected to be in the year 1995.

## Answers

**1** 36   **2(a)** $x^4 - 5x^2y^2 + xy^3 + 2y^4$   **(b)** $x^2 - xy - y^2$   **3(a)** $4y(2x-1)(4x^2 + 2x + 1)$
**(b)** $(3a - b)(2u + v)$   **4(a)** $\dfrac{2(x-2)(x-3y)}{3(x+2)}$   **(b)** $\dfrac{4x-1}{(2x+1)(2x-1)(x-1)}$
**5(a)** No solution   **(b)** $-1, \frac{7}{3}$   **6(a)** $y < \frac{22}{5}$
**(b)** $x \leq -\frac{1}{2}$ or $x \geq 2$   **7(a)** $\frac{1}{27}$   **(b)** $\dfrac{x^8}{y^{20}}$
**8(a)** $2x^2 y\sqrt[3]{2xy^2}$   **(b)** $5x^2 y\sqrt{y}$   **9(a)** $9\sqrt{3}$   **(b)** $7 + \sqrt{15}$   **10(a)** $\dfrac{5\sqrt{7}}{7}$
**(b)** $\dfrac{a^2 - 2a\sqrt{b} + b}{a^2 - b}$   **11(a)** $9 - 2i$   **(b)** $-1 + 10i$   **(c)** $41 - 11i$   **(d)** $-\frac{7}{29} - \frac{26}{29}i$
**12(a)** $\dfrac{3 - \sqrt{7}i}{4}, \dfrac{3 + \sqrt{7}i}{4}$   **(b)** $-3, -1, 1, 3$   **(c)** No solution   **(d)** 11
**13** $3 - 2\sqrt{2}, 3 + 2\sqrt{2}$   **14** $\dfrac{3\sqrt{5}}{2}$   **15** center $= (3, -4)$, radius $= 7$   **16(a)** $y = \dfrac{2}{3}x + 3$
**(b)** $\dfrac{x}{2} + \dfrac{y}{5} = 1$   **(c)** $y + 2 = \dfrac{3}{4}(x - 3)$   **17(a)** below   **(b)** parallel   **18(a)** $(3, 1)$
**(b)** $(2, 1, -1)$   **19(a)** 10   **(b)** 53   **20(a)** $\frac{8}{7}$   **(b)** 1   **21(a)** $2^4 = 16$   **(b)** $c^b = x$
**22(a)** $\log_b x^2 y^3$   **(b)** $\log_c \dfrac{t}{\sqrt{u}}$   **23(a)** 1.1133   **(b)** 0.2519   **(c)** 1.2921
**24(a)** 200   **(b)** 0.002   **25** 256,805

## CHAPTER 9  TESTS

### Chapter Test A

**1** Write each statement in logarithmic form.
   (a) $2^9 = 512$
   (b) $3^4 = 81$
   (c) $10^{-3} = 0.001$

**2** Write each statement in exponential form.
   (a) $\log_2 32 = 5$
   (b) $\log_5 125 = 3$
   (c) $\log_7 \frac{1}{49} = -2$

**3** Determine $x$ in each equation.
   (a) $\log_x 625 = 4$
   (b) $\log_2 \frac{1}{8} = x$
   (c) $\log_{32} x = \frac{1}{5}$
   (d) $\log_b x = 0$

**4** Find the solution of each equation.
   (a) $2^{x-1} = 8$
   (b) $9(3^{1-x}) = 81$
   (c) $\log_3(2x - 1) = 4$
   (d) $\log_5(7x - 2) = 2$

5  Evaluate each expression.
   (a) $\log_3 \frac{1}{9}$
   (b) $\log_5 \frac{1}{125}$
   (c) $\log_7 7\sqrt{7}$
   (d) $\log_9 27\sqrt{3}$

6  Simplify each expression.
   (a) $\log_3 \frac{17}{2} + \log_3 \frac{5}{34}$
   (b) $\log_a x^5 - 2 \log_a x^2$
   (c) $\log_a(x^2 - 16) - \log_a(x - 4)$
   (d) $3 \log_a x + 5 \log_a y^2$

7  Find the solution of equation $\log_2(x^2 - 9) - \log_2(x + 3) = 1$.

8  Evaluate each expression by using a table of common logarithms.
   (a) log 3.21
   (b) log 0.665
   (c) antilog 2.1818
   (d) antilog (−3.6162)

9  Evaluate each expression by using a calculator.
   (a) log 27.8
   (b) antilog 2.0891
   (c) ln 18.5
   (d) $(4.51)^{3.2}$

10 Use a calculator and the properties of logarithms to solve each equation.
   (a) $2^x = 7$
   (b) $e^{3x-1} = 15$

11 If $500 is invested at 8% annually compounded quarterly, how much money is accumulated after 4 years?

---

## Solutions

1  Since $b^x = c$ is equivalent to $x = \log_b c$, then
   (a) $2^9 = 512$ is equivalent to $9 = \log_2 512$
   (b) $3^4 = 81$ is equivalent to $4 = \log_3 81$
   (c) $10^{-3} = 0.001$ is equivalent to $-3 = \log_{10} 0.001$

2  Since $\log_b c = x$ is equivalent to $b^x = c$, then
   (a) $\log_2 32 = 5$ is equivalent to $2^5 = 32$
   (b) $\log_5 125 = 3$ is equivalent to $5^3 = 125$
   (c) $\log_7 \frac{1}{49} = -2$ is equivalent to $7^{-2} = \frac{1}{49}$

3  (a) $\log_x 625 = 4$
   $x^4 = 625$ and $x > 0$
   $x = \sqrt[4]{625}$
   $x = 5$

   (b) $\log_2 \frac{1}{8} = x$
   $2^x = \frac{1}{8}$
   $2^x = 2^{-3}$
   $x = -3$

   (c) $\log_{32} x = \frac{1}{5}$
   $x = 32^{1/5}$
   $x = \sqrt[5]{32}$
   $x = 2$

   (d) $\log_b x = 0$
   $b^0 = x$
   $1 = x$

4  (a) $2^{x-1} = 8$
   $2^{x-1} = 2^3$
   $x - 1 = 3$
   $x = 4$

   (b) $9(3^{1-x}) = 81$
   $3^2(3^{1-x}) = 3^4$
   $3^{3-x} = 3^4$
   $3 - x = 4$
   $x = -1$

   (c) $\log_3(2x - 1) = 4$
   $2x - 1 = 3^4$
   $2x - 1 = 81$
   $2x = 82$
   $x = 41$

   (d) $\log_5(7x - 2) = 2$
   $7x - 2 = 5^2$
   $7x - 2 = 25$
   $7x = 27$
   $x = \frac{27}{7}$

5  (a) $\log_3 \frac{1}{9} = x$
   $3^x = \frac{1}{9}$
   $3^x = 3^{-2}$
   $x = -2$

   (b) $\log_5 \frac{1}{125} = x$
   $5^x = \frac{1}{125}$
   $5^x = 5^{-3}$
   $x = -3$

   (c) $\log_7 7\sqrt{7} = x$
   $7^x = 7\sqrt{7}$
   $7^x = 7(7^{1/2})$
   $7^x = 7^{3/2}$
   $x = \frac{3}{2}$

   (d) $\log_9 27\sqrt{3} = x$
   $9^x = 27\sqrt{3}$
   $(3^2)^x = 3^3 3^{1/2}$
   $3^{2x} = 3^{7/2}$
   $2x = \frac{7}{2}$
   $x = \frac{7}{4}$

6  (a) $\log_3 \frac{17}{2} + \log_3 \frac{5}{34} = \log_3 (\frac{17}{2})(\frac{5}{34}) = \log_3 \frac{5}{4}$

   (b) $\log_a x^5 - 2 \log_a x^2 = \log_a x^5 - \log_a (x^2)^2 = \log_a \frac{x^5}{x^4} = \log_a x$

   (c) $\log_a(x^2 - 16) - \log_a(x - 4) = \log_a \frac{x^2 - 16}{x - 4} = \log_a(x + 4)$

   (d) $3 \log_a x + 5 \log_a y^2 = \log_a x^3 + \log_a (y^2)^5 = \log_a x^3 y^{10}$

7 $\log_2(x^2 - 9) - \log_2(x + 3) = 1$

$$\log_2 \frac{x^2 - 9}{x + 3} = 1$$

$$\log_2(x - 3) = 1$$
$$x - 3 = 2$$
$$x = 5$$

8 (a) $\log 3.21 = 0.5065 + 0 = 0.5065$

(b) $\log 0.665 = \log 6.65 + (-1) = 0.8228 + (-1) = -0.1772$

(c) antilog $2.1818 = x$, or
$\log x = 2.1818 = 0.1818 + 2$
$= \log 1.52 + \log 10^2$
$= \log(1.52 \times 10^2)$
$x = 152$

(d) antilog $(-3.6162) = $ antilog$[0.3838 + (-4)] = x$
or $\log x = 0.3838 + (-4)$
$= \log 2.42 + \log 10^{-4}$
$= \log(2.42 \times 10^{-4})$
$x = 0.000242$

9 (a) Enter 27.8, press the $\boxed{\log}$ key to obtain 1.4440.

(b) Enter 10, press the $\boxed{y^x}$ key, enter 2.0891, and press the $\boxed{=}$ key to obtain 122.77.

(c) Enter 18.5, press the $\boxed{\ln}$ key to obtain 2.9178.

(d) Enter 4.51, press the $\boxed{y^x}$ key, enter 3.2, and press the $\boxed{=}$ key to obtain 123.98.

10 (a) $2^x = 7$
$\log 2^x = \log 7$
$x \log 2 = \log 7$
$x = \dfrac{\log 7}{\log 2}$
$x = 2.807$

(b) $e^{3x-1} = 15$
$\ln e^{3x-1} = \ln 15$
$(3x - 1) \ln e = \ln 15$
$3x - 1 = \ln 15$ $\quad (\ln e = 1)$
$3x = 1 + \ln 15$
$x = \dfrac{1 + \ln 15}{3}$
$x = 1.236$

11 $S = P\left(1 + \dfrac{r}{t}\right)^{nt}$

$= 500\left(1 + \dfrac{0.08}{4}\right)^{4(4)} = 500(1.02)^{16}$

$= 686.39$

The amount accumulated is $686.39.

## Chapter Test B

*Multiple Choice:* Select the *one* correct answer for each of the following questions.

1 $\log_2 8 = 3$ is equivalent to _____.
(a) $2^{-3} = 8$   (b) $2^3 = 8$   (c) $2^3 = -8$   (d) none of these

2 $\log_b x^2 = 2$ is equivalent to _____.
(a) $x^2 = b^2$   (b) $x^2 = -b^2$   (c) $x^{-2} = b^2$   (d) none of these

3 $\left(\frac{4}{25}\right)^{-3/2} = \frac{125}{8}$ is equivalent to _____.
(a) $\log_4 \frac{125}{8} = -\frac{3}{2}$   (b) $\log_{4/25} \frac{125}{8} = -\frac{3}{2}$   (c) $\log_5 \frac{125}{8} = -\frac{3}{2}$   (d) none of these

4 $a^c = b$ is equivalent to _____.
(a) $\log_a c = b$   (b) $\log_a b = -c$   (c) $\log_b a = c$   (d) $\log_a b = c$

5 The base $b$ in $\log_b 16 = \frac{2}{3}$ is _____.
(a) 16   (b) $-16$   (c) 64   (d) none of these

6 $\log_4 64 = $ _____.
(a) $-3$   (b) 42   (c) 3   (d) none of these

7 $\log_{1/5} 25 = $ _____.
(a) 2   (b) $-5$   (c) 5   (d) $-2$

**8** The solution of $\log_2(x - 3) = 5$ is _____.
 (a) $2^5$   (b) 35   (c) 32   (d) none of these

**9** Which of the following statements is true?
 (a) $\log_b x_1 \cdot \log_b x_2 = \log_b x_1 + \log_b x_2$
 (b) $\log_b(x_1 x_2) = \log_b x_1 \cdot \log_b x_2$
 (c) $\log_b x^p = p \log_b x$
 (d) $\dfrac{\log_b x_1}{\log_b x_2} = \log_b x_1 - \log_b x_2$

**10** If $\log_{10} 2 = 0.3010$ and $\log_{10} 3 = 0.4771$, then $\log_{10} 6 =$ _____.
 (a) 0.7781   (b) 0.1761   (c) 2.8713   (d) none of these

**11** If $\log_{10} 2 = 0.3010$, then $\log_2 8 =$ _____.
 (a) 0.9030   (b) $(0.3010)^3$   (c) 3   (d) none of these

**12** The solution of $2^x = 256$ is _____.
 (a) 3   (b) 8   (c) $3^8$   (d) $8^3$

**13** Expressing $5 \log_a x + 3 \log_a y$ as a single logarithm, we have _____.
 (a) $\log_a x^5 \cdot \log_a y^3$   (b) $15 \log_a xy$   (c) $(\log_a xy)^5$   (d) $\log_a x^5 y^3$

**14** $\log_2 \sqrt[5]{xy}$ is equal to _____.
 (a) $\frac{1}{3}(\log_2 x + \log_2 y)$
 (b) $\sqrt[5]{(\log_2 x - \log_2 y)}$
 (c) $\frac{1}{5}(\log_2 x - \log_2 y)$
 (d) $\frac{1}{5}(\log_2 x + \log_2 y)$

**15** The solution of $\log_3 x + \log_3(x - 4) = \log_3 5$ is _____.
 (a) 5   (b) $-1$   (c) $-5$   (d) 1

**16** The solution of $\log_{10}(x^2 - 121) - \log_{10}(x + 11) = 1$ is _____.
 (a) 11   (b) 121   (c) 21   (d) none of these

**17** The solution of $\log_3(x + 1) + \log_3(x + 3) = 1$ is _____.
 (a) $-4$   (b) 0   (c) 3   (d) 4

**18** Using the tables of logarithms, log 53,900 is _____.
 (a) 4.7316   (b) 4.3716   (c) 4.6173   (d) 4.6137

**19** Using the tables of logarithms, antilog$[0.5105 + (-2)]$ is _____.
 (a) 0.0324   (b) 0.00324   (c) 0.324   (d) 3.204

**20** Using a calculator, ln 29.36 is _____.
 (a) 1.4678   (b) 5.6349   (c) 3.3796   (d) 2.2909

**21** Using a calculator, antilog$(-1.3459)$ is _____.
 (a) $-22.18$   (b) 0.0451   (c) 0.1290   (d) $-0.1290$

**22** If $3^x = 7$, then $x =$ _____.
 (a) $\dfrac{\log 7}{\log 3}$   (b) $\dfrac{\log 3}{\log 7}$   (c) $\log \dfrac{7}{3}$   (d) $\log \dfrac{3}{7}$

**23** If $1,000 was invested at a nominal annual interest rate of 12% compounded quarterly, in how many years would the original investment be doubled?
 (a) 8.3 years   (b) 7.6 years   (c) 6.8 years   (d) 5.9 years

---

**Answers**

| 1 b | 2 a | 3 b | 4 d | 5 c | 6 c | 7 d | 8 b | 9 c | 10 a |
| 11 a | 12 b | 13 d | 14 d | 15 a | 16 c | 17 b | 18 a | 19 a | |
| 20 c | 21 b | 22 a | 23 d | | | | | | |

# 10 Functions and Related Curves

In this chapter we consider one of the most useful and universal concepts in mathematics—functions. After completing the appropriate sections, the student should be able to:

1. Apply the definitions of functions.
2. Determine the domain and range of a given function.
3. Graph functions.
4. Apply the definitions of direct and inverse variation.
5. Graph linear and quadratic functions.
6. Graph exponential and logarithmic functions.
7. Graph the conic sections.
8. Solve systems containing quadratic equations in two variables.

## 10.1 Functions

### SEMIPROGRAMMED PROBLEMS

In problems 1–4, find the value indicated for each function.

| | |
|---|---|
| 2, −5 <br> 3, −8 <br> 0, 1 | **1** $f(2), f(3), f(0)$ if $f(x) = -3x + 1$ <br> $f(2) = -3(\_\_\_) + 1 = \_\_\_$ <br> $f(3) = -3(\_\_\_) + 1 = \_\_\_$ <br> $f(0) = -3(\_\_\_) + 1 = \_\_\_$ |
| 0 <br><br> $\frac{1}{9}, 27$ <br><br> $\frac{1}{3}, 9$ <br><br> 3, 1 <br><br> 15, $\frac{1}{5}$ | **2** $f\left(\frac{1}{9}\right), f\left(\frac{1}{3}\right), f(3), f(15)$ if $f(x) = \frac{3}{x}$, $x \neq \_\_\_$. <br> $f\left(\frac{1}{9}\right) = \dfrac{3}{\_\_\_} = \_\_\_$ <br> $f\left(\frac{1}{3}\right) = \dfrac{3}{\_\_\_} = \_\_\_$ <br> $f(3) = \dfrac{3}{\_\_\_} = \_\_\_$ <br> $f(15) = \dfrac{3}{\_\_\_} = \_\_\_$ |
| 10, 0 <br> 8, 6 <br> 6, 8 | **3** $f(10), f(8), f(6)$ if $f(x) = \sqrt{100 - x^2}$ <br> $f(10) = \sqrt{100 - (\_\_\_)^2} = \_\_\_$ <br> $f(8) = \sqrt{100 - (\_\_\_)^2} = \_\_\_$ <br> $f(6) = \sqrt{100 - (\_\_\_)^2} = \_\_\_$ |

| | |
|---|---|
| 2, 5, 5<br>−7, −4, 4<br>−3, 0, 0 | **4** $h(2), h(-7), h(-3)$ if $h(x) = \|x + 3\|$<br>$h(2) = \|\underline{\phantom{xx}} + 3\| = \|\underline{\phantom{xx}}\| = \underline{\phantom{xx}}$<br>$h(-7) = \|\underline{\phantom{xx}} + 3\| = \|\underline{\phantom{xx}}\| = \underline{\phantom{xx}}$<br>$h(-3) = \|\underline{\phantom{xx}} + 3\| = \|\underline{\phantom{xx}}\| = \underline{\phantom{xx}}$ |

In problems 5–8, find the domain of the function defined by each equation.

| | |
|---|---|
| real<br>real numbers | **5** $f(x) = -2x + 3$<br>Since the expression is defined for all _____ values of $x$, the domain is the set of _____. |
| 0<br>2, 2 | **6** $g(x) = \dfrac{4}{x - 2}$<br>Since the expression is defined for all values of $x$ for which $x - 2 \neq$ _____ or for $x \neq$ _____, the domain is the set of all real numbers except _____. |
| 0, 6<br>6 | **7** $h(x) = \sqrt{6 - x}$<br>The expression $\sqrt{6 - x}$ is defined for all values of $x$ for which $6 - x \geq$ _____ or $x \leq$ _____. Therefore, the domain of $h$ consists of all real numbers less than or equal to _____. |
| 0<br>1 | **8** $f(x) = \dfrac{3}{\sqrt{x - 1}}$<br>Since the expression $\dfrac{3}{\sqrt{x - 1}}$ is defined for all values of $x$ for which $x - 1 >$ _____, the domain of $f$ consists of all real numbers greater than _____. |

In problems 9 and 10, sketch the graph of each function.

| | |
|---|---|
| real numbers<br>$x + 2$<br>1, 2 | **9** $f(x) = x + 2$<br>The domain of $f$ is the set of all _____. The graph of the function $f$ is the graph of the equation $y =$ _____, which is a straight line with slope _____ and $y$ intercept _____. The graph is |

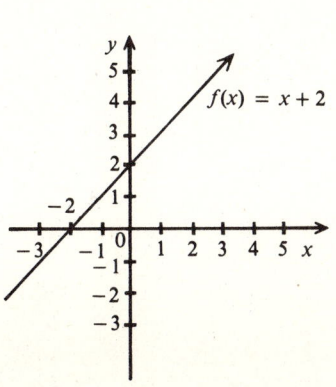

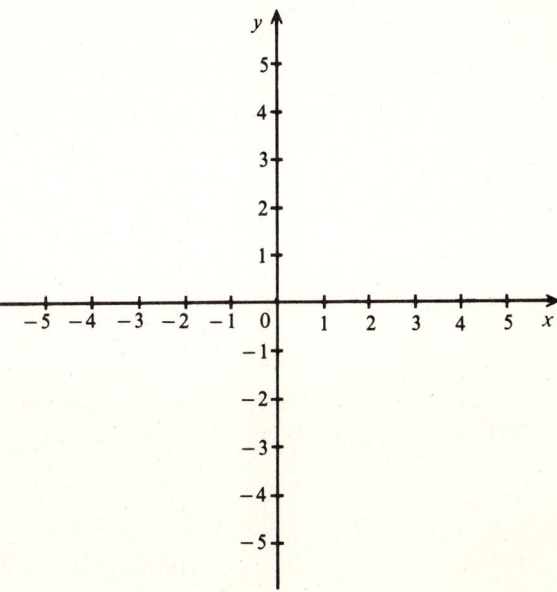

**10** $f(x) = |x - 2|$

The domain of $f$ is the set of all _____. The graph of the function $f$ is the graph of the equation $y =$ _____. This is equivalent to

$$y = \begin{cases} x - 2 & \text{for } x - 2 \geq 0 \text{ or } x \geq \underline{\quad} \\ \underline{\hspace{2cm}} & \text{for } x - 2 < 0 \text{ or } x < \underline{\quad} \end{cases}$$

Graphing $y = x - 2$ for $x \geq 2$ and $y =$ _____ for $x < 2$, we obtain the graph of $f$. The graph is

real numbers
$|x - 2|$

2
$-(x - 2), 2$
$2 - x$

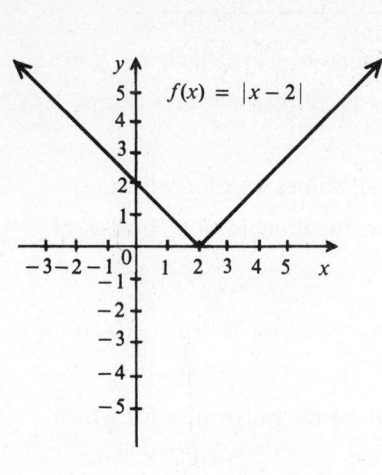
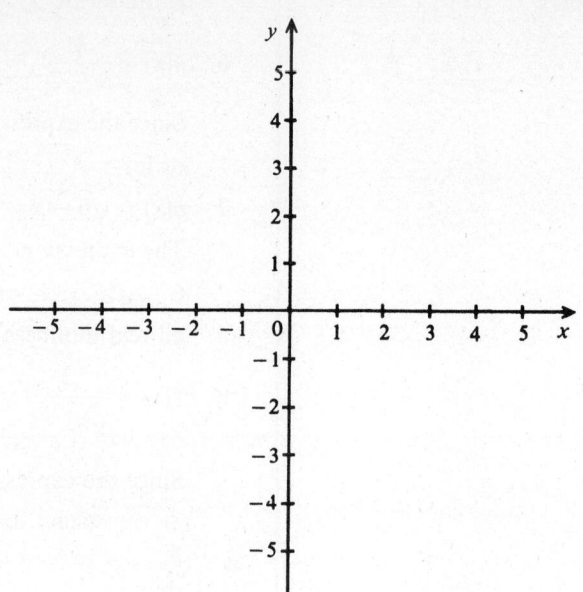

In problems 11 and 12, use the given pair of functions to find the given expressions.

**11** $f(x) = 4 - x$ and $g(x) = 2x + 1$

$f(x) + g(x) =$ _____.

$f(x) - g(x) =$ _____.

$f(x) \cdot g(x) =$ _____.

$\dfrac{f(x)}{g(x)} =$ _____.

$x + 5$

$3 - 3x$

$4 + 7x - 2x^2$

$\dfrac{4 - x}{2x + 1}$

**12** $f(x) = 2x$ and $g(x) = 3x - 5$

$f(x) + g(x) =$ _____.

$f(x) - g(x) =$ _____.

$f(x) \cdot g(x) =$ _____.

$\dfrac{f(x)}{g(x)} =$ _____.

$5x - 5$

$5 - x$

$6x^2 - 10x$

$\dfrac{2x}{3x - 5}$

In problem 13, find the difference quotient of $f$ at $x$ for $f(x) = 7x + 11$.

**13** $\dfrac{f(x + h) - f(x)}{h} = \dfrac{[7(x + h) + 11] - (\underline{\hspace{2cm}})}{h}$

$= \dfrac{\overline{\hspace{2cm}}}{h} =$ _____.

$7x + 11$

$7h, 7$

## SOLUTIONS TO SELECTED ODD PROBLEMS  Section 10.1, text pages 460–462

**1**  $f(x) = 3x + 1$
$f(1) = 3(1) + 1$
$\quad = 4$

**5**  $f(x) = 3x + 1$
$f(0) = 3(0) + 1$
$\quad = 1$

**9**  $f(x) = 3x + 1$
$[f(4)]^2 = [3(4) + 1]^2$
$\quad = [13]^2$
$\quad = 169$

**13**  $g(x) = \sqrt{16 - x^2}$
$g(4) = \sqrt{16 - (4)^2}$
$\quad = \sqrt{16 - 16}$
$\quad = 0$

**17**  $g(x) = \sqrt{16 - x^2}$
$g(\sqrt{7}) = \sqrt{16 - (\sqrt{7})^2}$
$\quad = \sqrt{16 - 7}$
$\quad = 3$

**21**  $h(x) = |x - 2|$
$h(-3) = |-3 - 2|$
$\quad = |-5|$
$\quad = 5$

**25**  $h(x) = |x - 2|$
$h(-a) = |-a - 2|$

**29**  $f(x) = \dfrac{1}{x}$

$\dfrac{1}{x}$ is defined for all real numbers except 0. Thus, the domain of $f$ is all real numbers except 0.

**33**  $f(x) = \sqrt{2 - x}$
$\sqrt{2 - x}$ represents a real number when $2 - x \geq 0$ or $x \leq 2$. Thus, the domain of $f$ is all real numbers $x$ such that $x \leq 2$.

**37**  The graph of $f(x) = x + 1$ is the graph of the equation $y = x + 1$, whose $y$ intercept is 1 and whose $x$ intercept is $-1$.

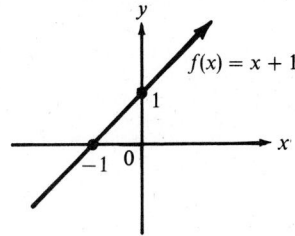

**41**  $\dfrac{f(x + h) - f(x)}{h} = \dfrac{[3(x + h) + 2] - (3x + 2)}{h}$
$\quad = \dfrac{3x + 3h + 2 - 3x - 2}{h}$
$\quad = \dfrac{3h}{h} = 3$

**45**  $f(x) + g(x) = (2x - 1) + (3x + 5)$
$\quad = 5x + 4$

**49**  $f(x) \cdot g(x) = (2x - 1)(3x + 5)$
$\quad = 6x^2 + 7x - 5$

**53**  $f(x) = 2x - 1$
$f[g(x)] = 2[g(x)] - 1$
$\quad = 2(3x + 5) - 1$
$\quad = 6x + 10 - 1$
$\quad = 6x + 9$

**57**  Any vertical line would intersect the graph in exactly one point. Therefore, the curve is the graph of a function.

**61**  Some vertical lines would intersect the graph in two points. Therefore, it is not the graph of a function.

**234** CHAPTER 10 FUNCTIONS AND RELATED CURVES

# 10.2 Variation

## SEMIPROGRAMMED PROBLEMS

In problems 1 and 2, express $y$ as a function of $x$.

$y = kx$
3, 2
$y = 2x$

**1** $y$ is directly proportional to $x$, and $y = 6$ when $x = 3$. Since $y$ is directly proportional to $x$, write the equation _____. Then $6 = k(\_\_\_)$ and $k = \_\_\_$. The equation that expresses the relationship is written as _____.

$y = kx$
3, $\frac{8}{3}$
$y = \frac{8}{3}x$

**2** $y$ is directly proportional to $x$, and $y = 8$ when $x = 3$. Since $y$ is directly proportional to $x$, write the equation _____. Then $8 = k(\_\_\_)$ and $k = \_\_\_$. The equation that expresses the relationship is written as _____.

In problems 3 and 4, assume that $y$ is directly proportional to $x^n$, and $y = 6$ when $x = 2$. Find $y$ for the given values of $x$ and $n$.

$y = kx^3$
2, $\frac{3}{4}$
$y = \frac{3}{4}x^3$
5, $\frac{375}{4}$

**3** $x = 5$ and $n = 3$
Since $y$ is directly proportional to $x^3$, write the equation _____. Then $6 = k(\_\_\_)^3$ and $k = \_\_\_$. The equation that expresses the relationship is written as _____. To obtain $y$ when $x = 5$, write $y = \frac{3}{4}(\_\_\_)^3$, so that $y = \_\_\_$.

$y = kx^2$
2, $\frac{3}{2}$
$y = \frac{3}{2}x^2$
$-2, 6$

**4** $x = -2$ and $n = 2$
Since $y$ is directly proportional to $x^2$, write the equation _____. Then $6 = k(\_\_\_)^2$ and $k = \_\_\_$. The equation that expresses the relationship is written as _____. To obtain $y$ when $x = -2$, write $y = \frac{3}{2}(\_\_\_)^2$, so that $y = \_\_\_$.

$d = kt^2$, 1, 16

$d = 16t^2$
4, 256 feet

**5** The distance in feet that an object will fall from rest is directly proportional to the square of the time in seconds. A body will fall 16 feet the first second. How far will it fall in 4 seconds? Let $d =$ the distance in feet, and $t =$ the time in seconds. Since $d$ is directly proportional to $t^2$, write the equation _____. Then $16 = k(\_\_\_)^2$ and $k = \_\_\_$. The equation that expresses the relationship between the distance and the time is written as _____. To obtain the distance $d$ that the body will fall in 4 seconds, write $d = 16(\_\_\_)^2$ or $d = \_\_\_$.

$y = \frac{k}{x}$, 3, 24

$y = \frac{24}{x}$

4, 6

**6** $y$ is inversely proportional to $x$, and $y = 8$ when $x = 3$. Express $y$ as a function of $x$. Find $y$ when $x = 4$. Since $y$ is inversely proportional to $x$, write the equation _____. Then $8 = \frac{k}{\_\_\_}$ and $k = \_\_\_$.

Therefore, the equation that expresses the relationship is _____.

To obtain $y$ when $x = 4$, we write $y = \frac{24}{\_\_\_}$ and obtain $y = \_\_\_$.

**7** If $y$ is inversely proportional to $x^2$, and $y = 5$ when $x = 2$, find $y$ when $x = 3$. Since $y$ is inversely proportional to $x^2$, write the equation

$y = \dfrac{k}{x^2}$, 2, 20

$y = \dfrac{20}{x^2}$

3, $\dfrac{20}{9}$

$p = \dfrac{k}{v}$

140, 5,600

$p = \dfrac{5,600}{v}$

100, 56

$I = \dfrac{k}{x^2}$, 4, 128

16, $\tfrac{1}{2}$

$z = kx^2y^2$
4, 2, 6
$z = 6x^2y^2$
6, 3, 1,944

$z = \dfrac{kx^3}{\sqrt{y}}$

6, 2

$z = \dfrac{2x^3}{\sqrt{y}}$

3, 9

———. Then $5 = \dfrac{k}{(\underline{\phantom{x}})^2}$ and $k = \underline{\phantom{xx}}$. The equation that expresses the relationship is ———. To obtain $y$ when $x = 3$, we write $y = \dfrac{20}{(\underline{\phantom{x}})^2}$ and obtain $y = \underline{\phantom{xx}}$.

8   When air is pumped into a tire, the pressure required varies inversely as the volume. If the pressure is 40 pounds when the volume is 140 cubic inches, what is the pressure when the volume is 100 cubic inches? Let $p$ = the pressure in pounds, and $v$ = the volume in cubic inches. Since $p$ is inversely proportional to $v$, write the equation ———. If $p = 40$ when $v = 140$, then $40 = \dfrac{k}{\underline{\phantom{x}}}$ and $k = \underline{\phantom{xx}}$. The equation that expresses the relationship is ———. To obtain $p$ when $v = 100$, we write $p = \dfrac{5,600}{\underline{\phantom{x}}}$ and obtain $p = \underline{\phantom{xx}}$.

9   The intensity of light upon a surface varies inversely as the square of the distance of the surface from the source of the light. If the intensity of light on a surface from a lamp 4 feet away is 8 candela, what would the intensity be if the lamp were moved to a distance of 16 feet from the surface? Let $I$ = the intensity in candela and $x$ = the distance in feet. Since $I$ is inversely proportional to $x^2$, write the equation ———. Since $I = 8$ when $x = 4$, then $8 = \dfrac{k}{(\underline{\phantom{x}})^2}$ and $k = \underline{\phantom{xx}}$. To obtain the intensity $I$ when the lamp is 16 feet from the surface, write $I = \dfrac{128}{(\underline{\phantom{x}})^2}$ to obtain $I = \underline{\phantom{xx}}$.

10   If $z$ is directly proportional to the product of $x^2$ and $y^2$, and $z = 384$ when $x = 4$ and $y = 2$, find $z$ when $x = 6$ and $y = 3$. Since $z$ varies jointly as $x^2$ and $y^2$, write the equation ————. To find $k$, use $384 = k(\underline{\phantom{x}})^2(\underline{\phantom{x}})^2$. Then $k = \underline{\phantom{xx}}$. The equation that expresses the relationship is ————. To obtain $z$ when $x = 6$ and $y = 3$, $z = 6x^2y^2$ becomes $z = 6(\underline{\phantom{x}})^2(\underline{\phantom{x}})^2 = \underline{\phantom{xx}}$.

11   If $z$ varies directly as $x^3$ and inversely as $\sqrt{y}$, and $z = 108$ when $x = 6$ and $y = 16$, find $z$ when $x = 3$ and $y = 36$. Since $z$ varies directly as $x^3$ and inversely as $\sqrt{y}$, write the equation ———. The constant $k$ can be determined from the equation $108 = \dfrac{k(\underline{\phantom{x}})^3}{\sqrt{16}}$, or $k = \underline{\phantom{xx}}$. The equation that expresses the relationship is ———. To obtain $z$ when $x = 3$ and $y = 36$, $z = \dfrac{2x^3}{\sqrt{y}}$ becomes $z = \dfrac{2(\underline{\phantom{x}})^3}{\sqrt{36}} = \underline{\phantom{xx}}$.

# 236 CHAPTER 10 FUNCTIONS AND RELATED CURVES

**SOLUTIONS TO SELECTED ODD PROBLEMS**  Section 10.2, text pages 466–467

**1**  Let $y = f(x) = kx$
$8 = k4$
$k = 2$
so that $y = f(x) = 2x$.
$f(x + 2) = 2(x + 2) = 2x + 4$
$f(2) + f(3) = 2(2) + 2(3) = 10$
$\dfrac{f(x + h) - f(x)}{h} = \dfrac{2(x + h) - 2x}{h} = \dfrac{2h}{h} = 2$

**5**  $y = kx^3$
$3 = k(1)^3$
$3 = k$
$y = 3x^3$

**9**  $y = \dfrac{k}{x^2}$
$9 = \dfrac{k}{(2)^2}$
$36 = k$
$y = \dfrac{36}{x^2}$
$y = \dfrac{36}{(3)^2}$
$y = 4$

**13**  $y = \dfrac{k}{x^3}$
$3 = \dfrac{k}{(4)^3}$
$192 = k$
$y = \dfrac{192}{x^3}$

**17**  $S = kx^2$
$54 = k \cdot 3^2$
$6 = k$
$S = 6x^2$
$S = 6(12^2)$
$\quad = 864$
Therefore, the surface is 864 square inches.

## 10.3 Linear and Quadratic Functions

### SEMIPROGRAMMED PROBLEMS

In problems 1–5, find the domain, the range, and the slope and sketch the graph of each linear function.

**1**  $f(x) = -2x + 3$

ℝ (real numbers)

ℝ (real numbers), $-2$

straight

$0, 3$

$\tfrac{3}{2}, 0$

The domain of $f$ is the set _____. The range of $f$ is the set _____. The slope $m =$ \_\_\_\_\_.

Since the graph of a linear function is a _____ line, it is enough to determine the graph by two points.

If $x = 0$, then $f(0) = -2(\underline{\quad}) + 3 = \underline{\quad}$.
If $x = \tfrac{3}{2}$, then $f(\tfrac{3}{2}) = -2(\underline{\quad}) + 3 = \underline{\quad}$.
The graph is

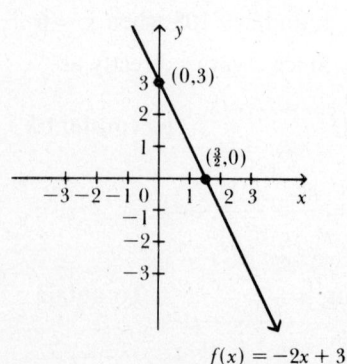

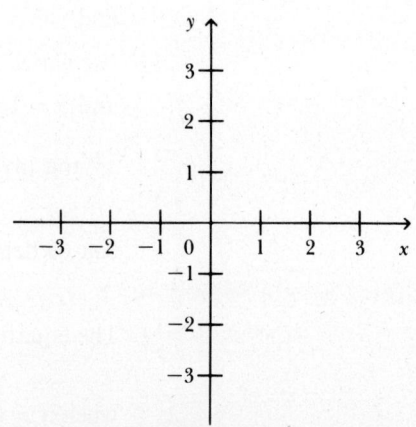

ℝ (real numbers)
ℝ (real numbers), 1
0, −2
2, 0

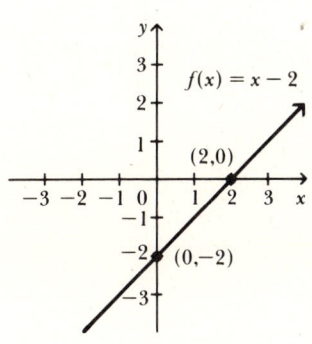

$f(x) = mx + b$

$1 - 3, -3$

$7$

$-3x + 7, \mathbb{R}$

$\mathbb{R}$

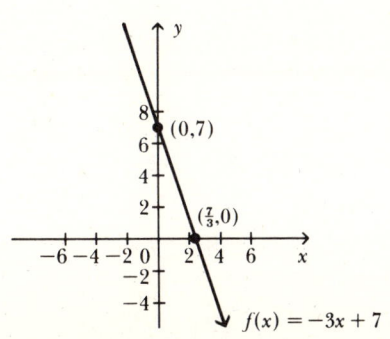

**2** $f(x) = x - 2$

The domain of $f$ is the set _____. The range of $f$ is the set _____. The slope $m = $ ____.
If $x = 0$, then $f(0) = $ ____ $- 2 = $ ____.
If $x = 2$, then $f(2) = $ ____ $- 2 = $ ____.
The graph is

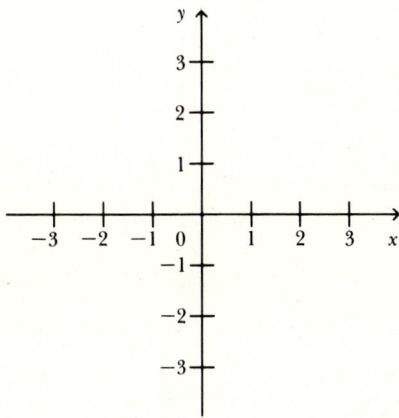

**3** Suppose that $f$ is a linear function whose graph contains $(3, -2)$ and $(1, 4)$, and $y = f(x)$. Since $(3, -2)$ and $(1, 4)$ both lie on the same line, they must satisfy the equation _____. Since
$$m = \frac{y_2 - y_1}{x_2 - x_1} = \frac{4 + 2}{\rule{1cm}{0.4pt}} = \underline{\hspace{1cm}},$$
we have $f(x) = -3x + b$. Using the point $(3, -2)$, the equation becomes $-2 = -3(3) + b$ or $b = $ ____. That is, $f(x) = $ _____. The domain is the set ____. The range is the set ____. The graph is

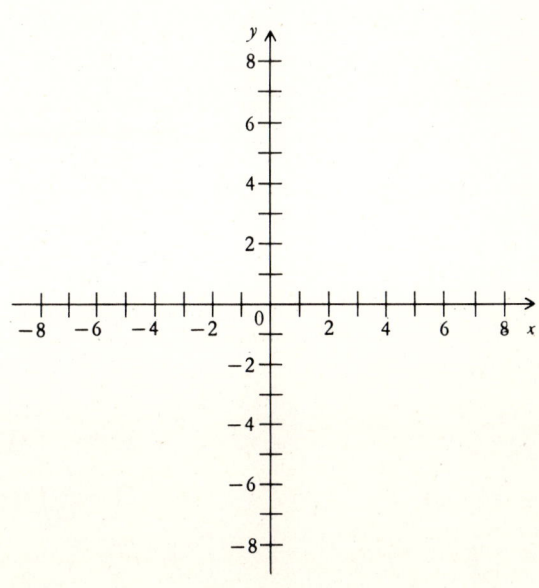

**238** CHAPTER 10 FUNCTIONS AND RELATED CURVES

**4** $f(x) = -1$

The domain of $f$ is the set _____. The range of $f$ is _____. The graph is

ℝ, {−1}

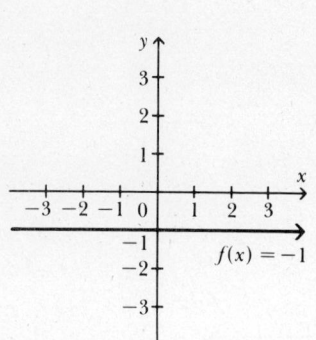
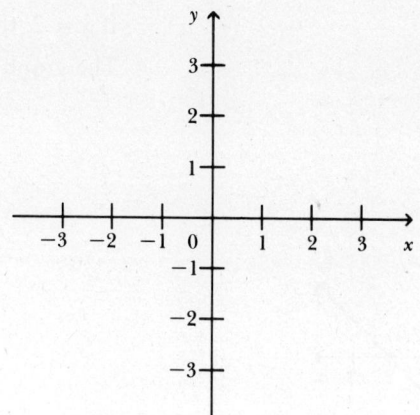

**5** $f(x) = 2$

The domain of $f$ is the set _____. The range of $f$ is _____. The graph is

ℝ {2}

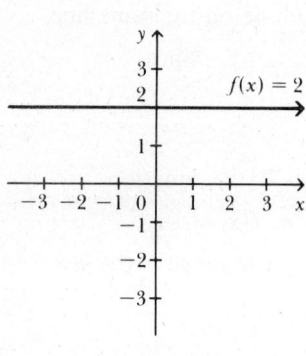
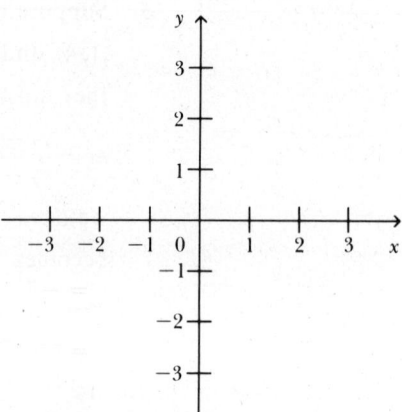

In problems 6–10, find the domain, the range, the $x$ and $y$ intercepts, and the vertex of each quadratic function $f$. Sketch the graph of $f$.

**6** $f(x) = 6 - x - x^2$

The domain of $f$ is the set _____. The $x$ intercepts are _____ and _____. The $y$ intercept is _____. Here, $a =$ _____, $b =$ _____, and $c =$ _____, so that the graph of $f$ opens _____ and

$$h = -\frac{b}{2a} = -\frac{\phantom{xx}}{2(-1)} = \underline{\phantom{xx}}$$

$$k = f\left(-\frac{b}{2a}\right) = \frac{4ac - b^2}{4a} = \frac{4(-1)(6) - (\phantom{xx})^2}{4(-1)} = \underline{\phantom{xx}}, \text{ so that}$$

$(h, k) =$ _____. The graph is

ℝ (real numbers)
−3, 2, 6
−1, −1, 6
downward

−1, −½

−1, 25/4

(−½, 25/4)

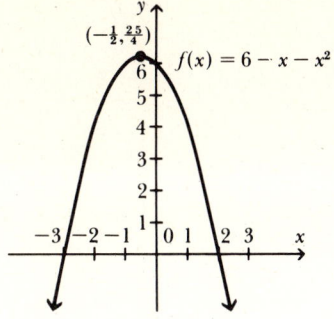

$\le \frac{25}{4}$

$\mathbb{R}$ (real numbers)
$-3, 1, -3, 1$
$2, -3$, upward

$2, -1$

$1, -3, 2, -4$

$(-1, 4)$

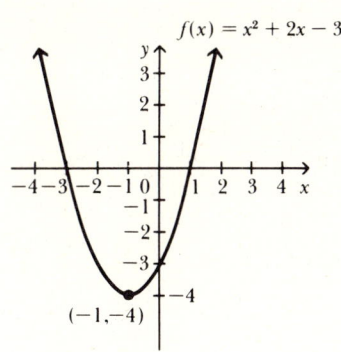

$\ge -4$

$\mathbb{R}$ (real numbers)
$-1, 2, 2, -1$
$1, 2$, downward

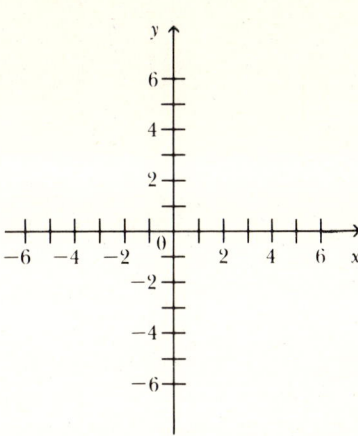

From the graph we see that the range of $f$ is the set of all real numbers _____.

7  $f(x) = x^2 + 2x - 3$
The domain is the set _____. The $x$ intercepts are ____ and ____. The $y$ intercept is ____. Here, $a = $ ____, $b = $ ____, and $c = $ ____, so that the graph of $f$ opens _____ and
$h = -\dfrac{b}{2a} = -\dfrac{\_\_}{2(1)} = \_\_.$
$k = f\left(-\dfrac{b}{2a}\right) = \dfrac{4ac - b^2}{4a} = \dfrac{4(\_\_)(\_\_) - (\_\_)^2}{4(1)} = \_\_,$
so that $(h, k) = $ _____. The graph is

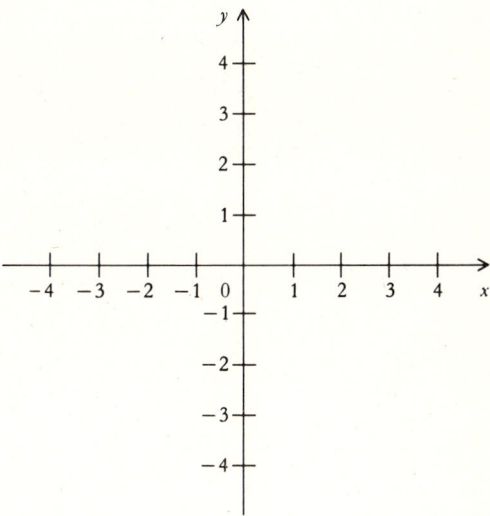

From the graph we can see that the range of $f$ is the set of all real numbers _____.

8  $f(x) = 2 + x - x^2$
The domain is the set _____. The $x$ intercepts are ____ and ____. The $y$ intercept is ____. Here, $a = $ ____, $b = $ ____, and $c = $ ____, so that the graph of $f$ opens _____

1, ½

2, 1, 9/4

(½, 9/4)

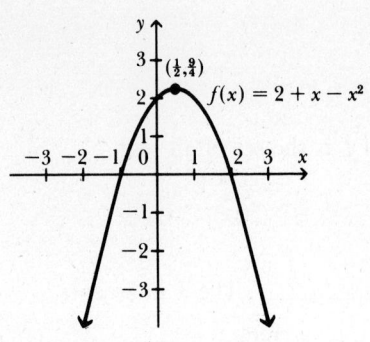

≤ 9/4

ℝ (real numbers)

1, 1, 1, −2

1, upward

1, 1

1, −2, 0

(1, 0)

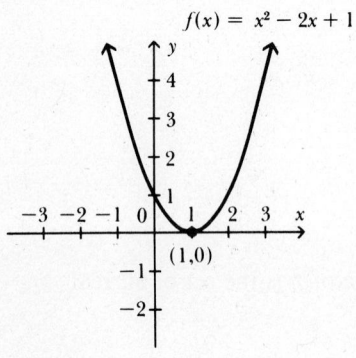

≥ 0

and

$$h = -\frac{b}{2a} = -\frac{\rule{1cm}{0.15mm}}{2(-1)} = \rule{1cm}{0.15mm}.$$

$$k = \frac{4ac - b^2}{4a} = \frac{4(-1)(\rule{0.5cm}{0.15mm}) - (\rule{0.5cm}{0.15mm})^2}{4(-1)} = \rule{1cm}{0.15mm},$$

so that $(h, k) = \rule{1cm}{0.15mm}$. The graph is

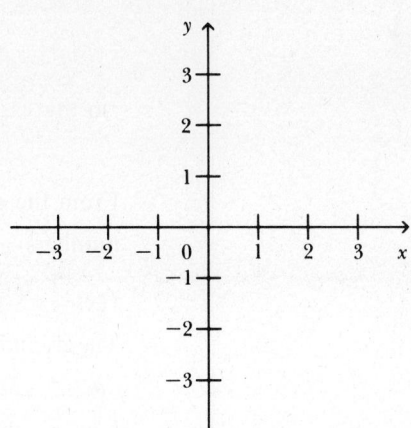

From the graph we can see that the range is the set of all real numbers \rule{1cm}{0.15mm}.

9  $f(x) = x^2 - 2x + 1$

The domain of $f$ is the set \rule{2cm}{0.15mm}. The $x$ intercept is \rule{1cm}{0.15mm}. The $y$ intercept is \rule{1cm}{0.15mm}. Here, $a = $ \rule{1cm}{0.15mm}, $b = $ \rule{1cm}{0.15mm}, and $c = $ \rule{1cm}{0.15mm}, so that the graph opens \rule{1cm}{0.15mm} and

$$h = -\frac{b}{2a} = -\frac{-2}{2(\rule{0.5cm}{0.15mm})} = \rule{1cm}{0.15mm}.$$

$$k = \frac{4ac - b^2}{4a} = \frac{4(1)(\rule{0.5cm}{0.15mm}) - (\rule{0.5cm}{0.15mm})^2}{4(1)} = \rule{1cm}{0.15mm},$$

so that $(h, k) = \rule{1cm}{0.15mm}$. The graph is

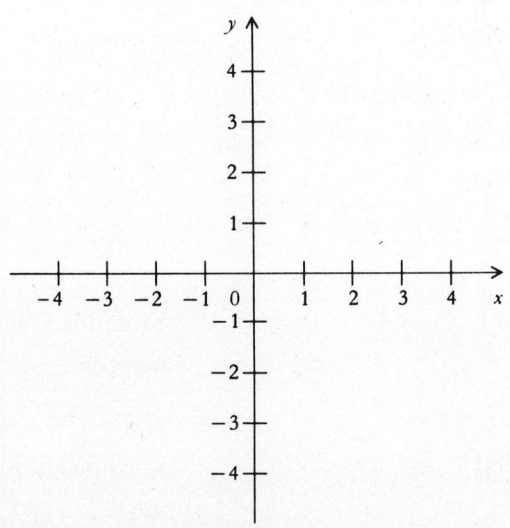

The range is the set of all real numbers \rule{1cm}{0.15mm}.

ℝ (real numbers)
complex
4, 1, 0, 4
upward

0, 0

4, 0, 4

(0, 4)

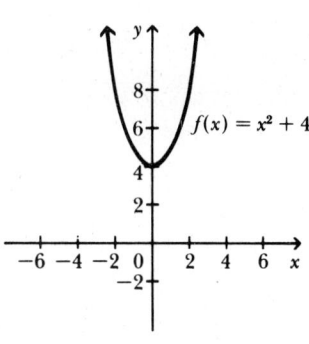

≥4

**10** $f(x) = x^2 + 4$

The domain of $f$ is the set _____. There are no $x$ intercepts, since $x^2 + 4 = 0$ will give _____ solutions. The $y$ intercept is _____. Here, $a = $ _____, $b = $ _____, and $c = $ _____, so that the graph of $f$ opens _____ and

$$h = -\frac{b}{2a} = -\frac{\underline{\phantom{xx}}}{2(1)} = \underline{\phantom{xx}}.$$

$$k = \frac{4ac - b^2}{4a} = \frac{4(1)(\underline{\phantom{xx}}) - (\underline{\phantom{xx}})^2}{4(1)} = \underline{\phantom{xx}},$$

so that $(h, k) = $ _____. The graph is

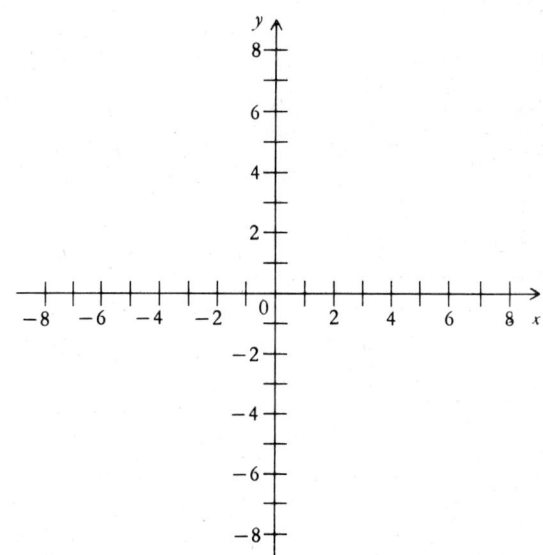

The range is the set of all real numbers _____.


## SOLUTIONS TO SELECTED ODD PROBLEMS   Section 10.3, text pages 474–475

**1** $f(x) = -3x + 5$

The graph of $f$ is the graph of $y = -3x + 5$. The graph is a line with slope $m = -3$. The $y$ intercept = 5, and the $x$ intercept = $\frac{5}{3}$. From the graph, we see that the domain is the set of all real numbers, and that the range is the set of all real numbers.

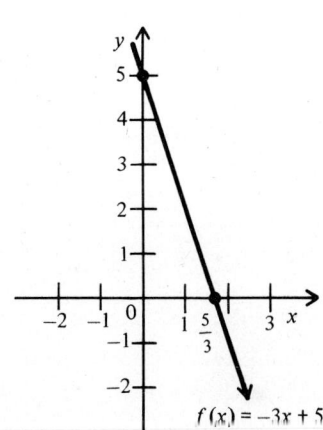

**5** $f(x) = -1$

The graph of $f$ is the graph of $y = -1$, which is a horizontal line with $y$ intercept $= -1$. From the graph, we see that the domain is the set of all real numbers, and that the range is the set $\{-1\}$.

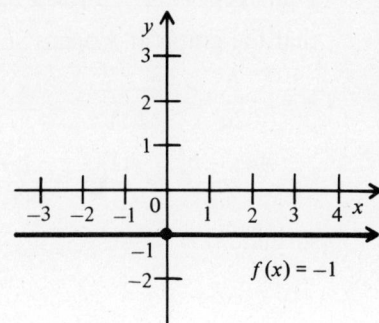

**9** $f(x) = 4x$

The graph of $f$ is the graph of $y = 4x$, which is a line with both $x$ and $y$ intercepts $= 0$, and whose slope is $m = 4$. From the graph, we see that the domain is the set of all real numbers, and that the range is the set of all real numbers.

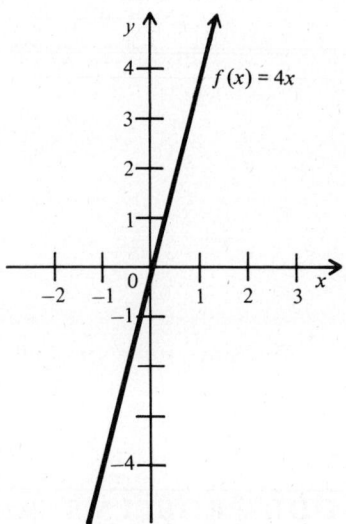

**13** $f(0) = 4$ means $y = 4$ when $x = 0$.
$f(3) = 0$ means $y = 0$ when $x = 3$.
Substituting these values in the general linear function $y = f(x) = mx + b$, we have the system

$$\begin{cases} 4 = m(0) + b \\ 0 = m(3) + b \end{cases}$$

whose solution is $b = 4$ and $m = -\frac{4}{3}$. Therefore, $f(x) = -\frac{4}{3}x + 4$.

**17** For $f(x) = mx + b$,
$2f(x) = 2(mx + b) = 2mx + 2b$
$f(2x) = m(2x) + b = 2mx + b$
so that $2f(x) = f(2x)$
if $2mx + 2b = 2mx + b$
or $b = 0$. Thus, $f(x) = mx$.

21. $f(x) = 2x^2$
Here, $a = 2$, $b = 0$, and $c = 0$,
so $h = -\dfrac{b}{2a} = -\dfrac{0}{2(2)} = 0$, and
$k = f\left(-\dfrac{b}{2a}\right) = \dfrac{4ac - b^2}{4a} = \dfrac{4(2)(0) - (0)^2}{4(2)} = 0$.
The vertex is $(h, k) = (0, 0)$.
Set $x = 0$, so $y = f(0) = 0$ is the $y$ intercept.
Set $y = 0$, so $0 = 2x^2$
or $x = 0$ is the $x$ intercept.

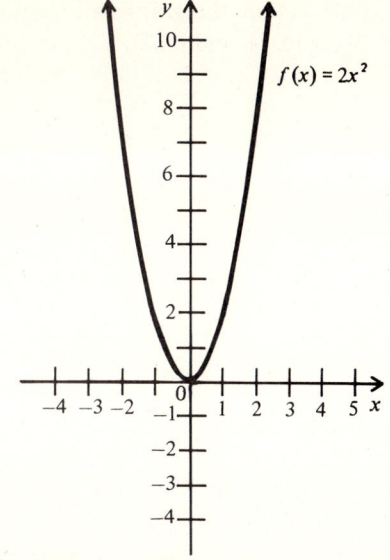

The domain is the set of all real numbers, and the range is the set of all real numbers $y \geq 0$.

25. $f(x) = -x^2 - 2x - 1$
$a = -1$, $b = -2$, $c = -1$
$h = -\dfrac{b}{2a} = -\dfrac{(-2)}{2(-1)} = -1$
$k = \dfrac{4ac - b^2}{4a} = \dfrac{4(-1)(-1) - (-2)^2}{4(-1)} = 0$
The vertex is $(h, k) = (-1, 0)$.
Set $x = 0$, $y = f(0) = -(0)^2 - 2(0) - 1 = -1$ is the $y$ intercept.
Set $y = 0$, and $-x^2 - 2x - 1 = 0$
$x^2 + 2x + 1 = 0$
$(x + 1)^2 = 0$
$x = -1$ is the $x$ intercept.

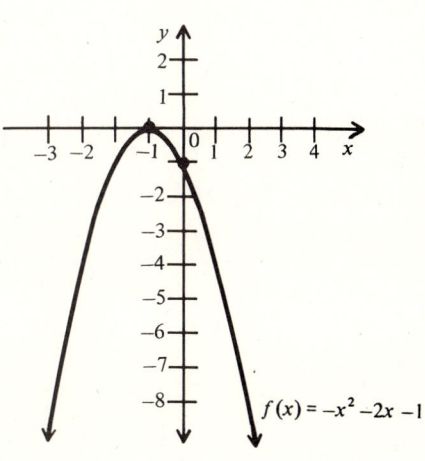

The domain is the set of all real numbers, and the range is the set of all real numbers $y \leq 0$.

29. $f(x) = 2x^2 - 3x$
$a = 2$, $b = -3$, $c = 0$
$h = -\dfrac{b}{2a} = -\dfrac{(-3)}{2(2)} = \dfrac{3}{4}$
$k = \dfrac{4ac - b^2}{4a} = \dfrac{4(2)(0) - (-3)^2}{4(2)} = -\dfrac{9}{8}$
The vertex is $(h, k) = \left(\dfrac{3}{4}, -\dfrac{9}{8}\right)$.
Set $x = 0$, $y = f(0) = 2(0)^2 - 3(0) = 0$ is the $y$ intercept.
Set $y = 0$, and $2x^2 - 3x = 0$
$x(2x - 3) = 0$
$x = 0$ or $x = \dfrac{3}{2}$
are the $x$ intercepts.

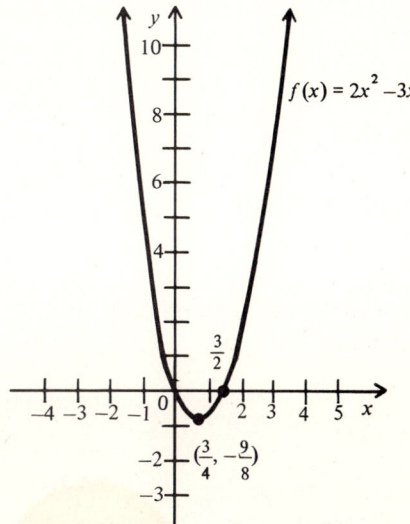

The domain is the set of all real numbers, and the range is the set of all real numbers $y \geq -\dfrac{9}{8}$.

**244** CHAPTER 10 FUNCTIONS AND RELATED CURVES

**33** Since $a = 1$ for each function, the graphs all open upward and their vertices are found by setting $x = 0$ to obtain $(0, -2)$, $(0, -1)$, $(0, 1)$, and $(0, 2)$.

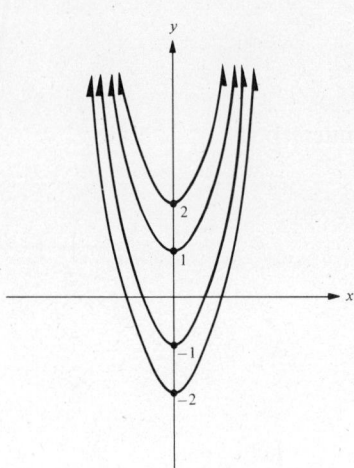

## 10.4 Exponential and Logarithmic Functions

### SEMIPROGRAMMED PROBLEMS

In problems 1–3, find $f(0)$, $f(-1)$, and $f(1)$.

|  | **1** $f(x) = 2^x$ |
|---|---|
| $^0, 1$ | $f(0) = 2^{\rule{1cm}{0.15mm}} = \rule{1cm}{0.15mm}$ |
| $^{-1}, \frac{1}{2}$ | $f(-1) = 2^{\rule{1cm}{0.15mm}} = \rule{1cm}{0.15mm}$ |
| $^1, 2$ | $f(1) = 2^{\rule{1cm}{0.15mm}} = \rule{1cm}{0.15mm}$ |
|  | **2** $f(x) = 5^x$ |
| $^0, 1$ | $f(0) = 5^{\rule{1cm}{0.15mm}} = \rule{1cm}{0.15mm}$ |
| $^{-1}, \frac{1}{5}$ | $f(-1) = 5^{\rule{1cm}{0.15mm}} = \rule{1cm}{0.15mm}$ |
| $^1, 5$ | $f(1) = 5^{\rule{1cm}{0.15mm}} = \rule{1cm}{0.15mm}$ |
|  | **3** $f(x) = (\frac{1}{3})^x$ |
| $^0, 1$ | $f(0) = (\frac{1}{3})^{\rule{1cm}{0.15mm}} = \rule{1cm}{0.15mm}$ |
| $^{-1}, 3$ | $f(-1) = (\frac{1}{3})^{\rule{1cm}{0.15mm}} = \rule{1cm}{0.15mm}$ |
| $^1, \frac{1}{3}$ | $f(1) = (\frac{1}{3})^{\rule{1cm}{0.15mm}} = \rule{1cm}{0.15mm}$ |

In problems 4–7, graph each function. Indicate the domain and the range.

**4** $f(x) = 2^x$

The graph can be drawn by considering specific values of $x$, say, $-2, -1, 0, 1, 2$, to determine $f(-2), f(-1), f(0), f(1),$ and $f(2)$. Thus,

$\frac{1}{4}, \frac{1}{2}, 1, 2$  $f(-2) = \rule{1cm}{0.15mm}, f(-1) = \rule{1cm}{0.15mm}, f(0) = \rule{1cm}{0.15mm}, f(1) = \rule{1cm}{0.15mm}$, and

$4$  $f(2) = \rule{1cm}{0.15mm}$. Use these points to graph $f$. The graph is

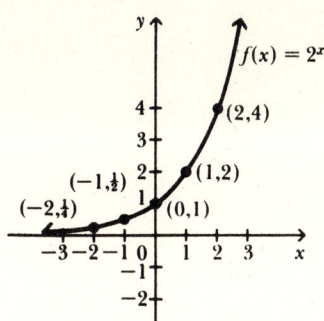

real numbers ℝ
positive

The domain of $f$ is the set of all _____. The range of $f$ is the set of all _____ real numbers.

5  $f(x) = 2^{-x}$

We can locate specific points of the graph by considering specific values of $x$, say, $-2, -1, 0, 1, 2$, to determine $f(-2)$, $f(-1)$, $f(0)$, $f(1)$, and $f(2)$. Thus, $f(-2) =$ ____, $f(-1) =$ ____, $f(0) =$ ____, $f(1) =$ ____, and $f(2) =$ ____. Use these points to graph $f$. The graph is

4, 2, 1
$\frac{1}{2}, \frac{1}{4}$

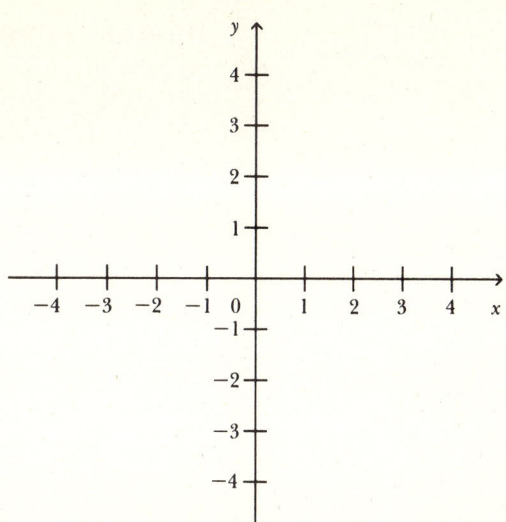

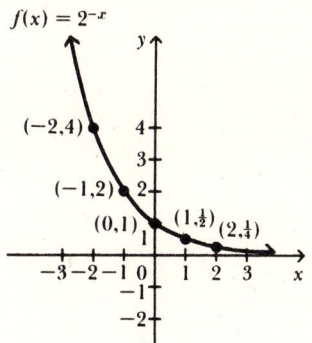

real numbers ℝ
positive

The domain is the set of all _____. The range is the set of all _____ real numbers.

6  $f(x) = -3^x$

Consider specific values of $x$, say, $-2, -1, 0, 1, 2$, to determine $f(-2)$, $f(-1)$, $f(0)$, $f(1)$, and $f(2)$. We obtain $f(-2) =$ ____, $f(-1) =$ ____, $f(0) =$ ____, $f(1) =$ ____, and $f(2) =$ ____. Use these points to

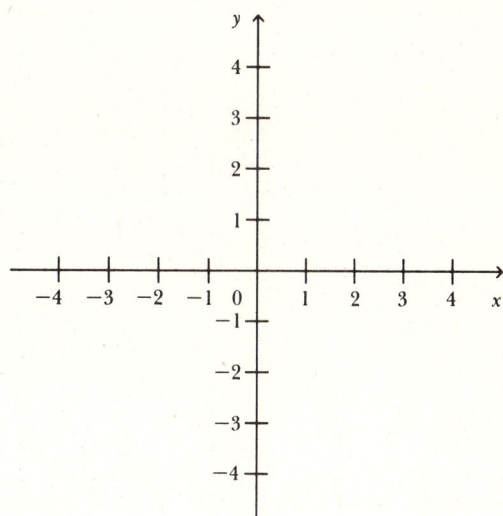

$-\frac{1}{9}, -\frac{1}{3}$
$-1, -3, -9$

**246** CHAPTER 10 FUNCTIONS AND RELATED CURVES

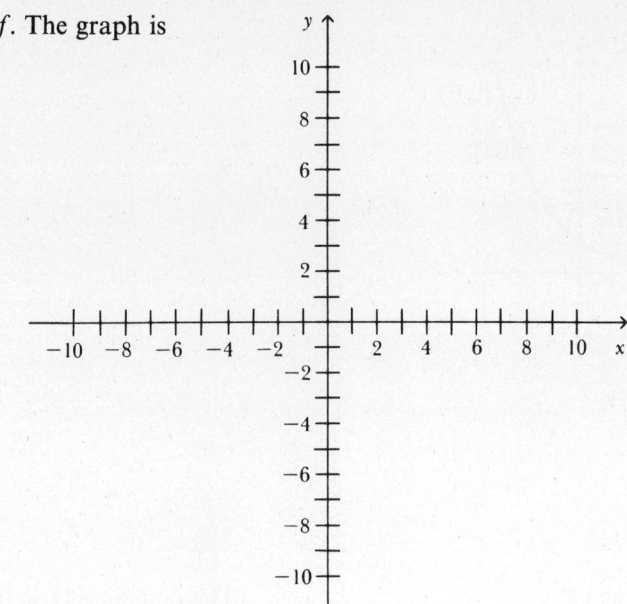

graph $f$. The graph is

real numbers $\mathbb{R}$
negative

The domain is the set of all _____. The range is the set of all _____ real numbers.

**7** $f(x) = -3^{-x}$

Consider specific values of $x$, say, $-2, -1, 0, 1, 2$, to determine $f(-2) = $ _____, $f(-1) = $ _____, $f(0) = $ _____, $f(1) = $ _____, and $f(2) = $ _____. Use these points to graph $f$. The graph is

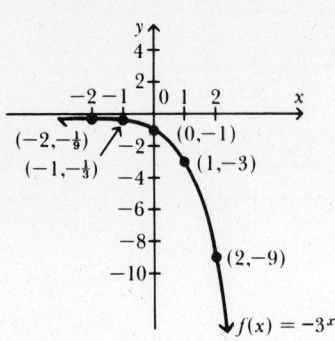

$-9, -3, -1, -\frac{1}{3}$
$-\frac{1}{9}$

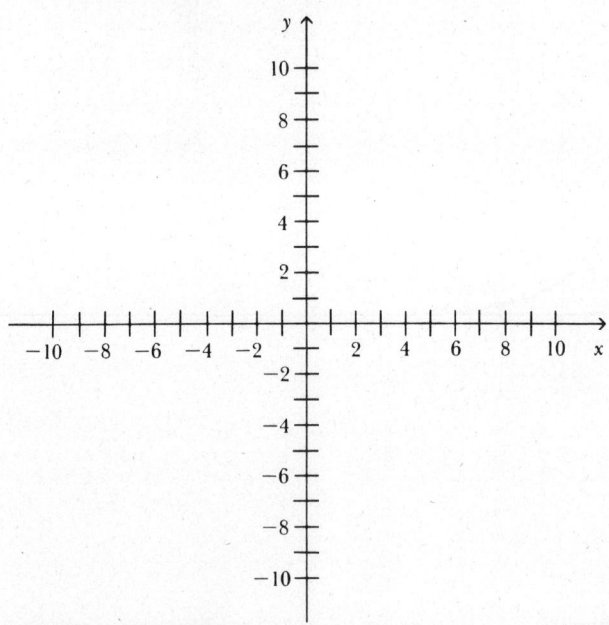

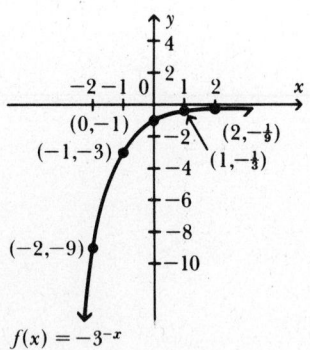

real numbers $\mathbb{R}$
negative

The domain is the set of all _____. The range is the set of all _____ real numbers.

In problems 8–10, graph each function. Indicate the domain and the range.

**8** $f(x) = \log_3 x$

Using the equivalent equation $x = 3^{f(x)}$, and considering specific values of $f(x)$, say, $-2, -1, 0,$ and $1$, we obtain $x = 3^{-2} = $ _____,

$\frac{1}{9}$

$\tfrac{1}{3}$, 1, 3

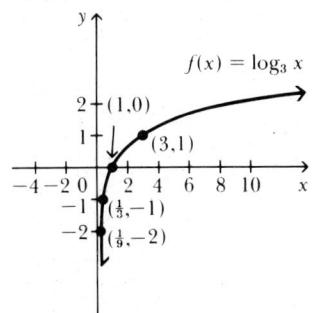

positive

real numbers $\mathbb{R}$

$(\tfrac{1}{3})^{f(x)}$

$\tfrac{1}{9}$

$\tfrac{1}{3}$, 1, 3

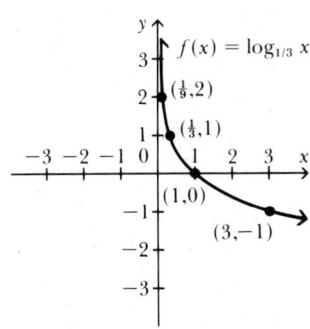

positive

real numbers $\mathbb{R}$

$-f(x)$

$3^{-f(x)}$

$\tfrac{1}{9}, \tfrac{1}{3}$

1, 3

---

$x = 3^{-1} = $ _____, $x = 3^0 = $ _____, and $x = 3^1 = $ _____. Use these points to graph $f$. The graph is

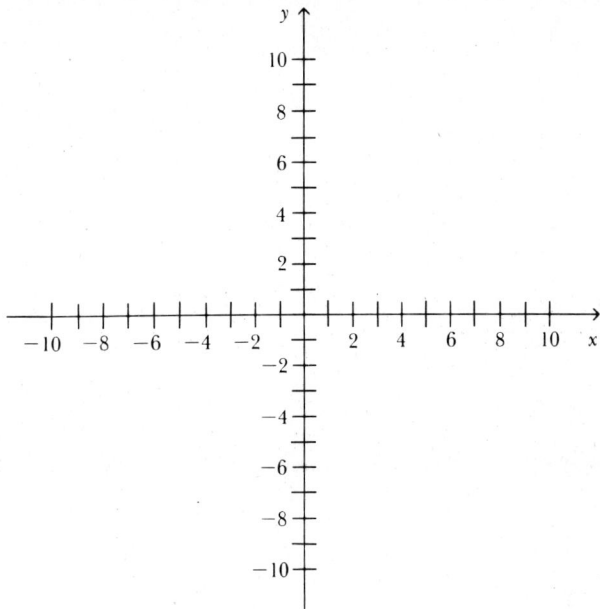

From the graph we see that the domain is the set of all _____ real numbers. The range is the set of all _____.

9  $f(x) = \log_{1/3} x$

Here, the equivalent equation is $x = $ _____. Consider specific values of $f(x)$, say, 2, 1, 0, and $-1$, to obtain $x = (\tfrac{1}{3})^2 = $ _____, $x = (\tfrac{1}{3})^1 = $ _____, $x = (\tfrac{1}{3})^0 = $ _____, and $x = (\tfrac{1}{3})^{-1} = $ _____. Use these points to graph $f$. The graph is

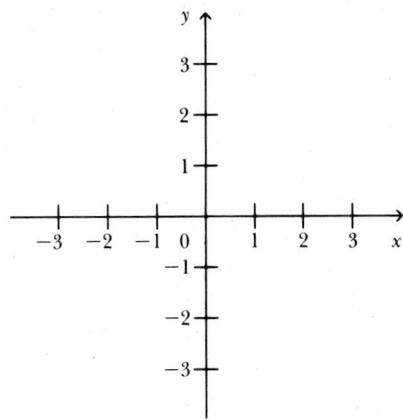

The domain is the set of all _____ real numbers.
The range is the set of all _____.

10  $f(x) = -\log_3 x$

Rewrite this equation as _____ $= \log_3 x$, so that the equivalent equation is $x = $ _____. Consider specific values of $f(x)$, say, 2, 1, 0, and $-1$, to obtain $x = 3^{-2} = $ _____, $x = 3^{-1} = $ _____, $x = 3^0 = $ _____, and $x = 3^{-(-1)} = $ _____. Use these points to graph $f$. The

**248** CHAPTER 10 FUNCTIONS AND RELATED CURVES

graph is

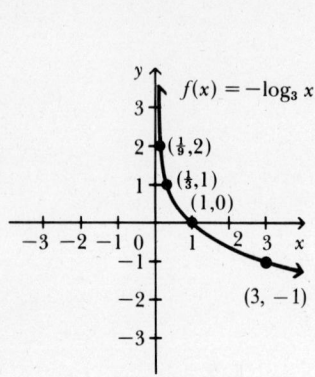

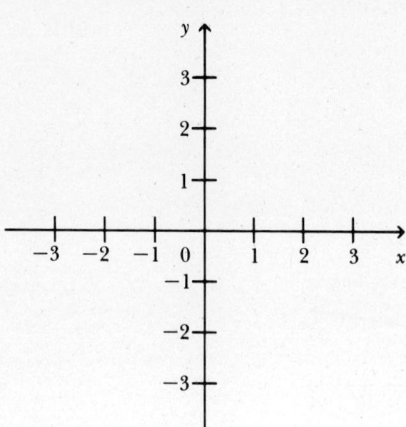

positive
real numbers ℝ

The domain is the set of all _____ real numbers.
The range is the set of all _____.

## SOLUTIONS TO SELECTED ODD PROBLEMS   Section 10.4, text pages 481–482

1

| $x$ | $y = 4^x$ |
|---|---|
| $-3$ | $4^{-3} = \frac{1}{64}$ |
| $-2$ | $4^{-2} = \frac{1}{16}$ |
| $-1$ | $4^{-1} = \frac{1}{4}$ |
| $0$ | $4^0 = 1$ |
| $1$ | $4^1 = 4$ |
| $2$ | $4^2 = 16$ |

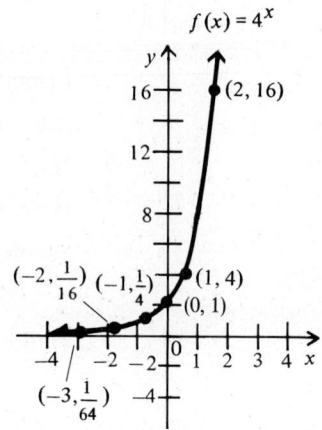

From the graph, we see that the domain is the set of all real numbers, and that the range is the set of all positive real numbers.

5

| $x$ | $y = -(\frac{1}{3})^x$ |
|---|---|
| $-2$ | $-(\frac{1}{3})^{-2} = -9$ |
| $-1$ | $-(\frac{1}{3})^{-1} = -3$ |
| $0$ | $-(\frac{1}{3})^0 = -1$ |
| $1$ | $-(\frac{1}{3})^1 = -\frac{1}{3}$ |
| $2$ | $-(\frac{1}{3})^2 = -\frac{1}{9}$ |

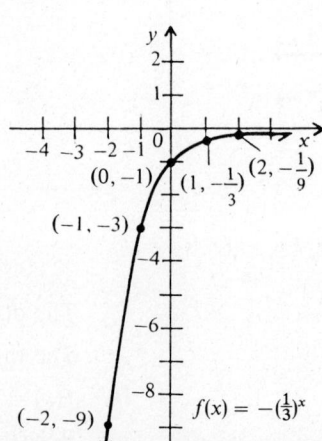

From the graph, we see that the domain is the set of all real numbers, and that the range is the set of all negative real numbers.

| 9 | $x$ | $y = 5(3^x)$ |
|---|---|---|
| | $-1$ | $5(3^{-1}) = \frac{5}{3}$ |
| | $0$ | $5(3^0) = 5$ |
| | $1$ | $5(3^1) = 15$ |

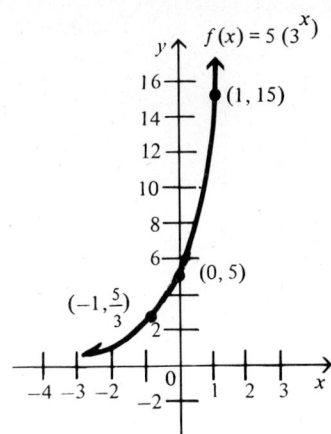

From the graph, we see that the domain is the set of all real numbers, and that the range is the set of all positive real numbers.

13  $f(x) = b^x$ where $b > 0$
$16 = b^2$
$b = \sqrt{16} = 4$

17  $f(x) = b^x$
$f(u - v) = b^{u-v}$
$f(u) \div f(v) = b^u \div b^v$
Since $b^u \div b^v = b^{u-v}$, then $f(u - v) = f(u) \div f(v)$.

21  Let $y = \log_{1/2} x$
so $x = (\frac{1}{2})^y$

| $x = (\frac{1}{2})^y$ | $y$ |
|---|---|
| $(\frac{1}{2})^{-1} = 2$ | $-1$ |
| $(\frac{1}{2})^0 = 1$ | $0$ |
| $(\frac{1}{2})^1 = \frac{1}{2}$ | $1$ |
| $(\frac{1}{2})^2 = \frac{1}{4}$ | $2$ |
| $(\frac{1}{2})^3 = \frac{1}{8}$ | $3$ |

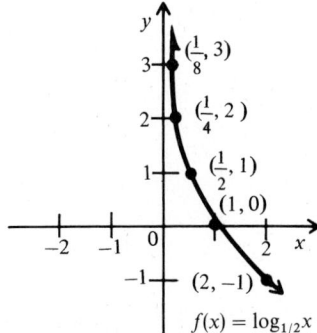

From the graph, we see that the domain is the set of all positive real numbers, and that the range is the set of all real numbers.

25  $f(x) = \log_b x$
$3 = \log_b 8$
$b^3 = 8 = 2^3$
$b = 2$

29  $f(x) = \log_b x$
$\frac{1}{2} = \log_b 3$
$b^{1/2} = 3$
$b = 3^2 = 9$

33  $N = 225e^{0.02t}$
$= 225e^{0.02(15)}$
$= 225e^{0.3}$
$= 304$

## 10.5  Graphs of Special Curves—Conic Sections

### SEMIPROGRAMMED PROBLEMS

In problems 1 and 2, sketch the graph of each circle and find its radius.

1  $x^2 + y^2 = 4$
This equation can be rewritten as $x^2 + y^2 = (\_\_\_)^2$, so that its radius $r = \_\_\_$.

2
2

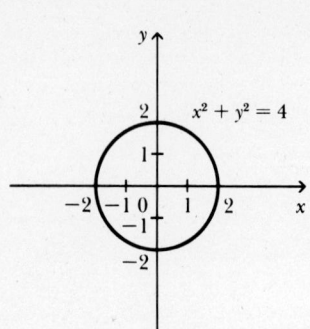

9, 3, 3

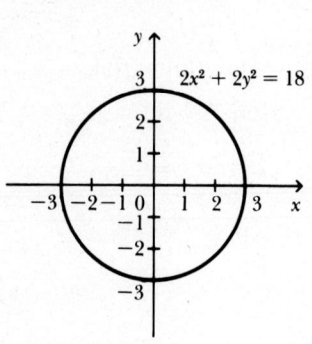

25, 9, 5, 3
(0, −3)

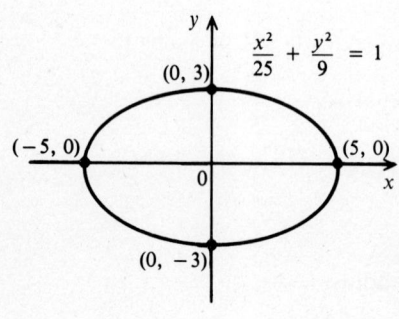

The graph is

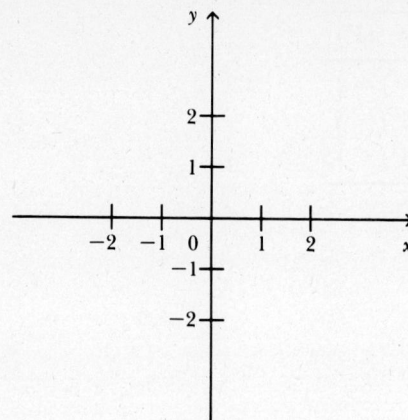

2. $2x^2 + 2y^2 = 18$

Dividing both sides of this equation by 2, we obtain $x^2 + y^2 = $ _____, or $x^2 + y^2 = ($____$)^2$. Thus, the radius $r = $ _____. The graph is

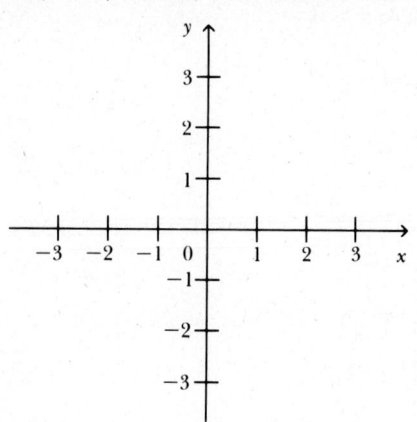

In problems 3–5, sketch the graph of each ellipse and find its vertices.

3. $\dfrac{x^2}{25} + \dfrac{y^2}{9} = 1$

Here $a^2 = $ _____ and $b^2 = $ _____, so that $a = $ _____ and $b = $ _____. The vertices are (5, 0), (−5, 0), (0, 3), and _____. The graph is

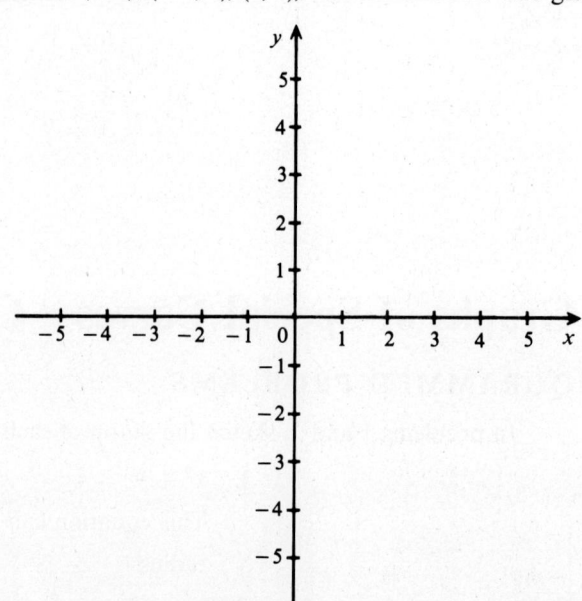

1

1, 1

(0, −1), (0, 1)

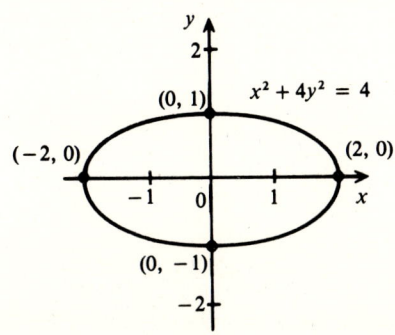

36

9, 9, 3

(0, −3), (0, 3)

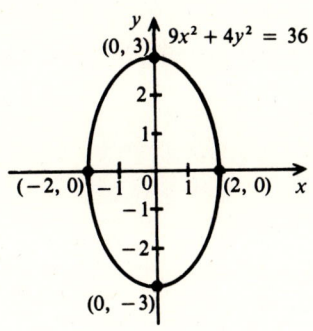

**4** $x^2 + 4y^2 = 4$

Dividing both sides of the equation by 4, we obtain $\dfrac{x^2}{4} + \dfrac{y^2}{\underline{\phantom{xx}}} = 1$, so that $a^2 = 4$, $b^2 = $ \_\_\_\_\_, $a = 2$, and $b = $ \_\_\_\_\_. The vertices are $(-2, 0)$, $(2, 0)$, _____, and _____. The graph is

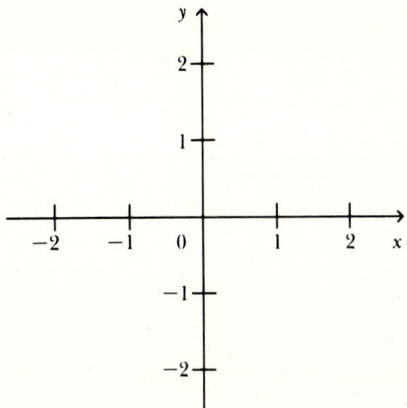

**5** $9x^2 + 4y^2 = 36$

Dividing both sides of the equation by \_\_\_\_\_, we obtain $\dfrac{x^2}{4} + \dfrac{y^2}{\underline{\phantom{xx}}} = 1$, so that $a^2 = 4$, $b^2 = $ \_\_\_\_\_, $a = 2$, and $b = $ \_\_\_\_\_. The vertices are $(-2, 0)$, $(2, 0)$, _____, and \_\_\_\_\_. The graph is

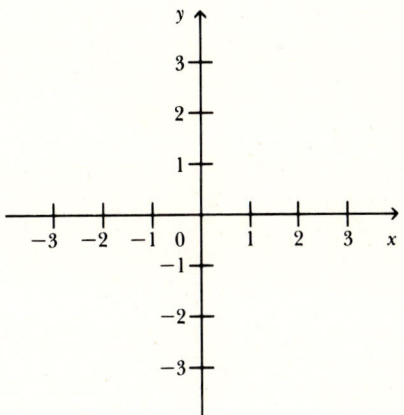

In problems 6 and 7, sketch the graph of each parabola.

$\frac{1}{16}$

right

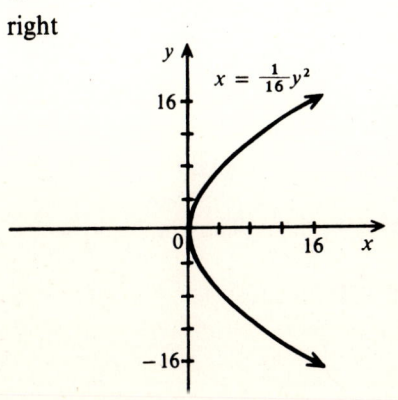

**6** $x = \frac{1}{16}y^2$

Here, $a = $ \_\_\_\_\_, a positive number, so that the graph opens to the \_\_\_\_\_. The graph is

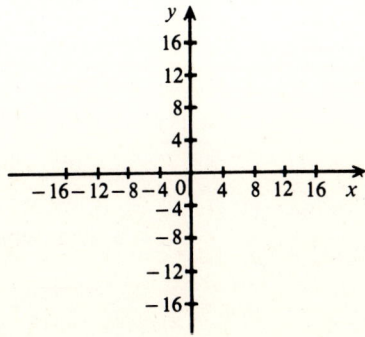

**252** CHAPTER 10 FUNCTIONS AND RELATED CURVES

$-\frac{1}{8}y^2$

$-\frac{1}{8}$, left

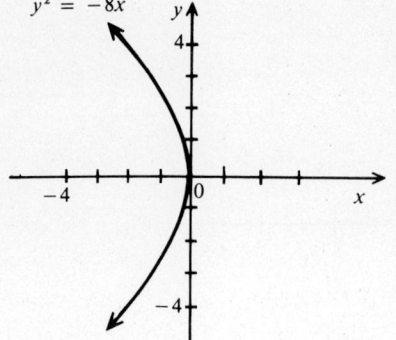

**7** $y^2 = -8x$

Solving the given equation for $x$, we have $x =$ _____, so that $a =$ _____. Hence, the graph opens to the _____. The graph is

In problems 8 and 9, sketch the graph of each hyperbola and find its vertices.

**8** $16x^2 - 9y^2 = 144$

144

9, 9, 3

$(-3, 0), (3, 0)$

Dividing both sides of the equation by _____, we obtain $\dfrac{x^2}{\_\_\_} - \dfrac{y^2}{16} = 1$, so that $a^2 =$ _____ or $a =$ _____. The graph is a hyperbola with horizontal transverse axis and vertices _____. The graph is

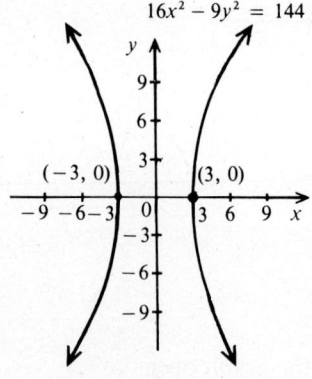

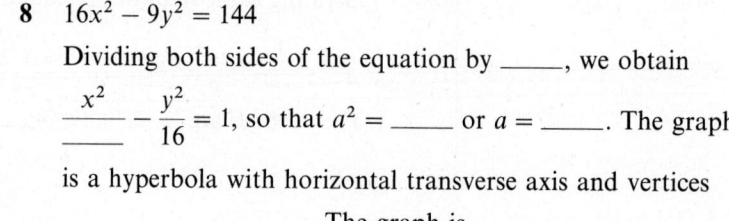

**9** $4y^2 - x^2 = 4$

4

1, 1, 1

$(0, -1), (0, 1)$

Dividing both sides of the equation by _____, we obtain $\dfrac{y^2}{\_\_\_} - \dfrac{x^2}{4} = 1$, so that $b^2 =$ _____ or $b =$ _____. The graph is a hyperbola with vertical transverse axis and vertices _____. The graph is

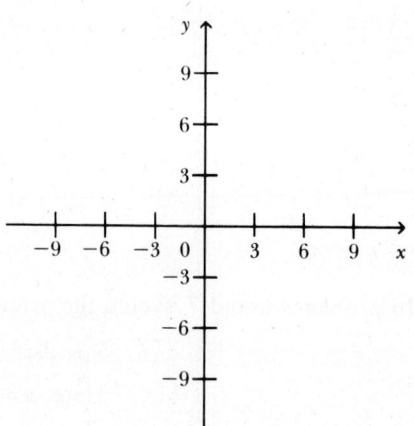

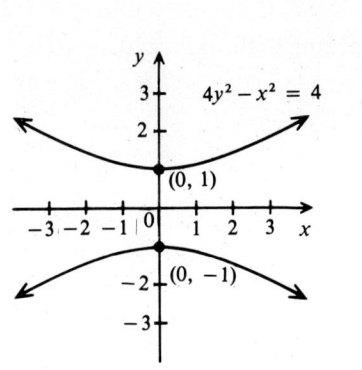
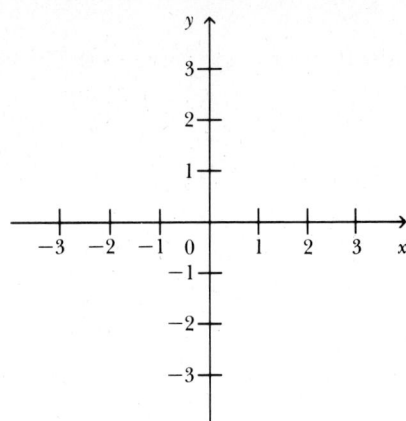

## SOLUTIONS TO SELECTED ODD PROBLEMS  Section 10.5, text page 490

1. $x^2 + y^2 = 9$
$x^2 + y^2 = 3^2$
radius $= 3$

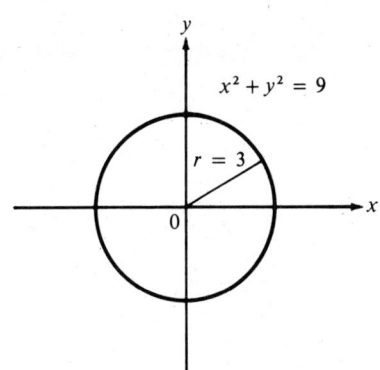

5. $\dfrac{x^2}{3} + \dfrac{y^2}{3} = 2$

$3\left(\dfrac{x^2}{3} + \dfrac{y^2}{3}\right) = 3(2)$

$x^2 + y^2 = 6$
$x^2 + y^2 = (\sqrt{6})^2$
radius $= \sqrt{6}$

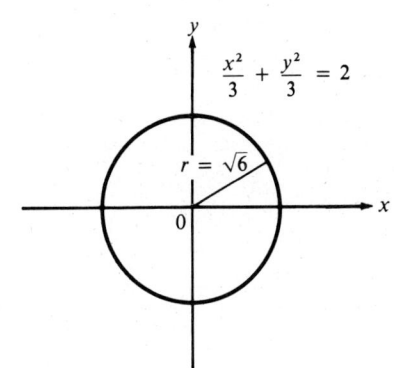

9. $25x^2 + 4y^2 = 100$

$\dfrac{25x^2}{100} + \dfrac{4y^2}{100} = \dfrac{100}{100}$

$\dfrac{x^2}{4} + \dfrac{y^2}{25} = 1$

$a^2 = 4, b^2 = 25$
$a = 2, \ b = 5$
The vertices are $(-2, 0), (2, 0), (0, 5),$ and $(0, -5)$.

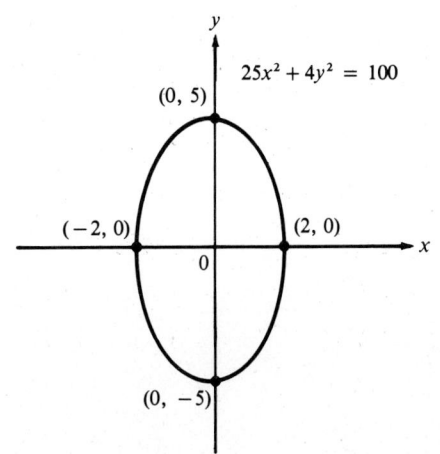

**13**    $x = 2y^2$

The form of the equation is $x = ay^2$ with $a > 0$. Therefore, its graph is a parabola opening to the right with its vertex at the point of origin.

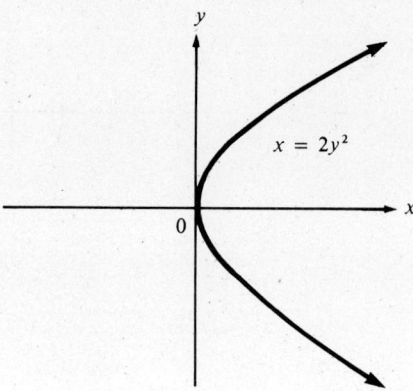

**17**    $4x = 3y^2$
$x = \frac{3}{4}y^2$

since $\frac{3}{4}$ is positive, the graph is a parabola opening to the right.

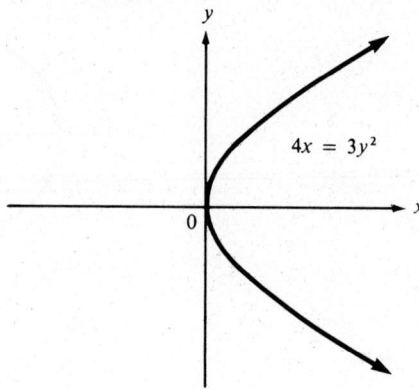

**21**    $\dfrac{y^2}{4} - \dfrac{x^2}{16} = 1$

$b^2 = 4, a^2 = 16$
$b = 2, \quad a = 4$

The graph has a vertical transverse axis with vertices $(0, -2)$ and $(0, 2)$.

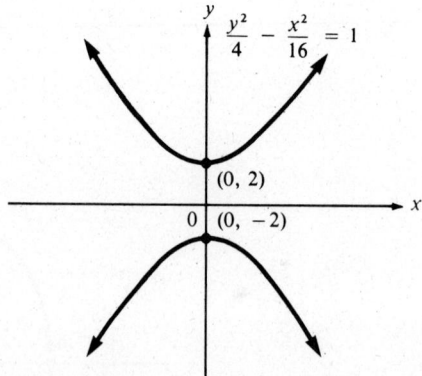

**25**    $y^2 - x^2 = 1$

or $\dfrac{y^2}{1} - \dfrac{x^2}{1} = 1$

so $b^2 = 1, a^2 = 1$
$b = 1, \quad a = 1$

The graph is a hyperbola with a vertical transverse axis and vertices $(0, -1)$ and $(0, 1)$.

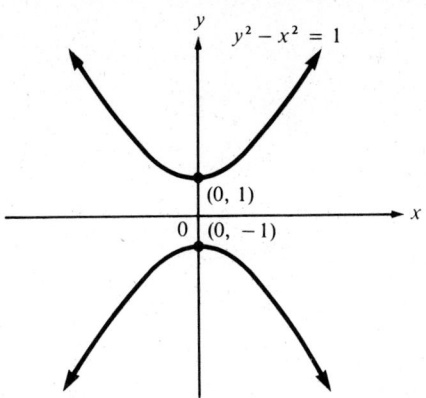

**29**   $x^2 - 4y = 0$
$x^2 = 4y$
$y = \tfrac{1}{4}x^2$
The graph is a parabola with its vertex at the origin, and since $\tfrac{1}{4} > 0$, it opens upward.

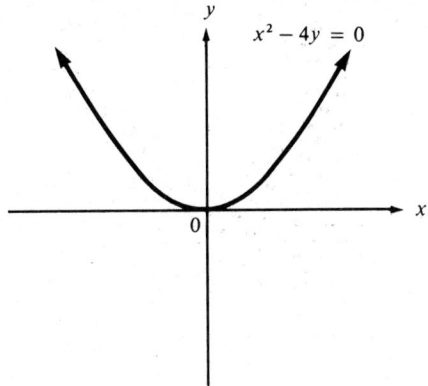

**33**   $36x^2 - 9y^2 = 1$

$$\dfrac{x^2}{\dfrac{1}{36}} - \dfrac{y^2}{\dfrac{1}{9}} = 1$$

$a^2 = \dfrac{1}{36},\ b^2 = \dfrac{1}{9}$

$a = \dfrac{1}{6},\ b = \dfrac{1}{3}$

The graph is a hyperbola with a horizontal transverse axis and vertices $\left(-\dfrac{1}{6}, 0\right)$ and $\left(\dfrac{1}{6}, 0\right)$.

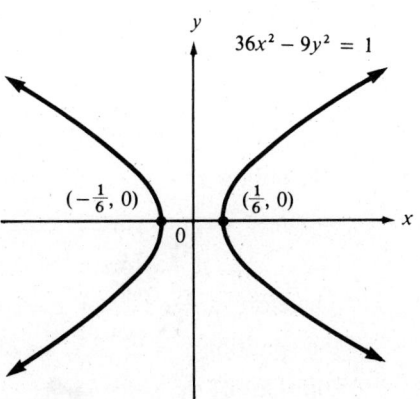

**37** $4x^2 + 4y^2 = 9$

$$\frac{4x^2}{4} + \frac{4y^2}{4} = \frac{9}{4}$$

$$x^2 + y^2 = \left(\frac{3}{2}\right)^2$$

The graph is a circle with center at the origin and radius $= \frac{3}{2}$.

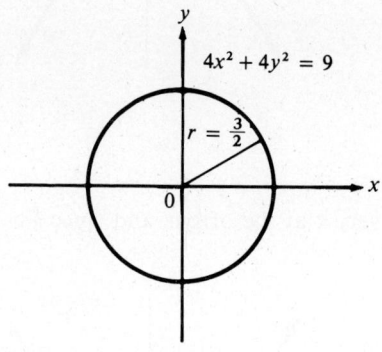

**41** If we use the form

$$\frac{x^2}{a^2} + \frac{y^2}{b^2} = 1$$

with $a = \dfrac{480}{2} = 240$

and $b = \dfrac{280}{2} = 140$,

we have $\dfrac{x^2}{240^2} + \dfrac{y^2}{140^2} = 1$.

## 10.6 Systems Containing Quadratic Equations

### SEMIPROGRAMMED PROBLEMS

In problems 1–7, solve each system of equations and check the solutions by sketching the graphs of the equations and approximating the points of intersection.

**1** $\begin{cases} x + y = 23 \\ x^2 + y^2 = 277 \end{cases}$

$y = 23 - x$

$23 - x$

$x^2 - 23x + 126 = 0$

$x - 9, x - 14, x - 9, x - 14$

$9, 14$

$14, 9$

$(9, 14), (14, 9)$

Solve the equation $x + y = 23$ explicitly for $y$ in terms of $x$ and obtain _____. Substitute $23 - x$ for $y$ in the equation $x^2 + y^2 = 277$ and obtain $x^2 + ($ _____ $)^2 = 277$. This equation is written in standard form as _____. Then $x^2 - 23x + 126 = ($ _____ $)($ _____ $) = 0$, so that _____ $= 0$ or _____ $= 0$. Therefore, $x = $ _____ or $x = $ _____. Replacing $x$ by 9 or $x$ by 14 in the equation $y = 23 - x$, we obtain $y = $ _____ or $y = $ _____. The solutions of the system are _____. The graph is

10.6 SEMIPROGRAMMED PROBLEMS 257

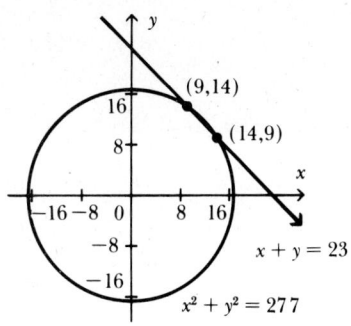

$y = x - 7$

$x - 7$
$2x^2 - 14x - 120 = 0$
$x - 12, x + 5, -5, 12$

$-12, 5$
$(-5, -12), (12, 5)$

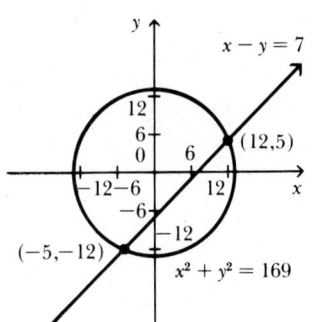

$\dfrac{8 - 3y}{2}$

$\dfrac{8 - 3y}{2}$
$y^2 - 16y + 28 = 0$
$y - 2, y - 14, 2, 14$

2, 1

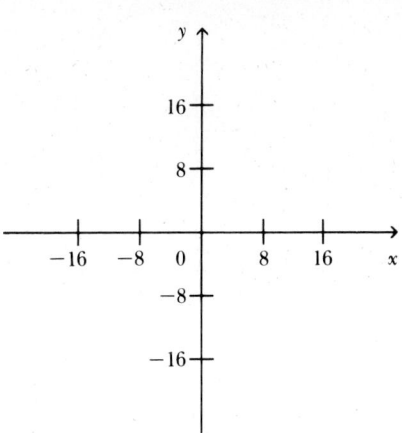

2  $\begin{cases} x - y = 7 \\ x^2 + y^2 = 169 \end{cases}$

Solve $x - y = 7$ explicitly for $y$ and obtain _____.
Substitute $x - 7$ for $y$ in $x^2 + y^2 = 169$ and obtain
$x^2 + (\underline{\phantom{xx}})^2 = 169$. This equation is written in standard
form as _____, so that $x^2 - 7x - 60 = (\underline{\phantom{xx}})(\underline{\phantom{xx}}) = 0$. That is, $x = \underline{\phantom{xx}}$ or $x = \underline{\phantom{xx}}$.
Replacing $x$ by $-5$ or $x$ by $12$ in the equation $y = x - 7$, we obtain
$y = \underline{\phantom{xx}}$ or $y = \underline{\phantom{xx}}$. The solutions of the system are
_____. The graph is

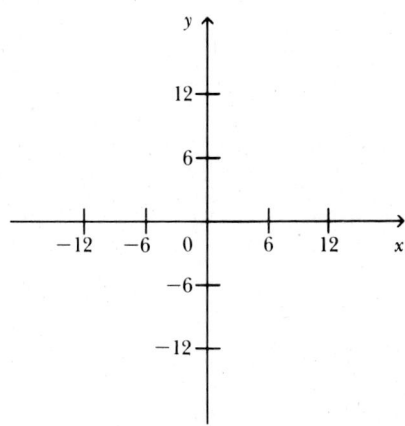

3  $\begin{cases} 2x + 3y = 8 \\ 2x^2 - 3y^2 = -10 \end{cases}$

Solve $2x + 3y = 8$ explicitly for $x$ and obtain $x = \underline{\phantom{xx}}$.

Then, substitute $\dfrac{8 - 3y}{2}$ for $x$ in $2x^2 - 3y^2 = -10$, and obtain

$2\left(\underline{\phantom{xx}}\right)^2 - 3y^2 = -10$. This equation is written in standard

form as _____, so that $y^2 - 16y + 28 = (\underline{\phantom{xx}})(\underline{\phantom{xx}}) = 0$. Then $y = \underline{\phantom{xx}}$ or $y = \underline{\phantom{xx}}$.

When $y = 2$, $x = \dfrac{8 - 3(\underline{\phantom{xx}})}{2} = \underline{\phantom{xx}}$.

14, −17

(1, 2), (−17, 14)

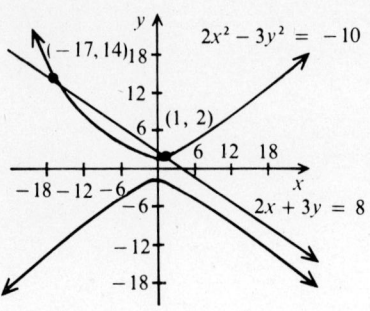

$12x^2 + 21y^2 = 96$

$-12x^2 + 44y^2 = 164$

$65y^2 = 260$

4, −2, 2

−2, −1, 1

2, −1, 1

(1, 2), (−1, 2), (1, −2), (−1, −2)

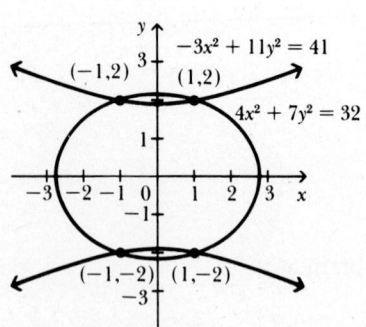

$2x^2 + 4y^2 = 44$

$3y^2 = 27, y^2 = 9$

−3, 3

When $y = 14$,

$$x = \frac{8 - 3(\underline{\phantom{xx}})}{2} = \underline{\phantom{xx}}.$$

Hence, the solutions are _____. The graph is

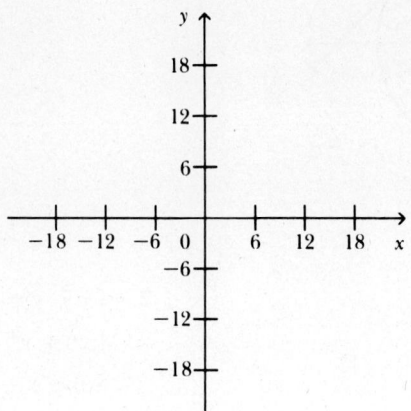

4 $\begin{cases} 4x^2 + 7y^2 = 32 & (1) \\ -3x^2 + 11y^2 = 41 & (2) \end{cases}$

Multiplying equation (1) by 3 and equation (2) by 4, we obtain

$\begin{cases} \underline{\phantom{xxxxxxxxxxxx}} \\ \underline{\phantom{xxxxxxxxxxxx}} \end{cases}$

Adding the two equations, we obtain _____ or

$y^2 = $ ____, so that $y = $ ____ or $y = $ ____.

When we substitute −2 for $y$ in equation (1), we obtain

$4x^2 + 7(\underline{\phantom{xx}})^2 = 32$, so that $x = $ ____ or $x = $ ____.

When we substitute 2 for $y$ in equation (1), we obtain

$4x^2 + 7(\underline{\phantom{xx}})^2 = 32$, so that $x = $ ____ or $x = $ ____.

The solutions of the system are _____.

The graph is

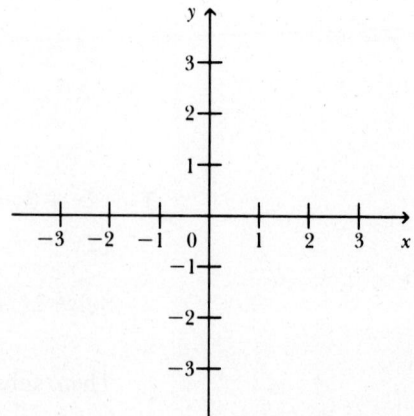

5 $\begin{cases} x^2 + 2y^2 = 22 & (1) \\ 2x^2 + y^2 = 17 & (2) \end{cases}$

Multiplying equation (1) by 2, we obtain _____. (3)

Subtracting equation (2) from equation (3), we obtain

_____ or _____, so that

$y = $ ____ or $y = $ ____.

−3
−2, 2
3
−2, 2
(2, 3), (−2, 3), (2, −3), (−2, −3)

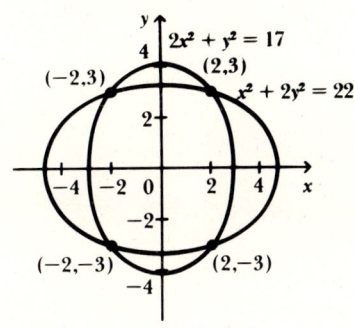

$x^2 = 57 - 9y^2$
$57 - 9y^2 + y^2 = 25$
−2, 2
−2
$-\sqrt{21}, \sqrt{21}$

2, $-\sqrt{21}, \sqrt{21}$
$(-\sqrt{21}, 2), (-\sqrt{21}, -2),$
$(\sqrt{21}, 2), (\sqrt{21}, -2)$

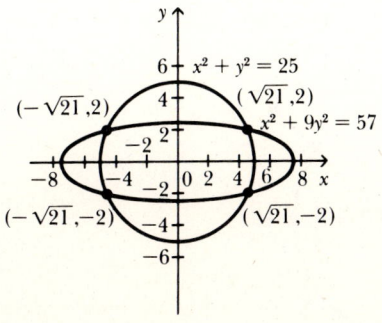

Substituting −3 for $y$ in equation (1), we obtain $x^2 + 2(\underline{\phantom{xx}})^2 = 22$, so that $x = \underline{\phantom{xx}}$ or $x = \underline{\phantom{xx}}$.
Substituting 3 for $y$ in equation (1), we obtain $x^2 + 2(\underline{\phantom{xx}})^2 = 22$, so that $x = \underline{\phantom{xx}}$ or $x = \underline{\phantom{xx}}$.
Hence, the solutions are $\underline{\phantom{xxxxxxxxxxxxxxxxxxxx}}$. The graph is

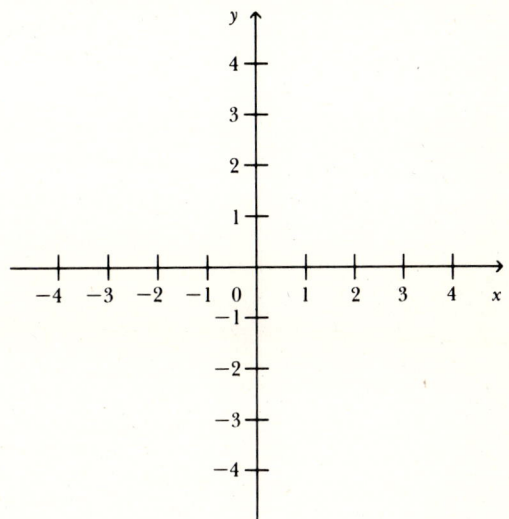

6 $\begin{cases} x^2 + 9y^2 = 57 & (1) \\ x^2 + y^2 = 25 & (2) \end{cases}$

Solving equation (1) explicitly for $x^2$, we obtain $\underline{\phantom{xxxxxxxxx}}$. Substituting $57 - 9y^2$ for $x^2$ in equation (2), we obtain $\underline{\phantom{xxxxxxxxx}}$, so that $-8y^2 = -32$ or $y^2 = 4$.
Therefore, $y = \underline{\phantom{xx}}$ or $y = \underline{\phantom{xx}}$.
Substituting −2 for $y$ in equation (2), we obtain $x^2 + (\underline{\phantom{xx}})^2 = 25$, so that $x = \underline{\phantom{xxx}}$ or $x = \underline{\phantom{xxx}}$.
Now, substituting 2 for $y$ in equation (2), we obtain $x^2 + (\underline{\phantom{xx}})^2 = 25$, so that $x = \underline{\phantom{xxx}}$ or $x = \underline{\phantom{xxx}}$.
The solutions are $\underline{\phantom{xxxxxxxxxx}}$
$\underline{\phantom{xxxxxxxxx}}$. The graph is

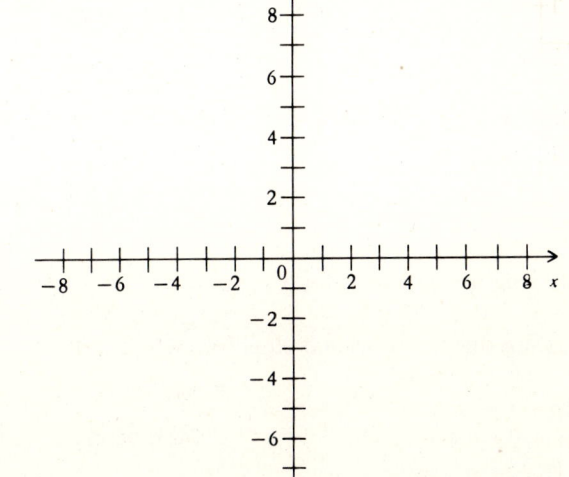

**260** CHAPTER 10 FUNCTIONS AND RELATED CURVES

$9x^2 + 9y^2 = 36$

$16y^2 = 189$, $y^2 = \frac{189}{16}$

$\frac{-3\sqrt{21}}{4}, \frac{3\sqrt{21}}{4}$

$x^2 = \frac{-125}{16}$

$\frac{-5\sqrt{5}i}{4}, \frac{5\sqrt{5}i}{4}$

$x^2 = \frac{-125}{16}, \frac{-5\sqrt{5}i}{4}, \frac{5\sqrt{5}i}{4}$

$\left(\frac{5\sqrt{5}i}{4}, \frac{3\sqrt{21}}{4}\right)$,

$\left(\frac{5\sqrt{5}i}{4}, \frac{-3\sqrt{21}}{4}\right)$,

$\left(\frac{-5\sqrt{5}i}{4}, \frac{3\sqrt{21}}{4}\right)$,

$\left(\frac{-5\sqrt{5}i}{4}, \frac{-3\sqrt{21}}{4}\right)$

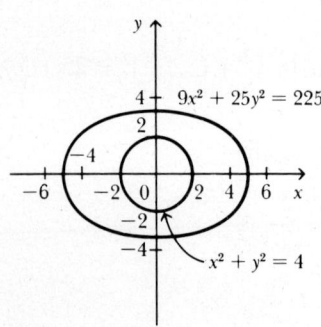

7 $\begin{cases} 9x^2 + 25y^2 = 225 & (1) \\ x^2 + y^2 = 4 & (2) \end{cases}$

Multiplying equation (2) by 9, we obtain _____. (3)

Subtracting equation (3) from equation (1), we obtain

_____ or _____.

Then $y =$ _____ or $y =$ _____.

Substituting $\frac{-3\sqrt{21}}{4}$ for $y$ in equation (2), we obtain _____,

so that $x =$ _____ or $x =$ _____.

Now, substituting $\frac{3\sqrt{21}}{4}$ for $y$ in equation (2), we obtain

_____, so that $x =$ _____ or $x =$ _____.

The solutions are _____

_____

_____

_____. The graph is

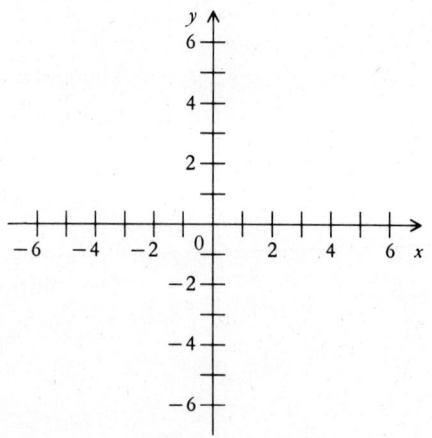

## SOLUTIONS TO SELECTED ODD PROBLEMS   Section 10.6, text pages 495–496

**1** From the graph, we see that there are two solutions.

$\begin{cases} x - y = 1 \\ x^2 + y^2 = 5 \end{cases}$

We solve the first equation for $x$.

$x = y + 1$

Substituting this in the second equation, we obtain

$(y + 1)^2 + y^2 = 5$

$y^2 + 2y + 1 + y^2 = 5$

$2y^2 + 2y - 4 = 0$

$y^2 + y - 2 = 0$

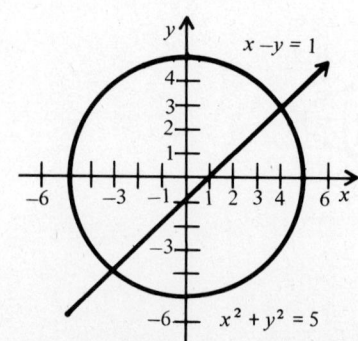

$(y + 2)(y - 1) = 0$, so that
$y = -2 \quad$ or $\quad y = 1$
$x = -2 + 1 \quad\quad x = 1 + 1$
$\phantom{x} = -1 \quad\quad\quad\quad = 2$
Therefore, the solutions are $(-1, -2)$ and $(2, 1)$.

**5**  From the graph, we see that there are no real solutions.
$\begin{cases} 3x + 2y = 1 \\ 3x^2 - y^2 = -4 \end{cases}$

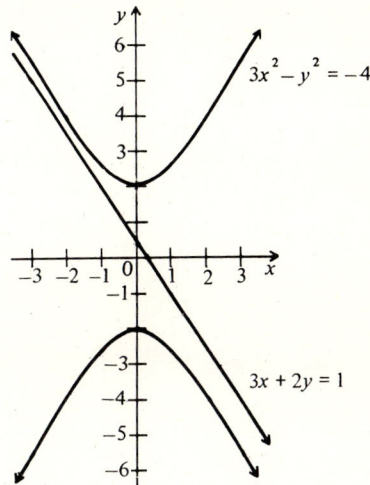

$3x + 2y = 1$
$\quad 2y = 1 - 3x$
$\quad\phantom{2}y = \dfrac{1 - 3x}{2}$

So $3x^2 - \left(\dfrac{1 - 3x}{2}\right)^2 = -4$

$12x^2 - (9x^2 - 6x + 1) = -16$
$\quad\quad\quad 3x^2 + 6x + 15 = 0$
$\quad\quad\quad\phantom{3}x^2 + 2x + 5 = 0$
$\quad\quad\quad\phantom{3}x^2 + 2x + 1 = -4$
$\quad\quad\quad\quad\quad (x + 1)^2 = -4$
$\quad\quad\quad\quad\quad\phantom{(x + 1)^2} x + 1 = \pm\sqrt{-4}$
$\quad\quad\quad\quad\quad\quad\quad\quad x = -1 \pm 2i$

For $x = -1 - 2i$, $y = \dfrac{1 - 3(-1 - 2i)}{2} = \dfrac{4 + 6i}{2} = 2 + 3i$

For $x = -1 + 2i$, $y = \dfrac{1 - 3(-1 + 2i)}{2} = \dfrac{4 - 6i}{2} = 2 - 3i$

Therefore, the complex solutions are $(-1 + 2i, 2 - 3i)$ and $(-1 - 2i, 2 + 3i)$.

**9**  $\begin{cases} 2x + 3y = 7 \\ x^2 + y^2 + 4y + 4 = 0 \end{cases}$

$2x + 3y = 7$
$\quad 2x = 7 - 3y$
$\quad\phantom{2}x = \dfrac{7 - 3y}{2}$

$\left(\dfrac{7 - 3y}{2}\right)^2 + y^2 + 4y + 4 = 0$
$49 - 42y + 9y^2 + 4y^2 + 16y + 16 = 0$
$\quad\quad\quad\quad 13y^2 - 26y + 65 = 0$
$\quad\quad\quad\quad\phantom{13}y^2 - 2y + 5 = 0$
$\quad\quad\quad\quad\phantom{13}y^2 - 2y + 1 = -4$
$\quad\quad\quad\quad\quad\quad (y - 1)^2 = -4$
$\quad\quad\quad\quad\quad\quad\phantom{(y - 1)^2}y - 1 = \pm\sqrt{-4}$
$\quad\quad\quad\quad\quad\quad\quad\quad\quad y = 1 \pm 2i$

For $y = 1 - 2i$, $x = \dfrac{7 - 3(1 - 2i)}{2}$
$\quad\quad\quad\quad\quad\quad x = 2 + 3i$

For $y = 1 + 2i$, $x = \dfrac{7 - 3(1 + 2i)}{2}$
$\quad\quad\quad\quad\quad\quad x = 2 - 3i$

The solutions are $(2 + 3i, 1 - 2i)$ and $(2 - 3i, 1 + 2i)$.

**13**  $\begin{cases} x - y^2 = 0 \\ x^2 + 2y^2 = 24 \end{cases}$

$x - y^2 = 0$
$\quad\phantom{x - } x = y^2$
$x^2 + 2(x) = 24$
$x^2 + 2x - 24 = 0$
$(x + 6)(x - 4) = 0$
$x + 6 = 0 \quad$ or $\quad x - 4 = 0$
$\quad x = -6 \quad\quad\quad\quad x = 4$
For $x = 4$, $y^2 = 4$
$\quad\quad\quad\quad\phantom{x = 4,} y = \pm 2$
For $x = -6$, $y^2 = -6$
$\quad\quad\quad\quad\phantom{x = -6,} y = \pm\sqrt{-6}$
$\quad\quad\quad\quad\quad\phantom{x = -6, y} = \pm\sqrt{6}i$

The solutions are $(4, 2)$, $(4, -2)$, $(-6 - \sqrt{6}i)$, and $(-6, \sqrt{6}i)$.

**17** $\begin{cases} x^2 + 9y^2 = 33 \\ x^2 + y^2 = 25 \end{cases}$ (subtract)

$$8y^2 = 8$$
$$y^2 = 1$$
$$y = \pm 1$$
$$x^2 + 1 = 25$$
$$x^2 = 24$$
$$x = \pm\sqrt{24}$$
$$x = \pm 2\sqrt{6}$$

The solutions are $(-2\sqrt{6}, 1)$, $(-2\sqrt{6}, -1)$, $(2\sqrt{6}, 1)$, and $(2\sqrt{6}, -1)$.

**21** $\begin{cases} 2x^2 - 3y^2 = 20 \\ x^2 + 2y = 20 \end{cases}$

or multiplying the second equation by 2,

$\begin{cases} 2x^2 - 3y^2 = 20 \\ 2x^2 + 4y = 40 \end{cases}$ (subtract)

$$3y^2 + 4y = 20$$
$$3y^2 + 4y - 20 = 0$$
$$(3y + 10)(y - 2) = 0$$
$$3y + 10 = 0 \quad \text{or} \quad y - 2 = 0$$
$$y = -\frac{10}{3} \qquad y = 2$$

For $y = -\frac{10}{3}$, $x^2 + 2\left(-\frac{10}{3}\right) = 20$

$$x^2 = \frac{80}{3}$$
$$x = \pm\sqrt{\frac{80}{3}}$$
$$= \pm\frac{4\sqrt{15}}{3}$$

For $y = 2$, $x^2 + 2(2) = 20$
$$x^2 = 16$$
$$x = \pm 4$$

The solutions are $(4, 2)$, $(-4, 2)$, $\left(\frac{4\sqrt{15}}{3}, -\frac{10}{3}\right)$, and $\left(-\frac{4\sqrt{15}}{3}, -\frac{10}{3}\right)$.

**25.** $\begin{cases} x^2 + 4y = 8 \\ x^2 + y^2 = 5 \end{cases}$ (subtract)

$$y^2 - 4y = -3$$
$$y^2 - 4y + 3 = 0$$
$$(y - 1)(y - 3) = 0$$
$$y - 1 = 0 \quad \text{or} \quad y - 3 = 0$$
$$y = 1 \qquad y = 3$$

For $y = 1$, $x^2 + 4(1) = 8$
$$x^2 = 4$$
$$x = \pm 2$$

For $y = 3$, $x^2 + 4(3) = 8$
$$x^2 = -4$$
$$x = \pm 2i$$

The solutions are $(2i, 3)$, $(-2i, 3)$, $(-2, 1)$, and $(2, 1)$.

**29.** $\begin{cases} x^2 + y^2 = 25 \\ (x - 5)^2 + y^2 = 9 \end{cases}$ (subtract)

$$x^2 - (x - 5)^2 = 16$$
$$x^2 - (x^2 - 10x + 25) = 16$$
$$10x - 25 = 16$$
$$x = \frac{41}{10}$$

$$\left(\frac{41}{10}\right)^2 + y^2 = 25$$
$$y^2 = 25 - \frac{1681}{100}$$
$$y^2 = \frac{819}{100}$$
$$y = \pm\sqrt{\frac{819}{100}}$$
$$y = \pm\frac{3\sqrt{91}}{10}$$

The solutions are $\left(\frac{41}{10}, \frac{3\sqrt{91}}{10}\right)$ and $\left(\frac{41}{10}, -\frac{3\sqrt{91}}{10}\right)$.

**33** Let $x$ = amount lent
$y$ = interest rate
Using the formula $I = Prt$, we have the system:

$$\begin{cases} 170 = x\left(\dfrac{y}{100}\right)(1) \\ 238 = x\left(\dfrac{y+1}{100}\right)(1) \end{cases}$$

or $\begin{cases} xy = 17{,}000 \\ xy + x = 23{,}800 \end{cases}$ (subtract)

$\phantom{xx}x = 6{,}800$

so $6{,}800y = 17{,}000$
$\phantom{xxxx}y = 2.5$

The amount lent was $6,800, and the interest rate was 2.5%.

## CUMULATIVE REVIEW PROBLEM SET  Chapters 1–10

**1** Evaluate $3x^2 - 2xy - 5y^3$ for $x = -1$, $y = -2$.

**2** Perform the indicated operations.
  (a) $(x^2 - 7)(x^2 + 7)$
  (b) $(3x^2 - 2x + 7) - (x^2 - 3x - 4)$
  (c) $(x^4 - 3x^3 + 7x - 5) \div (x - 1)$

**3** Factor each expression.
  (a) $x^2(y + 2z) - 4(y + 2z)$
  (b) $8x^2 - 2x - 15$
  (c) $x^4 - 27x$

**4** Perform the indicated operations.
  (a) $\dfrac{4y^2 - 64}{2y^2 - 8y} \div \dfrac{y + 4}{y - 4}$
  (b) $\dfrac{7}{x^2 + 2x - 3} + \dfrac{3}{x^2 + 8x + 15}$

**5** Simplify $\dfrac{\dfrac{2}{x} - \dfrac{3}{y}}{\dfrac{4y}{x} - \dfrac{9x}{y}}$.

**6** Solve each equation.
  (a) $3(y - 2) - 2(3 - 2y) = 7 - (5 - 4y)$
  (b) $\dfrac{2}{x^2 - 4} + \dfrac{4}{x + 2} = \dfrac{x}{x^2 - 4}$

**7** Solve the formula $A = P + Prt$ for $P$.

**8** Solve each inequality.
  (a) $7 + 2x - 5(2x - 3) \le 4x + 13$
  (b) $|2x - 3| > 4$

**9** Graph each solution on a number line.
  (a) $\{x \mid x \ge -3\} \cap \{x \mid x \le 4\}$
  (b) $\{x \mid x > 5\} \cup \{x \mid x \le 0\}$

**10** Rewrite each expression without negative exponents and simplify.
  (a) $\left[\dfrac{x^4 y^{-3} z^{-2}}{(xyz)^{-1}}\right]^{-1}$
  (b) $\left(\dfrac{-x^3 y^{-6/5}}{8}\right)^{-5/3}$

**11** Rationalize the denominator of each expression.
  (a) $\dfrac{4}{\sqrt{28}}$
  (b) $\dfrac{\sqrt{x} + 3}{\sqrt{x} - 2}$

**12** Perform the indicated operations.
  (a) $(3 - 2i) - (2 - i)$
  (b) $(4 + 5i)(6 - 2i)$

**13** Solve each equation.
  (a) $2x^2 - 3x + 7 = 0$
  (b) $\sqrt{5x - 4} = 2 - \sqrt{x}$
  (c) $y^{2/3} = 16$

**14** A rectangle whose dimensions are 15 feet and 20 feet has its width and length increased by the same amount. If the area of the new rectangle is 414 square feet, find the amount of the increase of the length and width.

**15** Solve each inequality and graph its solution set.

(a) $2x^2 + x - 1 < 0$

(b) $\dfrac{x - 3}{2x + 5} \geq 0$

**16** Find the distance between the points $(\tfrac{1}{2}, 3)$ and $(2, -1)$.

**17** Find the center and radius of the circle $x^2 + y^2 - 4x + 10y + 3 = 0$.

**18** Find the slope of the line through the points $(-1, 3)$ and $(4, -5)$ and write the equation of this line in slope-intercept form.

**19** Solve each system.

(a) $\begin{cases} 2x - 3y = 2 \\ x + 2y = 8 \end{cases}$

(b) $\begin{cases} 3x + y - z = 6 \\ x + 2y + 3z = 2 \\ 2x - y + 4z = -4 \end{cases}$

**20** Sketch the graph of each system.

(a) $\begin{cases} 2x - y < 4 \\ 3x + 4y \geq 12 \end{cases}$

(b) $\begin{cases} x \geq 0, y \geq 0 \\ 2x + 7y \leq 14 \\ 5x + 2y \leq 10 \end{cases}$

**21** Solve each equation.

(a) $4^{2-3x} = 8^{x-1}$

(b) $\log_3(2x - 1) = 3$

**22** Use the properties of logarithms to write each expression as a single logarithm.

(a) $3 \log_b x + 2 \log_b y^3$

(b) $\log \sqrt{x} - \log 3$

**23** Let $f(x) = \sqrt{25 - x^2}$ and find:

(a) $f(-5)$

(b) $[f(0)]^2$

**24** Find the domain of each function.

(a) $g(x) = \dfrac{3}{x^2 - 1}$

(b) $h(x) = \dfrac{2}{\sqrt{1 - 3x}}$

**25** Express $y$ as a function of $x$; that is, $y = f(x)$, if:

(a) $y$ is directly proportional to $x$, and $y = 30$ when $x = 6$.

(b) $y$ is inversely proportional to $x$, and $y = 11$ when $x = 3$.

**26** Find the vertex and the $x$ and $y$ intercepts of the graph of $f(x) = 2x^2 + x - 3$. Also, give the domain and range of $f$.

**27** Let $f(x) = 3^x$ and find:

(a) $f(-2)$

(b) $f(0)$

(c) $f(3)$

**28** Let $g(x) = \log_8 x$ and find:

(a) $g(1)$

(b) $g(16)$

(c) $g(\tfrac{1}{2})$

**29** Solve $\begin{cases} x^2 + y^2 = 29 \\ y^2 - 3x^2 = 13 \end{cases}$ algebraically.

**30** Use a system of equations to solve the given problem. A rectangle has an area of 21 square meters and a perimeter of 20 meters. Find the length and width of the rectangle.

---

## Answers

**1** 4  **2(a)** $x^4 - 49$  **(b)** $2x^2 + x + 11$  **(c)** $x^3 - 2x^2 - 2x + 5$  **3(a)** $(x - 2)(x + 2)(y + 2z)$

**(b)** $(2x - 3)(4x + 5)$  **(c)** $x(x - 3)(x^2 + 3x + 9)$  **4(a)** $\dfrac{2(y - 4)}{y}$  **(b)** $\dfrac{10x + 32}{(x - 1)(x + 3)(x + 5)}$

CHAPTER 10 TESTS    265

5  $\dfrac{1}{2y + 3x}$    6(a)  $\dfrac{14}{3}$    (b) no solution    7  $P = \dfrac{A}{1 + rt}$    8(a)  $x \geq \dfrac{3}{4}$

(b)  $x < -\dfrac{1}{2}$ or $x > \dfrac{7}{2}$    9(a)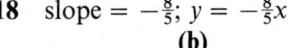

(b)     10(a)  $\dfrac{y^2 z}{x^5}$    (b)  $-\dfrac{32y^2}{x^5}$    11(a)  $\dfrac{2\sqrt{7}}{7}$

(b)  $\dfrac{x + 5\sqrt{x} + 6}{x - 4}$    12(a)  $1 - i$    (b)  $34 + 22i$    13(a)  $\dfrac{3 - \sqrt{47}i}{4}, \dfrac{3 + \sqrt{47}i}{4}$    (b)  4

(c)  $-64, 64$    14  3 feet    15(a)  $-1 < x < \tfrac{1}{2}$

(b)  $x < -\tfrac{5}{2}$ or $x \geq 3$    16  $\dfrac{\sqrt{73}}{2}$

17  center = $(2, -5)$; radius = $\sqrt{26}$    18  slope = $-\tfrac{8}{5}$; $y = -\tfrac{8}{5}x + \tfrac{7}{5}$    19(a)  $(4, 2)$    (b)  $(1, 2, -1)$

20(a) 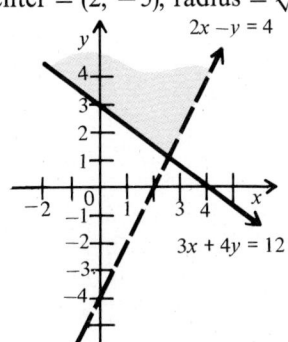   (b)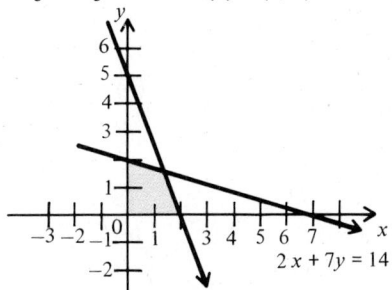

21(a)  $\tfrac{7}{9}$    (b)  14    22(a)  $\log_b x^3 y^6$    (b)  $\log \dfrac{\sqrt{x}}{3}$    23(a)  0    (b)  25

24(a)  all real $x$ except $\pm 1$    (b)  all real $x < \tfrac{1}{3}$    25(a)  $y = 5x$    (b)  $y = \dfrac{33}{x}$

26  The vertex is $(-\tfrac{1}{4}, -\tfrac{25}{8})$; $x$ intercepts are $-\tfrac{3}{2}$ and 1; $y$ intercept is $-3$; the domain = all real numbers; the range is all real numbers $y \geq -\tfrac{25}{8}$.

27(a)  $\tfrac{1}{9}$    (b)  1    (c)  27    28(a)  0    (b)  $\tfrac{4}{3}$    (c)  $-\tfrac{1}{3}$

29  $(-2, -5), (-2, 5), (2, -5),$ and $(2, 5)$    30  length = 7 meters; width = 3 meters

## CHAPTER 10  TESTS

### Chapter Test A

1  Let $f(x) = 3x + 2$, $g(x) = \sqrt{9 - x^2}$, and $h(x) = |x - 7|$. Find the following values.
  (a)  $f(4)$    (b)  $g(\sqrt{5})$    (c)  $h(3)$    (d)  $g(0)$    (e)  $h(9)$    (f)  $f(-2)$

2  Find the domain of the function determined by each equation.
  (a)  $f(x) = -2x + 7$    (b)  $g(x) = \dfrac{1}{x - 2}$    (c)  $h(x) = \sqrt{3x - 1}$

3  Let $f(x) = 2x + 3$ and $g(x) = -x + 4$. Find the following values and simplify.
  (a)  $f(x) + g(x)$    (b)  $\dfrac{f(x)}{g(x)}$    (c)  $\dfrac{g(x + h) - g(x)}{h}$

4  Express $y$ as a function of $x$ if:
  (a)  $y$ is directly proportional to $x$, and $y = 4$ when $x = 8$.
  (b)  $y$ is inversely proportional to $x^2$, and $y = 2$ when $x = 2$.

**5** Let $f(x) = x^2 + 2x + 2$. Find:
  (a) $f(-2)$     (b) $f(0)$     (c) $f(2)$     (d) $f(4)$

**6** Let $f(x) = x^2 + x - 2$. Find:
  (a) $x$ intercepts     (b) $y$ intercept     (c) vertex
  (d) domain     (e) range

**7** Find the equation of the linear function whose $y$ intercept is 2 and which contains the point (3, 5).

**8** For each function, sketch the graph and find the domain and the range.
  (a) $f(x) = 3^x$     (b) $f(x) = \log_3 x$

**9** For each equation, identify and sketch the graph of the given conic.
  (a) $x^2 + y^2 = 1$     (b) $y^2 = -12x$     (c) $4x^2 + 9y^2 = 36$     (d) $4y^2 - 9x^2 = 36$

**10** Solve each system of nonlinear equations and illustrate your solution graphically.
  (a) $\begin{cases} x + y = 5 \\ x^2 + y^2 = 13 \end{cases}$     (b) $\begin{cases} x^2 + 2y^2 = 29 \\ 2x^2 + y^2 = 46 \end{cases}$

---

## Solutions

**1** (a) $f(x) = 3x + 2$
$f(4) = 3(4) + 2$
$= 14$

(b) $g(x) = \sqrt{9 - x^2}$
$g(\sqrt{5}) = \sqrt{9 - (\sqrt{5})^2}$
$= \sqrt{4}$
$= 2$

(c) $h(x) = |x - 7|$
$h(3) = |3 - 7|$
$= |-4|$
$= 4$

(d) $g(0) = \sqrt{9 - 0^2}$
$= \sqrt{9}$
$= 3$

(e) $h(9) = |9 - 7|$
$= |2|$
$= 2$

(f) $f(-2) = 3(-2) + 2$
$= -6 + 2$
$= -4$

**2** (a) $f(x) = -2x + 7$
$-2x + 7$ is defined for all real numbers. Therefore, the domain of $f$ is the set of all real numbers.

(b) $g(x) = \dfrac{1}{x - 2}$
$x - 2 = 0$ when $x = 2$. Therefore, the domain of $g$ is the set of all real numbers except 2.

(c) $h(x) = \sqrt{3x - 1}$
$\sqrt{3x - 1}$ represents a real number when $3x - 1 \geq 0$ or $x \geq \frac{1}{3}$. Therefore, the domain of $h$ is the set of all real numbers $x$ such that $x \geq \frac{1}{3}$.

**3** (a) $f(x) + g(x) = (2x + 3) + (-x + 4) = x + 7$

(b) $\dfrac{f(x)}{g(x)} = \dfrac{2x + 3}{-x + 4}$

(c) $\dfrac{g(x + h) - g(x)}{h} = \dfrac{[-(x + h) + 4] - (-x + 4)}{h}$
$= \dfrac{-x - h + 4 + x - 4}{h}$
$= \dfrac{-h}{h} = -1$

**4** (a) $y = kx$
$4 = k(8)$
$k = \frac{4}{8} = \frac{1}{2}$
Therefore, $y = \frac{1}{2}x$.

(b) $y = \dfrac{k}{x^2}$
$2 = \dfrac{k}{(2)^2}$
$k = 2(2)^2 = 8$
Therefore, $y = \dfrac{8}{x^2}$.

5  $f(x) = x^2 + 2x + 2$
   (a) $f(-2) = (-2)^2 + 2(-2) + 2 = 4 - 4 + 2 = 2$
   (b) $f(0) = (0)^2 + 2(0) + 2 = 0 + 0 + 2 = 2$
   (c) $f(2) = (2)^2 + 2(2) + 2 = 4 + 4 + 2 = 10$
   (d) $f(4) = (4)^2 + 2(4) + 2 = 16 + 8 + 2 = 26$

6  $f(x) = x^2 + x - 2$
   (a) $x^2 + x - 2 = 0$
       $(x + 2)(x - 1) = 0$
       $x + 2 = 0 \quad x - 1 = 0$
       $x = -2 \quad x = 1$
   (b) $f(0) = 0^2 + 0 - 2$
            $= -2$
   (c) $a = 1, b = 1, c = -2$
       $h = -\dfrac{b}{2a} = -\dfrac{1}{2(1)} = -\dfrac{1}{2}$
       $k = \dfrac{4ac - b^2}{4a} = \dfrac{4(1)(-2) - (1)^2}{4(1)} = -\dfrac{9}{4}$
       $(h, k) = \left(-\dfrac{1}{2}, -\dfrac{9}{4}\right)$
   (d) $x^2 + x - 2$ is defined for all real numbers. Therefore, the domain of $f$ is the set of all real numbers.
   (e) The graph of $f$ is a parabola that opens upward, since $a = 1$ is positive. Since the vertex is the point $\left(-\tfrac{1}{2}, -\tfrac{9}{4}\right)$, the range is the set of all real numbers $y \geq -\tfrac{9}{4}$.

7  $y = f(x) = mx + b$ with $b = 2$, so that $y = mx + 2$.
   Since $(3, 5)$ is on the line, $5 = m(3) + 2$
   $3m = 3$
   $m = 1$.
   Therefore, $f(x) = x + 2$.

8  (a)

| $x$ | $y = 3^x$ |
|---|---|
| $-2$ | $3^{-2} = \tfrac{1}{9}$ |
| $-1$ | $3^{-1} = \tfrac{1}{3}$ |
| $0$ | $3^0 = 1$ |
| $1$ | $3^1 = 3$ |

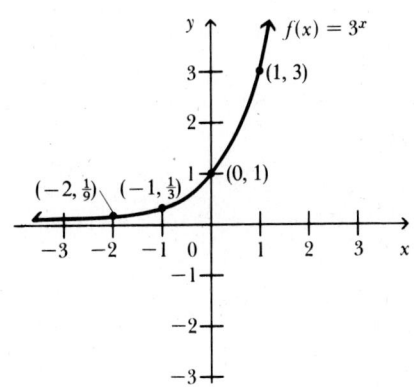

From the graph, domain: all real numbers; range: all positive real numbers

(b) $y = \log_3 x$
    $x = 3^y$

| $x$ | $y$ |
|---|---|
| $\tfrac{1}{27}$ | $-3$ |
| $\tfrac{1}{9}$ | $-2$ |
| $\tfrac{1}{3}$ | $-1$ |
| $1$ | $0$ |
| $3$ | $1$ |

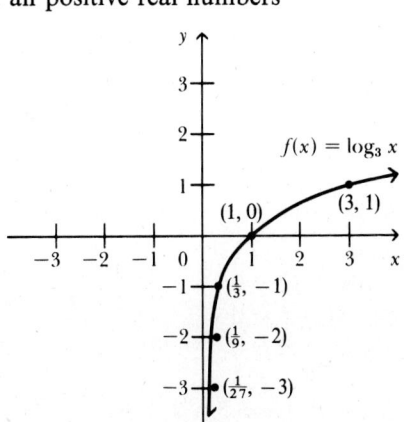

From the graph, domain: all positive real numbers; range: all real numbers

**268** CHAPTER 10 FUNCTIONS AND RELATED CURVES

9  (a)  $x^2 + y^2 = 1$ or $x^2 + y^2 = 1^2$
a circle with center at the origin and radius = 1

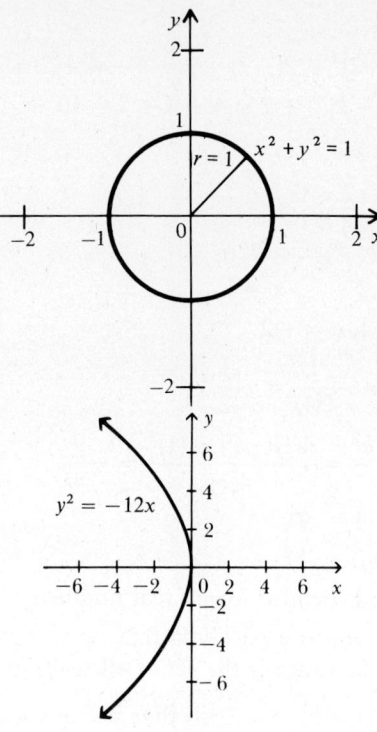

(b)  $y^2 = -12x$
a parabola with vertex at the origin and that opens to the left, since $a = -12 < 0$

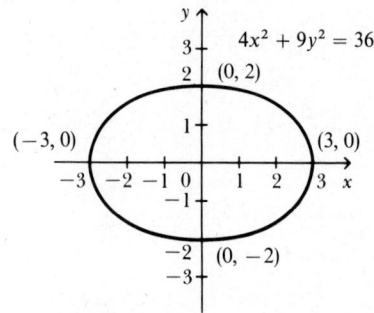

(c)  $4x^2 + 9y^2 = 36$

$\dfrac{4x^2}{36} + \dfrac{9y^2}{36} = \dfrac{36}{36}$

$\dfrac{x^2}{9} + \dfrac{y^2}{4} = 1$

$a = 3 \quad b = 2$

an ellipse with vertices $(-3, 0)$, $(3, 0)$, $(0, -2)$, and $(0, 2)$

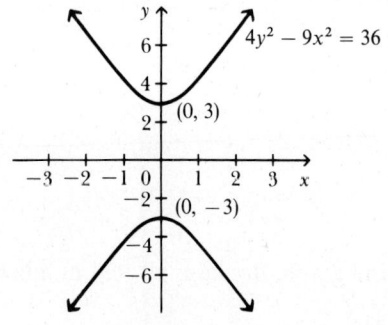

(d)  $4y^2 - 9x^2 = 36$

$\dfrac{4y^2}{36} - \dfrac{9x^2}{36} = \dfrac{36}{36}$

$\dfrac{y^2}{9} - \dfrac{x^2}{4} = 1$

$b = 3$ and $a = 2$
a hyperbola with vertical transverse axis and vertices $(0, 3)$ and $(0, -3)$

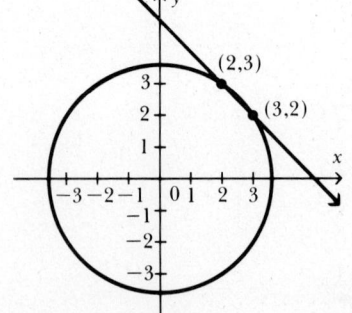

10  (a)  $\begin{cases} x + y = 5 \\ x^2 + y^2 = 13 \end{cases}$

$x + y = 5$
$y = 5 - x$

$\quad\quad\quad x^2 + (5 - x)^2 = 13$
$\quad\quad x^2 + 25 - 10x + x^2 = 13$
$\quad\quad\quad\quad 2x^2 - 10x + 12 = 0$
$\quad\quad\quad\quad\quad x^2 - 5x + 6 = 0$
$\quad\quad\quad\quad\quad (x - 2)(x - 3) = 0$
$\quad\quad\quad\quad\quad\quad x = 2$ or $x = 3$

For $x = 2$, $y = 5 - x = 5 - 2 = 3$
For $x = 3$, $y = 5 - x = 5 - 3 = 2$
The solutions are (2, 3) and (3, 2).

**(b)** $\begin{cases} x^2 + 2y^2 = 29 \\ 2x^2 + y^2 = 46 \end{cases}$ $\xrightarrow{\text{multiply by 2}}$ $\begin{cases} 2x^2 + 4y^2 = 58 \\ 2x^2 + y^2 = 46 \end{cases}$ (subtract)

$$3y^2 = 12$$
$$y^2 = 4$$
$$y = \pm 2$$

For $y = \pm 2$, $x^2 + 2(\pm 2)^2 = 29$
$$x^2 = 21$$
$$x = \pm\sqrt{21}$$

The solutions are $(-\sqrt{21}, 2)$, $(-\sqrt{21}, -2)$, $(\sqrt{21}, 2)$, $(\sqrt{21}, -2)$.

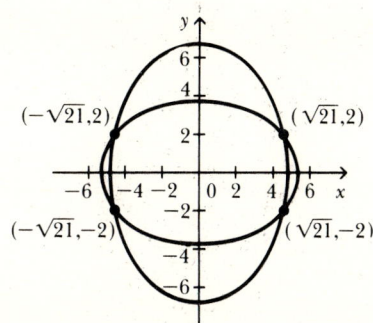

## Chapter Test B

*Multiple Choice:* Select the *one* correct answer for each of the following questions.

**1** If $f(x) = -3x + 1$, then $f(-3) = $ _____.
(a) $-8$    (b) $10$    (c) $-10$    (d) none of these

**2** If $f(x) = \sqrt{16 - x^2}$, then $f(-4) = $ _____.
(a) $0$    (b) $4$    (c) $16$    (d) not defined

**3** If $f(x) = 2 - x$, then $f(3 + a) = $ _____.
(a) $a - 2$    (b) $3 - a$    (c) $1 - a$    (d) $-1 - a$

**4** If $f(x) = \sqrt{100 - x^2}$, then $f(8) - f(6) = $ _____.
(a) $6$    (b) $8$    (c) $-2$    (d) $2$

**5** If $h(x) = |x - 8|$, then $h(3) = $ _____.
(a) $-5$    (b) $11$    (c) $-11$    (d) $5$

**6** Of the following, _____ is the graph of a function.

(a)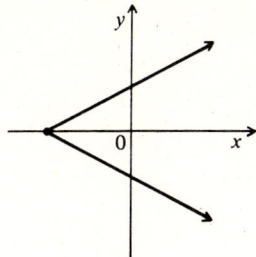

(b)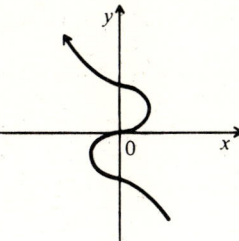

(c)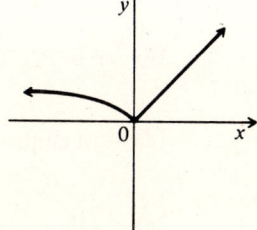

(d)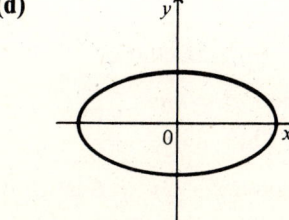

# CHAPTER 10  FUNCTIONS AND RELATED CURVES

7  The domain of the function $f(x) = \dfrac{1}{x-1}$ is _____.
   - (a) all real numbers
   - (b) all positive numbers
   - (c) all real numbers except 0
   - (d) all real numbers except 1

8  The domain of the function $h(x) = \sqrt{1-2x}$ is the set of all real numbers $x$ such that _____.
   - (a) $x \leq \tfrac{1}{2}$
   - (b) $x \geq \tfrac{1}{2}$
   - (c) $x < \tfrac{1}{2}$
   - (d) $x > \tfrac{1}{2}$

9  If $y$ is directly proportional to $x$, and $y = 6$ when $x = 3$, then $y = $ _____.
   - (a) $-2x$
   - (b) $\dfrac{2}{x}$
   - (c) $2x$
   - (d) $3x$

10  If $y$ is inversely proportional to $x$, and $y = 8$ when $x = 3$, then we express $y$ as a function of $x$ as _____.
   - (a) $y = \dfrac{2}{x}$
   - (b) $y = \dfrac{-24}{x}$
   - (c) $y = \dfrac{24}{x}$
   - (d) $y = 24x$

11  If $y$ is inversely proportional to $x^2$, and $y = 5$ when $x = 2$, then we express $y$ as a function of $x$ as _____.
   - (a) $y = 20x$
   - (b) $y = -20x^2$
   - (c) $y = \dfrac{-20}{x^2}$
   - (d) $y = \dfrac{20}{x^2}$

12  The graph of a linear function $f(x) = mx + b$ is a _____.
   - (a) parabola
   - (b) straight line
   - (c) ellipse
   - (d) none of these

13  The slope of the linear function $f(x) = -3x + 2$ is _____.
   - (a) $-\tfrac{3}{2}$
   - (b) $-\tfrac{2}{3}$
   - (c) $-3$
   - (d) $2$

14  The $x$ intercepts of the quadratic function $f(x) = 8x - x^2$ are _____.
   - (a) 2 and 0
   - (b) 0 and 8
   - (c) $-2$ and 0
   - (d) 0 and $-8$

15  The vertex of the quadratic function $f(x) = x^2 - x - 2$ is the point _____.
   - (a) $(1, -2)$
   - (b) $(-1, 2)$
   - (c) $(0, -2)$
   - (d) $(\tfrac{1}{2}, -\tfrac{9}{4})$

16  The range of the function $f(x) = x^2 - x - 2$ is the set of all real $y$ such that _____.
   - (a) $y \geq -\tfrac{9}{4}$
   - (b) $y \leq -\tfrac{9}{4}$
   - (c) $y \geq -2$
   - (d) $y \leq -2$

17  The domain of the function $f(x) = 2^x$ is the set of _____.
   - (a) all positive numbers
   - (b) all real numbers
   - (c) all negative numbers
   - (d) integers

18  The domain of the function $h(x) = \log_3 x$ is the set of _____.
   - (a) all positive numbers
   - (b) all real numbers
   - (c) all negative numbers
   - (d) integers

19  If $f(x) = 9^x$, then $f(\tfrac{1}{2}) = $ _____.
   - (a) $-\tfrac{9}{2}$
   - (b) $\tfrac{9}{2}$
   - (c) $-3$
   - (d) $3$

20  If $h(x) = \log_3 x$, then $h(81) = $ _____.
   - (a) 4
   - (b) 27
   - (c) 2
   - (d) $81^3$

21  The graph of the equation $x^2 + y^2 = 16$ is _____.
   - (a) a line
   - (b) a parabola
   - (c) an ellipse
   - (d) a circle

22  The graph of the equation $2x^2 + 3y^2 = 12$ is _____.
   - (a) a circle
   - (b) an ellipse
   - (c) a line
   - (d) a hyperbola

23  The graph of the equation $y^2 = x$ is _____.
   - (a) a circle
   - (b) a line
   - (c) a parabola
   - (d) an ellipse

24  The graph of the equation $2x^2 - 3y^2 = 6$ is _____.
   - (a) a circle
   - (b) a parabola
   - (c) a hyperbola
   - (d) a line

**25** The solutions of $\begin{cases} x + y = 23 \\ x^2 + y^2 = 277 \end{cases}$ are _____.

(a) (9, 14), (14, 9)  
(b) (−9, 14), (−14, 9)  
(c) (−9, −14), (−14, −9)  
(d) no real solutions

**26** The solutions of $\begin{cases} x + y = 12 \\ 2x^2 + 2y^2 = 6 \end{cases}$ are _____.

(a) (3, 9), (9, 3)  
(b) (−3, 9), (−9, 3)  
(c) (−3, −9), (−9, −3)  
(d) no real solutions

## Answers

| 1 | b | 2 | a | 3 | d | 4 | c | 5 | d | 6 | c | 7 | d | 8 | a | 9 | c | 10 | c |
|---|---|---|---|---|---|---|---|---|---|---|---|---|---|---|---|---|---|----|---|
| 11 | d | 12 | b | 13 | c | 14 | b | 15 | d | 16 | a | 17 | b | 18 | a | 19 | d | | |
| 20 | a | 21 | d | 22 | b | 23 | c | 24 | c | 25 | a | 26 | d | | | | | | |

# 11 Topics in Algebra

In this chapter we study certain functions whose domains are the set of positive integers. Such functions are called sequences. Other topics covered are series and the binomial theorem. After completing the appropriate sections, the student should be able to:

1. Write the terms of a given sequence.
2. Write the terms of a given arithmetic sequence.
3. Find the sum of the terms of an arithmetic sequence.
4. Write the terms of a given geometric sequence.
5. Find the sum of the terms of a geometric sequence.
6. Employ the summation notation.
7. Apply the binomial theorem.

## 11.1 Sequences

### SEMIPROGRAMMED PROBLEMS

In problems 1 and 2, write the first five terms

1. $a_n = (-1)^n$

$-1, 1$      $a_1 = (-1)^1 = \underline{\phantom{xx}}$    $a_2 = (-1)^2 = \underline{\phantom{xx}}$

$-1, -1, 1$     $a_3 = (-1)^3 = \underline{\phantom{xx}}$    $a_4 = (\underline{\phantom{xx}})^4 = \underline{\phantom{xx}}$

$-1$           $a_5 = (-1)^5 = \underline{\phantom{xx}}$

$-1, 1, -1, 1, -1$    The first five terms of the sequence are $\underline{\phantom{xxxxxxxxxxxx}}$.

2. $a_n = 3 - \dfrac{1}{n}$

$2, \frac{5}{2}$      $a_1 = 3 - \frac{1}{1} = \underline{\phantom{xx}}$    $a_2 = 3 - \frac{1}{2} = \underline{\phantom{xx}}$

$\frac{8}{3}, \frac{11}{4}$     $a_3 = 3 - \frac{1}{3} = \underline{\phantom{xx}}$    $a_4 = 3 - \frac{1}{4} = \underline{\phantom{xx}}$

$\frac{14}{5}$         $a_5 = 3 - \frac{1}{5} = \underline{\phantom{xx}}$

$2, \frac{5}{2}, \frac{8}{3}, \frac{11}{4}, \frac{14}{5}$    The first five terms of the sequence are $\underline{\phantom{xxxxxxxxxxxx}}$.

In problems 3 and 4, write the next five terms and the $n$th term of the arithmetic sequence whose first two terms are shown.

3. $2, 4, \ldots$

$2$              The common difference $d = 4 - 2 = \underline{\phantom{xx}}$. The next five terms of the

$6, 8, 10, 12, 14$    sequence are $\underline{\phantom{xx}}$, $\underline{\phantom{xx}}$, $\underline{\phantom{xx}}$, $\underline{\phantom{xx}}$, and $\underline{\phantom{xx}}$. The first term

$2, 2, 2 + (n-1)2$    $a_1 = \underline{\phantom{xx}}$, $d = \underline{\phantom{xx}}$, so $a_n = a_1 + (n-1)d = \underline{\phantom{xxxxxxxx}} =$

$2n$            $\underline{\phantom{xx}}$.

## 11.1 SEMIPROGRAMMED PROBLEMS

**4** 5, 9, ...
The common difference $d = 9 - 5 =$ ____. The next five terms of the sequence are ____, ____, ____, ____, and ____. The first term $a_1 =$ ____, $d =$ ____, so $a_n = a_1 + (n-1)d =$ ____ = ____.

4
13, 17, 21, 25, 29
5, 4, 5 + (n − 1)4
4n + 1

In problems 5 and 6, find the indicated term of each arithmetic sequence.

**5** The ninth term of 4, 9, 14, ...
Notice that $a_1 =$ ____ and compute $d =$ ____. Using $a_n = a_1 + (n-1)d$, with $n = 9$, we have $a_9 = 4 + (9-1)5 =$ ____.

4, 5
44

**6** The twenty-fifth term of 4, 13, 22, ...
First observe that $a_1 =$ ____ and compute $d =$ ____. Using $a_n = a_1 + (n-1)d$, and taking $n = 25$, we have $a_{25} = 4 + (25-1)($____$) =$ ____.

4, 9

9, 220

In problems 7–9, find the common ratio and the next three terms in each geometric sequence.

**7** 2, 8, 32, ...
To find the common ratio, write $r = \dfrac{8}{\rule{1cm}{0.4pt}} =$ ____. The fourth term is equal to $32($____$) =$ ____. The fifth term is equal to $128($____$) =$ ____. The sixth term is equal to $512($____$) =$ ____. Therefore, the geometric sequence is ____.

2, 4
4, 128
4, 512, 4
2,048
2, 8, 32, 128, 512, 2,048, ...

**8** 8, 4, 2, ...
To find the common ratio, write $r = \dfrac{\rule{1cm}{0.4pt}}{8} =$ ____. The fourth term is equal to $2($____$) =$ ____. The fifth term is equal to $1($____$) =$ ____. The sixth term is equal to $\frac{1}{2}($____$) =$ ____. The geometric sequence is ____.

4, $\frac{1}{2}$
$\frac{1}{2}$, 1
$\frac{1}{2}, \frac{1}{2}, \frac{1}{2}, \frac{1}{4}$
8, 4, 2, 1, $\frac{1}{2}, \frac{1}{4}$, ...

**9** $6, -4, \frac{8}{3}, \ldots$
To find the common ratio, write $r = \dfrac{-4}{\rule{1cm}{0.4pt}} =$ ____. The fourth term is equal to $\frac{8}{3}($____$) =$ ____. The fifth term is equal to $-\frac{16}{9}($____$) =$ ____. The sixth term is equal to $\frac{32}{27}($____$) =$ ____. The geometric sequence is ____.

6, $-\frac{2}{3}$
$-\frac{2}{3}, -\frac{16}{9}$
$-\frac{2}{3}, \frac{32}{27}, -\frac{2}{3}, -\frac{64}{81}$
$6, -4, \frac{8}{3}, -\frac{16}{9}, \frac{32}{27}, -\frac{64}{81}$

In problems 10–12, find the indicated term of each geometric sequence.

**10** The ninth term of 2, 6, 18, ...
$a_1 =$ ____, $r =$ ____, $n =$ ____.
Using the formula $a_n = a_1 r^{n-1}$ and substituting for $a_1$, $r$, and $n$, we obtain
$a_9 = 2(3)^{9-1} = 2(3)^{\rule{0.5cm}{0.4pt}} =$ ____.

2, 3, 9

8, 13,122

1, 2, 8

7, 128

5, 2, 6

5, 160

**11** The eighth term of 1, 2, 4, ...

$a_1 = $ _____, $r = $ _____, $n = $ _____.

Using the formula $a_n = a_1 r^{n-1}$ and substituting for $a_1$, $r$, and $n$, we obtain

$a_8 = 1(2)^{\phantom{xx}} = $ _____.

**12** The sixth term of 5, 10, 20, ...

$a_1 = $ _____, $r = $ _____, $n = $ _____.

Using the formula $a_n = a_1 r^{n-1}$ and substituting for $a_1$, $r$, and $n$, we obtain

$a_6 = 5(2)^{\phantom{xx}} = $ _____.

## SOLUTIONS TO SELECTED ODD PROBLEMS   Section 11.1, text pages 504–505

**1**  $a_n = \dfrac{n(n+2)}{2}$

$a_1 = \dfrac{1(1+2)}{2} = \dfrac{3}{2}$

$a_2 = \dfrac{2(2+2)}{2} = 4$

$a_3 = \dfrac{3(3+2)}{2} = \dfrac{15}{2}$

$a_4 = \dfrac{4(4+2)}{2} = 12$

$a_5 = \dfrac{5(5+2)}{2} = \dfrac{35}{2}$

**5**  $a_n = (-1)^n + 3$
$a_1 = (-1)^1 + 3 = -1 + 3 = 2$
$a_2 = (-1)^2 + 3 = 1 + 3 = 4$
$a_3 = (-1)^3 + 3 = -1 + 3 = 2$
$a_4 = (-1)^4 + 3 = 1 + 3 = 4$
$a_5 = (-1)^5 + 3 = -1 + 3 = 2$

**9**  $12 - 7 = 5$
$17 - 12 = 5$
$22 - 17 = 5$
An arithmetic progression with $d = 5$.

**13**  $6.9 - 5.7 = 1.2$
$8.1 - 6.9 = 1.2$
$9.3 - 8.1 = 1.2$
An arithmetic progression with $d = 1.2$.

**17**  $(4a + 20b) - (a + 24b) = 3a - 4b$
and
$(7a + 16b) - (4a + 20b) = 3a - 4b$
Therefore, $d = 3a - 4b$
and $a_1 = a + 24b$.
Using $a_n = a_1 + (n-1)d$
$a_6 = (a + 24b) + (6-1)(3a - 4b)$
$\phantom{a_6} = (a + 24b) + (15a - 20b)$
$\phantom{a_6} = 16a + 4b$
$a_9 = (a + 24b) + (9-1)(3a - 4b)$
$\phantom{a_9} = (a + 24b) + (24a - 32b) = 25a - 8b$.

**21**  $\dfrac{-2}{1} = -2$

$\dfrac{4}{-2} = -2$

A geometric progression with $r = -2$.

**25**  $\dfrac{-6}{9} = -\dfrac{2}{3}$

$\dfrac{4}{-6} = -\dfrac{2}{3}$

A geometric progression with $r = -\dfrac{2}{3}$.

**29**  $\dfrac{16}{32} = \dfrac{1}{2} = r$
$a_1 = 32$
$a_n = a_1 r^{n-1}$
$a_5 = 32(\tfrac{1}{2})^{5-1}$
$\phantom{a_5} = 2$

33  $\frac{12}{6} = 2 = r$
    $a_1 = 6$
    $a_n = a_1 r^{n-1}$
    $a_6 = 6(2)^{6-1} = 6(2)^5$
        $= 192$
    $a_{10} = 6(2)^{10-1} = 6(2)^9$
        $= 3{,}072$

37  $1 \div 2 = \frac{1}{2} = r$
    $a_1 = 2$
    $a_n = a_1 r^{n-1}$
    $\frac{1}{16} = 2(\frac{1}{2})^{n-1}$
    $\frac{1}{32} = (\frac{1}{2})^{n-1}$
    $(\frac{1}{2})^5 = (\frac{1}{2})^{n-1}$
    $5 = n - 1$
    $n = 6$
    Therefore, $\frac{1}{16}$ is the sixth term.

## 11.2 Series

### SEMIPROGRAMMED PROBLEMS

In problems 1–4, find the numerical values of each finite sum.

1  $\sum_{k=1}^{4} k$

The expanded form is

$1 + 2 + 3 + 4$

$\sum_{k=1}^{4} k = \underline{\hspace{2cm}}.$

10

The finite sum is equal to _____.

2  $\sum_{i=0}^{3} (\frac{1}{3})^i$

The expanded form is

$1 + \frac{1}{3} + \frac{1}{9} + \frac{1}{27}$

$\sum_{i=0}^{3} (\frac{1}{3})^i = \underline{\hspace{2cm}}.$

$\frac{40}{27}$

The finite sum is equal to _____.

3  $\sum_{k=1}^{3} (4k^2 + 3k)$

The expanded form is

$[4(3)^2 + 3(3)]$

$\sum_{k=1}^{3} (4k^2 + 3k) = [4(1)^2 + 3(1)] + [4(2)^2 + 3(2)] + \underline{\hspace{2cm}}$

22, 45

$= 7 + \underline{\hspace{1cm}} + \underline{\hspace{1cm}}$

74

$= \underline{\hspace{1cm}}.$

4  $\sum_{i=3}^{6} i(i-2)$

The expanded form is

$6(6-2)$

$\sum_{i=3}^{6} i(i-2) = 3(3-2) + 4(4-2) + 5(5-2) + \underline{\hspace{2cm}}$

15, 24

$= 3 + 8 + \underline{\hspace{1cm}} + \underline{\hspace{1cm}}$

50

$= \underline{\hspace{1cm}}.$

In problems 5 and 6, find the sum of the first $n$ terms of the arithmetic sequence with first term $a_1$ and common difference $d$.

5  $n = 20$, $a_1 = 3$, and $d = 2$

Substituting these values in the formula for $S_n$, we have

20, 2

$S_{20} = \frac{20}{2}[6 + (\underline{\hspace{1cm}} - 1)\underline{\hspace{1cm}}]$ so that

440

$S_{20} = \underline{\hspace{1cm}}.$

# CHAPTER 11 TOPICS IN ALGEBRA

**6** $n = 30$, $a_1 = -40$, and $d = \frac{3}{4}$

30, 2(−40)

Substituting, we obtain

$$S_{30} = \frac{\rule{1cm}{0.4pt}}{2}\left[\rule{1cm}{0.4pt} + (30-1)\frac{3}{4}\right], \text{ so that}$$

$-\frac{3,495}{4}$

$$S_{30} = \rule{2cm}{0.4pt}.$$

In problems 7 and 8, certain elements of an arithmetic sequence are given. Find the indicated unknown.

**7** $a_9$ and $S_9$ if $a_1 = 37$, $d = -5$

Using $a_n = a_1 + (n-1)d$, we obtain

−3

$a_9 = 37 + (9-1)(-5) = \rule{1cm}{0.4pt}.$

Using $S_n = \frac{n}{2}(a_1 + a_n)$, we have

153

$S_9 = \frac{9}{2}[37 + (-3)] = \rule{1cm}{0.4pt}.$

**8** $a_1$, $d$, and $S_{15}$ if $a_4 = 9$ and $a_{15} = 31$

Using $a_n = a_1 + (n-1)d$, we have

$a_1 + 14d$

$9 = a_1 + (4-1)d$ and $31 = \rule{1cm}{0.4pt}.$

3, 2

Solving for $a_1$ and $d$, we obtain $a_1 = \rule{1cm}{0.4pt}$ and $d = \rule{1cm}{0.4pt}.$

Using $S_n = \frac{n}{2}(a_1 + a_n)$, we have

31, 255

$S_{15} = \frac{15}{2}(3 + \rule{1cm}{0.4pt}) = \rule{1cm}{0.4pt}.$

In problems 9 and 10, find the indicated sum for each geometric sequence.

**9** $S_{10}$ if $a_1 = 6$, $r = 2$

Using the formula $S_n = \dfrac{a_1 - a_1 r^n}{1 - r}$, we obtain

$^{10}$, 6,138

$$S_{10} = \frac{6 - 6(2)^{\rule{0.5cm}{0.4pt}}}{1 - 2} = \rule{1cm}{0.4pt}.$$

**10** $S_6$ if $a_1 = 4$ and $r = \frac{3}{2}$

Using the formula $S_n = \dfrac{a_1 - a_1 r^n}{1 - r}$, we write

$\frac{665}{8}$

$$S_6 = \frac{4 - 4(\frac{3}{2})^6}{1 - \frac{3}{2}} = \rule{1cm}{0.4pt}.$$

In problems 11 and 12, find the indicated element in each geometric sequence.

**11** $a_1$ and $a_7$ if $r = 2$ and $S_7 = 1{,}397$

Using the formula $S_n = \dfrac{a_1 - a_1 r^n}{1 - r}$, $r \neq 1$, we obtain

1,397, 128

$$\rule{1cm}{0.4pt} = \frac{a_1 - a_1(\rule{0.5cm}{0.4pt})}{1 - 2}, \text{ so that}$$

$-127a_1$, 11

$-1{,}397 = \rule{1cm}{0.4pt}$ or $a_1 = \rule{1cm}{0.4pt}.$

Using $a_n = a_1 r^{n-1}$, we obtain

$^6$, 704

$a_7 = 11(2)^{\rule{0.5cm}{0.4pt}} = \rule{1cm}{0.4pt}.$

$\frac{1}{64}$, $^{n-1}$

7

$\frac{127}{192}$

**12**   $n$ and $S_n$ if $a_1 = \frac{1}{3}$, $r = \frac{1}{2}$, and $a_n = \frac{1}{192}$

Using the formula $a_n = a_1 r^{n-1}$, we obtain $\frac{1}{192} = \frac{1}{3}(\frac{1}{2})^{n-1}$ or

\_\_\_\_ $= (\frac{1}{2})^{n-1}$, so that $(\frac{1}{2})^6 = (\frac{1}{2})$\_\_\_\_. Therefore, $n - 1 = 6$ or

$n = $ \_\_\_\_. Using $S_n = \dfrac{a_1 - a_1 r^n}{1 - r}$, with $a_1 = \frac{1}{3}$, $r = \frac{1}{2}$, and $n = 7$, we

obtain

$$S_7 = \frac{\frac{1}{3} - \frac{1}{3}(\frac{1}{2})^7}{1 - \frac{1}{2}} = \underline{\quad}.$$

## SOLUTIONS TO SELECTED ODD PROBLEMS   Section 11.2, text pages 511–512

**1**   $\displaystyle\sum_{k=1}^{5} k = 1 + 2 + 3 + 4 + 5 = 15$

**5**   $\displaystyle\sum_{k=2}^{5} 2^{k-2} = 2^0 + 2^1 + 2^2 + 2^3$
$= 1 + 2 + 4 + 8 = 15$

**9**   $\displaystyle\sum_{i=1}^{4} \frac{i}{i+1} = \frac{1}{1+1} + \frac{2}{2+1} + \frac{3}{3+1} + \frac{4}{4+1}$
$= \frac{1}{2} + \frac{2}{3} + \frac{3}{4} + \frac{4}{5} = \frac{163}{60}$

**13**   $\displaystyle\sum_{k=1}^{5} \frac{1}{k(k+1)} = \frac{1}{1(1+1)} + \frac{1}{2(2+1)} + \frac{1}{3(3+1)} + \frac{1}{4(4+1)} + \frac{1}{5(5+1)}$
$= \frac{1}{2} + \frac{1}{6} + \frac{1}{12} + \frac{1}{20} + \frac{1}{30} = \frac{5}{6}$

**17**   $S_n = \dfrac{n}{2}[2a_1 + (n-1)d]$

$S_8 = \dfrac{8}{2}\left[2(-5) + (8-1)\dfrac{3}{7}\right]$

$S_8 = 4[-10 + 3]$

$S_8 = -28$

**21**   $a_n = a_1 + (n-1)d$

$a_{10} = 6 + (10-1)3$

$a_{10} = 33$

$S_n = \dfrac{n}{2}(a_1 + a_n)$

$S_{10} = \dfrac{10}{2}(6 + 33)$

$S_{10} = 195$

**25**   $S_n = \dfrac{n}{2}(a_1 + a_n)$

$1{,}200 = \dfrac{n}{2}(27 + 48)$

$2{,}400 = 75n$

$32 = n$

$S_n = \dfrac{n}{2}[2a_1 + (n-1)d]$

$1{,}200 = \dfrac{32}{2}[2(27) + (32-1)d]$

$1{,}200 = 16[54 + 31d]$

$1{,}200 = 864 + 496d$

$336 = 496d$

$\dfrac{21}{31} = d$

**29**   $S_n = \dfrac{a_1(1 - r^n)}{1 - r}$

$S_{12} = \dfrac{-4[(1 - (-2)^{12})]}{1 - (-2)} = \dfrac{-4(1 - 4{,}096)}{3}$

$S_{12} = \dfrac{-4(-4{,}095)}{3} = 5{,}460$

**33**

$$S_n = \frac{a_1(1-r^n)}{1-r}$$

$$26 = \frac{2(1-r^3)}{1-r}$$

$$13 = \frac{1-r^3}{1-r}$$

$$13 = \frac{(1-r)(1+r+r^2)}{1-r}$$

$$13 = 1 + r + r^2$$

$$r^2 + r - 12 = 0$$

$$(r+4)(r-3) = 0$$

$$r = -4 \text{ or } r = 3$$

**37**

$$a_n = a_1 r^{n-1}$$

$$\frac{1}{8} = a_1\left(-\frac{1}{2}\right)^{5-1}$$

$$\frac{1}{8} = a_1\left(-\frac{1}{2}\right)^4$$

$$\frac{1}{8} = a_1\left(\frac{1}{16}\right)$$

$$2 = a_1$$

$$a_9 = (2)\left(-\frac{1}{2}\right)^{9-1}$$

$$a_9 = (2)\left(-\frac{1}{2}\right)^8$$

$$a_9 = 2\left(\frac{1}{256}\right)$$

$$a_9 = \frac{1}{128}$$

$$S_8 = \frac{2[1-(-\frac{1}{2})^8]}{1-(-\frac{1}{2})}$$

$$S_8 = \frac{2(1-\frac{1}{256})}{\frac{3}{2}}$$

$$S_8 = \frac{2(\frac{255}{256})}{\frac{3}{2}}$$

$$S_8 = \frac{85}{64}$$

## 11.3 The Binomial Theorem

**SEMIPROGRAMMED PROBLEMS**

In problems 1–3, simplify each expression.

5!, 6

5!, 3 · 2 · 1, 7

$(n-1)!, \dfrac{1}{(n+1)n}$

**1** $\dfrac{6!}{5!} = \dfrac{6 \cdot 5!}{\underline{\phantom{5!}}} = \underline{\phantom{000}}$

**2** $\dfrac{7!}{5!3!} = \dfrac{7 \cdot 6 \underline{\phantom{0}}}{5!} \dfrac{\underline{\phantom{0}}}{3!} = \dfrac{7.6}{\underline{\phantom{000}}} = \underline{\phantom{000}}$

**3** $\dfrac{(n-1)!}{(n+1)!} = \dfrac{(n-1)!}{(n+1)(n)(\underline{\phantom{000}})} = \underline{\phantom{000}}$

In problems 4–7, expand each binomial.

**4** $(2x+y)^4$

$(2x)^4, 16x^4$

The first term of the expansion is $\underline{\phantom{000}} = \underline{\phantom{000}}$.

$\dfrac{4}{1!}, 32x^3y$

The second term is $\underline{\phantom{0}}(2x)^3 y = \underline{\phantom{000}}$.

$2!, 24x^2y^2$

The third term is $\dfrac{(4\cdot 3)(2x)^2 y^2}{\underline{\phantom{0}}} = \underline{\phantom{000}}$.

$3!, 8xy^3$

The fourth term is $\dfrac{4\cdot 3\cdot 2}{\underline{\phantom{0}}}(2x)y^3 = \underline{\phantom{000}}$.

11.3 SEMIPROGRAMMED PROBLEMS 279

| | |
|---|---|
| $y^4$ | The fifth term is ____. |
| | Therefore, the expansion of $(2x+y)^4 =$ |
| $16x^4 + 32x^3y + 24x^2y^2 + 8xy^3 + y^4$ | _____. |
| | 5  $\left(2 + \dfrac{x}{3}\right)^5$ |
| $2^5$, 32 | The first term of the expansion is ____ = ____. |
| $\frac{80}{3}x$ | The second term is $\dfrac{5(2)^4\left(\frac{x}{3}\right)}{1!} =$ ____. |
| 4, $\frac{80}{9}x^2$ | The third term is $\dfrac{5(\underline{\phantom{xx}})(2)^3\left(\frac{x}{3}\right)^2}{2!} =$ ____. |
| 3, $\frac{40}{27}x^3$ | The fourth term is $\dfrac{(5)(4)(\underline{\phantom{xx}})(2)^2\left(\frac{x}{3}\right)^3}{3!} =$ ____. |
| 4, $\frac{10}{81}x^4$ | The fifth term is $\dfrac{(5)(4)(3)(2)(2)\left(\frac{x}{3}\right)^{\underline{\phantom{x}}}}{4!} =$ ____. |
| $\left(\frac{x}{3}\right)^5, \dfrac{x^5}{243}$ | The sixth term is $\underline{\phantom{xxxx}} = \underline{\phantom{xx}}$. |
| | Therefore, the expansion is |
| $\dfrac{40x^3}{27} + \dfrac{10x^4}{81} + \dfrac{x^5}{243}$ | $\left(2 + \dfrac{x}{3}\right)^5 = 32 + \dfrac{80x}{3} + \dfrac{80x^2}{9} +$ ____. |
| | 6  $(3x^2 - y)^4$ |
| $-y$ | Write $(3x^2 - y)^4$ in the form of $(a+b)^n$ to obtain $[3x^2 + (\underline{\phantom{xx}})]^4$. |
| $(3x^2)^4$, $81x^8$ | The first term of the expansion is ____ = ____. |
| $-y$, $-108x^6y$ | The second term is $\dfrac{4(3x^2)^3(\underline{\phantom{xx}})^1}{1!} =$ ____. |
| 3, $54x^4y^2$ | The third term is $\dfrac{4(\underline{\phantom{xx}})(3x^2)^2(-y)^2}{2!} =$ ____. |
| $3!$, $-12x^2y^3$ | The fourth term is $\dfrac{4(3)(2)(3x^2)(-y)^3}{\underline{\phantom{xx}}} =$ ____. |
| $4$, $y^4$ | The fifth term is $(-y)^{\underline{\phantom{x}}} =$ ____. |
| | Hence, |
| $54x^4y^2 - 12x^2y^3 + y^4$ | $(3x^2 - y)^4 = 81x^8 - 108x^6y +$ ____. |
| | 7  $\left(x^2 - \dfrac{1}{2x}\right)^6$ |
| $-\dfrac{1}{2x}$ | Write $\left(x^2 - \dfrac{1}{2x}\right)^6$ in the form $(a+b)^n$ to obtain $\left[x^2 + (\underline{\phantom{xx}})\right]^6$. |
| 6, $x^{12}$ | The first term of the expansion is $(x^2)^{\underline{\phantom{x}}} =$ ____. |
| $x^2$, $-3x^9$ | The second term is $\dfrac{6(\underline{\phantom{xx}})^5\left(-\frac{1}{2x}\right)}{1!} =$ ____. |
| $x^2$, $\frac{15}{4}x^6$ | The third term is $\dfrac{6(5)(\underline{\phantom{xx}})^4\left(-\frac{1}{2x}\right)^2}{2!} =$ ____. |

**280** CHAPTER 11 TOPICS IN ALGEBRA

4, $-\frac{5}{2}x^3$

3, $\frac{15}{16}$

2, $-\frac{3}{16x^3}$

6, $\frac{1}{64x^6}$

$\frac{15}{16} - \frac{3}{16x^3} + \frac{1}{64x^6}$

The fourth term is $\dfrac{6(5)(\underline{\phantom{xx}})(x^2)^3\left(-\dfrac{1}{2x}\right)^3}{3!} = \underline{\phantom{xxxx}}$.

The fifth term is $\dfrac{6(5)(4)(\underline{\phantom{xx}})(x^2)^2\left(-\dfrac{1}{2x}\right)^4}{4!} = \underline{\phantom{xx}}$.

The sixth term is $\dfrac{6(5)(4)(3)(\underline{\phantom{xx}})(x^2)\left(-\dfrac{1}{2x}\right)^5}{5!} = \underline{\phantom{xx}}$.

The seventh term is $\left(-\dfrac{1}{2x}\right)^{\underline{\phantom{xx}}} = \underline{\phantom{xx}}$.

Hence,

$$\left(x^2 - \frac{1}{2x}\right)^6 = x^{12} - 3x^9 + \frac{15}{4}x^6 - \frac{5}{2}x^3 + \underline{\phantom{xxxxxxxx}}.$$

In problems 8 and 9, write the first four terms of the expansion and simplify.

**8** $(x + y)^7$

$x^7$

The first term is $\underline{\phantom{xx}}$.

7, 7

The second term is $\dfrac{\overline{\phantom{xx}}}{1!}x^6y = \underline{\phantom{xx}}x^6y$.

6, 21

The third term is $\dfrac{7(\underline{\phantom{xx}})x^5y^2}{2!} = \underline{\phantom{xx}}x^5y^2$.

$\dfrac{7 \cdot 6 \cdot 5}{3!}x^4y^3$, $35x^4y^3$

The fourth term is $\underline{\phantom{xxxxxxxx}} = \underline{\phantom{xxxx}}$.

**9** $(1 + 2x)^{13}$

1

The first term is $\underline{\phantom{xx}}$.

1!, 26x

The second term is $\dfrac{13(1)^{12}(2x)}{\underline{\phantom{xx}}} = \underline{\phantom{xx}}$.

12, 312$x^2$

The third term is $\dfrac{13(\underline{\phantom{xx}})(1)^{11}(2x)^2}{2!} = \underline{\phantom{xx}}$.

$\dfrac{(13)(12)(11)(1)^{10}(2x)^3}{3!}$, $2{,}288x^3$

The fourth term is $\underline{\phantom{xxxxxxxx}} = \underline{\phantom{xxxx}}$.

In problems 10–12, find only the indicated term for each expression and simplify.

**10** The sixth term of $(x^2 + 2y)^{12}$

In the expansion, $n = 12$, and since the sixth term is being written $k = \underline{\phantom{xx}}$, use

6

$$u_k = \frac{n(n-1)(n-2)\cdots(n-k+2)}{(k-1)!}a^{n-k+1}b^{k-1}$$

$x^2$, $2y$

where $a = \underline{\phantom{xx}}$ and $b = \underline{\phantom{xx}}$. Therefore,

25,344$x^{14}y^5$

$$u_6 = \frac{12 \cdot 11 \cdot 10 \cdot 9 \cdot 8}{(6-1)!}(x^2)^7(2y)^5 = \underline{\phantom{xxxx}}.$$

**11** The term that involves $x^7$ in the expansion of $(2 - x)^{12}$

12, 7, 2

In this expansion, $n = \underline{\phantom{xx}}$, $k = \underline{\phantom{xx}}$, $a = \underline{\phantom{xx}}$, and

$-x$

$b = \underline{\phantom{xx}}$. Use the formula

$-32x^7$

$792, -25{,}344x^7$

$$u_{k+1} = \frac{n!}{k!(n-k)!} a^{n-k}b^k \text{ to obtain}$$

$$u_8 = \frac{12!}{7!(12-7)!}(2)^5(-x)^7 = \frac{12 \cdot 11 \cdot 10 \cdot 9 \cdot 8}{5 \cdot 4 \cdot 3 \cdot 2 \cdot 1}(\underline{\phantom{xxx}}) \text{ or}$$

$$u_8 = (\underline{\phantom{xxx}})(-32x^7) = \underline{\phantom{xxxxx}}.$$

**12** The fourth term of $\left(3x - \dfrac{y}{6}\right)^9$

In this expansion $n = 9$, and since the fourth term is being

$4, 3x, \dfrac{-y}{6}$

written, $k = \underline{\phantom{xx}}$. Also, $a = \underline{\phantom{xx}}$ and $b = \underline{\phantom{xx}}$. Use

$$u_k = \frac{n(n-1)(n-2)\cdots(n-k+2)}{(k-1)!}a^{n-k+1}b^{k-1} \text{ to obtain}$$

$3!, -\dfrac{567}{2}x^6 y^3$

$$u_4 = \frac{9 \cdot 8 \cdot 7}{\underline{\phantom{xx}}}(3x)^6\left(\frac{-y}{6}\right)^3 = \underline{\phantom{xxxxx}}.$$

## SOLUTIONS TO SELECTED ODD PROBLEMS  Section 11.3, text page 517

**1** $\dfrac{4!}{6!} = \dfrac{4 \cdot 3 \cdot 2 \cdot 1}{6 \cdot 5 \cdot 4 \cdot 3 \cdot 2 \cdot 1} = \dfrac{1}{30}$

**5** $\dfrac{3! \cdot 8!}{4! \cdot 7!} = \dfrac{(3 \cdot 2 \cdot 1) \cdot (\overset{2}{8} \cdot 7 \cdot 6 \cdot 5 \cdot 4 \cdot 3 \cdot 2 \cdot 1)}{(4 \cdot 3 \cdot 2 \cdot 1) \cdot (7 \cdot 6 \cdot 5 \cdot 4 \cdot 3 \cdot 2 \cdot 1)} = 2$

**9** $\dfrac{(n+1)!}{(n-3)!} = \dfrac{(n+1)(n)(n-1)(n-2)(n-3)(n-4)\cdots(1)}{(n-3)(n-4)\cdots(1)}$

$= (n+1)(n)(n-1)(n-2)$

**13** $(x^2 + 4y^2)^3 = (x^2)^3 + \dfrac{3}{1!}(x^2)^2(4y^2) + \dfrac{3 \cdot 2}{2!}(x^2)(4y^2)^2 + \dfrac{3 \cdot 2 \cdot 1}{3!}(4y^2)^3$

$= x^6 + 12x^4 y^2 + 48x^2 y^4 + 64y^6$

**17** $\left(2 + \dfrac{x}{y}\right)^5 = 2^5 + \dfrac{5}{1!}(2^4)\left(\dfrac{x}{y}\right) + \dfrac{5 \cdot 4}{2!}(2^3)\left(\dfrac{x}{y}\right)^2 + \dfrac{5 \cdot 4 \cdot 3}{3!}(2^2)\left(\dfrac{x}{y}\right)^3 + \dfrac{5 \cdot 4 \cdot 3 \cdot 2}{4!}(2)\left(\dfrac{x}{y}\right)^4 + \dfrac{5 \cdot 4 \cdot 3 \cdot 2 \cdot 1}{5!}\left(\dfrac{x}{y}\right)^5$

$= 32 + 80\left(\dfrac{x}{y}\right) + 80\left(\dfrac{x}{y}\right)^2 + 40\left(\dfrac{x}{y}\right)^3 + 10\left(\dfrac{x}{y}\right)^4 + \left(\dfrac{x}{y}\right)^5$

**21** $(y^2 - 2x)^5 = (y^2)^5 + \dfrac{5}{1!}(y^2)^4(-2x) + \dfrac{5 \cdot 4}{2!}(y^2)^3(-2x)^2 + \dfrac{5 \cdot 4 \cdot 3}{3!}(y^2)^2(-2x)^3$

$+ \dfrac{5 \cdot 4 \cdot 3 \cdot 2}{4!}(y^2)(-2x)^4 + \dfrac{5 \cdot 4 \cdot 3 \cdot 2 \cdot 1}{5!}(-2x)^5$

$= y^{10} - 10y^8 x + 40y^6 x^2 - 80y^4 x^3 + 80y^2 x^4 - 32x^5$

**25** $\left(\sqrt{\dfrac{x}{2}} + 2y\right)^7$, the first four terms

$\left(\sqrt{\dfrac{x}{2}} + 2y\right)^7 = \left[\left(\dfrac{x}{2}\right)^{1/2} + 2y\right]^7$

$= \left[\left(\dfrac{x}{2}\right)^{1/2}\right]^7 + \dfrac{7}{1!}\left[\left(\dfrac{x}{2}\right)^{1/2}\right]^6 (2y) + \dfrac{7 \cdot 6}{2!}\left[\left(\dfrac{x}{2}\right)^{1/2}\right]^5 (2y)^2 + \dfrac{7 \cdot 6 \cdot 5}{3!}\left[\left(\dfrac{x}{2}\right)^{1/2}\right]^4 (2y)^3$

$= \left(\dfrac{x}{2}\right)^{7/2} + 14\left(\dfrac{x}{2}\right)^3 y + 84\left(\dfrac{x}{2}\right)^{5/2} y^2 + 280\left(\dfrac{x}{2}\right)^2 y^3$

**282** CHAPTER 11 TOPICS IN ALGEBRA

**29** $(a - 2b^2)^{11}$, the first five terms

$$(a - 2b^2)^{11} = a^{11} + \frac{11}{1!}(a)^{10}(-2b^2) + \frac{11 \cdot 10}{2!}(a)^9(-2b^2)^2$$
$$+ \frac{11 \cdot 10 \cdot 9}{3!}(a)^8(-2b^2)^3 + \frac{11 \cdot 10 \cdot 9 \cdot 8}{4!}(a)^7(-2b^2)^4$$
$$= a^{11} - 22a^{10}b^2 + 220a^9b^4 - 1{,}320a^8b^6 + 5{,}280a^7b^8$$

**33** $(a^3 - a^2)^9$, the first five terms

$$(a^3 - a^2)^9 = (a^3)^9 + \frac{9}{1!}(a^3)^8(-a^2) + \frac{9 \cdot 8}{2!}(a^3)^7(-a^2)^2$$
$$+ \frac{9 \cdot 8 \cdot 7}{3!}(a^3)^6(-a^2)^3 + \frac{9 \cdot 8 \cdot 7 \cdot 6}{4!}(a^3)^5(-a^2)^4$$
$$= a^{27} - 9a^{26} + 36a^{25} - 84a^{24} + 126a^{23}$$

**37** $\left(2x^2 - \frac{a^2}{3}\right)^9$, seventh term

$a = 2x^2, \; b = -\frac{a^2}{3}, \; n = 9$ and $k = 7$

$$u_k = \frac{n(n-1)(n-2)\cdots(n-k+2)}{(k-1)!} a^{n-k+1} b^{k-1}$$

$$u_7 = \frac{9(9-1)(9-2)(9-3)(9-4)(9-7+2)}{(7-1)!} (2x^2)^{9-7+1} \left(-\frac{a^2}{3}\right)^{7-1}$$

$$= \frac{9 \cdot 8 \cdot 7 \cdot 6 \cdot 5 \cdot 4}{6!}(2x^2)^3 \left(-\frac{a^2}{3}\right)^6$$

$$= \frac{224}{243} x^6 a^{12}$$

## CUMULATIVE REVIEW PROBLEM SET  Chapters 1–11

**1** Evaluate $2x^2 + 2xy - 3y^3$ for $x = -1, y = -2$.

**2** Perform the indicated operations.
 (a) $(x^2 - xy + 2y^2)(x + 3y)$
 (b) $(3x^3 - 2x^2 + 5) - (x^3 + x^2 - 3x + 2) + (3x^2 + 5x - 1)$

**3** Factor the following expressions completely.
 (a) $2x^3y - 10x^2y^2 - 12xy^3$
 (b) $15ay^2 - bx + 15ax - by^2$

**4** Determine the values of the variable for which $\dfrac{2x + 3}{x^2 - 2x - 15}$ is not defined.

**5** Perform the indicated operations and simplify the result.
 (a) $\dfrac{x^2 - x - 2}{x^2 - x - 6} \cdot \dfrac{x^2 + 2x}{x^2 - 2x}$
 (b) $\dfrac{3y}{3y^2 + 5y - 2} - \dfrac{2y}{2y^2 + 5y + 2} + \dfrac{y}{6y^2 + y - 1}$

**6** Solve each equation.
 (a) $\dfrac{12}{x^2 - 25} - \dfrac{1}{x + 5} = \dfrac{2}{x - 5}$
 (b) $ac + d = bc + 2$, solve for $c$

**7** Susan is 4 years older than Paul. If the sum of their ages in 6 years is 34, what are their ages now?

**8** Solve each absolute value inequality and graph each solution set.
 (a) $|2x - 3| < 5$
 (b) $|x + 7| \geq 9$

**9** Rewrite each expression so that it contains positive exponents, and simplify.
 (a) $\dfrac{1}{3^{-1} + 2^{-2}}$
 (b) $\left(\dfrac{x^{-3}y^4}{x^{-5}y^{-3}}\right)^{-2}$

10. Simplify the following expression and write the answer in scientific notation.
$$\frac{(8.4 \times 10^5) \cdot (3.2 \times 10^{-3})}{(1.6 \times 10^{-6}) \cdot (4 \times 10^4)}$$

11. Find the value of each expression.
   (a) $\sqrt[3]{-125}$
   (b) $-4^{5/2}$

12. Use the properties of radicals to simplify for $x \geq 0$ and $y \geq 0$.
   (a) $\sqrt{3xy^3}\sqrt{12xy}$
   (b) $\sqrt[4]{\frac{64x^5y}{4x}}$

13. Perform the indicated operations and simplify the result.
   (a) $5\sqrt{12} - 2\sqrt{75}$
   (b) $(2\sqrt{3} + 1)(\sqrt{3} - 4)$

14. Rationalize the denominator of $\dfrac{\sqrt{y}}{2\sqrt{x} + \sqrt{y}}$.

15. Solve each equation.
   (a) $3x^2 - 8x + 5 = 0$
   (b) $\sqrt{5x+1} + \sqrt{x-2} = 5$

16. Solve each inequality and graph its solution set.
   (a) $2x^2 + 9x - 5 \geq 0$
   (b) $\dfrac{3x-1}{2x+5} \leq 0$

17. Two bicycle riders leave the same point and travel at right angles to each other, each moving at a uniform speed. One rides 2 miles per hour faster than the other. After 2 hours, they are 20 miles apart. Find the speed of each bicyclist.

18. For $P = (3, 5)$ and $Q = (-1, 7)$, find:
   (a) the distance between $P$ and $Q$
   (b) the midpoint of $\overline{PQ}$
   (c) the slope of $\overline{PQ}$

19. Find the $x$ intercept, the $y$ intercept, and the slope of the line $3x - 4y = 12$.

20. Find the equation of each line $L$ if:
   (a) $L$ contains $P = (-2, 3)$ and slope $= 4$
   (b) $L$ contains $P = (4, 5)$ and is perpendicular to the line $2x + 3y - 6 = 0$

21. Complete the following statement by inserting "above" or "below" in the blank: The graph of the inequality $y < 3x - 4$ is the half-plane that lies _____ the line $y = 3x - 4$.

22. If Cramer's rule is used to solve the system $\begin{cases} 3x - 4y = 7 \\ 2x + y = 4 \end{cases}$ then find the value of
   (a) $D$
   (b) $D_x$
   (c) $D_y$

23. Sketch the graph of the system of inequalities $\begin{cases} 2x - 3y \leq 6 \\ 4x + 5y > 20 \end{cases}$.

24. Write $3^x = 15$ in logarithmic form.

25. Use the properties of logarithms to write each expression as a sum or difference of multiples of logarithms.
   (a) $\log_3 4x^2$ for $x > 0$
   (b) $\log_2 \dfrac{x^5}{7}$ for $x > 0$

26. If $f(x) = 2x + 3$ and $g(x) = |x - 1|$, find:
   (a) $f(-3)$
   (b) $g(-7)$
   (c) $f(x + 2)$

27. Solve the following system of equations algebraically.
$$\begin{cases} x + y^2 = 10 \\ y^2 - x^2 = 8 \end{cases}$$

28. Find the tenth term and $S_{10}$ for the arithmetic progression $-1, 2, 5, \ldots$.

**29** Find the fifth term and $S_5$ for the geometric progression $1, -\frac{1}{2}, \frac{1}{4}, \ldots$.

**30** Use the binomial theorem to expand $(x + 2y)^6$.

## Answers

**1** 30  **2(a)** $x^3 + 2x^2y - xy^2 + 6y^3$  **(b)** $2x^3 + 8x + 2$  **3(a)** $2xy(x - 6y)(x + y)$
**(b)** $(15a - b)(x + y^2)$  **4** $-3, 5$  **5(a)** $\dfrac{x + 1}{x - 3}$  **(b)** $\dfrac{y^2 + 7y}{(3y - 1)(y + 2)(2y + 1)}$  **6(a)** $\frac{7}{3}$
**(b)** $c = \dfrac{2 - d}{a - b}$  **7** 9, 13  **8(a)** $-1 < x < 4$

**(b)** $x \leq -16$ or $x \geq 2$   **9(a)** $\frac{12}{7}$  **(b)** $\dfrac{1}{x^4 y^{14}}$

**10** $4.2 \times 10^4$  **11(a)** $-5$  **(b)** $-32$  **12(a)** $6xy^2$  **(b)** $2x\sqrt[4]{y}$  **13(a)** 0
**(b)** $2 - 7\sqrt{3}$  **14** $\dfrac{2\sqrt{xy} - y}{4x - y}$  **15(a)** $1, \frac{5}{3}$  **(b)** 3
**16(a)** $x \leq -5$ or $x \geq 2$   **(b)** $-\frac{5}{2} < x \leq \frac{1}{3}$

**17** 6 miles per hour, 8 miles per hour
**18(a)** $2\sqrt{5}$  **(b)** $(1, 6)$  **(c)** $-\frac{1}{2}$  **19** x intercept = 4, y intercept = $-3$, slope = $\frac{3}{4}$
**20(a)** $y - 3 = 4(x + 2)$  **(b)** $y - 5 = \frac{3}{2}(x - 4)$  **21** below  **22(a)** $D = 11$  **(b)** $D_x = 23$
**(c)** $D_y = -2$  **23** **24** $x = \log_3 15$

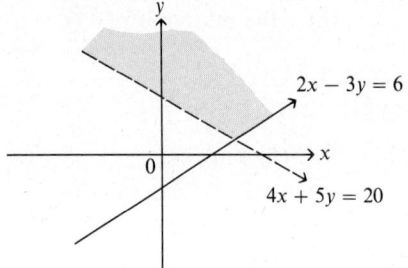

**25(a)** $\log_3 4 + 2\log_3 x$  **(b)** $5\log_2 x - \log_2 7$  **26(a)** $-3$  **(b)** 8  **(c)** $2x + 7$
**27** $(-2, -2\sqrt{3}), (-2, 2\sqrt{3}), (1, -3)$ and $(1, 3)$  **28** $a_{10} = 26, S_{10} = 125$  **29** $a_5 = \frac{1}{16}, S_5 = \frac{11}{16}$
**30** $x^6 + 12x^5y + 60x^4y^2 + 160x^3y^3 + 240x^2y^4 + 192xy^5 + 64y^6$

## CHAPTER 11 TESTS

### Chapter Test A

**1** Find the first five terms of each sequence.

  **(a)** $f(n) = (-1)^n 2^n$   **(b)** $f(n) = \dfrac{n + 1}{n}$   **(c)** $f(n) = \dfrac{n^2}{2^n}$

**2** Find the indicated term of each arithmetic progression.

  **(a)** $1, \frac{2}{3}, \frac{1}{3}, 0, -\frac{1}{3}, \ldots; a_{10}$   **(b)** If $a_1 = 40, a_n = 400, d = 2; n, S_n$

**3** Find the indicated term of each geometric progression.

  **(a)** $-2, -1, -\frac{1}{2}, -\frac{1}{4}, \ldots; a_{10}$   **(b)** $-\frac{2}{3}, -1, -\frac{3}{2}, \ldots; a_8, S_8$

4  If the sides of a right triangle are in arithmetic progression, let these sides be denoted by $a$, $a + d$, and $a + 2d$, with $d > 0$. Express $d$ in terms of $a$.

5  Find the numerical values of each finite sum.

(a) $\sum_{k=0}^{3} \dfrac{1}{2^k}$

(b) $\sum_{k=1}^{3} \dfrac{2^k}{3k - 1}$

6  Simplify each expression.

(a) $\dfrac{3! + 5!}{6!}$

(b) $\dfrac{(n + 1)!}{n!}$

7  Expand each expression by using the binomial theorem.

(a) $(x + y)^5$

(b) $(3x^2 - 2y)^3$

8  Find the indicated term in each binomial expansion and simplify the result.

(a) The seventh term of $(x - 2y)^{12}$

(b) The term involving $x^7$ of $(2 - x)^{12}$

---

## Solutions

1  (a) $f(1) = (-1)^1 2^1 = -2$
$f(2) = (-1)^2 2^2 = 4$
$f(3) = (-1)^3 2^3 = -8$
$f(4) = (-1)^4 2^4 = 16$
$f(5) = (-1)^5 2^5 = -32$

(b) $f(1) = \dfrac{1 + 1}{1} = 2$
$f(2) = \dfrac{2 + 1}{2} = \dfrac{3}{2}$
$f(3) = \dfrac{3 + 1}{3} = \dfrac{4}{3}$
$f(4) = \dfrac{4 + 1}{4} = \dfrac{5}{4}$
$f(5) = \dfrac{5 + 1}{5} = \dfrac{6}{5}$

(c) $f(1) = \dfrac{1^2}{2^1} = \dfrac{1}{2}$
$f(2) = \dfrac{2^2}{2^2} = 1$
$f(3) = \dfrac{3^2}{2^3} = \dfrac{9}{8}$
$f(4) = \dfrac{4^2}{2^4} = 1$
$f(5) = \dfrac{5^2}{2^5} = \dfrac{25}{32}$

2  (a) $\dfrac{2}{3} - 1 = -\dfrac{1}{3} = d$ and $a_1 = 1$
$a_n = a_1 + (n - 1)d$
$a_{10} = 1 + (10 - 1)\left(-\dfrac{1}{3}\right)$
$a_{10} = -2$

(b) $a_n = a_1 + (n - 1)d$
$400 = 40 + (n - 1)2$
$400 = 2n + 38$
$n = 181$
$S_n = \dfrac{n}{2}(a_1 + a_n)$
$S_n = \dfrac{181}{2}(40 + 400)$
$S_n = 39{,}820$

3  (a) $\dfrac{-1}{-2} = \dfrac{1}{2} = r$ and $a_1 = -2$
$a_n = a_1 r^{n-1}$
$a_{10} = (-2)\left(\dfrac{1}{2}\right)^{10-1}$
$a_{10} = -\dfrac{1}{256}$

(b) $\dfrac{-1}{-\frac{2}{3}} = \dfrac{3}{2} = r$ and $a_1 = -\dfrac{2}{3}$
$a_n = a_1 r^{n-1}$
$a_8 = \left(-\dfrac{2}{3}\right)\left(\dfrac{3}{2}\right)^{8-1} = -\dfrac{729}{64}$
$S_n = \dfrac{a_1(1 - r^n)}{1 - r}$
$S_8 = \dfrac{\left(-\frac{2}{3}\right)\left[1 - \left(\frac{3}{2}\right)^8\right]}{1 - \frac{3}{2}}$
$S_8 = \dfrac{4}{3}\left(1 - \dfrac{6{,}561}{256}\right)$
$S_8 = -\dfrac{6{,}305}{192}$

**4** 
$$(a + 2d)^2 = (a + d)^2 + a^2$$
$$a^2 + 4ad + 4d^2 = a^2 + 2ad + d^2 + a^2$$
$$3d^2 + 2ad - a^2 = 0$$
$$(3d - a)(d + a) = 0$$
$$3d - a = 0 \quad d + a = 0$$
$$d = \frac{a}{3} \qquad d = -a$$

Since $d > 0$, then $d = \frac{a}{3}$.

**5** (a) $\sum_{k=0}^{3} \frac{1}{2^k} = \frac{1}{2^0} + \frac{1}{2^1} + \frac{1}{2^2} + \frac{1}{2^3}$

$= \frac{1}{1} + \frac{1}{2} + \frac{1}{4} + \frac{1}{8}$

$= \frac{8 + 4 + 2 + 1}{8} = \frac{15}{8}$

(b) $\sum_{k=1}^{3} \frac{2^k}{3k-1} = \frac{2^1}{3(1)-1} + \frac{2^2}{3(2)-1} + \frac{2^3}{3(3)-1}$

$= \frac{2}{2} + \frac{4}{5} + \frac{8}{8}$

$= 1 + \frac{4}{5} + 1 = \frac{14}{5}$

**6** (a) $\dfrac{3! + 5!}{6!} = \dfrac{3! + 5 \cdot 4 \cdot (3!)}{6 \cdot 5 \cdot 4 \cdot (3!)}$

$= \dfrac{3!(1 + 5 \cdot 4)}{3!(6 \cdot 5 \cdot 4)}$

$= \dfrac{21}{120} = \dfrac{7}{40}$

(b) $\dfrac{(n+1)!}{n!} = \dfrac{(n+1) \cdot n!}{n!} = n + 1$

**7** (a) $(x + y)^5 = x^5 + \dfrac{5}{1!} x^4 y + \dfrac{5 \cdot 4}{2!} x^3 y^2 + \dfrac{5 \cdot 4 \cdot 3}{3!} x^2 y^3 + \dfrac{5 \cdot 4 \cdot 3 \cdot 2}{4!} xy^4 + \dfrac{5 \cdot 4 \cdot 3 \cdot 2 \cdot 1}{5!} y^5$

$= x^5 + 5x^4 y + 10x^3 y^2 + 10x^2 y^3 + 5xy^4 + y^5$

(b) $(3x^2 - 2y)^3 = (3x^2)^3 + \dfrac{3}{1!}(3x^2)^2(-2y) + \dfrac{3 \cdot 2}{2!}(3x^2)(-2y)^2 + \dfrac{3 \cdot 2 \cdot 1}{3!}(-2y)^3$

$= 27x^6 - 54x^4 y + 36x^2 y^2 - 8y^3$

**8** (a) Substituting $a = x$, $b = -2y$, $n = 12$, and $k = 7$

in the formula $u_k = \dfrac{n(n-1)(n-2) \cdots (n-k+2)}{(k-1)!} a^{n-k+1} b^{k-1}$

we obtain $u_7 = \dfrac{12(11)(10)(9)(8)(7)}{6!} x^6(-2y)^6 = 59{,}136 x^6 y^6$

(b) Substituting $a = 2$, $b = -x$, $n = 12$, and $k = 7$

in the formula $u_{k+1} = \dfrac{n!}{k!(n-k)!} a^{n-k} b^k$

we obtain the term involving $x^7$ as follows:

$u_{7+1} = \dfrac{12!}{7!(12-7)!} (2)^{12-7}(-x)^7$

$= -\dfrac{12 \cdot 11 \cdot 10 \cdot 9 \cdot 8}{5 \cdot 4 \cdot 3 \cdot 2 \cdot 1} 2^5 x^7$

$= -25{,}344 x^7$.

# Chapter Test B

*Multiple Choice:* Select the *one* correct answer for each of the following questions.

1. A sequence is a function whose domain is _____.
   - (a) the set of rational numbers
   - (b) the set of integers
   - (c) the set of real numbers
   - (d) the set of positive integers

2. The first five terms of the sequence $\left\{\dfrac{(-1)^n}{2^n}\right\}$ are _____.
   - (a) $\frac{1}{2}, \frac{1}{3}, \frac{1}{4}, \frac{1}{5}, \frac{1}{6}$
   - (b) $\frac{1}{2}, -\frac{1}{4}, \frac{1}{8}, -\frac{1}{16}, \frac{1}{32}$
   - (c) $-\frac{1}{2}, \frac{1}{4}, -\frac{1}{8}, \frac{1}{16}, -\frac{1}{32}$
   - (d) $1, -\frac{1}{2}, \frac{1}{4}, -\frac{1}{16}, \frac{1}{32}$

3. The $n$th term of the arithmetic progression 2, 4, 6, ... is _____.
   - (a) $2 + (2n - 1)2$
   - (b) $2n$
   - (c) $1 + (n - 1)2$
   - (d) none of these

4. The ninth term of the arithmetic progression 4, 9, 14, ... is _____.
   - (a) 44
   - (b) 45
   - (c) 54
   - (d) none of these

5. If $a_1 = 2$ and $d = 4$, the sum $S_{20}$ of the arithmetic progression is _____.
   - (a) $-18$
   - (b) 795
   - (c) 25
   - (d) 800

6. The sixth term of the geometric progression $6, -4, \frac{8}{3}, \ldots$ is _____.
   - (a) $-\frac{64}{81}$
   - (b) $\frac{32}{27}$
   - (c) $\frac{64}{81}$
   - (d) $-\frac{81}{64}$

7. The ninth term of the geometric progression 2, 6, 18, ... is _____.
   - (a) 13
   - (b) 13,122
   - (c) $2^9$
   - (d) none of these

8. If $a_1 = 6$ and $r = 2$, then the sum of the first 10 terms of the geometric progression is _____.
   - (a) $2^{10}$
   - (b) 12
   - (c) 6
   - (d) 6,138

9. The value of $\sum_{k=1}^{4} (k^2 + 2k)$ is _____.
   - (a) 50
   - (b) 82
   - (c) 24
   - (d) none of these

10. The expression $1 + \frac{1}{2} + \frac{1}{4} + \frac{1}{8} + \frac{1}{16}$ is written in sigma notation as _____.
    - (a) $\sum_{k=1}^{4} \dfrac{1}{2^k}$
    - (b) $\sum_{k=1}^{5} \dfrac{1}{2^k}$
    - (c) $\sum_{k=0}^{4} \left(\dfrac{1}{2}\right)^k$
    - (d) none of these

11. Of the following, statement _____ is true.
    - (a) $\sum_{k=0}^{49} (k+1)^2 = \sum_{k=1}^{50} k^2$
    - (b) $\sum_{k=1}^{16} k^2 = \left(\sum_{k=1}^{16} k\right)^2$
    - (c) $\sum_{k=1}^{n} (k+1)^2 = \left(\sum_{k=1}^{n} k^2\right) + n$
    - (d) $\sum_{k=0}^{4} \left(\dfrac{1}{2}\right)^k = 32$

12. The value of $\dfrac{6!}{5!}$ is _____.
    - (a) 5
    - (b) $6!$
    - (c) 6
    - (d) none of these

13. The value of $\dfrac{7!}{5!3!}$ is _____.
    - (a) $3 \cdot 2 \cdot 1$
    - (b) 7
    - (c) 51
    - (d) none of these

14. The value of $\dfrac{6! + 4!}{4! + 5!}$ is _____.
    - (a) $\frac{31}{6}$
    - (b) $6!4!$
    - (c) $-\frac{25}{4}$
    - (d) $\frac{5}{24}$

## CHAPTER 11  TOPICS IN ALGEBRA

15  The value of $\dfrac{(n+1)!}{n!}$ is _____.
   (a) $(n+1)!$  (b) $n$  (c) $n+1$  (d) none of these

16  The second term of the expansion $\left(2 + \dfrac{x}{3}\right)^5$ is _____.
   (a) 32  (b) $-\dfrac{80x}{3}$  (c) $\dfrac{3}{80}x$  (d) $\dfrac{80x}{3}$

17  The third term of the expansion $(2x + y)^4$ is _____.
   (a) $24x^2y^2$  (b) $-24x^2y^3$  (c) $24x^2$  (d) none of these

18  The sixth term of the expansion $\left(x^2 - \dfrac{1}{2x}\right)^6$ is _____.
   (a) $\dfrac{3}{16x^3}$  (b) $\dfrac{16}{x^3}$  (c) $-\dfrac{3}{16x^3}$  (d) none of these

19  The term that involves $x^4$ in the expansion of $(y^2 + 2x)^{12}$ is _____.
   (a) $7{,}920y^{16}x^4$  (b) $9{,}720x^4y^1$  (c) $-7{,}920x^4y^{18}$  (d) none of these

20  The term that involves $x^7$ in the expansion of $(2 - x)^{12}$ is _____.
   (a) $25{,}344x^7$  (b) $34{,}452x^7$  (c) $-25{,}344x^7$  (d) none of these

---

## Answers

1  d   2  c   3  b   4  a   5  d   6  a   7  b   8  d   9  a   10  c
11 a   12 c   13 b   14 a   15 c   16 d   17 a   18 c   19 a
20 c

# 12 Geometry

In this chapter we review the basic concepts of geometry. After completing the appropriate sections, the student should be able to:

1. Apply the terminology and notations associated with rays, line segments, and lines.
2. Identify angles by the standard classifications.
3. Identify the various standard polygons.
4. Compute the area of particular polygons.
5. Compute the perimeter of particular polygons.
6. Compute the volume and surface area of particular solids.

## 12.1 Basic Elements of Geometry

### SEMIPROGRAMMED PROBLEMS

In problems 1–3, give the appropriate symbol for the indicated ray, line segment, or line.

$\overline{CD}$ (or $\overline{DC}$)

1. The line segment joining two points $C$ and $D$ on the same line is labeled as _____.

$\overrightarrow{FH}$

2. The ray that is the portion of the line containing $F$ and $H$ and that starts at $F$ and continues indefinitely through $H$ is labeled as _____.

$\overleftrightarrow{AB}$ (or $AB$)

3. The line determined by the two points $A$ and $B$ is labeled as _____.

In problems 4 and 5, give the rays that are the sides of the given angle.

$\overrightarrow{FA}$ and $\overrightarrow{FE}$

4. $\angle AFE$
   The sides are _____.

$\overrightarrow{OC}$ and $\overrightarrow{OD}$

5. $\angle COD$
   The sides are _____.

In problems 6–11, complete each statement.

less
180°
38°, 52°
123°, 57°

6. An acute angle is _____ than 90°.
7. An obtuse angle is more than 90° and less than _____.
8. If $\angle A = 38°$, its complementary angle = 90° − _____ = _____.
9. If $\angle B = 123°$, its supplementary angle = 180° − _____ = _____.

equal

10. If two parallel lines are cut by a transversal, then the corresponding angles are _____.

parallel

11. If alternate interior angles formed by a transversal intersecting two lines are equal, then the lines are _____.

289

## SOLUTIONS TO SELECTED ODD PROBLEMS    Section 12.1, text pages 525–528

1  (a) [diagram: ray from A through B, C]     $\overrightarrow{AB} \cap \overrightarrow{BC} = \overrightarrow{BC}$

   (b) [diagram: rays from A to B and A to C]     $\overrightarrow{AB} \cap \overrightarrow{AC} = \text{point } A$

   (c) [diagram: rays through C, B, A]     $\overleftarrow{CA} \cap \overrightarrow{AC} = \overline{CA}$

5   $\angle 1 + \angle 2 + \angle 3 = 180°$
Since $\angle 1 + \angle 3 = 90°$ and
$\angle 2 + 90° = 180°$,
$\angle 2 = 90°$.

9  (a) vertical angles by definition
   (b) complementary angles, since $\angle 3 + \angle 4 = 90°$
   (c) adjacent angles, since $\angle 1 = \angle COA$ and $\angle 2 = \angle AOE$, and they share a common side $\overline{OA}$
   (d) supplementary angles, since $\angle 4 + \angle 5 = \angle DOB + \angle BOC = \angle DOC$, which is a straight angle
   (e) complementary angles, since $\angle 1 = \angle 4$ (vertical angles) and $\angle 3$ and $\angle 4$ are complementary angles

## 12.2 Polygons

### SEMIPROGRAMMED PROBLEMS

In problems 1–5, complete each statement.

| | |
|---|---|
| trapezoid | 1  A _____ is a quadrilateral with one pair of opposite sides parallel. |
| parallelogram | 2  A _____ is a quadrilateral with each pair of opposite sides parallel and equal. |
| rhombus | 3  A _____ is a parallelogram with all its sides equal. |
| rectangle | 4  A _____ is a parallelogram in which each of the angles is a right angle. |
| square | 5  A _____ is a rectangle with all four sides equal. |

In problems 6–8, determine if the given triangles are congruent for the given information.

6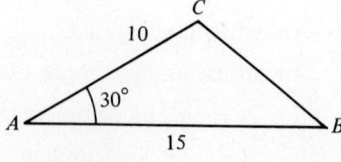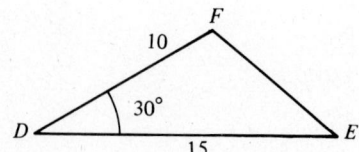

|$\overline{DF}$| 
|$\overline{DE}$| 
$\angle D$ 
$\triangle DEF$, SAS

Since $|\overline{AC}| = $ _____ $= 10$,
$|\overline{AB}| = $ _____ $= 15$, and
$\angle A = $ _____ $= 30°$,
then $\triangle ABC \cong$ _____ by _____.

## 12.2 SOLUTIONS TO SELECTED ODD PROBLEMS    291

7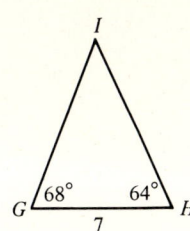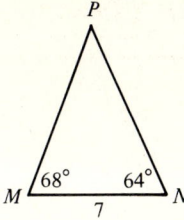

∠M  
|MN|  
∠N  
△MNP, ASA

Since ∠G = ___ = 68°,  
|GH| = ___ = 7, and  
∠H = ___ = 64°,  
then △GHI ≅ ___ by ___.

8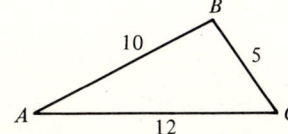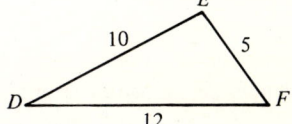

|DE|  
|DF|  
|EF|  
△DEF, SSS

Since |AB| = ___ = 10,  
|AC| = ___ = 12, and  
|BC| = ___ = 5,  
then △ABC ≅ ___ by ___.

In problems 9 and 10, determine the length of the indicated side of a triangle if △ABC ~ △DEF.

9  $\overline{AB}$, if $|\overline{BC}| = 5$, $|\overline{DE}| = 7$ and $|\overline{EF}| = 10$

Also, $\overline{AB}$ corresponds to $\overline{DE}$, and $\overline{BC}$ corresponds to $\overline{EF}$.

Since △ABC ~ △DEF, then

$\dfrac{|BC|}{|EF|}$

$\dfrac{|AB|}{|DE|} = $ ___ . Substitute

$\dfrac{5}{10}$

$\dfrac{|AB|}{7} = $ ___ , so that

5

$10|\overline{AB}| = 7($ ___ $)$. Thus

35, $\dfrac{7}{2}$

$|\overline{AB}| = \dfrac{\phantom{00}}{10} = $ ___ .

10  $\overline{DF}$, if $|\overline{AB}| = 15$, $|\overline{AC}| = 12$, and $|\overline{DE}| = 45$

Also, $\overline{AB}$ corresponds to $\overline{DE}$, and $\overline{AC}$ corresponds to $\overline{DF}$.

Since △ABC ~ △DEF, then

$\dfrac{|DE|}{|AB|}$

$\dfrac{|DF|}{|AC|} = $ ___ . Substitute.

$\dfrac{45}{15}$, 3

$\dfrac{|DF|}{12} = $ ___ = ___ , so that

3, 36

$|\overline{DF}| = 12($ ___ $) = $ ___ .

**SOLUTIONS TO SELECTED ODD PROBLEMS**  Section 12.2, text pages 534–537

1  ∠1 = ∠2 and ∠3 = ∠4  
AC is the included side between ∠1 and ∠4 of △II and between ∠2 and ∠3 of △I. Therefore, △I ≅ △II by ASA.

5  Since $\overline{DE} \parallel \overline{AB}$, then ∠CDE = ∠CBA and ∠CED = ∠CAB. (Alternate interior angles determined by a transversal of parallel lines are equal.) Also, it is given that $|\overline{DE}| = |\overline{AB}|$. Therefore, △ABC ≅ △CDE by ASA.

**292** CHAPTER 12 GEOMETRY

9 (a) $\dfrac{|CE|}{|AE|} = \dfrac{|BD|}{|AD|}$

$\dfrac{|CE|}{4} = \dfrac{5}{3}$

$|CE| = \dfrac{20}{3}$

(b) $\dfrac{|AE|}{|AC|} = \dfrac{|AD|}{|AB|}$ and $|AB| = |AD| + |BD|$

$\dfrac{|AE|}{10} = \dfrac{2}{2+3} = \dfrac{2}{5}$

$|AE| = \dfrac{20}{5} = 4$

(c) $\dfrac{|AB|}{|AD|} = \dfrac{|BC|}{|DE|}$

$\dfrac{|AB|}{4} = \dfrac{30}{10} = 3$

$|AB| = 12$

(d) $\dfrac{|AD|}{|BD|} = \dfrac{|AE|}{|EC|}$ and $|AD| = |EC|$

$\dfrac{|AD|}{4} = \dfrac{9}{|AD|}$

$|AD|^2 = 36$

$|AD| = 6$

13 Since each triangle contains an angle of 25° and a right angle, then two angles of one triangle equal two angles of the other triangle and they are similar.

17 $\dfrac{x+1}{x} = \dfrac{4}{3}$

$4x = 3x + 3$

$x = 3$

21 $A, B, C, D$ are midpoints of the sides of the quadrilateral $EFGH$.
In $\triangle EGH$, $\overline{DC} \parallel \overline{EG}$ and $|\overline{DC}| = \tfrac{1}{2}|\overline{EG}|$ (from problem 19).
In $\triangle EFG$, $\overline{AB} \parallel \overline{EG}$ and $|\overline{AB}| = \tfrac{1}{2}|\overline{EG}|$.
Therefore, $\overline{DC} \parallel \overline{AB}$ and $|\overline{DC}| = |\overline{AB}|$.
In $\triangle EFH$, $\overline{AD} \parallel \overline{HF}$ and $|\overline{AD}| = \tfrac{1}{2}|\overline{HF}|$.
In $\triangle FGH$, $\overline{BC} \parallel \overline{HF}$ and $|\overline{BC}| = \tfrac{1}{2}|\overline{HF}|$.
Therefore, $\overline{BC} \parallel \overline{AD}$ and $|\overline{BC}| = |\overline{AD}|$.
Thus, $ABCD$ is a parallelogram.

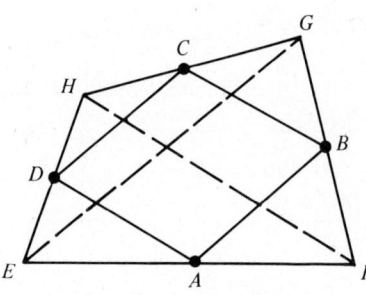

## 12.3 Areas of Plane Figures

### SEMIPROGRAMMED PROBLEMS

In problems 1–12, find the area of the plane figure having the given dimensions.

1 a rectangle whose length is 12 meters and width is 8 meters

12, 8   $\ell =$ _____ meters and $w =$ _____ meters, so that

  $A = \ell w$

12, 8   $A = (\underline{\phantom{xx}})(\underline{\phantom{xx}})$

96   $A =$ _____.

96 square meters   The area is _____.

2 a rectangle whose length is 7.3 feet and width is 5.8 feet

$\ell w$   The formula for the area of a rectangle is $A =$ _____. Here,

7.3, 5.8   $\ell =$ _____ feet and $w =$ _____ feet, so that

7.3, 5.8   $A = (\underline{\phantom{xx}})(\underline{\phantom{xx}})$

42.34   $A =$ _____.

42.34 square feet   The area is _____.

3 a square whose side is 3.5 yards

$s^2$   The formula for the area of a square is $A =$ _____. Here,

3.5   $s =$ _____, so that

| | |
|---|---|
| 3.5 | $A = (\_\_\_\_)^2$ |
| 12.25 | $A = \_\_\_\_$. |
| 12.25 square yards | The area is _____. |
| | 4  a square whose side is $4\frac{1}{3}$ inches |
| $4\frac{1}{3}$ | Here, $s = \_\_\_\_$, so that |
| $4\frac{1}{3}$ | $A = (\_\_\_\_)^2$ |
| $18\frac{7}{9}$ | $A = \_\_\_\_$. |
| $18\frac{7}{9}$ square inches | The area is _____. |
| | 5  a parallelogram whose base is 10 inches and height is 6.3 inches |
| $bh$ | The formula for the area of a parallelogram is $A = \_\_\_\_$. Here, |
| 10, 6.3 | $b = \_\_\_\_$ and $h = \_\_\_\_$, so that |
| 10, 6.3 | $A = (\_\_\_\_)(\_\_\_\_)$ |
| 63 | $A = \_\_\_\_$. |
| 63 square inches | The area is _____. |
| | 6  a parallelogram whose base is $3\frac{2}{5}$ yards and height is $2\frac{2}{3}$, yards |
| $\frac{17}{5}, \frac{8}{3}$ | Here, $b = \_\_\_\_$ and $h = \_\_\_\_$, so that |
| $\frac{17}{5}, \frac{8}{3}$ | $A = (\_\_\_\_)(\_\_\_\_)$ |
| $9\frac{1}{15}$ | $A = \_\_\_\_$. |
| $9\frac{1}{15}$ square yards | The area is _____. |
| | 7  a triangle whose base is 10 feet and height is 7 feet |
| $\frac{1}{2}$ | The formula for the area of a triangle is $A = (\_\_\_\_)bh$. Here, |
| 10, 7 | $b = \_\_\_\_$ and $h = \_\_\_\_$, so that |
| 10, 7 | $A = \frac{1}{2}(\_\_\_\_)(\_\_\_\_)$ |
| 35 | $A = \_\_\_\_$. |
| 35 square feet | The area is _____. |
| | 8  a triangle whose base is 4.3 centimeters and height is 5.8 centimeters |
| | $A = \frac{1}{2}bh$ |
| 4.3, 5.8 | $A = \frac{1}{2}(\_\_\_\_)(\_\_\_\_)$ |
| 12.47 | $A = \_\_\_\_$ |
| 12.47 square centimeters | The area is _____. |
| | 9  a trapezoid whose bases are 4 feet and 6 feet and whose height is 3 feet |
| | The formula for the area of a trapezoid is $A = \frac{1}{2}h(b_1 + b_2)$. Here, |
| 6 | $A = \frac{1}{2}(3)(4 + \_\_\_\_)$ |
| 10 | $= \frac{3}{2}(\_\_\_\_)$ |
| 15 | $= \_\_\_\_$. |
| 15 square feet | The area is _____. |
| | 10  a trapezoid whose height is 4.6 meters and whose bases are 3.7 meters and 5.5 meters |
| 4.6, 5.5 | $A = \frac{1}{2}(\_\_\_\_)(3.7 + \_\_\_\_)$ |
| 9.2 | $A = (2.3)(\_\_\_\_)$ |
| 21.16 | $A = \_\_\_\_$ |
| 21.16 square meters | The area is _____. |

**294** CHAPTER 12 GEOMETRY

| | |
|---|---|
| $r$ | **11** a circle whose radius is 11 centimeters |
| 11 | The formula for the area of a circle is $A = \pi(\_\_\_)^2$. Here, |
| 11 | $r = \_\_\_$, so that |
| 121, 380.13 | $A = \pi(\_\_\_)^2$ |
| 380.13 square centimeters | $= \pi(\_\_\_) \approx \_\_\_$ |
| | The area is approximately _____. |
| | **12** a circle whose radius is $3\frac{1}{2}$ feet |
| $3\frac{1}{2}$ | Here, $r = \_\_\_$, so that |
| $\frac{7}{2}$ | $A = \pi(\_\_\_)^2$ |
| $\frac{49}{4}$, 38.48 | $= \pi(\_\_\_) \approx \_\_\_$. |
| 38.48 square feet | The area is approximately _____. |

### SOLUTIONS TO SELECTED ODD PROBLEMS  Section 12.3, text page 541

**1**  $A = \ell w$
$A = (5)(3)$
$A = 15$
Area = 15 square inches

**5**  $A = \ell w$
$A = (7.3)(5.6)$
$A = 40.88$
Area = 40.88 square feet

**9**  $A = s^2$
$A = (6.3)^2$
$A = 39.69$
Area = 39.69 square feet

**13**  $A = bh$
$A = (5)(3.4)$
$A = 17$
Area = 17 square feet

**17**  $A = bh$
$A = (4\frac{1}{3})(6)$
$A = 26$
Area = 26 square inches

**21**  $A = \frac{1}{2}bh$
$A = \frac{1}{2}(5.1)(7.3)$
$A = 18.615$
Area = 18.615 square meters

**25**  $A = \frac{1}{2}h(b_1 + b_2)$
$A = \frac{1}{2}(5)(7 + 9)$
$A = 40$
Area = 40 square feet

**29**  $A = \frac{1}{2}h(b_1 + b_2)$
$A = \frac{1}{2}(6\frac{1}{3})(7\frac{1}{2} + 5\frac{1}{4})$
$A = \frac{1}{2}(\frac{19}{3})(\frac{51}{4})$
$A = 40\frac{3}{8}$
Area = $40\frac{3}{8}$ square feet

**33**  $A = \pi r^2$
$A = \pi(3.1)^2$
$A = 30.19$ (using $\pi = 3.1416$)
Area = 30.19 square meters

## 12.4 Perimeters of Plane Figures

### SEMIPROGRAMMED PROBLEMS

In problems 1–10, find the perimeter of the given plane figure.

| | |
|---|---|
| | **1** a rectangle whose length is 12 feet and width is 9 feet |
| 12, 9 | The formula is $P = 2\ell + 2w$. Here, $\ell = \_\_\_$ and $w = \_\_\_$, so that |
| 12, 9 | $P = 2(\_\_\_) + 2(\_\_\_)$ |
| 42 | $P = \_\_\_$. |
| 42 feet | The perimeter is _____. |
| | **2** a rectangle whose length is $5\frac{1}{6}$ yards and width is $4\frac{1}{3}$ yards |
| $2\ell + 2w$ | $P = _____$ |
| $2(4\frac{1}{3})$ | $P = 2(5\frac{1}{6}) + \_\_\_\_$ |
| 19 | $P = \_\_\_$ |
| 19 yards | The perimeter is _____. |

12.4 SEMIPROGRAMMED PROBLEMS    295

|   |   |
|---|---|
| 12 | 3  a square whose side is 12 inches |
| 12 | The formula is $P = 4s$. Here, $s = $ _____, so that |
| 48 | $P = 4(\_\_\_)$ |
| 48 inches | $P = \_\_\_$. |
|   | The perimeter is _____. |
| $\frac{8}{3}$ | 4  a square whose side is $2\frac{2}{3}$ meters |
| $\frac{8}{3}$ | Here, $s = $ _____, so that |
| $\frac{32}{3}$ | $P = 4(\_\_\_)$ |
| $10\frac{2}{3}$ meters | $P = \_\_\_$. |
|   | The perimeter is _____. |
|   | 5  a parallelogram whose length is 14 inches and width is 11 inches |
| $2w$ | The formula is $P = 2\ell + $ _____, so that |
| $2(11)$ | $P = 2(14) + $ _____ |
| 50 | $P = \_\_\_$. |
| 50 inches | The perimeter is _____. |
|   | 6  a parallelogram whose length is $5\frac{1}{3}$ feet and width is $3\frac{1}{2}$ feet. |
| $5\frac{1}{3}, 3\frac{1}{2}$ | $P = 2(\_\_\_) + 2(\_\_\_)$ |
| $17\frac{2}{3}$ | $P = \_\_\_$. |
| $17\frac{2}{3}$ feet | The perimeter is _____. |
|   | 7  a triangle whose sides are 5 centimeters, 7 centimeters, and 9 centimeters |
|   | The formula is $P = a + b + c$, where $a$, $b$, and $c$ represent the three sides. Thus, |
| 9 | $P = 5 + 7 + $ _____ |
| 21 | $P = \_\_\_$. |
| 21 centimeters | The perimeter is _____. |
|   | 8  a triangle whose sides are 3.4 meters, 5.7 meters, and 7.2 meters |
|   | $P = a + b + c$ |
| 5.7 | $P = 3.4 + $ _____ $+ 7.2$ |
| 16.3 | $P = \_\_\_$ |
| 16.3 meters | The perimeter is _____. |
|   | 9  a trapezoid whose sides are 16 yards, 12 yards, 7 yards, and 7 yards |
|   | The formula is $P = a + b + c + d$, so that |
| 7 | $P = 16 + 12 + 7 + $ _____ |
| 42 | $P = \_\_\_$. |
| 42 yards | The perimeter is _____. |
|   | 10  a trapezoid whose sides are 5 meters, 5.7 meters, 8.9 meters, and 15.9 meters |
|   | $P = a + b + c + d$ |
| 15.9 | $P = 5 + 5.7 + 8.9 + $ _____ |
| 35.5 | $P = \_\_\_$ |
| 35.5 meters | The perimeter is _____. |

In problems 11 and 12, find the circumference of each circle.

**11** a circle whose radius is 10 centimeters

The formula is $C = 2\pi r$, so that

$C = 2\pi(10) =$ _____     20π

$C \approx$ _____.     62.83

The circumference is approximately _____.     62.83 centimeters

**12** a circle whose radius is 4.5 meters

$C = 2\pi r$

$C = 2\pi(\_\_\_) =$ _____     4.5, 9π

$C \approx$ _____     28.27

The circumference is approximately _____.     28.27 meters

## SOLUTIONS TO SELECTED ODD PROBLEMS  Section 12.4, text page 545

**1**  $P = 2\ell + 2w$
 $= 2(5) + 2(3)$
 $= 16$
 Perimeter = 16 inches

**5**  $P = 2\ell + 2w$
 $= 2(7.3) + 2(5.6)$
 $= 25.8$
 Perimeter = 25.8 feet

**9**  $P = 4s$
 $= 4(6.3)$
 $= 25.2$
 Perimeter = 25.2 feet

**13**  $P = 2\ell + 2w$
 $= 2(5) + 2(3.4)$
 $= 16.8$
 Perimeter = 16.8 feet

**17**  $P = 2\ell + 2w$
 $= 2(6) + 2(4\frac{1}{3})$
 $= 20\frac{2}{3}$
 Perimeter = $20\frac{2}{3}$ inches

**21**  $P = a + b + c$
 $= 5.6 + 4.9 + 7.2$
 $= 17.7$
 Perimeter = 17.7 meters

**25**  $P = a + b + c + d$
 $= 10 + 4 + 6 + 4$
 $= 24$
 Perimeter = 24 inches

**29**  $P = a + b + c + d$
 $= \frac{1}{2} + \frac{1}{4} + \frac{1}{3} + \frac{1}{4}$
 $= 1\frac{1}{3}$
 Perimeter = $1\frac{1}{3}$ feet

**33**  $C = 2\pi r$
 $= 2\pi(5.2)$
 $= 10.4\pi$
 $\simeq 32.67$
 Circumference = 32.67 inches

## 12.5  Volumes and Surface Areas

### SEMIPROGRAMMED PROBLEMS

In problems 1 and 2, find the volume and the total surface area of the rectangular box having the given dimensions.

**1**  $\ell = 7$ inches, $w = 5$ inches, and $h = 4$ inches

The volume is given by the formula $V = \ell wh$, so that

$V = (7)(5)(\_\_\_) =$ _____.     4, 140

The volume is _____.     140 cubic inches

The total surface area is given by the formula

$S = 2(\ell w + \ell h + wh)$, so that

$S = 2[(7)(5) + (7)(4) + _____] =$ _____.     (5)(4), 166

The total surface area is _____.     166 square inches

**2**  $\ell = 3.4$ meters, $w = 4.7$ meters, and $h = 2.6$ meters

$V = \ell wh$

$= (3.4)(4.7)(\_\_\_) =$ _____     2.6, 41.55

The volume is _____.     41.55 cubic meters

## 12.5 SEMIPROGRAMMED PROBLEMS

| | |
|---|---|
| (4.7)(2.6) | $S = 2(\ell w + \ell h + wh)$ |
| 74.08 | $= 2[(3.4)(4.7) + (3.4)(2.6) + \underline{\hspace{1cm}}]$ |
| 74.08 square meters | $= \underline{\hspace{1cm}}$ |
| | The total surface area is _____. |

In problems 3 and 4, find the lateral surface area, the total surface area, and the volume of the right circular cylinder having the given dimensions.

| | |
|---|---|
| | **3**    $r = 2$ feet and $h = 4$ feet |
| | The lateral surface area is given by the formula $LS = 2\pi rh$, so that |
| 4, $16\pi$ | $LS = 2\pi(2)(\underline{\hspace{0.5cm}}) = \underline{\hspace{0.5cm}}$. |
| $16\pi$ square feet | The lateral surface area is _____. |
| | The total surface area is given by the formula $S = 2\pi r^2 + 2\pi rh$, so that |
| 2, $24\pi$ | $S = 2\pi(\underline{\hspace{0.5cm}})^2 + 2\pi(2)(4) = \underline{\hspace{0.5cm}}$. |
| $24\pi$ square feet | The total surface area is _____. |
| | The volume is given by the formula $V = \pi r^2 h$, so that |
| 2, 4, $16\pi$ | $V = \pi(\underline{\hspace{0.5cm}})^2(\underline{\hspace{0.5cm}}) = \underline{\hspace{0.5cm}}$. |
| $16\pi$ cubic feet | The volume is _____. |
| | **4**    $r = 3.2$ inches and $h = 7.1$ inches |
| | $LS = 2\pi rh$ |
| 3.2, 7.1, $45.44\pi$ | $= 2\pi(\underline{\hspace{0.5cm}})(\underline{\hspace{0.5cm}}) = \underline{\hspace{0.5cm}}$ |
| $45.44\pi$ square inches | The lateral surface area is _____. |
| | $S = 2\pi r^2 + 2\pi rh$ |
| 3.2, 3.2, 7.1 | $S = 2\pi(\underline{\hspace{0.5cm}})^2 + 2\pi(\underline{\hspace{0.5cm}})(\underline{\hspace{0.5cm}})$ |
| $65.92\pi$ | $= \underline{\hspace{0.5cm}}$ |
| $65.92\pi$ square inches | The total surface area is _____. |
| | $V = \pi r^2 h$ |
| 3.2, 7.1, $72.7\pi$ | $= \pi(\underline{\hspace{0.5cm}})^2(\underline{\hspace{0.5cm}}) = \underline{\hspace{0.5cm}}$ |
| $72.7\pi$ cubic inches | The volume is _____. |

In problems 5 and 6, find the lateral surface area and the volume of the regular pyramid having the given dimensions.

| | |
|---|---|
| | **5**    a regular square pyramid with a base edge of 6 inches, a height of 4 inches, and a slant height of 5 inches |
| | The formula for the lateral surface area is $LS = \frac{1}{2}P\ell$. Here, |
| 6, 24, 5 | $P = 4(\underline{\hspace{0.5cm}}) = \underline{\hspace{0.5cm}}$ and $\ell = \underline{\hspace{0.5cm}}$, so that |
| 24, 60 | $LS = \frac{1}{2}(\underline{\hspace{0.5cm}})5 = \underline{\hspace{0.5cm}}$. |
| 60 square inches | The lateral surface area is _____. |
| 6, 36 | The formula for the volume is $V = \frac{1}{3}Ah$. Here, $A = (\underline{\hspace{0.5cm}})^2 = \underline{\hspace{0.5cm}}$, so that |
| 36, 48 | $V = \frac{1}{3}(\underline{\hspace{0.5cm}})4 = \underline{\hspace{0.5cm}}$. |
| 48 cubic inches | The volume is _____. |
| | **6**    a regular triangular pyramid with a base edge of 10 feet, a base area of $25\sqrt{3}$ square feet, a height of 8 feet, and a slant height of 8.5 feet |

**298** CHAPTER 12 GEOMETRY

10, 30

30, 8.5, 127.5

127.5 square feet

$25\sqrt{3}$, 8, $\dfrac{200\sqrt{3}}{3}$

$\dfrac{200\sqrt{3}}{3}$ cubic feet

The perimeter of the base is $P = 3(\_\_\_) = \_\_\_$, so that
$LS = \frac{1}{2}P\ell$
$= \frac{1}{2}(\_\_\_)(\_\_\_) = \_\_\_.$
The lateral surface area is _____.

$V = \frac{1}{3}Ah$, so that $V = \frac{1}{3}(_____)(\_\_\_\_) = _____.$

The volume is _____.

In problems 7 and 8, find the lateral surface area and the volume of the right circular cone having the given dimensions.

**7** a right circular cone with a radius of 5 centimeters, height of 12 centimeters, and a slant height of 13 centimeters

The formula for the lateral surface area is $LS = \pi r \ell$, so that

5, 13, 65π

65π square centimeters

$LS = \pi(\_\_\_)(\_\_\_) = \_\_\_.$
The lateral surface area is _____.
The formula for the volume is $V = \frac{1}{3}\pi r^2 h$, so that

5, 12, 100π

100π cubic centimeters

$V = \frac{1}{3}\pi(\_\_\_)^2(\_\_\_) = \_\_\_.$
The volume is _____.

**8** a right circular cone with a radius of $2\sqrt{10}$ meters, a height of 9 meters, and a slant height of 11 meters
$LS = \pi r \ell$

$2\sqrt{10}$, 11, $22\sqrt{10}\pi$

$22\sqrt{10}\pi$ square meters

$= \pi(_____)(\_\_\_\_) = _____$
The lateral surface area is _____.
$V = \frac{1}{3}\pi r^2 h$, so that

$2\sqrt{10}$, 9, 120π

120π cubic meters

$V = \frac{1}{3}\pi(_____)^2(\_\_\_\_) = _____.$
The volume is _____.

In problems 9 and 10, find the surface area and the volume of the sphere having the given radius.

**9** a sphere with a radius of 2 inches
The formula for the surface area is $S = 4\pi r^2$, so that

2, 16π

16π square inches

$S = 4\pi(\_\_\_)^2 = \_\_\_.$
The surface area is _____.
The formula for the volume is $V = \frac{4}{3}\pi r^3$, so that

2, $\dfrac{32\pi}{3}$

$\dfrac{32\pi}{3}$ cubic inches

$V = \frac{4}{3}\pi(\_\_\_)^3 = \_\_\_.$

The volume is _____.

**10** a sphere with a radius of 3.1 meters
$S = 4\pi r^2$, so that

3.1, 38.44π

38.44π square meters

$S = 4\pi(\_\_\_)^2 = \_\_\_.$
The surface area is _____.

3.1, 39.72$\pi$

39.72$\pi$ cubic meters

$V = \frac{4}{3}\pi r^3$, so that

$V = \frac{4}{3}\pi(\underline{\phantom{xx}})^3 = \underline{\phantom{xxxxxx}}$.

The volume is \underline{\phantom{xxxxxxxxxxxxxxxxxxxxxxxx}}.

## SOLUTIONS TO SELECTED ODD PROBLEMS   Section 12.5, text pages 550–551

**1**   $V = \ell w h$
    $= 5(4)(3) = 60$
The volume is 60 cubic inches.

$S = 2(\ell w + \ell h + wh)$
$= 2[5(4) + 5(3) + 4(3)] = 94$
The total surface area is 94 square inches.

**5**   $V = \ell w h$
    $= 11.9(10.3)(7.5) = 919.275$
The volume is 919.275 cubic feet.

$S = 2(\ell w + \ell h + wh)$
$= 2[11.9(10.3) + 11.9(7.5) + 10.3(7.5)]$
$= 578.14$
The total surface area is 578.14 square feet.

**9**   $V = \pi r^2 h$
    $= \pi(4.1)^2(11.3)$
    $= 189.953\pi$
The volume is 189.953$\pi$ cubic meters.
$S = 2\pi rh + 2\pi r^2$
$= 2\pi(4.1)(11.3) + 2\pi(4.1)^2$
$= 126.28\pi$
The total surface area is 126.28$\pi$ square meters.

$LS = 2\pi rh$
$= 2\pi(4.1)(11.3)$
$= 92.66\pi$
The lateral surface area is 92.66$\pi$ square meters.

**13**   $A = s^2$
     $= 10^2 = 100$
   $V = \frac{1}{3}Ah$
     $= \frac{1}{3}(100)(12) = 400$
The volume is 400 cubic inches.

$P = 4s$
$= 4(10) = 40$
$LS = \frac{1}{2}P\ell$
$= \frac{1}{2}(40)(13) = 260$
The lateral surface area is 260 square inches.

**17**   $V = \frac{1}{3}Ah$
     $= \frac{1}{3}(247.8)(9) = 743.4$
The volume is 743.4 cubic inches.

$P = 5s$
$= 5(12) = 60$
$LS = \frac{1}{2}P\ell$
$= \frac{1}{2}(60)(12.2) = 366$
The lateral surface area is 366 square inches.

**21**   $V = \frac{1}{3}\pi r^2 h$
     $= \frac{1}{3}\pi(9.2)^2(6.9)$
     $= 194.672\pi$
The volume is 194.672$\pi$ cubic feet.

$LS = \pi r \ell$
$= \pi(9.2)(11.5)$
$= 105.8\pi$
The lateral surface area is 105.8$\pi$ square feet.

**25**   $V = \frac{4}{3}\pi r^3$
     $= \frac{4}{3}\pi(4)^3 = \frac{256\pi}{3}$
The volume is $\frac{256\pi}{3}$ cubic feet.

$S = 4\pi r^2$
$= 4\pi(4)^2$
$= 64\pi$
The surface area is 64$\pi$ square feet.

**29**   $V = \frac{4}{3}\pi r^3$
     $= \frac{4}{3}\pi(10)^3$
     $= \frac{4000\pi}{3}$
The volume is $\frac{4000\pi}{3}$ cubic inches.

$S = 4\pi r^2$
$= 4\pi(10)^2$
$= 400\pi$
The surface area is 400$\pi$ square inches.

# CUMULATIVE REVIEW PROBLEM SET   Chapters 1–12

1. Evaluate $2x^2 - 3xy^2 - 2y^3$ for $x = -1$, $y = -2$.

2. Perform the indicated operations.
   (a) $(x^2 - 3x + 2) + (5x - 3x^2 - 2) - (4 - 3x - 5x^2)$
   (b) $(x + 2y)(x^2 - xy - y^2)$
   (c) $(x^4 - 3x^2 + 7x - 5) \div (x - 1)$

3. Factor each expression completely.
   (a) $x^3 - 16x$
   (b) $x^3 - 64$
   (c) $a^2x^2 - 2a^2xy - 35a^2y^2$

4. Perform the indicated operations and simplify.
   (a) $\dfrac{z^2 - 4w^2}{z^2 - 4zw + 3w^2} \div \dfrac{3z + 6w}{z^2 - 9w^2}$
   (b) $\dfrac{3z + 1}{8z^2 + 2z - 1} - \dfrac{z - 1}{4z^2 + 11z - 3} + \dfrac{z + 1}{2z^2 + 7z + 3}$
   (c) $\dfrac{\dfrac{1}{x+1} + \dfrac{1}{x}}{\dfrac{1}{x} - \dfrac{1}{x+1}}$

5. Solve each equation.
   (a) $\dfrac{z}{z - 1} + 3 = \dfrac{4z}{z + 2}$
   (b) $|1 - 3x| = |-5|$
   (c) $3x^2 - 12x - 1 = 0$
   (d) $x^4 - 5x^2 + 6 = 0$
   (e) $\sqrt{5y - 4} - 1 = \sqrt{2y + 1}$

6. Solve each inequality.
   (a) $4 - 3(1 + 2x) \le 3x - (4 - x)$
   (b) $|2x - 3| < 7$
   (c) $|3x - 4| \ge 5$

7. Rewrite each expression so that it contains only positive exponents and simplify.
   (a) $\left(\dfrac{u^2 v^{-1/2}}{8v^{-3}}\right)^{-2/3}$
   (b) $\left(\dfrac{8x^{-6}}{125y^{12}}\right)^{-1/3} \left(\dfrac{25x^4}{4y^{-8}}\right)^{-1/2}$

8. Perform the indicated operations and simplify.
   (a) $\sqrt{5}(\sqrt{15} + \sqrt{35})$
   (b) $\sqrt{8x} + \sqrt{50x^3}$, $x > 0$

9. Rationalize the denominator of each fraction.
   (a) $\dfrac{7}{\sqrt[3]{4}}$
   (b) $\dfrac{\sqrt{x} + 2}{\sqrt{x} - 3}$

10. Perform the indicated operations and write the results in the form $a + bi$.
    (a) $(-3 + 4i) + (7 - 2i)$
    (b) $(6 - 5i) - (-3 + 2i)$
    (c) $(2 - 3i)(4 + 2i)$
    (d) $\dfrac{1 + 4i}{3 - 2i}$

11. A chemist has 100 milliliters of a solution that contains a 40% concentration of acid. How many milliliters of pure acid must be added in order to increase the concentration to 50%?

12. Find the center and the radius of the circle $x^2 + y^2 - 6x + 2y + 1 = 0$.

13. Find the slope-intercept form of the equation of a line that contains the point $(-1, 3)$ and is perpendicular to the line $3x - 2y = 7$.

14. The half-plane that lies below the graph of the line $2x - 3y = 6$ is the graph of the inequality $2x - 3y < 6$. True or false?

CUMULATIVE REVIEW PROBLEM SET  CHAPTERS 1–12

**15** Use Cramer's rule to solve the following system.
$$\begin{cases} 3x - 4y = -2 \\ -x + 2y = 0 \end{cases}$$

**16** Sketch the graph of the system of inequalities.
$$\begin{cases} -x + y \geq 1 \\ 2x + 3y \geq 6 \end{cases}$$

**17** Write each exponential equation in equivalent logarithmic form.
 (a) $4^3 = 64$  (b) $z^c = b$

**18** Use the properties of logarithms to write each expression as a single logarithm.
 (a) $3 \log_7 x + 2 \log_7 y$  (b) $\frac{1}{2} \log_a u - 3 \log_a v$

**19** Let $f(x) = |1 - x^2|$. Find:
 (a) $f(0)$  (b) $f(-1)$  (c) $f(4)$

**20** Find the domain of $f(x) = \sqrt{4 - x}$.

**21** Express $y$ as a function of $x$ if $y$ is inversely proportional to $x$, and $y = 8$ when $x = \frac{1}{2}$.

**22** Let $f(x) = 2x^2 - x - 6$. Find the $x$ and $y$ intercepts and the vertex of the graph of $f$. Also, determine the domain and the range of $f$.

**23** Let $f(x) = 2^x$ and $g(x) = \log_4 x$. Find:
 (a) $f(3)$  (b) $f(-2)$  (c) $g(32)$  (d) $g(\frac{1}{8})$

**24** Find the eleventh term and $S_{11}$ for the arithmetic progression $-2, -\frac{3}{2}, -1, -\frac{1}{2}, \ldots$.

**25** Evaluate each sum.
 (a) $\sum_{k=1}^{4} k(k-1)$  (b) $\sum_{k=2}^{5} \frac{1}{k(2k-3)}$

**26** How many lines are determined by four points if:
 (a) three of the points are collinear  (b) no three points are collinear

**27** If $\triangle ABC \sim \triangle DEF$ and $\angle A = \angle D$, $\angle B = \angle E$, find $|\overline{EF}|$ if $|\overline{AB}| = 7$, $|\overline{BC}| = 5$, and $|\overline{DE}| = 14$.

**28** Find the perimeter and the area of each of the following geometric figures.
 (a) A rectangle whose length is 5 feet and width is 3 feet
 (b) A triangle whose sides are 7 inches, 8 inches, and 12 inches and whose height is 4.5 inches when the 12-inch side is the base

**29** Find the volume and the lateral surface area of each of the following geometric figures.
 (a) A right circular cylinder whose radius is 4 meters and height is 7 meters
 (b) A regular square pyramid with a base edge of 6 feet, a height of $\sqrt{7}$ feet, and a slant height of 4 feet

**30** Find the volume and surface area of a sphere whose radius is 9 centimeters.

---

## Answers

**1** 30  **2(a)** $3x^2 + 5x - 4$  **(b)** $x^3 + x^2y - 3xy^2 - 2y^3$  **(c)** $x^3 + x^2 - 2x + 5$

**3(a)** $x(x-4)(x+4)$  **(b)** $(x-4)(x^2 + 4x + 16)$  **(c)** $a^2(x - 7y)(x + 5y)$  **4(a)** $\dfrac{(z - 2w)(z + 3w)}{3(z - w)}$

**(b)** $\dfrac{5z^2 + 14z + 3}{(4z - 1)(2z + 1)(z + 3)}$  **(c)** $2x + 1$  **5(a)** $\frac{2}{3}$  **(b)** $-\frac{4}{3}, 2$  **(c)** $\dfrac{6 \pm \sqrt{39}}{3}$

**(d)** $\pm\sqrt{2}, \pm\sqrt{3}$  **(e)** 4  **6(a)** $x \geq \frac{1}{2}$  **(b)** $-2 < x < 5$  **(c)** $x \leq -\frac{1}{3}$ or $x \geq 3$

**7(a)** $\dfrac{4}{u^{4/3} v^{5/3}}$  **(b)** 1  **8(a)** $5\sqrt{3} + 5\sqrt{7}$  **(b)** $(5x + 2)\sqrt{2x}$  **9(a)** $\dfrac{7\sqrt[3]{2}}{2}$

302   CHAPTER 12   GEOMETRY

(b) $\dfrac{x + 5\sqrt{x} + 6}{x - 9}$   10(a)  $4 + 2i$   (b)  $9 - 7i$   (c)  $14 - 8i$   (d)  $-\tfrac{7}{13} + \tfrac{14}{13}i$

11   20 milliliters   12  $(3, -1); 3$   13  $y = -\tfrac{2}{3}x + \tfrac{7}{3}$   14  false   15  $(-2, -1)$

16 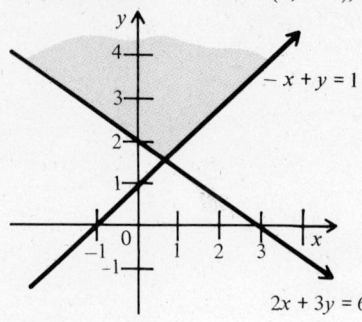   17(a)  $3 = \log_4 64$   (b)  $c = \log_z b$

18(a)  $\log_7 x^3 y^2$   (b)  $\log_a \dfrac{\sqrt{u}}{v^3}$   19(a)  1   (b)  0   (c)  15

20   all real numbers $x$ such that $x \leq 4$   21   $y = \dfrac{4}{x}$

22   $x$ intercepts are $-\tfrac{3}{2}, 2$; $y$ intercept is $-6$; the vertex is $(\tfrac{1}{4}, -\tfrac{49}{8})$; the domain is $\mathbb{R}$; and the range is the set of all real numbers $y \geq -\tfrac{49}{8}$.   23(a)  8   (b)  $\tfrac{1}{4}$   (c)  $\tfrac{5}{2}$   (d)  $-\tfrac{3}{2}$   24  $a_{11} = 3; S_{11} = \tfrac{11}{2}$
25(a)  20   (b)  $\tfrac{869}{1260}$   26(a)  4   (b)  6   27  10   28(a)  $P = 16$ feet; $A = 15$ square feet
(b)  $P = 27$ inches; $A = 27$ square inches   29(a)  $V = 112\pi$ cubic meters; $LS = 56\pi$ square meters
(b)  $V = 12\sqrt{7}$ cubic feet; $LS = 48$ square feet   30  $V = 972\pi$ cubic centimeters; $A = 324\pi$ square centimeters

## CHAPTER 12   TESTS

### Chapter Test A

1  In Figure 1, identify by number the angle whose sides are the rays:
   (a)  $\overrightarrow{AC}$ and $\overrightarrow{AB}$
   (b)  $\overrightarrow{BC}$ and $\overrightarrow{BA}$

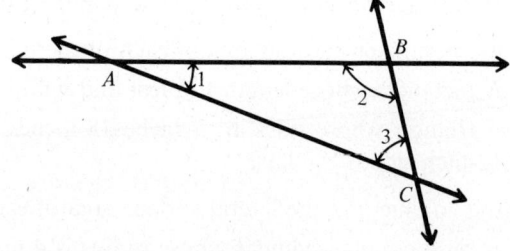

*Figure 1*

2  In Figure 2, identify by number the pairs of angles that are:
   (a)  vertical angles
   (b)  adjacent angles

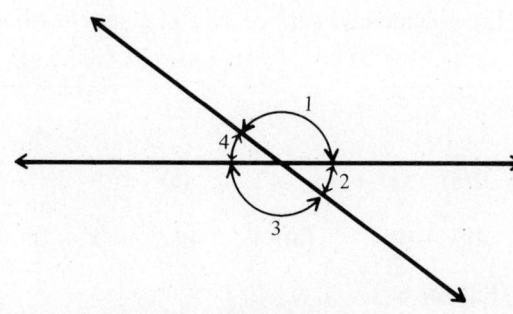

*Figure 2*

3  If $\angle A = 37°$, find the size of:
   (a)  the complement of $\angle A$   (b)  the supplement of $\angle A$

4   In Figure 3, $\ell \parallel m$ and $\angle 3 = 56°$. Find:
   (a)  $\angle 1$
   (b)  $\angle 2$

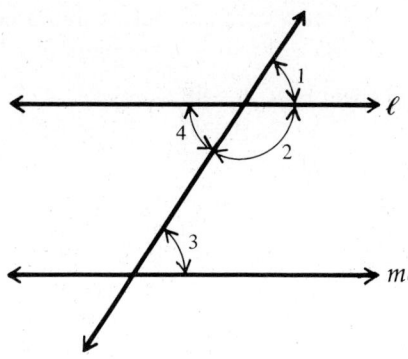

**Figure 3**

5   In Figure 4, ABCD is a rectangle. Show that $\triangle ABC \cong \triangle ADC$.

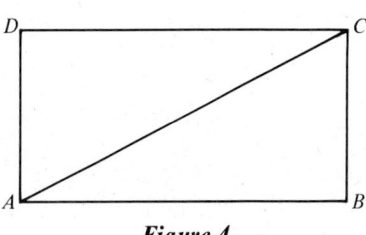

**Figure 4**

6   In Figure 5, $\overline{DE} \parallel \overline{AB}$ and $|\overline{AD}| = |\overline{DC}|$. If $|\overline{DE}| = 5$ feet, find $|\overline{AB}|$.

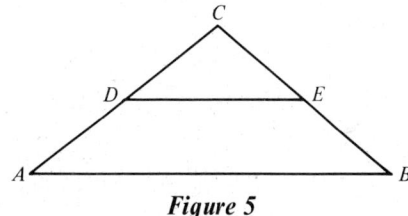

**Figure 5**

7   Find the area of the given plane figures.
   (a)  A rectangle whose length is 11.3 centimeters and width is 7.6 centimeters
   (b)  A trapezoid whose bases are 10 feet and 12 feet and whose height is 7 feet
   (c)  A circle whose radius is 6.5 meters

8   Find the perimeter of the given plane figures.
   (a)  A square whose side is 7.2 centimeters
   (b)  A triangle whose sides are $3\frac{1}{2}$ yards, $4\frac{1}{4}$ yards, and $5\frac{2}{3}$ yards
   (c)  A parallelogram whose length is 8.3 meters and width is 4.9 meters

9   Find the volume and the lateral surface area of the given solid.
   (a)  A right circular cylinder whose radius is 3 inches and height is $6\frac{1}{3}$ inches
   (b)  A regular triangular pyramid with a base edge of 12 inches, a base area of $36\sqrt{3}$ square inches, a height of 15 inches, and a slant height of 15.5 inches
   (c)  A right circular cone whose radius is 10 centimeters, height is 24 centimeters, and slant height is 26 centimeters

10  Find the volume and surface area of the sphere whose radius is 6 inches.

---

## Solutions

1   (a)  $\overrightarrow{AC}$ and $\overrightarrow{AB}$ intersect at A, so that they form $\angle 1$.
   (b)  $\overrightarrow{BC}$ and $\overrightarrow{BA}$ intersect at B, so that they form $\angle 2$.

2   (a)  Nonadjacent angles formed by two intersecting lines are called vertical angles. Here, the pairs of vertical angles are $\angle 2$ and $\angle 4$, and $\angle 1$ and $\angle 3$.

(b) $\angle 1$ and $\angle 2$ share a common vertex and a common side, so they are adjacent angles. Also, $\angle 2$ and $\angle 3$, $\angle 3$ and $\angle 4$, and $\angle 4$ and $\angle 1$ are pairs of adjacent angles.

3 (a) The complement of $\angle A = 90° - \angle A$
$= 90° - 37° = 53°$

(b) The supplement of $\angle A = 180° - \angle A$
$= 180° - 37° = 143°$

4 (a) Since $\ell \| m$, $\angle 1 = \angle 3$, since corresponding angles determined by a transversal of parallel lines are equal. Thus, $\angle 1 = 56°$.

(b) $\angle 4 = \angle 3 = 56°$, since alternate interior angles determined by a transversal of parallel lines are equal. Since $\angle 2 + \angle 4 = 180°$, $\angle 2 + 56° = 180°$ or $\angle 2 = 124°$.

5 Since $ABCD$ is a rectangle, $|\overline{AD}| = |\overline{BC}|$, $|\overline{AB}| = |\overline{DC}|$, and $\angle D = \angle B = 90°$, so that $\triangle ABC \cong \triangle ADC$ by SAS.

6 Since $\overline{DE} \| \overline{AB}$, $\angle BAC = \angle EDC$. Also, $\angle C$ is a common angle, so that $\triangle ABC \sim \triangle DEC$. Therefore, $\dfrac{|\overline{AB}|}{|\overline{DE}|} = \dfrac{|\overline{AC}|}{|\overline{DC}|}$, where $|\overline{AC}| = |\overline{AD}| + |\overline{DC}| = 2|\overline{DC}|$. Thus, $\dfrac{|\overline{AB}|}{5} = \dfrac{2|\overline{DC}|}{|\overline{DC}|} = 2$ or $|\overline{AB}| = 10$.

7 (a) $A = \ell w$
$= (11.3)(7.6)$
$= 85.88$
Area = 85.88 square centimeters

(b) $A = \frac{1}{2}h(b_1 + b_2)$
$= \frac{1}{2}(7)(10 + 12)$
$= 77$
Area = 77 square feet

(c) $A = \pi r^2$
$= \pi(6.5)^2$
$= 42.25\pi$
Area = $42.25\pi$ square meters

8 (a) $P = 4s$
$= 4(7.2)$
$= 28.8$
Perimeter = 28.8 centimeters

(b) $P = a + b + c$
$= 3\frac{1}{2} + 4\frac{1}{4} + 5\frac{2}{3}$
$= 13\frac{5}{12}$
Perimeter = $13\frac{5}{12}$ yards

(c) $P = 2\ell + 2w$
$= 2(8.3) + 2(4.9)$
$= 26.4$
Perimeter = 26.4 meters

9 (a) $V = \pi r^2 h$
$= \pi(3)^2(6\frac{1}{3})$
$= 57\pi$
Volume = $57\pi$ cubic inches

$LS = 2\pi rh$
$= 2\pi(3)(6\frac{1}{3})$
$= 38\pi$
Lateral surface area = $38\pi$ square inches

(b) $V = \frac{1}{3}Ah$
$= \frac{1}{3}(36\sqrt{3})(15)$
$= 180\sqrt{3}$
Volume = $180\sqrt{3}$ cubic inches

$LS = \frac{1}{2}Pl^*$
$= \frac{1}{2}(36)(15.5)$
$= 279$
Lateral surface area = 279 square inches

*$P = 3s$
$= 3(12) = 36$

(c) $V = \frac{1}{3}\pi r^2 h$
$= \frac{1}{3}\pi(10)^2(24)$
$= 800\pi$
Volume = $800\pi$ cubic centimeters

$LS = \pi r \ell$
$= \pi(10)(26)$
$= 260\pi$
Lateral surface area = $260\pi$ square centimeters

10 $V = \frac{4}{3}\pi r^3$
$= \frac{4}{3}\pi(6)^3$
$= 288\pi$
Volume = $288\pi$ cubic inches

$A = 4\pi r^2$
$= 4\pi(6)^2$
$= 144\pi$
Area = $144\pi$ square inches

## Chapter Test B

*Multiple Choice:* Select the *one* correct answer for each of the following questions.

1 The line segment joining the points $A$ and $B$ is labeled as _____.

(a) $AB$  (b) $\overrightarrow{AB}$  (c) $\overline{AB}$  (d) $\overleftrightarrow{AB}$

2  The point of intersection of the two sides of an angle is called the _____.
   (a) vertex   (b) initial side   (c) terminal side   (d) ray

3  A right angle is an angle of _____.
   (a) 180°   (b) 90°   (c) 270°   (d) 360°

4  Nonadjacent angles formed by two intersecting lines are called _____ angles.
   (a) complementary   (b) supplementary   (c) vertical   (d) corresponding

5  Two angles the sum of whose measures is 90° are called _____ angles.
   (a) supplementary   (b) complementary   (c) corresponding   (d) adjacent

6  A quadrilateral with one pair of opposite sides parallel is called a _____.
   (a) square   (b) rectangle   (c) parallelogram   (d) trapezoid

7  Two triangles that can be made to coincide are called _____ triangles.
   (a) congruent   (b) similar   (c) isosceles   (d) equilateral

8  Two triangles that have equal angles and corresponding sides that are proportional are called _____ triangles.
   (a) congruent   (b) similar   (c) isosceles   (d) equilateral

9  The area of a rectangle with sides whose lengths are 8 inches and 6 inches is _____.
   (a) 48 inches   (b) 28 inches   (c) 28 square inches   (d) 48 square inches

10 The area of a trapezoid with bases of 5 meters and 4 meters and a height of 3 meters is _____.
   (a) 27 square meters   (b) 24 square meters   (c) 16 square meters   (d) 13.5 square meters

11 The area of a circle with a radius of 5 feet is _____.
   (a) $25\pi$ feet   (b) $10\pi$ feet   (c) $25\pi$ square feet   (d) $10\pi$ square feet

12 The perimeter of a rectangle with a length of 7 yards and a width of 3.5 yards is _____.
   (a) 10.5 yards   (b) 21 yards   (c) 17.5 yards   (d) 15.5 yards

13 The perimeter of a triangle with sides that are 4 centimeters, 5 centimeters, and 7 centimeters is _____.
   (a) 16 square centimeters   (b) 16 centimeters
   (c) 14 centimeters   (d) 14 square centimeters

14 The circumference of a circle with a radius of 4 feet is _____.
   (a) $8\pi$ feet   (b) $16\pi$ feet   (c) $4\pi$ feet   (d) $64\pi$ feet

15 The volume of a rectangular box with dimensions that are 5 meters, 7 meters, and 4 meters is _____.
   (a) 140 cubic meters   (b) 166 cubic meters   (c) 83 cubic meters   (d) 100 cubic meters

16 The lateral surface area of a right circular cylinder with a radius of 4 inches and a height of 8 inches is _____.
   (a) $96\pi$ square inches   (b) $128\pi$ square inches
   (c) $64\pi$ square inches   (d) $80\pi$ square inches

17 The total surface area of the right circular cylinder in Problem 16 is _____.
   (a) $96\pi$ square inches   (b) $128\pi$ square inches
   (c) $64\pi$ square inches   (d) $80\pi$ square inches

18 The volume of a regular square pyramid with a base edge of 6 feet, a height of $\sqrt{7}$ feet, and a slant height of 4 feet is _____.
   (a) $36\sqrt{7}$ cubic feet   (b) 8 cubic feet   (c) 72 cubic feet   (d) $12\sqrt{7}$ cubic feet

**306** CHAPTER 12 GEOMETRY

19  The lateral surface area of a right circular cone with a radius of 3 feet, a height of 4 feet, and a slant height of 5 feet is _____.

    (a)   $15\pi$ square feet    (b)   $12\pi$ square feet    (c)   $45\pi$ square feet    (d)   $36\pi$ square feet

20  The volume of a sphere with a radius of 6 feet is _____.

    (a)   $144\pi$ cubic feet    (b)   $72\pi$ cubic feet    (c)   $288\pi$ cubic feet    (d)   $216\pi$ cubic feet

---

*Answers*

| 1 c | 2 a | 3 b | 4 c | 5 b | 6 d | 7 a | 8 b | 9 d | 10 d |
| 11 c | 12 b | 13 b | 14 a | 15 a | 16 c | 17 a | 18 d | 19 a | |
| 20 c | | | | | | | | | |

# Appendix A     TABLES

## TABLE I   COMMON LOGARITHMS

| n | 0 | 1 | 2 | 3 | 4 | 5 | 6 | 7 | 8 | 9 |
|---|---|---|---|---|---|---|---|---|---|---|
| 10 | 0000 | 0043 | 0086 | 0128 | 0170 | 0212 | 0253 | 0294 | 0334 | 0374 |
| 11 | 0414 | 0453 | 0492 | 0531 | 0569 | 0607 | 0645 | 0682 | 0719 | 0755 |
| 12 | 0792 | 0828 | 0864 | 0899 | 0934 | 0969 | 1004 | 1038 | 1072 | 1106 |
| 13 | 1139 | 1173 | 1206 | 1239 | 1271 | 1303 | 1335 | 1367 | 1399 | 1430 |
| 14 | 1461 | 1492 | 1523 | 1553 | 1584 | 1614 | 1644 | 1673 | 1703 | 1732 |
| 15 | 1761 | 1790 | 1818 | 1847 | 1875 | 1903 | 1931 | 1959 | 1987 | 2014 |
| 16 | 2041 | 2068 | 2095 | 2122 | 2148 | 2175 | 2201 | 2227 | 2253 | 2279 |
| 17 | 2304 | 2330 | 2355 | 2380 | 2405 | 2430 | 2455 | 2480 | 2504 | 2529 |
| 18 | 2553 | 2577 | 2601 | 2625 | 2648 | 2672 | 2695 | 2718 | 2742 | 2765 |
| 19 | 2788 | 2810 | 2833 | 2856 | 2878 | 2900 | 2923 | 2945 | 2967 | 2989 |
| 20 | 3010 | 3032 | 3054 | 3075 | 3096 | 3118 | 3139 | 3160 | 3181 | 3201 |
| 21 | 3222 | 3243 | 3263 | 3284 | 3304 | 3324 | 3345 | 3365 | 3385 | 3404 |
| 22 | 3424 | 3444 | 3464 | 3483 | 3502 | 3522 | 3541 | 3560 | 3579 | 3598 |
| 23 | 3617 | 3636 | 3655 | 3674 | 3692 | 3711 | 3729 | 3747 | 3766 | 3784 |
| 24 | 3802 | 3820 | 3838 | 3856 | 3874 | 3892 | 3909 | 3927 | 3945 | 3962 |
| 25 | 3979 | 3997 | 4014 | 4031 | 4048 | 4065 | 4082 | 4099 | 4116 | 4133 |
| 26 | 4150 | 4166 | 4183 | 4200 | 4216 | 4232 | 4249 | 4265 | 4281 | 4298 |
| 27 | 4314 | 4330 | 4346 | 4362 | 4378 | 4393 | 4409 | 4425 | 4440 | 4456 |
| 28 | 4472 | 4487 | 4502 | 4518 | 4533 | 4548 | 4564 | 4579 | 4594 | 4609 |
| 29 | 4624 | 4639 | 4654 | 4669 | 4683 | 4698 | 4713 | 4728 | 4742 | 4757 |
| 30 | 4771 | 4786 | 4800 | 4814 | 4829 | 4843 | 4857 | 4871 | 4886 | 4900 |
| 31 | 4914 | 4928 | 4942 | 4955 | 4969 | 4983 | 4997 | 5011 | 5024 | 5038 |
| 32 | 5051 | 5065 | 5079 | 5092 | 5105 | 5119 | 5132 | 5145 | 5159 | 5172 |
| 33 | 5185 | 5198 | 5211 | 5224 | 5237 | 5250 | 5263 | 5276 | 5289 | 5302 |
| 34 | 5315 | 5328 | 5340 | 5353 | 5366 | 5378 | 5391 | 5403 | 5416 | 5428 |
| 35 | 5441 | 5453 | 5465 | 5478 | 5490 | 5502 | 5514 | 5527 | 5539 | 5551 |
| 36 | 5563 | 5575 | 5587 | 5599 | 5611 | 5623 | 5635 | 5647 | 5658 | 5670 |
| 37 | 5682 | 5694 | 5705 | 5717 | 5729 | 5740 | 5752 | 5763 | 5775 | 5786 |
| 38 | 5798 | 5809 | 5821 | 5832 | 5843 | 5855 | 5866 | 5877 | 5888 | 5899 |
| 39 | 5911 | 5922 | 5933 | 5944 | 5955 | 5966 | 5977 | 5988 | 5999 | 6010 |
| 40 | 6021 | 6031 | 6042 | 6053 | 6064 | 6075 | 6085 | 6096 | 6107 | 6117 |
| 41 | 6128 | 6138 | 6149 | 6160 | 6170 | 6180 | 6191 | 6201 | 6212 | 6222 |
| 42 | 6232 | 6243 | 6253 | 6263 | 6274 | 6284 | 6294 | 6304 | 6314 | 6325 |
| 43 | 6335 | 6345 | 6355 | 6365 | 6375 | 6385 | 6395 | 6405 | 6415 | 6425 |
| 44 | 6435 | 6444 | 6454 | 6464 | 6474 | 6484 | 6493 | 6503 | 6513 | 6522 |
| 45 | 6532 | 6542 | 6551 | 6561 | 6571 | 6580 | 6590 | 6599 | 6609 | 6618 |
| 46 | 6628 | 6637 | 6646 | 6656 | 6665 | 6675 | 6684 | 6693 | 6702 | 6712 |
| 47 | 6721 | 6730 | 6739 | 6749 | 6758 | 6767 | 6776 | 6785 | 6794 | 6803 |
| 48 | 6812 | 6821 | 6830 | 6839 | 6848 | 6857 | 6866 | 6875 | 6884 | 6893 |
| 49 | 6902 | 6911 | 6920 | 6928 | 6937 | 6946 | 6955 | 6964 | 6972 | 6981 |

| n  | 0    | 1    | 2    | 3    | 4    | 5    | 6    | 7    | 8    | 9    |
|----|------|------|------|------|------|------|------|------|------|------|
| 50 | 6990 | 6998 | 7007 | 7016 | 7024 | 7033 | 7042 | 7050 | 7059 | 7067 |
| 51 | 7076 | 7084 | 7093 | 7101 | 7110 | 7118 | 7126 | 7135 | 7143 | 7152 |
| 52 | 7160 | 7168 | 7177 | 7185 | 7193 | 7202 | 7210 | 7218 | 7226 | 7235 |
| 53 | 7243 | 7251 | 7259 | 7267 | 7275 | 7284 | 7292 | 7300 | 7308 | 7316 |
| 54 | 7324 | 7332 | 7340 | 7348 | 7356 | 7364 | 7372 | 7380 | 7388 | 7396 |
| 55 | 7404 | 7412 | 7419 | 7427 | 7435 | 7443 | 7451 | 7459 | 7466 | 7474 |
| 56 | 7482 | 7490 | 7497 | 7505 | 7513 | 7520 | 7528 | 7536 | 7543 | 7551 |
| 57 | 7559 | 7566 | 7574 | 7582 | 7589 | 7597 | 7604 | 7612 | 7619 | 7627 |
| 58 | 7634 | 7642 | 7649 | 7657 | 7664 | 7672 | 7679 | 7686 | 7694 | 7701 |
| 59 | 7709 | 7716 | 7723 | 7731 | 7738 | 7745 | 7752 | 7760 | 7767 | 7774 |
| 60 | 7782 | 7789 | 7796 | 7803 | 7810 | 7818 | 7825 | 7832 | 7839 | 7846 |
| 61 | 7853 | 7860 | 7868 | 7875 | 7882 | 7889 | 7896 | 7903 | 7910 | 7917 |
| 62 | 7924 | 7931 | 7938 | 7945 | 7952 | 7959 | 7966 | 7973 | 7980 | 7987 |
| 63 | 7993 | 8000 | 8007 | 8014 | 8021 | 8028 | 8035 | 8041 | 8048 | 8055 |
| 64 | 8062 | 8069 | 8075 | 8082 | 8089 | 8096 | 8102 | 8109 | 8116 | 8122 |
| 65 | 8129 | 8136 | 8142 | 8149 | 8156 | 8162 | 8169 | 8176 | 8182 | 8189 |
| 66 | 8195 | 8202 | 8209 | 8215 | 8222 | 8228 | 8235 | 8241 | 8248 | 8254 |
| 67 | 8261 | 8267 | 8274 | 8280 | 8287 | 8293 | 8299 | 8306 | 8312 | 8319 |
| 68 | 8325 | 8331 | 8338 | 8344 | 8351 | 8357 | 8363 | 8370 | 8376 | 8382 |
| 69 | 8388 | 8395 | 8401 | 8407 | 8414 | 8420 | 8426 | 8432 | 8439 | 8445 |
| 70 | 8451 | 8457 | 8463 | 8470 | 8476 | 8482 | 8488 | 8494 | 8500 | 8506 |
| 71 | 8513 | 8519 | 8525 | 8531 | 8537 | 8543 | 8549 | 8555 | 8561 | 8567 |
| 72 | 8673 | 8579 | 8585 | 8591 | 8597 | 8603 | 8609 | 8615 | 8621 | 8627 |
| 73 | 8633 | 8639 | 8645 | 8651 | 8657 | 8663 | 8669 | 8675 | 8681 | 8686 |
| 74 | 8692 | 8698 | 8704 | 8710 | 8716 | 8722 | 8727 | 8733 | 8739 | 8745 |
| 75 | 8751 | 8756 | 8762 | 8768 | 8774 | 8779 | 8785 | 8791 | 8797 | 8802 |
| 76 | 8808 | 8814 | 8820 | 8825 | 8831 | 8837 | 8842 | 8848 | 8854 | 8859 |
| 77 | 8865 | 8871 | 8876 | 8882 | 8887 | 8893 | 8899 | 8904 | 8910 | 8915 |
| 78 | 8921 | 8927 | 8932 | 8938 | 8943 | 8949 | 8954 | 8960 | 8965 | 8971 |
| 79 | 8976 | 8982 | 8987 | 8993 | 8998 | 9004 | 9009 | 9015 | 9020 | 9025 |
| 80 | 9031 | 9036 | 9042 | 9047 | 9053 | 9058 | 9063 | 9069 | 9074 | 9079 |
| 81 | 9085 | 9090 | 9096 | 9101 | 9106 | 9112 | 9117 | 9122 | 9128 | 9133 |
| 82 | 9138 | 9143 | 9149 | 9154 | 9159 | 9165 | 9170 | 9175 | 9180 | 9186 |
| 83 | 9191 | 9196 | 9201 | 9206 | 9212 | 9217 | 9222 | 9227 | 9232 | 9238 |
| 84 | 9243 | 9248 | 9253 | 9258 | 9263 | 9269 | 9274 | 9279 | 9284 | 9289 |
| 85 | 9294 | 9299 | 9304 | 9309 | 9315 | 9320 | 9325 | 9330 | 9335 | 9340 |
| 86 | 9345 | 9350 | 9355 | 9360 | 9365 | 9370 | 9375 | 9380 | 9385 | 9390 |
| 87 | 9395 | 9400 | 9405 | 9410 | 9415 | 9420 | 9425 | 9430 | 9435 | 9440 |
| 88 | 9445 | 9450 | 9455 | 9460 | 9465 | 9469 | 9474 | 9479 | 9484 | 9489 |
| 89 | 9494 | 9499 | 9504 | 9509 | 9513 | 9518 | 9523 | 9528 | 9533 | 9538 |
| 90 | 9542 | 9547 | 9552 | 9557 | 9562 | 9566 | 9571 | 9576 | 9581 | 9586 |
| 91 | 9590 | 9595 | 9600 | 9605 | 9609 | 9614 | 9619 | 9624 | 9628 | 9633 |
| 92 | 9638 | 9643 | 9647 | 9652 | 9657 | 9661 | 9666 | 9671 | 9675 | 9680 |
| 93 | 9685 | 9689 | 9694 | 9699 | 9703 | 9708 | 9713 | 9717 | 9722 | 9727 |
| 94 | 9731 | 9736 | 9741 | 9745 | 9750 | 9754 | 9759 | 9763 | 9768 | 9773 |
| 95 | 9777 | 9782 | 9786 | 9791 | 9795 | 9800 | 9805 | 9809 | 9814 | 9818 |
| 96 | 9823 | 9827 | 9832 | 9836 | 9841 | 9845 | 9850 | 9854 | 9859 | 9863 |
| 97 | 9868 | 9872 | 9877 | 9881 | 9886 | 9890 | 9894 | 9899 | 9903 | 9908 |
| 98 | 9912 | 9917 | 9921 | 9926 | 9930 | 9934 | 9939 | 9943 | 9948 | 9952 |
| 99 | 9956 | 9961 | 9965 | 9969 | 9974 | 9978 | 9983 | 9987 | 9991 | 9996 |

## TABLE II  POWERS AND ROOTS

| Number | Square | Square Root | Cube | Cube Root | Number | Square | Square Root | Cube | Cube Root |
|---|---|---|---|---|---|---|---|---|---|
| 1 | 1 | 1.000 | 1 | 1.000 | 51 | 2,601 | 7.141 | 132,651 | 3.708 |
| 2 | 4 | 1.414 | 8 | 1.260 | 52 | 2,704 | 7.211 | 140,608 | 3.733 |
| 3 | 9 | 1.732 | 27 | 1.442 | 53 | 2,809 | 7.280 | 148,877 | 3.756 |
| 4 | 16 | 2.000 | 64 | 1.587 | 54 | 2,916 | 7.348 | 157,464 | 3.780 |
| 5 | 25 | 2.236 | 125 | 1.710 | 55 | 3,025 | 7.416 | 166,375 | 3.803 |
| 6 | 36 | 2.449 | 216 | 1.817 | 56 | 3,136 | 7.483 | 175,616 | 3.826 |
| 7 | 49 | 2.646 | 343 | 1.913 | 57 | 3,249 | 7.550 | 185,193 | 3.849 |
| 8 | 64 | 2.828 | 512 | 2.000 | 58 | 3,364 | 7.616 | 195,112 | 3.871 |
| 9 | 81 | 3.000 | 729 | 2.080 | 59 | 3,481 | 7.681 | 205,379 | 3.893 |
| 10 | 100 | 3.162 | 1,000 | 2.154 | 60 | 3,600 | 7.746 | 216,000 | 3.915 |
| 11 | 121 | 3.317 | 1,331 | 2.224 | 61 | 3,721 | 7.810 | 226,981 | 3.936 |
| 12 | 144 | 3.464 | 1,728 | 2.289 | 62 | 3,844 | 7.874 | 238,328 | 3.958 |
| 13 | 169 | 3.606 | 2,197 | 2.351 | 63 | 3,969 | 7.937 | 250,047 | 3.979 |
| 14 | 196 | 3.742 | 2,744 | 2.410 | 64 | 4,096 | 8.000 | 262,144 | 4.000 |
| 15 | 225 | 3.873 | 3,375 | 2.466 | 65 | 4,225 | 8.062 | 274,625 | 4.021 |
| 16 | 256 | 4.000 | 4,096 | 2.520 | 66 | 4,356 | 8.124 | 287,496 | 4.041 |
| 17 | 289 | 4.123 | 4,913 | 2.571 | 67 | 4,489 | 8.185 | 300,763 | 4.062 |
| 18 | 324 | 4.243 | 5,832 | 2.621 | 68 | 4,624 | 8.246 | 314,432 | 4.082 |
| 19 | 361 | 4.359 | 6,859 | 2.668 | 69 | 4,761 | 8.307 | 328,509 | 4.102 |
| 20 | 400 | 4.472 | 8,000 | 2.714 | 70 | 4,900 | 8.367 | 343,000 | 4.121 |
| 21 | 441 | 4.583 | 9,261 | 2.759 | 71 | 5,041 | 8.426 | 357,911 | 4.141 |
| 22 | 484 | 4.690 | 10,648 | 2.802 | 72 | 5,184 | 8.485 | 373,248 | 4.160 |
| 23 | 529 | 4.796 | 12,167 | 2.844 | 73 | 5,329 | 8.544 | 389,017 | 4.179 |
| 24 | 576 | 4.899 | 13,824 | 2.884 | 74 | 5,476 | 8.602 | 405,224 | 4.198 |
| 25 | 625 | 5.000 | 15,625 | 2.924 | 75 | 5,625 | 8.660 | 421,875 | 4.217 |
| 26 | 676 | 5.099 | 17,576 | 2.962 | 76 | 5,776 | 8.718 | 438,976 | 4.236 |
| 27 | 729 | 5.196 | 19,683 | 3.000 | 77 | 5,929 | 8.775 | 456,533 | 4.254 |
| 28 | 784 | 5.292 | 21,952 | 3.037 | 78 | 6,084 | 8.832 | 474,552 | 4.273 |
| 29 | 841 | 5.385 | 24,389 | 3.072 | 79 | 6,241 | 8.888 | 493,039 | 4.291 |
| 30 | 900 | 5.477 | 27,000 | 3.107 | 80 | 6,400 | 8.944 | 512,000 | 4.309 |
| 31 | 961 | 5.568 | 29,791 | 3.141 | 81 | 6,561 | 9.000 | 531,441 | 4.327 |
| 32 | 1,024 | 5.657 | 32,768 | 3.175 | 82 | 6,724 | 9.055 | 551,368 | 4.344 |
| 33 | 1,089 | 5.745 | 35,937 | 3.208 | 83 | 6,889 | 9.110 | 571,787 | 4.362 |
| 34 | 1,156 | 5.831 | 39,304 | 3.240 | 84 | 7,056 | 9.165 | 592,704 | 4.380 |
| 35 | 1,225 | 5.916 | 42,875 | 3.271 | 85 | 7,225 | 9.220 | 614,125 | 4.397 |
| 36 | 1,296 | 6.000 | 46,656 | 3.302 | 86 | 7,396 | 9.274 | 636,056 | 4.414 |
| 37 | 1,369 | 6.083 | 50,653 | 3.332 | 87 | 7,569 | 9.327 | 658,503 | 4.431 |
| 38 | 1,444 | 6.164 | 54,872 | 3.362 | 88 | 7,744 | 9.381 | 681,472 | 4.448 |
| 39 | 1,521 | 6.245 | 59,319 | 3.391 | 89 | 7,921 | 9.434 | 704,969 | 4.465 |
| 40 | 1,600 | 6.325 | 64,000 | 3.420 | 90 | 8,100 | 9.487 | 729,000 | 4.481 |
| 41 | 1,681 | 6.403 | 68,921 | 3.448 | 91 | 8,281 | 9.539 | 753,571 | 4.498 |
| 42 | 1,764 | 6.481 | 74,088 | 3.476 | 92 | 8,464 | 9.592 | 778,688 | 4.514 |
| 43 | 1,849 | 6.557 | 79,507 | 3.503 | 93 | 8,649 | 9.644 | 804,357 | 4.531 |
| 44 | 1,936 | 6.633 | 85,184 | 3.530 | 94 | 8,836 | 9.695 | 830,584 | 4.547 |
| 45 | 2,025 | 6.708 | 91,125 | 3.557 | 95 | 9,025 | 9.747 | 857,375 | 4.563 |
| 46 | 2,116 | 6.782 | 97,336 | 3.583 | 96 | 9,216 | 9.798 | 884,736 | 4.579 |
| 47 | 2,209 | 6.856 | 103,823 | 3.609 | 97 | 9,409 | 9.849 | 912,673 | 4.595 |
| 48 | 2,304 | 6.928 | 110,592 | 3.634 | 98 | 9,604 | 9.899 | 941,192 | 4.610 |
| 49 | 2,401 | 7.000 | 117,649 | 3.659 | 99 | 9,801 | 9.950 | 970,299 | 4.626 |
| 50 | 2,500 | 7.071 | 125,000 | 3.684 | 100 | 10,000 | 10.000 | 1,000,000 | 4.642 |

# Appendix B  FORMULAS FROM GEOMETRY

## I  Plane Area

1 Circle $\quad A = \pi r^2$
2 Parallelogram $\quad A = bh$
3 Triangle $\quad A = \tfrac{1}{2}bh$
4 Square $\quad A = s^2$
5 Rectangle $\quad A = lw$
6 Trapezoid $\quad A = \tfrac{1}{2}h(b + b_1)$

## II  Surface Area

1 Sphere $\quad S = 4\pi r^2$
2 Cylinder $\quad S = 2\pi rh + 2\pi r^2, \quad LS = 2\pi rh$
3 Cone $\quad S = \pi r^2 + \pi r\sqrt{r^2 + h^2}, \quad LS = \pi r\sqrt{r^2 + h^2}$
4 Pyramid $\quad S = \tfrac{1}{2}Pl + A, \quad LS = \tfrac{1}{2}Pl$
5 Cube $\quad S = 6s$
6 Rectangular box $\quad S = 2A + Ph$

## III  Volumes

1 Cube $\quad V = s^3$
2 Rectangular box $\quad V = lwh$
3 Cylinder $\quad V = \pi r^2 h$
4 Pyramid $\quad V = \tfrac{1}{3}Ah$
5 Cone $\quad V = \tfrac{1}{3}\pi r^2 h$
6 Sphere $\quad V = \tfrac{4}{3}\pi r^3$

## IV  Circumference

1 Circle $\quad C = 2\pi r$